Computational Nanoscience

RSC Theoretical and Computational Chemistry Series

Editor in Chief:
Jonathan Hirst, *University of Nottingham, Nottingham, UK*

Series Editors:
Kenneth Jordan, *University of Pittsburgh, Pittsburgh, USA*
Carmay Lim, *Academia Sinica, Taipei, Taiwan*
Walter Thiel, *Max Planck Institute for Coal Research, Mülheim an der Ruhr, Germany*

Titles in the Series:
1: Knowledge-based Expert Systems in Chemistry: Not Counting on Computers
2: Non-Covalent Interactions: Theory and Experiment
3: Single-Ion Solvation: Experimental and Theoretical Approaches to Elusive Thermodynamic Quantities
4: Computational Nanoscience

How to obtain future titles on publication:
A standing order plan is available for this series. A standing order will bring delivery of each new volume immediately on publication.

For further information please contact:
Book Sales Department, Royal Society of Chemistry, Thomas Graham House, Science Park, Milton Road, Cambridge, CB4 0WF, UK
Telephone: +44 (0)1223 420066, Fax: +44 (0)1223 420247,
Email: books@rsc.org
Visit our website at http://www.rsc.org/Shop/Books/

Printed and bound in Great Britain by CPI Antony Rowe, Chippenham and Eastbourne

Computational Nanoscience

Edited by

Elena Bichoutskaia
School of Chemistry, University of Nottingham, Nottingham, UK

RSC Publishing

RSC Theoretical and Computational Chemistry Series No. 4

ISBN: 978-1-84973-133-1
ISSN: 2041-3181

A catalogue record for this book is available from the British Library

Published by The Royal Society of Chemistry,
Thomas Graham House, Science Park, Milton Road,
Cambridge CB4 0WF, UK

Registered Charity Number 207890

For further information see our web site at www.rsc.org

Preface

Nanoscience, in the view of the editor, is one of the most exciting modern scientific disciplines as it is concerned with materials and systems, which exhibit novel physical and chemical properties due to their nanoscale size. It stretches across the whole spectrum of science including medicine and health, physics, engineering and chemistry. It provides a deep understanding of the behaviour of matter at the scale of individual atoms and molecules which is a crucial step towards future applications of nanotechnology.

The *Computational Nanoscience* book addresses modern challenges in the computational science, within the context of the rapidly evolving field of nanotechnology. It satisfies the need for a comprehensive, yet concise and up-to-date, survey of new developments and applications presented by the world's leading academics. It documents major recent advances in scientific computation, mathematical models and theory development that specifically target the applications in nanotechnology. The remarkable improvements in both theoretical methods and computational techniques make it possible for modern computational nanoscience to achieve a new level of accuracy. Computational nanoscience is now a discipline capable of leading and guiding experimental efforts.

This book brings together prominent experts in the field to discuss current state-of-the-art research and highlight future challenges. The target readership is not solely theoreticians. Suitable for experimental researchers[1] and students[2] the book shows readers what computational nanoscience has achieved, and how it may be applied in their own work. The twelve chapters, cover topics which include the concepts behind recent breakthroughs in nanoscience, the development of cutting edge simulation tools, and a variety of new applications.

The editor wishes to thank the Royal Society of Chemistry for the invitation to edit this monograph, and the Engineering and Physical Sciences Research

RSC Theoretical and Computational Chemistry Series No. 4
Computational Nanoscience
Edited by Elena Bichoutskaia
© Royal Society of Chemistry 2011
Published by the Royal Society of Chemistry, www.rsc.org

Council (EPSRC) for the award of a Career Acceleration Fellowship. Most importantly, the editor is deeply grateful to the contributing authors whose highest quality research, scientific vision and commitment made this book such an excellent read.

Elena Bichoutskaia
Nottingham, UK

Contents

RSC Theoretical and Computational Chemistry Series No. 4
Computational Nanoscience
Edited by Elena Bichoutskaia
© Royal Society of Chemistry 2011
Published by the Royal Society of Chemistry, www.rsc.org

CHAPTER 1

Algorithms for Predicting the Physical Properties of Nanocrystals and Large Clusters

JAMES R. CHELIKOWSKY

Center for Computational Materials, Institute for Computational Engineering and Sciences, University of Texas at Austin, Austin, Texas 78712, USA; Departments of Physics and Chemical Engineering, University of Texas at Austin, Austin, Texas 78712, USA

1.1   Introduction

Quantum confinement offers one the opportunity to alter physical properties of matter without changing the chemical composition.[1,2] For example, quantum confinement can be used to tune the optical gap across the visible spectrum in CdSe nanocrystals[3] or it can be used to modify the magnetic moment of ferromagnetic clusters.[4]

However, constructing an efficacious algorithm for predicting the role of quantum confinement and its role in determining the properties of nanocrystals is a difficult task owing to the complexity of nanocrystals, which often contain thousands of atoms. Here, I will illustrate new algorithms for nanoscale systems. These algorithms are especially designed for highly parallel platforms. The nature of these algorithms will be discussed and applied to complex systems such as magnetic clusters, and doped semiconductor nanocrystals.

RSC Theoretical and Computational Chemistry Series No. 4
Computational Nanoscience
Edited by Elena Bichoutskaia
© Royal Society of Chemistry 2011
Published by the Royal Society of Chemistry, www.rsc.org

1.2   The Electronic Structure Problem

Many properties of materials can be predicted by a solution of the Kohn–Sham equation[5]:

$$\left(\frac{-\hbar^2 \nabla^2}{2m} + V_{\text{ion}}^{\text{p}} + V_{\text{H}} + V_{\text{xc}}\right)\psi_n = E_n \psi_n, \qquad (1.1)$$

where $V_{\text{ion}}^{\text{p}}$ is an ionic pseudopotential,[6,7] V_{H} is the Hartree or Coulomb potential, and V_{xc} is the exchange-correlation potential. The Hartree and exchange-correlation potentials can be determined from the electronic charge density. The density is given by

$$\rho(\vec{r}) = \sum_{n,\text{occup}} |\psi_n(\vec{r})|^2. \qquad (1.2)$$

The summation is over all occupied states. The Hartree potential is determined by

$$\nabla^2 V_{\text{H}}(\vec{r}) = -4\pi e^2 \rho(\vec{r}). \qquad (1.3)$$

This term can be interpreted as the electrostatic interaction of a point charge with the charge density of the system.

A physical interpretation of the exchange-correlation potential is more difficult as it has no classical analogue. This potential can be evaluated using a *local density approximation*. The central tenet of this approximation is that the total exchange-correlation energy may be written as a universal functional of the density:

$$E_{\text{xc}}[\rho] = \int \rho(\vec{r})\varepsilon_{\text{xc}}[\rho(\vec{r})]\mathrm{d}^3 r, \qquad (1.4)$$

where ε_{xc} is the exchange-correlation energy density. E_{xc} and ε_{xc} are to be interpreted as depending solely on the charge density. The exchange-correlation potential, V_{xc}, is then obtained as $V_{\text{xc}} = \delta E_{\text{xc}}[\rho]/\delta\rho$.

It is not difficult to solve the Kohn–Sham equation (eqn (1.1)) for an atom as the atomic charge density can be taken to be spherically symmetric. Thus, the Kohn–Sham problem reduces to solving a one-dimensional problem. The Hartree and exchange-correlation potentials can be iterated to form a self-consistent field. This atomic solution provides the input to construct a pseudopotential representing the effect of the core electrons and nucleus. The "ion-core" pseudopotential, $V_{\text{ion}}^{\text{p}}$, can be transferred to other systems such as molecules and nanocrystals.[6,7]

The Kohn–Sham equations represent a nonlinear, self-consistent eigenvalue problem. Typically, a solution is obtained by first approximating the Hartree and exchange-correlation potentials using a superposition of atomic charge densities. The Kohn–Sham equation is then solved using these approximate potentials. From the solution, new wavefunctions and charge densities are obtained and used to construct updated Hartree and exchange-correlation potentials. The process is repeated until the "input" and "output" potentials agree and a self-consistent solution is realised. At this point, the total electronic energy can be computed along with a variety of other electronic properties.[6,7] Once the Kohn–Sham equation is solved, the total electronic energy, E_T, of the system can be evaluated from

$$E_T = \sum_{n,occup} E_n - \frac{1}{2} \int V_H(\vec{r})\rho(\vec{r})d^3r \tag{1.5}$$
$$+ \int \rho(\vec{r})(\varepsilon_{xc}[\rho(\vec{r})] - V_{xc}[\rho(\vec{r})])d^3r.$$

The structural energy can be obtained by adding the ion-core electrostatic terms.[6,7] The forces can be obtained by taking the derivative of the energy with respect to position as from the Hellmann–Feynman theorem.

1.3 Algorithms for Solving the Kohn–Sham Equation

The Kohn–Sham equation as cast in eqn (1.1) can be solved using a variety of techniques. Often, the wavefunctions can be expanded in a basis such as plane waves or gaussians and the resulting secular equations can be solved using standard diagonalisation packages such as those found in VASP.[8]

Here, we focus on a different approach. We solve the Kohn–Sham equation without resort to an explicit basis.[9–13] We solve for the wavefunctions on a grid with a fixed domain, which encompasses the physical system of interest. The grid need not be uniform, but it greatly simplifies the problem if it is. The wavefunctions outside of the domain are required to vanish for confined systems or assume periodic boundary conditions for systems with translational symmetry. In contrast to methods employing an explicit basis, such boundary conditions are easily incorporated. In particular, real-space methods do not require the use of supercells for localised systems. As such, charged systems can easily be examined without considering any electrostatic divergences. The Hartree potential is also solved on the grid from eqn (1.3) using conjugate-gradient minimisation. For confined systems, the boundary condition for this Poisson equation can be obtained by first performing a multipole expansion of the electronic charge density, the Hartree potential on the boundary of the domain can then be evaluated using the multipoles. A few lowest-order multipoles are found to be sufficient to obtain an accurate Hartree potential.

Within a "real-space" approach, one can solve the eigenvalue problem using a finite element or finite difference approach.[10,11,13] We use a higher-order finite difference approach owing to its simplicity in implementation. The Laplacian operator can be expressed using

$$\left(\frac{\partial^2 \psi}{\partial x^2}\right)_{x_0} \approx \sum_{n=-N}^{N} C_i\psi(x_0 + nh, y, z), \tag{1.6}$$

where h is the grid spacing, N is the number of nearest grid points, and C_i are the coefficients for evaluating the required derivatives.[14] The error scales as $O(h^{2N+2})$.

Once the secular equation is created, the eigenvalue problem can be solved using iterative methods.[12,15,16] Typically, a method such as a preconditioned Davidson method can be used. This is a robust and efficient method, which never requires one to store the Hamiltonian matrix. In this chapter, we outline a method that avoids solving large-eigenvalue problems explicitly.[12] The method utilises a damped *Chebyshev polynomial filtered subspace iteration*. In this approach, only the initial iteration requires solving an eigenvalue problem, which can be handled by means of any available efficient eigensolver. This step is used to provide a good initial subspace (or good initial approximation to the wavefunctions). Because the subspace dimension is slightly larger than the number of wanted eigenvalues, the method does not require as much memory as standard restarted eigensolvers such as ARPACK and TRLan (Thick – Restart, Lanczos).[17,18] Moreover, the cost of orthogonalisation is much reduced as the filtering approach only requires a subspace with dimension slightly larger than the number of occupied states and orthogonalisation is performed only once per SCF iteration. In contrast, standard eigensolvers using restart usually require a subspace twice as large and the orthogonalisation and other costs related to updating the eigenvectors are much higher.

The main idea of the proposed method is to start with a good initial eigenbasis, $\{\psi_n\}$, corresponding to occupied states of the initial Hamiltonian, and then to improve adaptively the subspace by polynomial filtering. That is, at a given self-consistent step, a polynomial filter, $P_m(\mathcal{H})$, of order m is constructed for the current Hamiltonian $\mathcal{H}$. As the eigenbasis is updated, the polynomial will be different at each SCF step since $\mathcal{H}$ will change. The goal of the filter is to make the subspace spanned by $\{\hat{\psi}_n\} = P_m(H\mathcal{H})\{\psi_n\}$ approximate the eigensubspace corresponding to the occupied states of $\mathcal{H}$. There is no need to make the new subspace, $\{\hat{\psi}_n\}$, approximate the wanted eigensubspace of $\mathcal{H}$ to high accuracy at intermediate steps. Instead, the filtering is designed so that the new subspace obtained at each self-consistent iteration step will progressively approximate the wanted eigenspace of the final Hamiltonian when self-consistency is reached.

This can be efficiently achieved by exploiting the Chebyshev polynomials, C_m, for the polynomials P_m. Specifically, we wish to exploit the fast growth property outside of the [–1, 1] interval. To obtain a good filter for a given SCF step, one needs to provide a lower bound and an upper bound of the unoccupied spectrum of the current Hamiltonian $\mathcal{H}$. The lower bound can be readily obtained from the Ritz values computed from the previous step, and the upper bound can be inexpensively obtained by a very small number of (*e.g.*, 4 or 5) Lanczos steps.[12] Hence, the main cost of the filtering at each iteration is in computing the polynomial operation.

One can use an affine mapping, $l(t)$, to map interval [a, b] to the interval [–1, 1]:

$$l(t) = \frac{t - (a + b)/2}{(b - a)/2}. \tag{1.7}$$

The interval is chosen to encompass the energy interval that needs to be filtered out. The filtering operation can then be expressed as

$$\{\hat{\psi}_n\} = C_m(l(H\mathscr{H}))\{\psi_n\}. \tag{1.8}$$

This computation is accomplished by exploiting the convenient three-term recurrence property of Chebyshev polynomials:

$$C_0(l) = 1, \qquad C_1(l) = l,$$
$$C_{m+1}(l) = 2l\,C_m(l) - C_{m-1}(l). \tag{1.9}$$

An example of a damped Chebyshev polynomial as defined by eqns (1.7) and (1.9) is given in Figure 1.1, where we have taken the lower bound as $a = 0.2$ and the upper bound as $b = 2$. In this example, the filtering would enhance the eigenvalue components in the shaded region.

The filtering procedure for the self-consistent cycle is illustrated in Figure 1.2. Unlike traditional methods, the cycle only requires one explicit diagonalisation step. Instead of repeating this step again within the self-consistent loop, a filtering operation is used to create a new basis in which the desired eigensubspace is enhanced. After the new basis, $\{\hat{\psi}_n\}$, is formed, the basis is orthogonalised. The orthogonalisation step scales as the cube of the number of occupied states and as such this method is not an "order-n" method, *i.e.* it does not scale linearly with the number of occupied states. However, the prefactor is sufficiently small that the method is much faster than previous implementations of real-space methods.[12] As for the traditional methods, the self-consistent cycle is repeated until the "input" and "output" density is unchanged.

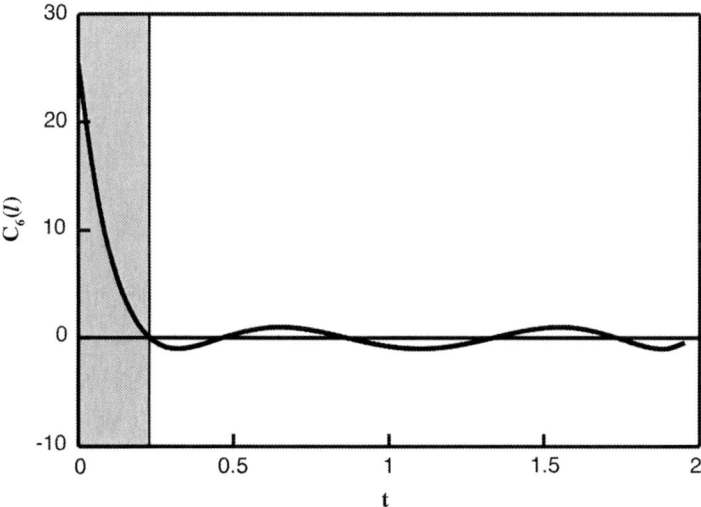

Figure 1.1 Example of a damped Chebyshev polynomial, C_6. The shaded area corresponds to eigenvalue spectrum regime that will be enhanced by the filtering operation (see text).

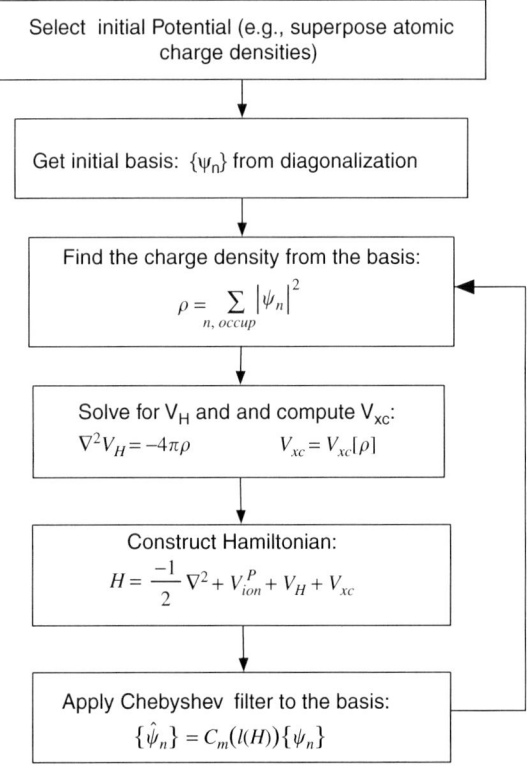

Figure 1.2   Schematic of the self-consistent cycle using Chebyshev filtering.

   The algorithm can be parallelised by partitioning the domain that encloses the physical system into continuous subdomains of grid points as implemented in the PARSEC code.[19] The number of subdomains is the same as the number of processors used in the calculation. A mapping can then be defined that assigns a subdomain to a processor. For optimal load balancing, different subdomains contain roughly the same number of grid points. The elements in the eigenvectors and potential arrays are indexed by grid points, hence a continuous block within the arrays can be distributed to different processors according to the defined mapping. As a result, the data required by the algorithm is completely distributed, and the Hamiltonian matrix is not stored explicitly. The Chebyshev filtering method requires matrix-vector operation between the Hamiltonian matrix and the eigenvectors. The operations of the local part of the ionic pseudopotential, Hartree and exchange-correlation potentials on the eigenvectors are local and therefore do not require communication. On the other hand, the Laplacian operator in eqn (1.6) and the nonlocal part of the ionic pseudopotential require communication with the processors that correspond to neighbouring blocks of grid points. The communication is done using the standard message passing interface (MPI) library.

Table 1.1 Comparison of the Chebyshev filtering method with ARPACK and TRLan. $H\times$ indicates the number of matrix-vector products, *SCFIts* indicates the number of iterations for a self-consistent soltion and the last entry is the *CPU* time in seconds.

Method	$H\times$	SCF Its	CPU(s)
Filtering	124 761	11	5947
ARPACK	142 047	10	62 026
TRLan	145 909	10	26 852

In Table 1.1, we compare the timings using the Chebyshev filtering method along with explicit diagonalisation solvers using the TRLan and the ARPACK. These timing are for a modest sized nanocrystal: $Si_{525}H_{276}$. The Hamiltonian size is 292 584 × 292 584 and 1194 eigenvalues were determined. The numerical runs were performed on the SGI Altix 3700 cluster at the Minnesota Supercomputing Institute. The CPU type is a 1.3-GHz Intel Madison processor. Although the number of matrix-vector products and SCF iterations is similar, the total time with filtering is over an order of magnitude faster compared to ARPACK, and a factor of better than four when compared to TRLan. Such improved timings are not limited to this particular example. We have also explored GaAs nanocrystal and large Fe clusters, including spin-dependent density functionals. In Figure 1.3, we

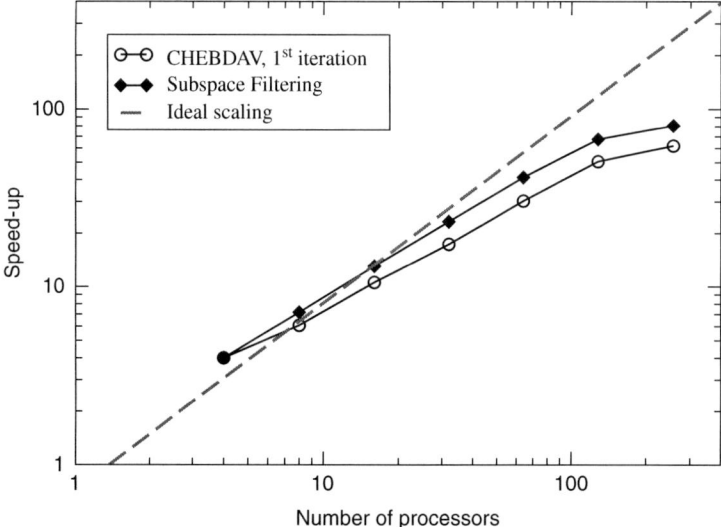

Figure 1.3 Scaling behaviour on Lonestar. Speed-up is the inverse of CPU time per processor for one SCF iteration step, the speed-up for the calculation with 4 processors is defined to be 4, the rest of the data points are with respect to the 4-processor calculation. The open circles correspond to the first diagonalisation step of the SCF loop, while the solid circles correspond to the subsequent Chebyshev subspace filtering step.

present the scaling behaviour of this algorithm on Lonestar, which is a TACC Dell Linux cluster containing 5840 computing cores. The scaling is studied using an iron cluster containing 43 iron atoms. The Hamiltonian matrix has a size of 110 245 and 258 eigenstates were calculated. We can see that the scaling starts linearly with the number of processors, but flattens off at around 128 processors for this particular system. The number of processors at which the saturation of the scaling occurs depends on the system size, a larger system size will need more processors to saturate the scaling.

1.4 Applications

1.4.1 The Electronic Properties of Si Nanocrystals

We illustrate the Chebyshev filtering method for a prototypical system: hydrogenated silicon nanocrystals. These nanocrystals, or quantum dots, are small fragments of the bulk in which the surface has been passivated by hydrogen atoms. In the case of silicon, the passivation is accomplished experimentally by capping the surface dangling bonds with hydrogen atoms.[20] These systems exhibit interesting changes as one approaches the nanoregime. For example, nanocrystals of silicon are expected to be optically active, whereas bulk crystals of silicon are not.[20,21]

The largest dot we examined contained over ten thousand atoms: $Si_{9041}H_{1860}$. This dot is approximately 7 nm in diameter. A ball and stick model for a much smaller quantum dot, approximately, 3 nm in diameter, is illustrated in Figure 1.4.

A solution of the Kohn–Sham equation (1.1) yields the energy levels or eigenstates for the quantum dot. If the dot is sufficiently large, this spectrum should approach the density of states of crystalline silicon. This is illustrated in Figure 1.5. The bulk density of states will reflect van Hove singularities in the energy-band topology, *i.e.* singularities associated with critical points when $\nabla_{\vec{k}} E(\vec{k})$ vanishes (where $E(\vec{k})$ is the energy band as a function of the wave-vector[22]). The generally good agreement between the density of states for the dot and crystal suggests that by ~ 7 nm the dot is of sufficient size to capture the van Hove singularities. (The comparison will not be exact because the quantum dot contains Si–H bonds in addition to the Si–Si bonds. In the dot eigenvalue spectrum, this results in the extra contributions near -5 eV.)

We can also examine the evolution of the ionisation potentials (I) and the electron affinities (A) for the quantum dot:

$$I = E(N - 1) - E(N)$$
$$A = E(N) - E(N + 1). \tag{1.10}$$

The difference between the ionisation potential and the electron affinity can be associated with the quasiparticle gap: $E_{qp} = I - A$. If the exciton (electron–hole)

Figure 1.4 Ball and stick model of a hydrogenated silicon quantum dot. The interior consists of a diamond fragment. The surface of the fragment is capped with hydrogen atoms.

interaction is small, this gap can be compared to the optical gap. However, for silicon quantum dots the exciton energy is believed to be on the order of ~ 1 eV for dots less than ~ 1 nm.

We can examine the scaling of the ionisation potential and electron affinity by assuming a simple scaling and fitting to the calculated values (shown in Figure 1.6):

$$I(D) = I_\infty + A/D^\alpha$$
$$A(D) = A_\infty + B/D^\beta, \tag{1.11}$$

where D is the dot diameter. A fit of these quantities results in $I_\infty = 4.5$ eV, $A_\infty = 3.9$ eV, $\alpha = 1.1$ and $\beta = 1.08$. The fit gives a quasiparticle gap of

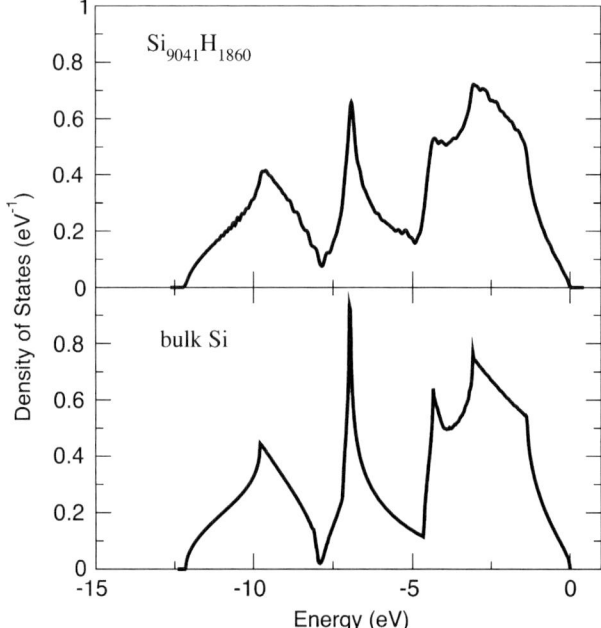

Figure 1.5   Eigenvalue spectrum for a large quantum dot of silicon: $Si_{9041}H_{1860}$ (top panel) compared to the density of states of crystalline silicon (bottom panel). The energy zero is taken to be the highest occupied level for the dot and the valence-band maximum for the crystal.

$E_{qp}(D \rightarrow \infty) = I_\infty - A_\infty = 0.6$ eV in the limit of an infinitely large dot. This value is in good agreement with the gap found for crystalline silicon using the local density approximation.[23] The gap is not in good agreement with experiment owing to the failure of density-functional theory to describe excited states.

A key aspect of this study is that the scaling of the ionisation potential and electron affinity for quantum dots ranging from silane (SiH_4) to dots with thousand of atoms can be examined. We not only verify the limiting value of the quasiparticle gap, we can ascertain how this limit is reached, *i.e.* how the ionisation potential and electron affinity scale with the size of the dot and what the relationship is between these quantities and the highest-occupied and lowest-empty energy levels.

1.4.2   The Vibrational Modes for Si Nanocrystals

Given the energy as a function of position, we can also examine structural and vibrational properties of nanocrystals. Vibrational properties are easier to describe as they converge more rapidly to the bulk values than do electronic states; however, because they involve small changes in energy with position, the wavefunctions need to be more highly converged.

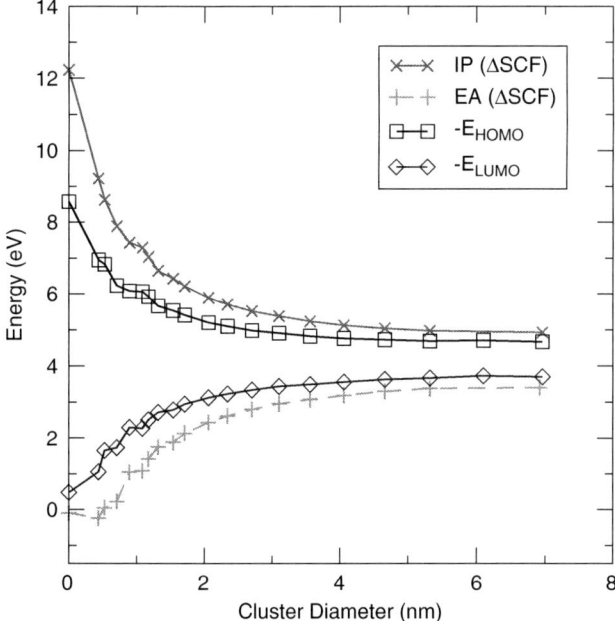

Figure 1.6 Evolution of the ionisation potential (IP) and electron affinity (EA) with quantum dot size. Also shown are the eigenvalue levels for the highest-occupied molecular orbital (HOMO) and the lowest-unoccupied molecular orbital (LUMO).

Owing to the localised nature of nanocrystals, it is feasible to predict vibrational modes by the direct force-constant method.[24] The dynamical matrix of the system is constructed by displacing all atoms one by one from their equilibrium positions along the Cartesian directions and finding the forces induced on the other atoms of the nanocrystal. We determine the forces using the Hellmann–Feynman theorem in real space[25] and employed a symmetrised form of the dynamical matrix expression.[26] The elements of the dynamical matrix, $D_{ij}^{\alpha\beta}$, are given by

$$D_{ij}^{\alpha\beta} = -\frac{1}{2} \left[\frac{F_i^{\alpha}(\{\mathbf{R}\} + d_j^{\beta}) - F_i^{\alpha}(\{\mathbf{R}\} - d_j^{\beta})}{2d_j^{\beta}} \right. \\ \left. + \frac{F_i^{\beta}(\{\mathbf{R}\} + d_j^{\alpha}) - F_j^{\beta}(\{\mathbf{R}\} - d_i^{\alpha})}{2d_i^{\alpha}} \right], \tag{1.12}$$

where F_j^{α} is the force on atom α in the direction j and $\{\mathbf{R}\} + d_j^{\beta}$ is the atomic configuration where only the atom β is displaced along j from its equilibrium position. The value of displacement was chosen to be 0.03 a.u. (1 a.u. = 0.5292 Å). The equilibrium structure was relaxed to satisfy the maximum forces below 5×10^{-5} Ry/au. For this accuracy, the grid spacing h is reduced to 0.4 a.u. as compared to a value of about 0.7 a.u., typically used for electronic properties.

The Chebyshev filtering algorithm is especially well suited for this procedure as the initial diagonalisation need not be repeated when the geometry changes are small.

The vibrational modes frequencies and corresponding eigenvectors can be obtained from the dynamical equation:

$$\sum_{\beta,k} \left[\omega^2 \delta_{\alpha\beta} \delta_{ik} - \frac{D_{ij}^{\alpha\beta}}{\sqrt{M_\alpha M_\beta}} \right] A_k^\beta = 0, \tag{1.13}$$

where M_α is the mass of atom labeled by α and the eigenvectors are given by the mode amplitudes, A_k^β.

We considered a series of Si nanocrystals: $Si_{29}H_{36}$, $Si_{66}H_{64}$, $Si_{87}H_{76}$, and $Si_{281}H_{172}$. The surfaces of the nanocrystals were passivated by hydrogen atoms.[27] We avoid any "one-fold" coordinated Si surface atoms, *i.e.*, the surface Si atoms are bonded to one or two hydrogen atoms at most. The interior of the nanocrystal assumes the diamond structure with the relaxed bond length of 2.31 Å.

In Figure 1.7, we illustrate the evolution of the vibrational density of states. The general features appear even in the smallest nanocrystal as the modes are dominated by nearest-neighbour interactions. The modes can be ascribed as follows. The lowest energy modes (below $\sim 200 \, cm^{-1}$) are Si-bond bending modes. Si-bond stretching modes occur at higher energies (below $\sim 600 \, cm^{-1}$). The dominant peak near 500 cm^{-1} corresponds to the crystalline peak transverse optical (TO) mode. The TO mode is Raman active in crystalline Si and is claimed to be size dependent[29,30]. The region above the Si–Si bond stretching modes is related to Si–H dominated vibrations such as librations and scissor motions. These modes are separated from the Si–H stretching modes by a gap from 1000 cm^{-1} to 2000 cm^{-1}.

In Figure 1.8, we compare the measured vibrational modes for crystalline silicon[28] to the modes for the largest nanocrystal that we have considered. We ignored the Si–H modes by limiting the comparison to modes below 600 cm^{-1}. The Van Hove singularities in the vibrational density of states have clearly evolved by a nanocrystal containing a few hundred atoms. A similar evolution of the electronic states would not occur until the size of the nanocrystal exceeds several thousand atoms.

1.4.3   Properties of Phosphorus-Doped Si Nanocrystals

In this section, we examine the electronic structures of phosphorus-doped Si nanocrystals. The doped Si nanocrystals are fragments of the bulk in which the surface has been passivated. The passivation is accomplished by capping the surface dangling bonds with hydrogen atoms. The doping is achieved by substituting the Si atom at the centre of the nanocrystal by a phosphorus atom. The largest nanocrystal we examined contained over six thousand

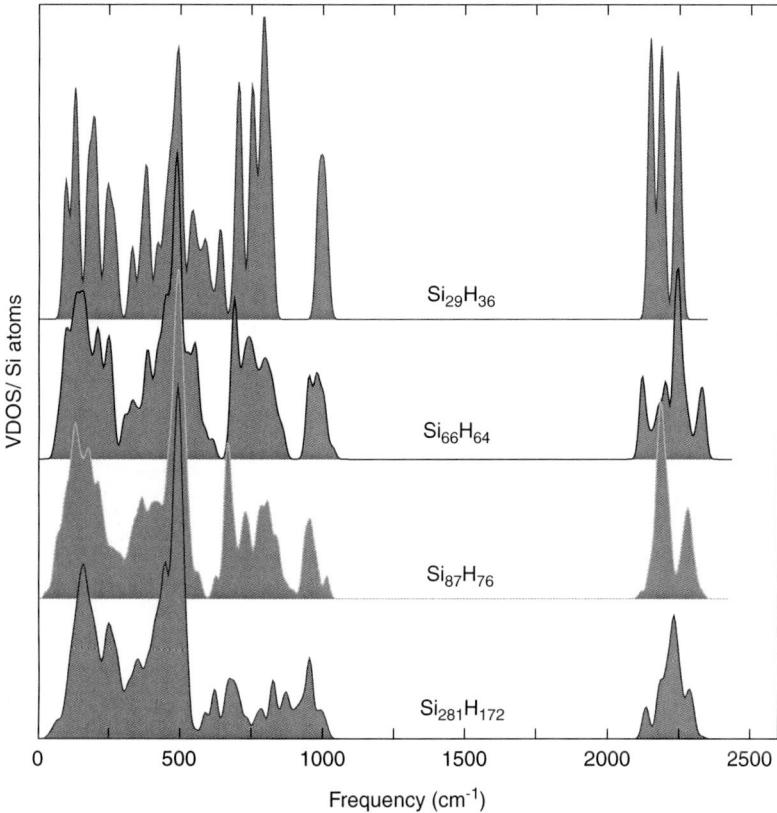

Figure 1.7   Evolution of vibrational modes in Si nanocrystals. For a better visualisation, the intensities of the spectra for each cluster were divided by the
number of Si atoms in the corresponding cluster.

atoms: $Si_{5011}H_{1276}P$. This nanocrystal is approximately 6 nm in diameter. A ball
and stick model for such quantum dots is illustrated in Figure 1.9.

The energy spectrum of a pure Si nanocrystal exhibits an energy gap E_G,
which increases when the size of the nanocrystal decreases as illustrated in
Figure 1.10(a). After doping with P, a defect energy level arises in the energy
gap. The electron occupying the defect energy level can be envisioned as a state
weakly bound by the defect P atom. We can examine the evolution of the
binding energy of the defect electron for the P-doped Si nanocrystal. The
binding energy can be defined as the energy required to ionise the P-doped Si
quantum dot by removing an electron (I_d) and the electron is subsequently
added to another pure Si nanocrystal (A_p) of the same size: $E_B = I_d - A_p$. As
shown in Figure 1.10(b), the ionisation energy of P-doped Si nanocrystal
depends weakly on size, and stays close to the bulk value of ~ 4 eV even for
nanoscale system. From the binding energy, the defect electron is found to be
more tightly bound as the dot size decreases. This is largely due to the weak
screening[31,32] in nanocrystals. By fitting to a power law, we find that the binding

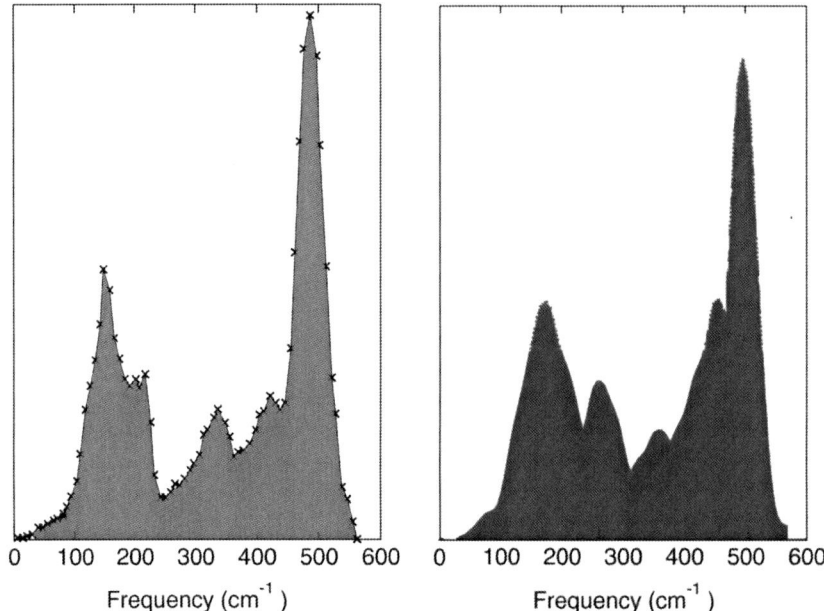

Figure 1.8 Vibrational modes in bulk silicon from experiment (left) and calculated modes for a nanocrystal ($Si_{281}H_{172}$). The experimental results are from ref. 28.

energy scales with the nanocrystal radius as $R^{-1.1}$ and approaches a bulk limit of 0.06 eV, which agrees well with the experimental measurement of 0.044 eV.

We can also study the isotropic hyperfine splitting (HFS) of the defect state. The HFS arises from the interaction between the electron spin of the defect state and the nucleus spin of the P atom, and can be calculated from the amplitude of the defect wavefunction at the P location. The isotropic hyperfine splitting of a wavefunction is given by

$$\text{HFS} = \frac{8\pi}{3} g_e \beta_e g_n \beta_n |\psi(\mathbf{0})|^2, \tag{1.14}$$

where g_e is the electron g factor, g_n is the g factor of the P nucleus, β_e is the electronic Bohr magneton, β_n is the nuclear magneton, and $|\psi(\mathbf{0})|^2$ is the spin density of the donor wavefunction at the nucleus site. Although the charge density in the core region is modified for calculations based on pseudopotentials, the amplitude of the charge density at the core can be corrected using Van de Walle and Blöchl method[33]:

$$|\psi(\mathbf{0})|^2 = |\tilde{\psi}(\mathbf{0})|^2 \left(\frac{|\phi_s(\mathbf{0})|^2}{|\tilde{\phi}_s(\mathbf{0})|^2} \right), \tag{1.15}$$

where $|\psi|^2$ denotes the all-electron defect charge density and $|\phi_s|^2$ denotes the charge density of the s-like partial wave of the defect atom. $|\tilde{\psi}|^2$ and $|\tilde{\phi}_s|^2$ are the

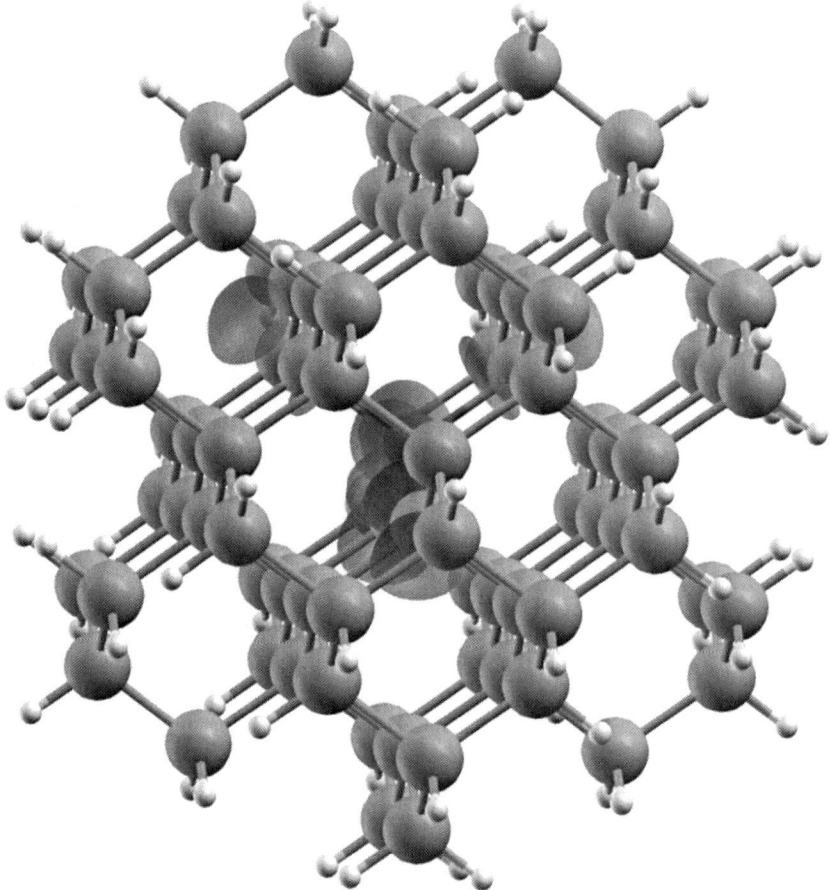

Figure 1.9 Ball and stick model of a phosphorus-doped silicon nanocrystal. The interior consists of a diamond fragment with the Si atom at the centre substituted by P. The surface of the fragment is capped with hydrogen atoms. A contour surface of the charge density of the defect wavefunction centred around the P atom is also illustrated.

pseudocounterparts of $|\tilde{\psi}|^2$ and $|\phi_s|^2$, respectively. Our calculated HFS based on this scheme are plotted in Figure 1.11. We can see that there is a very good agreement between experimental data[34] and our theoretical calculations. Even for nanocrystal with radius of 3 nm, the HFS is much higher than the bulk value of 42 G, and the HFS continues to increase as the nanocrystal decreases in size due to quantum confinement.

1.4.4 Hyperfine Splitting in Lithium-Doped ZnO Nanocrystals

ZnO is a wide-bandgap semiconductor that provides great prospects for optoelectronic and spintronic applications with various types of doping

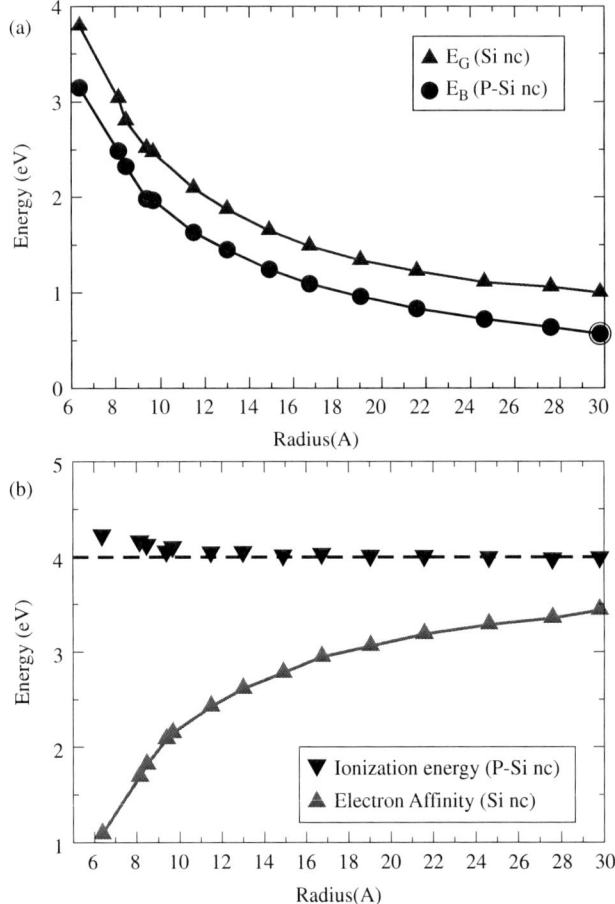

Figure 1.10   Electronic structures of P-doped Si nanocrystals. (a) The dependence on radius of the energy gap E_g in pure Si nanocrystal (▲) and binding energy E_B(•) in P-doped Si nanocrystal. (b) Ionisation energy of P-doped Si nanocrystal (▼) and electron affinity of a pure Si nanocrystal (▲) plotted as a function of radius.

materials.[35,36] In particular, Li is one of the promising dopants that reveals both p- and n-type conductivities.[35,37–41] Owing to its small atomic radius, Li can easily be introduced to form a substitutional, or interstitial defect in both bulk crystals and nanocrystals.[42]

Here, we present an analysis of the Li interstitial (Li_i) defect in ZnO nanocrystal. The Li_i-doped ZnO nanocrystals are constructed from a spherical fragment of the wurtzite structured ZnO. Each nanocrystal is chosen to have the same stoichiometry of Zn_nO_n. The nanocrystal surface is passivated with fictitious capping-layer atoms to remove any surface state from the energy gap.[43] We place the Li interstitial atom at the octahedral site[39,40] that is closest to the centre of a nanocrystal, as illustrated in Figure 1.12.

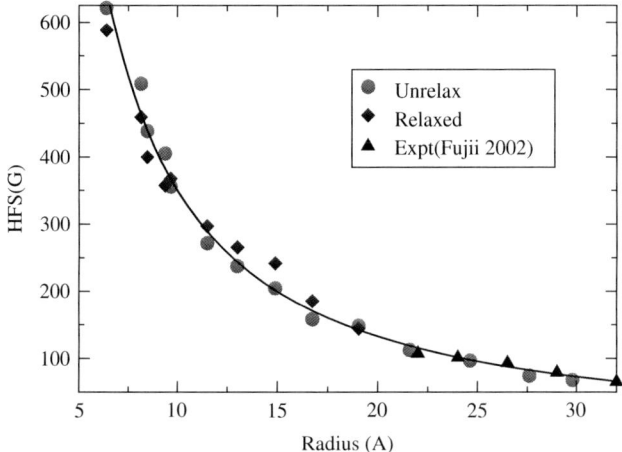

Figure 1.11 HFS of P-doped Si nanocrystal. Calculated HFS of P-doped Si quantum dot as a function of radius is plotted together with experimental data (▲) from ref. 34. Both calculation results corresponding to unrelaxed exact bulk-like geometry (●) and structures being relaxed to local energy minimum (◆) are shown in the figure.

We analyse the localisation shape of the defect wavefunction. The charge density of the donor wavefunction is calculated and depicted for a $(ZnO)_{87}Li$ nanocrystal (Figure 1.12). The wavefunction is not strongly localised at the defect atom. Instead, the defect state extends to the nearby oxygen atoms to form a strongly mixed state, leaving the Li atom partially ionised. The projection of the donor wavefunction inside a covalent radius of Li ($= 2\,\text{Å}$) was found to be less than 2% for all the Li_i-doped ZnO nanocrystals that we examined. It is very different from the defect charge density of P-doped Si nanocrystals in Figure 1.9, where the defect wavefunction was found to be localised at the impurity site. The drastic difference between the two systems illustrates the difference in nature between interstitial and substitutional defects.

A ramification of such a delocalised wavefunction is that the Van de Walle–Blöchl method to calculate HFS using pseudopotential is no longer applicable, as the theory is based on the defect wavefunction being localised on the defect atom. In Figure 1.13, we illustrate the results of our HFS calculations based on eqns (1.14) and (1.15). The values using the Van de Walle–Blöchl method are off by almost two orders of magnitude compared to the experiment.[44] Therefore, we generalise the Van de Walle–Blöchl method[45] and we find that HFS based on pseudopotentials can be calculated in general by

$$|\psi(\mathbf{0})| = |\tilde{\psi}(\mathbf{0})|\left(\frac{|\phi_s(\mathbf{0})|}{|\tilde{\phi}_s(\mathbf{0})|}(1-v)+v\right), \tag{1.16}$$

where v is a measure of the amount of off-site contributions and can be evaluated from first-principles calculations. Van de Walle–Blöchl's method

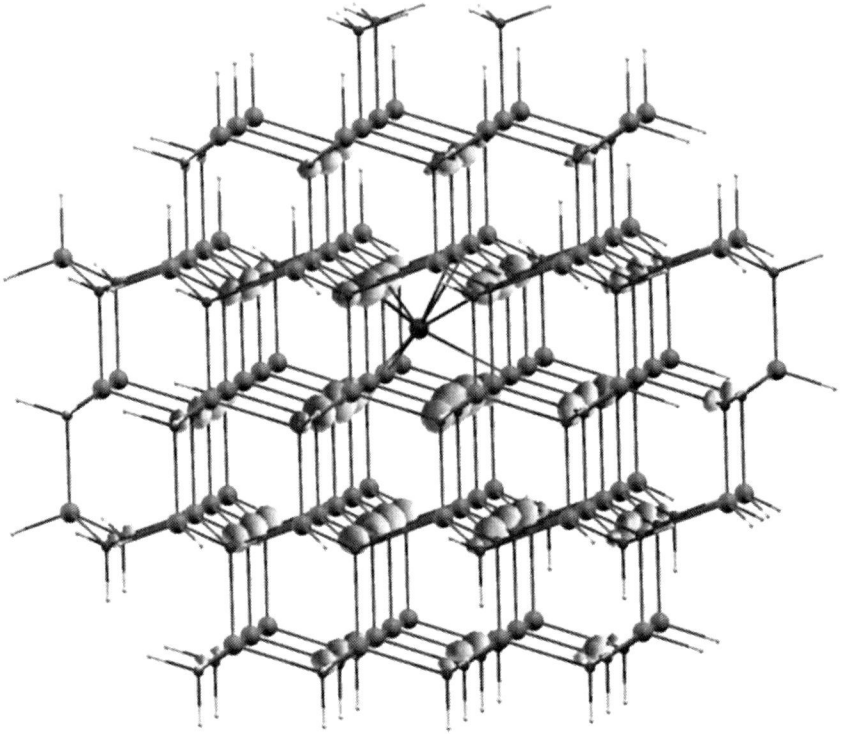

Figure 1.12   Charge-density isosurface plot of the defect wavefunction. Li atom
(center atom) is located at the octahedral site of the wurtzite structured
$(ZnO)_{87}$ nanocrystal. The grey isosurface indicates the diffuse nature of
the charge density. The isosurface corresponds to 40% of the maximum
amplitude. The large and small spheres represent zinc and oxygen atoms,
respectively. White spheres at the surface represent the capping atoms.

corresponds to one of the limiting cases where $v = 0$. By exploiting the fact that
v is relatively independent of the choice of pseudopotentials, v is evaluated to
be 0.95 ± 0.005 and is insensitive to the size of the nanocrystal. This value of
v is close to the limiting case ($v = 1$) where the defect atom becomes completely ionised. Our calculated values for HFS of the Li_i-doped ZnO nanocrystals using eqn (1.16) and the estimated off-site contribution $v = 0.95$
are plotted in Figure 1.13 and compared to experiment.[44] The agreement
between theory and experiment is dramatically improved by two orders of
magnitude.

1.4.5   Evolution of Magnetism in Iron Clusters

Ferromagnetism in bulk metals is reasonably well understood, the spontaneous
alignment of magnetic moments is caused by the exchange interaction mediated

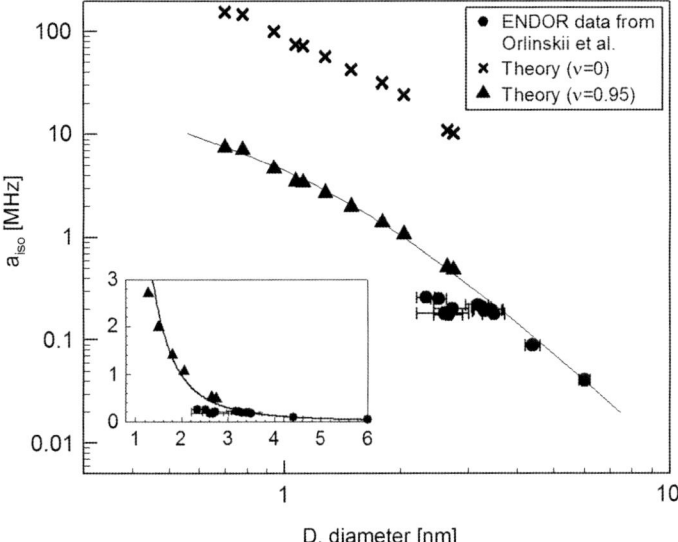

Figure 1.13   Log scale plot of Li$_i$ hyperfine splitting in ZnO nanocrystals as a function of nanocrystal size. The calculated splitting based on Van de Walle and Blöchl's scheme[33] ($\times$, $v=0$) differ by almost two orders of magnitude from the experimental data (•) at around $D=2.8$ nm. Theoretical results including the off-site contribution ($v=0.95$) are plotted as triangles (▲). Experimental data is taken from the ENDOR analysis by Orlinskii *et al.*[44] The solid line is a guide to the eye. The inset shows the same plot with linear scale axes.

by the itinerant electrons in the metal. For nanocrystals, electrons are confined and the mechanism of ferromagnetism is not well known. It is still an open problem of how the magnetic moment evolves from an atom *via* clusters to the bulk limit. This problem has been attracting considerable attention[46-56] owing to potential technological applications. Experimentally, it is found that the magnetic moment is suppressed as the clusters grow in size; however, the trend is not monotonic. On the theoretical side, studies based on first-principles calculations are still lacking to explain the observed phenomenon.

Here, we investigate the evolution of the magnetic moment in iron clusters containing 20 to 400 atoms. Three families of clusters with different geometries are considered in this study: icosahedral, body-centred cubic centred on an atom site, and body-centred cubic centred on the bridge between two neighbouring atoms. The magnetic moment is calculated as the expectation value of the total angular momentum:

$$M = \frac{\mu_B}{\hbar}\left[g_s\langle S_z\rangle + \langle L_z\rangle\right]$$
$$= \mu_B\left[\frac{g_s}{2}(n_\uparrow - n_\downarrow) + \frac{1}{\hbar}\langle L_z\rangle\right], \tag{1.17}$$

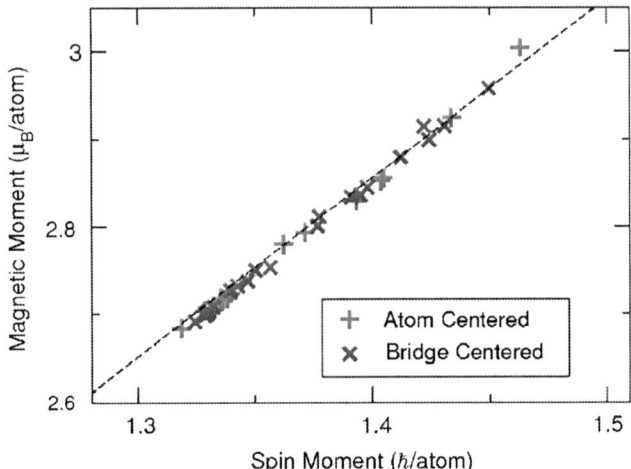

Figure 1.14   Magnetic moment *versus* spin moment calculated for the atom-centred
bcc (+) and bridge-centred bcc (×) iron clusters. The approximate ratio
is $M/\langle S_z \rangle = g_{\mathrm{eff}} = 2.04 \, \mu_\mathrm{B}/\hbar$.

where $g_\mathrm{s} = 2$ is the electron gyromagnetic ratio. The spin-orbit interaction is
included in the Kohn–Sham theory as a model potential

$$V_{\mathrm{so}} = -\xi \mathbf{L} \cdot \mathbf{S}, \tag{1.18}$$

where $\xi = 80$ meV/$\hbar^2$.[51] From Figure 1.14, the magnetic moment and spin
moment $\langle S_z \rangle$ has an approximate linear relationship throughout the range of
the size that we studied. We can therefore extract an effective gyromagnetic
ratio g_{eff} from this linear dependence, which is found to be 2.04 $\mu_\mathrm{B}/\hbar$. It is
somewhat smaller than the gyromagnetic ratio of 2.09 $\mu_\mathrm{B}/\hbar$ in bulk bcc iron.
The difference in ratios is probably due to an underestimation in the orbital
contribution, $\langle L_z \rangle$. In Figure 1.15, we show the size dependence of the
magnetic moment of the iron clusters for the three families of geometries.
The experimental data from Billas and collaborators[46] is also shown in the
figure.

   There is an overall suppression of magnetic moment as cluster size
increases, and the trend exhibits a nonmonotonic behaviour similar to
experimental observation. In the experiment, minima of magnetic moment
can be observed at sizes 45, 85, and 188. We find that these minima can be
associated to clusters with faceted surfaces. In the icosahedral family, a dip in
the magnetic moment can be found at a faceted cluster containing 55 iron
atoms, which is close to the first minimum measured experimentally. In the
atom-centred and bridge-centred bcc families, local minima in the magnetic
moment can be located around 65 to 92 iron atoms, which contain faceted
clusters as indicated in Figures 1.15(a) and (b). Clusters in this size range

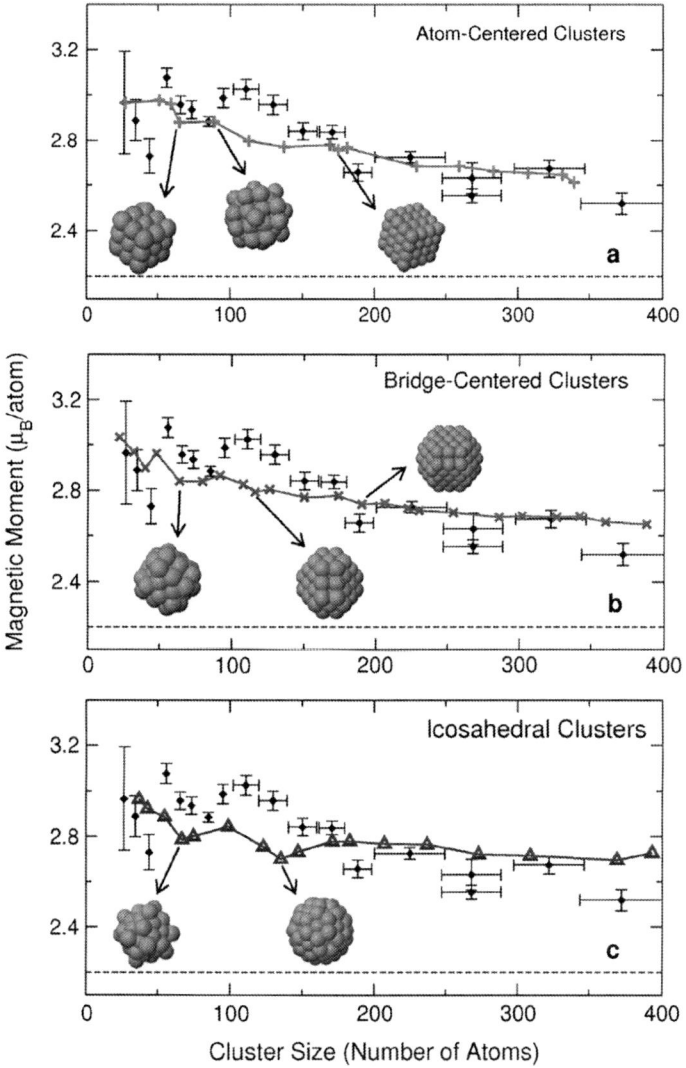

Figure 1.15 Calculated magnetic moments for clusters in the atom-centred ($+$, a), bridge-centred ($\times$, b), and icosahedral ($\triangle$, c) families. Experimental data[46] is shown in black diamonds with error bars. Some of the faceted and nonfaceted clusters are depicted next to their corresponding data points. The dashed lines indicate the value of the magnetic moment per atom in bulk iron.

include the second minimum of 85 observed in experiments. In addition, the 175 atom-centred and 173 bridge-centred iron clusters, close to the third minimum of 188 observed experimentally, are also faceted and produce smaller magnetic moment compared to clusters of similar size. The agreement between the calculated minima of magnetic moments and experimental data is not exact. This is probably because the magnetic moment was measured at 120 K,[46,48] which can induce vibrational modes or metastable configurations for the clusters.

Clusters with faceted surfaces have magnetic moments lower than other clusters with similar size owing to more effective hybridisation of *d* orbitals along the plane of the facets. Enhanced orbital hybridisation leads to a larger splitting in the energy levels, thus reducing the difference in electron population between the majority and minority spin channel, resulting in reduced spin polarisation and weaker magnetisation.

1.5 Conclusions

The algorithm presented in this chapter replaces the explicit eigenvalue calculations by an approximation of the wanted invariant subspace, obtained with the use of Chebyshev filters. Only the initial self-consistent field iteration requires solving an eigenvalue problem in order to provide a good initial subspace. In the remaining iterations, no iterative eigensolvers are involved. Instead, Chebyshev polynomials are used to refine the subspace. The subspace iteration at each step is easily five to ten times faster than solving a corresponding eigenproblem by the most efficient eigenalgorithms. Moreover, the subspace iteration reaches self-consistency within roughly the same number of steps as an eigensolver-based approach.

We illustrated this algorithm by applying it to hydrogenated silicon nanocrystals for both electronic and vibrational modes, doped semi-conductor nanocrystals and metallic iron clusters. The largest dot we examined contained over 10 000 atoms and was ~ 7 nm ($Si_{9041}H_{1860}$) in diameter. We examined the evolution of the electronic properties in the silicon nanocrystals, which we found to assume a bulk-like configuration for dots larger than ~ 5 nm. In addition, we obtained scaling relations for the ionisation potential, the electron affinity and the quasiparticle gap over the size regime of interest. We found the quasiparticle gap to approach the known bulk limit within density-functional theory. We did a similar calculation for vibrational modes, although the size of the nanocrystals is not as large, owing to computational issues, *i.e.* the need for more accurate forces, and that the vibrational modes converge to the bulk values more rapidly than the electronic states. A key aspect of our study is that we can examine the scaling of the electronic, vibrational and magnetic properties from an atom all the way up to a size that approaches the bulk limit. In particular, our calculations can simulate nanocrystals large enough that can be synthesised in experiments as illustrated by our studies on phosphorus-doped Si, Li-doped ZnO nanocrystals and the magnetic behaviour of Fe clusters. The good agreement between our calculated and experimental results demonstrates that our algorithm is efficient and accurate for electronic-structure prediction.

Acknowledgements

Computational work on nanoscale materials was supported in part by the US Department of Energy, Office of Basic Energy Sciences under grant No. DE-FG02-06ER46286. The National Science Foundation under DMR-0941645 supported the work on magnetic systems. Computational resources were provided in part by the National Energy Research Scientific Computing Center (NERSC), the Minnesota Supercomputing Institute (MSI) and the Texas Advanced Computing Center (TACC).

References

1. J. R. Chelikowsky, A. T. Zayak, T.-L. Chan, M. L. Tiago, Y. Zhou and Y. Saad, *J. Phys. Condens. Matter.*, 2009, **21**, 0642087.
2. A. P. Alivisatos, *Science*, 1996, **271**, 933.
3. C. Troparevsky, L. Kronik and J. R. Chelikowsky, *J. Chem. Phys.*, 2003, **119**, 2284.
4. M. L. Tiago, Y. Zhou, M. M. G. Alemany, Y. Saad and J. R. Chelikowsky, *Phys. Rev. Lett.*, 2006, **97**, 147201.
5. W. Kohn and L. J. Sham, *Phys. Rev.*, 1965, **140**, A1133.
6. J. R. Chelikowsky and M. L. Cohen, in *Handbook of Semiconductors*, ed. P. Landsberg, Elsevier, Amsterdam, v. 1. 1992, p. 59.
7. J. R. Chelikowsky, *J. Phys. D.*, 2000, **33**, R33.
8. G. Kresse and J. Furthmüller, *Phys. Rev. B*, 1996, **54**, 11169.
9. J. R. Chelikowsky, N. Troullier and Y. Saad, *Phys. Rev. Lett.*, 1994, **72**, 1240.
10. T. Torsti, T. Eirola, J. Enkovaara, T. Hakala, P. Havu, V. Havu, T. Höynälänmaa, J. Ignatius, M. Lyly, I. Makkonen, T. T. Rantala, J. Ruokolainen, K. Ruotsalainen, E. Räsänen, H. Saarikoski and M. J. Puska, *Phys. Status Solidi*, 2006, **243**, 1016.
11. L. Kronik, A. Makmal, M. L. Tiago, M. M. G. Alemany, M. Jain, X. Huang, Y. Saad and J. R. Chelikowsky, *Phys. Status Solidi (b)*, 2006, **243**, 1063.
12. Y. Zhou, Y. Saad, M. L. Tiago and J. R. Chelikowsky, *Phys. Rev. E.*, 2006, **74**, 066704.
13. T. L. Beck, *Rev. Mod. Phys.*, 2000, **74**, 1041.
14. B. Fornberg and D. M. Sloan, *Acta Numerica*, 1994, **94**, 203.
15. A. Stathopoulos, S. Öğüt, Y. Saad, J. Chelikowsky and H. Kim, *Comput. Sci. Eng.*, 2000, **2**, 19.
16. C. Bekas, Y. Saad, M. L. Tiago and J. R. Chelikowsky, *Comp. Phys. Comm.*, 2005, **171**, 175.
17. R. B. Lehoucq, D. C. SorensenC. Yang, *ARPACK Users Guide: Solution of Large Scale Eigenvalue Problems by Implicitly Restarted Arnoldi Methods*, SIAM, Philadelphia, 1998.
18. K. Wu, A. Canning, H. D. Simon and L. W. Wang, *J. Comp. Phys.*, 1999, **154**, 156.

19. http://parsec. ices. utexas. edu.
20. S. Furukawa and T. Miyasato, *Phys. Rev. B*, 1988, **38**, 5726.
21. S. Ogut and J. R. Chelikowsky, *Phys. Rev. B*, 1997, **55**, R4914.
22. J. R. Chelikowsky and M. L. Cohen, *Phys. Rev. B*, 1976, **14**, 556.
23. L. J. Sham and M. Schlüter, *Phys. Rev. Lett.*, 1983, **51**, 1888.
24. K. Parlinski, Z. Q. Li and Y. Kawazoe, *Phys. Rev. Lett.*, 1997, **78**, 4063.
25. M. M. M. Alemany, J. Manish, L. Kronik and J. R. Chelikowsky, *Phys. Rev. B*, 2004, **69**, 075101.
26. A. V. Postnikov, O. Pagès and J. Hugel, *Phys. Rev. B*, 2005, **71**, 115206.
27. X. Huang, E. Lindgren and J. R. Chelikowsky, *Phys. Rev. B*, 2005, **71**, 165328.
28. W. A. Kamitakahara, C. M. Soukoulis, H. R. Shanks, U. Buchenau and G. S. Grest, *Phys. Rev. B*, 1987, **36**, 6539.
29. X.-S. Zhao, Y.-R. Ge and X. Zhao, *J. Mater. Sci.*, 1998, **33**, 4267.
30. C. Meier, S.L.S.V.G. Kravets, H. Nienhaus, A. Lorke and H. Wiggers, *Physica E*, 2006, **32**, 155.
31. S. Öğüt, J. R. Chelikowsky and S. G. Louie, *Phys. Rev. Lett.*, 1997, **79**, 1770.
32. S. Öğüt, R. Burdick, Y. Saad and J. R. Chelikowsky, *Phys. Rev. Lett.*, 2003, **90**, 127401.
33. C. G. V. de Walle and P. E. Blöchl, *Phys. Rev. B*, 1993, **47**, 4244.
34. M. Fujii, A. Mimura, S. Hayashi, Y. Yamamoto and K. Murakami, *Phys. Rev. Lett.*, 2002, **89**, 206805.
35. O. Lopatiuk, L. Chernyak, A. Osinsky and J. Q. Xie, *Appl. Phys. Lett.*, 2005, **87**, 214110.
36. D. Karmakar, S. K. Mandal, R. M. Kadam, P. L. Paulose, A. K. Rajarajan, T. K. Nath, A. K. Das, I. Dasgupta and G. P. Das, *Phys. Rev. B*, 2007, **75**, 144404.
37. C. Park, S. Zhang and S.-H. Wei, *Phys. Rev. B*, 2002, **66**, 073202.
38. J. G. Lu, Y. Z. Zhang, Z. Z. Ye, Y. J. Zeng, H. P. He, L. P. Zhu, J. Y. Huang, L. Wang, J. Yuan, B. H. Zhao and X. H. Li, *Appl. Phys. Lett.*, 2006, **89**, 1112113.
39. M. G. Wardle, J. P. Goss and P. R. Briddon, *Phys. Rev. B*, 2005, **71**, 155205.
40. E.-C. Lee and K. J. Chang, *Phys. Rev. B*, 2004, **70**, 115210.
41. Y. J. Zeng, Z. Z. Ye, J. G. Lu, W. Z. Xu, L. P. Zhu and B. Zhao, *Appl. Phys. Lett.*, 2006, **89**, 042106.
42. A. Onodera, K. Yoshio, H. Satoh, H. Yamashita and N. Sakagami, *Jpn. J. Appl. Phys.*, 1998, **37**, 5315.
43. X. Huang, E. Lindgren and J. R. Chelikowsky, *Phys. Rev. B*, 2005, **71**, 165328.
44. S. B. Orlinskii, J. Schmidt, E. J. J. Groenen, P. G. Baranov, C. de Mello Donegá and A. Mei-jerink, *Phys. Rev. Lett.*, 2005, **94**, 097602.
45. H. Kwak, M. L. Tiago, T.-L. Chan and J. R. Chelikowsky, *Chem. Phys. Lett.*, 2008, **454**, 337.

46. I. M. L. Billas, J. A. Becker, A. Châtelain and W. A. de Heer, *Phys. Rev. Lett.*, 1993, **71**, 4067.
47. D. Gerion, A. Hirt, I. M. L. Billas, A. Châtelain and W. A. de Heer, *Phys. Rev. B*, 2000, **62**, 7491.
48. I. M. L. Billas, A. Châtelain and W. A. de Heer, *Science*, 1994, **265**, 1682.
49. J. A. Franco, A. Vega and F. Aguilera-Granja, *Phys. Rev. B*, 1999, **60**, 434.
50. A. Postnikov and P. Entel, *Phase Transit.*, 2004, **77**, 149.
51. O. Šipr, M. Košuth and H. Ebert, *Phys. Rev. B*, 2004, **70**, 174423.
52. A. Postnikov, P. Entel and J. M. Soler, *Eur. Phys. J. D*, 2003, **25**, 261.
53. R. Félix-Medina, J. Dorantes-Dávila and G. M. Pastor, *Phys. Rev. B*, 2003, **67**, 094430.
54. K. Edmonds, C. Binns, S. H. Baker, S. Thornton, C. Norris, J. B. Goedkoop, M. Finazzi and N. B. Brookes, *Phys. Rev. B*, 1999, **60**, 472.
55. G. M. Pastor, J. Dorantes-Dávila and K. H. Bennemann, *Phys. Rev. B*, 1989, **40**, 7642.
56. O. Diéguez, M. M. G. Alemany, C. Rey, P. Ordejón and L. J. Gallego, *Phys. Rev. B*, 2001, **63**, 205407.

CHAPTER 2

Rational Design of Mixed Nanoclusters: Metal Shells Supported and Shaped by Molecular Cores

FEDOR Y. NAUMKIN

Faculty of Science, UOIT, Oshawa, ON L1H 7K4, Canada

2.1 Introduction

Atomic and molecular clusters are the central objects of studies explosively expanding in recent decades due to the ever-increasing capability to produce and manipulate them. These nanometer-sized systems are, by definition, at the very heart of modern nanoscience and nanotechnology, and, due to their endless variety, have every chance of staying there in future. Scaling the size of a cluster of a given composition between the limits of a single atom/molecule and the corresponding bulk material implies variation of properties between those (very different) of a quantum and a classical entity. In particular, metal clusters can exhibit semiconductor or insulator properties;[1,2] gold, famous for chemical inertness in bulk quantities, becomes a versatile and efficient catalyst at cluster scale;[3] small metal clusters ("superhalogens") show electron affinities exceeding that of halogen atoms,[4] *etc.*

Cluster composition is another parameter strongly affecting cluster properties. For atoms or molecules of two different pure substances, A and B, the mixed cluster $A_m B_{n-m}$ of a fixed size n can apparently cover the whole interval

RSC Theoretical and Computational Chemistry Series No. 4
Computational Nanoscience
Edited by Elena Bichoutskaia
© Royal Society of Chemistry 2011
Published by the Royal Society of Chemistry, www.rsc.org

of properties between those of A_n and B_n with varying m. Further degrees of freedom are associated with more-than-binary systems. This is reflected by the quickly developing research in the area of doped (with both a nonmetal dopant atom in a metal host and *vice versa*)[5-7] and mixed (with multiple dopant atoms) clusters such as metcars[8] and nanoalloys (with dopant and host as different metals).[9]

Yet another factor is the cluster shape, which can considerably alter properties between different isomers of otherwise the same (in terms of size and composition) system. One relevant example of a system currently defining a whole new research area is graphene, an all-flat isomeric form of carbon (with a honecomb-shaped one-atom thick grid of atoms). Here, we could certainly recall other famous isomers of carbon like diamond, fullerenes, and carbon nanotubes, all with very different properties. By analogy with carbon species, metal cluster cages and nanotubes *versus* bulk crystalline solid are appropriate examples as well.[10,11]

An enormous multitude of property designs is accessible by combining the above factors, such as shape and size variations of clusters of mixed composition, for instance as in gold-covered fullerenes[12] and layered "onion" clusters.[13] Sometimes, one factor can spontaneously involve another. For instance, gold clusters switch from energetically preferred 2D to 3D geometry with increasing size (see, *e.g.*, ref. 14); doping a pure cluster with a different atom can induce shape change, such as originally planar (rhombic) metal tetra-atom (Al_4, Au_4) forming with the added carbon or silicon atom a tetrahedral methane-like structure[15-17] due to the dopant's inherent bond pattern, *etc.*

An extra dimension can be added when considering shape changes enabled by "dopants" forming no covalent bonds to the host cluster. For instance, water molecules are known to form hydrogen-bonded cages around inserted hydrophobic molecules such as saturated organics (methane, *etc.*). By analogy, we could imagine a metal cage (shell) surrounding a closed-shell atom or molecule (core) suspended inside it electrostatically, say *via* weak van der Waals or maybe (if a core–shell charge transfer occurs) appreciable Coulomb interaction. In particular, such an arrangement would facilitate exploring the otherwise inaccessible shape-related properties of a cage that would collapse without such a supporting "filler" inside. Moreover, using molecular cores of appropriate sizes and geometries, we could attempt designing metal-shell based nanosystems with desirable shapes and related properties. One recent example of a relevant system created experimentally is gold-covered polymer nanorods.[18]

In this work, we present and review recent studies of core–shell clusters with nonmetal atomic and molecular centres (C_n, CH_4, H_2) encapsulated into metal cluster shells (Au_n, Al_n, Mg_n). Both open- and closed-shell cores, bonded to the shell covalently (C_n) and electrostatically (CH_4, H_2), as well as the shells both stable by themselves as cages (with no core inside) and not stable are considered. The focus is on structures and stabilities, while charge distributions and some other electronic properties (excitation and ionisation energies, electron affinity) are also discussed.

2.2 General Aspects of Computational Procedure

All calculations have been carried out using the NWChem computational chemistry package,[19] and obtained structures visualised with the ViewMol3D molecular modelling software.[20] For relatively small C_nM_m ($n = 1$–3, $m = 1$–8, M = Au, Al) species an MP2 level of theory has been preferred, replaced due to the size of the system by DFT(B3LYP) for larger C_nAu_m ($n = 5$ and 10, $m = 12$ to 24) systems. The MP2 theory has also been used for $H_2@Mg_n$, while both approaches employed for $H_2@Au_{12}$ and $CH_4@Au_{12}$. These standard methods are implemented in NWChem.

Basis sets for atoms have varied from 6-31G* for carbon and hydrogen in larger C_nAu_m, $H_2@Au_{12}$ and $CH_4@Au_{12}$, to aug-cc-pVTZ for all atoms in small C_nAu_m, C_nAl_m and $H_2@Mg_n$. For gold, a standard choice has been the LANL2dz effective core potential (ECP) and associate basis set, with limited corrections using the Stuttgart's RSC (relativistic small-core) ECP and basis set. Details of all these bases are available at the PNL's Basis Set Exchange online facility.[21] In particular, the results for identical basis sets and methods used for larger C_nAu_m, $CH_4@Au_{12}$, and $H_2@Au_{12}$ allow a direct quantitative comparison of these systems. The choice of methods and basis sets is governed by a reasonable compromise between the accuracy judged from comparison to available experimental data or highly reliable calculated results for small fragments (such as ionisation energies and electron affinities for atoms, dissociation energies and equilibrium distances for diatoms),[22] and the acceptable time of calculations.

All-atom symmetry-unconstrained optimisation has been employed, and energy minima verified in terms of all-real vibrational frequencies. For each species, energies for both lowest and higher spin states are calculated, in order to properly establish the ground state multiplicity. For every system, multiple initial geometries have been tried. For the C_nAu_m and C_nAl_m species, it appears reasonable to start, in particular, with structures of metal-substituted C_nH_m molecules. The rationale behind this is the relations between the relative bond strengths (C–C > C–M > M–M), so that the integrity of carbon skeletons is preserved, while efficient carbon–metal interaction is enabled in the hydrocarbon-like structures. Hence, it is anticipated that such a starting point should lead to low-energy isomers. Besides, this allows considering species likely to be produced in experiments effectively *via* such a metal substitution, so that optimisation could mimic a subsequent process leading to realistic products. For the $H_2@Au_{12}$ and $CH_4@Au_{12}$ systems, a few possible orientations of a molecule inside a metal cluster cage, as well as a few different cage structures, are attempted.

It is possible that the resulting optimised geometries correspond to local energy minima. Finding the global minimum, *i.e.* lowest-energy isomer, is certainly important and desirable, while it is realised that identifying higher-energy isomers is also significant, especially if they may be stabilised by high enough energy barriers to be sufficiently long-living for practical purposes. An apparent support of their role, directly relevant to this work, is a variety of the isomers of virtually each hydrocarbon molecule, readily produced in experiments and having various useful applications. Thus, for instance, we should not aim only at graphite while suspecting possibility of diamond. One recent example related to the present work is the structure of the $SiAu_{16}$ cluster, predicted to have a higher-energy symmetric core–shell isomer $Si@Au_{16}$,[23] as well as a lower-energy isomer with Si attached to the surface of asymmetric Au_{16}.[24]

Another, related aspect is that the lower-energy species are usually easier to obtain experimentally, at least with currently established procedures. However, experimental methods develop quickly. Furthermore, if a higher-energy isomer promises (at least theoretically) to possess very attractive properties, then stronger efforts to find new, unusual ways to efficiently produce it will be fully justified, very likely made, and probably succeed. A recent illustration may be the experiments on generating carbon–gold clusters by laser vaporisation of gold-covered carbon rods,[25] producing carbon-dominated products with only a few (mainly one) gold atoms. One could interpret this outcome as a likely consequence of the target structure, with the gold layer evaporated first and being almost gone by the time carbon begins to vaporise. One possible attempt to overcome this could be to try also to invert the situation and enrich clusters with gold *via* irradiating carbon (graphite?) -covered gold targets. At least, the above considerations regarding relative bond strengths and their reflection in the stable system structure are consistent with these experimental results.

2.3   Results and Discussion

The presentation of results is organised as follows. First, systems with cores forming bonds to shell atoms are considered, in the general order of increasing size, beginning with the core. Secondly, saturated closed-shell cores non-covalently suspended inside the shell are introduced.

Hence, starting from a single carbon atom with a few gold atoms attached, we proceed to a carbon diatomic core, then look at the differences associated with replacing Au by Al atoms, and the core is then increased to a triatomic molecule. The next step involves return to auro-carbon systems, with the gold shell surrounding the carbon core represented by penta and deca-atomic (radical) species. In the following stage, the core is changed qualitatively, to a small hydrocarbon molecule (methane) as a closed-shell counterpart of the C_5 radical. Finally, the core is reduced to a very small (hydrogen) diatom, and then evolves into a hydrogen dimer, both encapsulated in cluster cages of magnesium atoms. The last case thus deals with more than one molecule inside a metal shell.

2.3.1   Small Gold Clusters Doped by Carbon

2.3.1.1   CAu_n (n = 2-6)

Adding a carbon atom to the gold diatom along or perpendicular to its axis produces two isomers, linear CAuAu and isosceles-triangular counterpart of methylene, respectively (Figure 2.1), similar to the $SiAu_2$ counterpart.[17] Gold clusters of three to several atoms are known to prefer planar structures composed of near-equilateral triangular Au_3 units fused by their sides (for example, Au_4 is a rhombus).[14] Accordingly, the C atom can be attached to the corner, side, or face of Au_3, producing planar Y-shaped and cyclic, and near-planar methyl-like isomers, respectively.

For both CAu_2 and CAu_3, the relative stability of isomers appears to be simply related to the number of stronger C–Au bonds, so that the energy for complete dissociation (into atoms) increases from linear (one bond) to triangular (two) CAu_2 and from corner- (one bond) to side- (two) to face-bound (three) CAu_3. In particular, for analogous $SiAu_3$, methyl-like isomer has also been found to be most stable.[17] The same trend is preserved for CAu_4, the tetrahedral (T_d) methane-like isomer being most stable, similar to $SiAu_4$,[17] followed by asymmetric trigonal bipyramidal (ATB), planar B-shaped and Φ-shaped, with four to one C–Au bonds, respectively, the first two isomers being identified previously as well.[16]

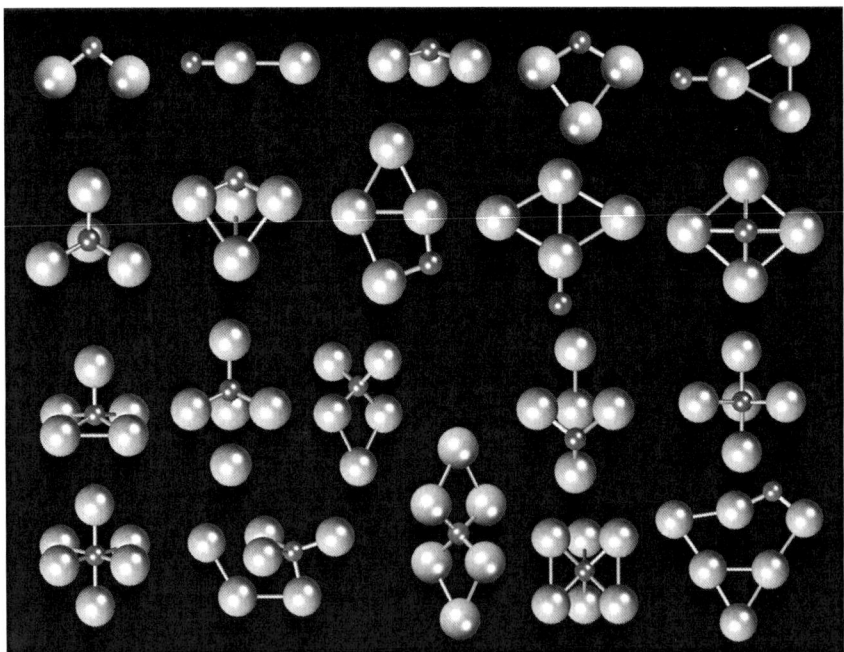

Figure 2.1 Optimised geometries of: the bent and linear isomers of CAu_2 (top row left), and the methyl-like, cyclic and Y-shaped isomers of CAu_3 (right); the methane-like, trigonal-bipyramidal, planar B- and Φ-shaped isomers, and transition state of CAu_4 (2nd row); the square-pyramidal, trigonal-bipyramidal, γ-shaped, and capped-ATB isomers, and transition state of CAu_5 (3rd row); the octahedral, capped-ATB, twisted-8-shaped, trigonal-prismatic, and planar isomers of CAu_6 (bottom row).

Starting from $n = 4$, adding the C atom transforms planar Au_n into 3D shells around the carbon core, at least for the most stable isomers of CAu_n (like the T_d isomer of CAu_4). Their higher-energy isomers can be obtained by attaching another Au atom to gold atoms of CAu_{n-1} (like the ATB isomer of CAu_4, with 4th Au attached to the Au_3 face of CAu_3) or C atom to Au_n (like the B- and Φ-isomers of CAu_4), see Figure 2.1. It is worth noting that adding C to the other corner of the Au_4 rhombus still leads to the Φ-isomer, thus effectively "rotating" the gold component.

The CAu_4 structure with C attached centrally to (bent) rhombic Au_4, predicted previously (with DFT) as an isomer,[16] is found (with MP2) to be a transition state relaxing to the T_d isomer, apparently due to an unfavourable C–Au bond pattern. However, the CAu_5 isomer predicted to be most stable in this work can be viewed as such a CAu_4 structure stabilised by attaching 5th Au to the carbon atom, on the opposite side (Figure 2.1). The resulting square-pyramidal CAu_5 species, analogous to that predicted for $SiAu_5$,[26] represents an unusual case of pentacoordinated carbon, also different from normally expected trigonal bipyramidal structure for such a coordination, as, *e.g.* for the CLi_5 counterpart.[27] In fact, the latter structure for CAu_5 is predicted to be a transition state relaxing to the square pyramid.

Three other isomers, in the order of increasing energy, can be obtained by attaching a 5th Au to the Au_3 face (adding three new Au–Au pair interactions in the resulting asymmetric trigonal bipyramid) or to the Au_2 side (two such interactions in the resulting γ-shaped isomer) of the T_d isomer of CAu_4, and to the Au_3 face of the (higher-energy) ATB isomer of CAu_4. Again, the relative stability of the CAu_5 isomers follows the number of C–Au bonds – from five to three. Starting from CAu_5, the similarity of at least some isomers to corresponding hydrocarbons is lost, as expected.

The most stable isomer of CAu_6 found in this work is octahedral, with hexacoordinated carbon atom at its centre (Figure 2.1), obtained by relaxing the above CAu_5 square pyramid with a 6th Au attached centrally to its base. This geometry is what is normally expected for such a coordination (found also, *e.g.*, for CLi_6[27]) which, however, is still rather unusual for carbon.

Two higher-energy isomers correspond to 6th Au attached to the Au_3 face (farther from C) of the above (higher-energy) ATB, and to the Au_2 side (opposite the 5th Au added before) of γ-shaped CAu_5, resulting in a twisted-8-shaped isomer of CAu_6. An isomer with a still higher energy is another unusual species with hexacoordinated carbon at the centre of a trigonal prism of the gold shell, produced *via* relaxing either the ATB or the square-pyramidal isomer of CAu_5 with 6th Au attached to its Au_3 face. The least-stable isomer of those studied here is obtained by attaching C atom to a side of the planar (near-equilateral triangular) Au_6. Here, the general trend of stability decreasing with the number of C–Au bonds (from six to four to two for the above isomers) is violated by the trigonal-prismatic isomer, indicating a higher strain for its bond pattern, with the AuCAu bond angles of less than 85°.

The above isomers of CAu_2 and CAu_3 are separated by energy differences of about 2 eV. For CAu_4 the gap between the lowest-energy and nearest next

Table 2.1 Characteristic energies (in eV) of the lowest-energy isomers of
CAu$_n$.

n	D_e ($\to C + n$ Au)	D_e ($\to CAu_{n-1} + Au$)	D_e ($\to C + Au_n$)	VE^{*a}	VIE	VEA
2	6.63	3.57	4.74	0.75	8.51	0.89
3	10.0	3.37	7.07	2.59	7.46	2.64
4	14.0	3.99	8.56	2.48	8.31	1.08
5	15.5	1.50	7.95	2.84	5.95	2.46
6	17.9	2.42	7.18	1.34	6.51	0.87

^{a}S = 0 → 1 for even n, S = 1/2 → 3/2 for odd n

isomer reduces to ≈ 1 eV, which recovers to $\approx 2 \pm 0.5$ eV for the subsequent
isomers. For CAu$_5$ and CAu$_6$, however, respectively two and three higher
energy isomers are within 0.3 ± 0.05 eV of the lowest-energy isomer.

The equilibrium distance R_e(C–Au) for the lowest-energy isomers found in this
work steadily increases from CAu to CAu$_6$, slowly up to $n = 4$ and then faster for
$n = 5, 6$. The overall variation falls within 15%, and the bond lengths are about
2 Å. In the lowest-energy CAu$_4$ and CAu$_6$ all R_e(C–Au) are equal, and in the
square-pyramidal CAu$_5$, the axial C–Au bond is slightly shorter (by ≈ 0.1 Å)
than the other four. The R_e(Au–Au) distance is longest in CAu$_3$, only slightly
shorter in CAu$_4$, and in CAu$_6$ recovers the value for CAu$_2$, varying within 7%. In
all systems, R_e(Au–Au) of about 3 Å is significantly longer than in isolated Au$_2$
(2.5 Å), consistent with dominant C–Au bonding.

For the lowest-energy isomers, the total dissociation energies D_e(CAu$_n$ →
C + n Au) per atom vary relatively weakly with n, peaking for $n = 4$ (Table 2.1).
The energies D_e(CAu$_n$ → CAu$_{n-1}$ + Au) for detachment of one Au atom from
CAu$_n$ show a slow increase of up to $n = 4$, and then a sharp drop for the
hypercoordinated carbon case with $n = 5$, followed by a significant increase for
$n = 6$. The highest relative stability of the methane-like CAu$_4$ is further confirmed
by the energies D_e(CAu$_n$ → C + Au$_n$) for removal of the C atom, with their faster
increase with n before, and a slower decrease after $n = 4$. In both cases the
remaining CAu$_{n-1}$ and Au$_n$ are reoptimised. Thus, from CAu to CAu$_4$ the stabilisation of the system with adding Au atoms accelerates with increasing n due to
more binding contributions: with each Au, one C–Au bond plus $n - 1$ Au–Au
interactions are added. In CAu$_4$ the valence of carbon is saturated, and the
system has a total of "magic" 8 valence electrons in its atoms. In CAu$_5$ and CAu$_6$,
the C–Au bonds are strained due to the hypercoordination of carbon and the
gold shell is overcrowded with atoms, slowing down the further stabilisation.

We can approximately partition the total dissociation energy D_e(CAu$_n$ →
C + n Au) using a simple model of effective D(C–Au) and D(Au–Au) pair
components. Then, for CAu$_4$:

$$D_e(CAu_4) \approx 4D(C - Au) + 6\ D(Au - Au).$$

Similarly, for the CAu$_3^*$ fragment with atoms frozen in the same positions as
in CAu$_4$:

$$D_e(CAu_3{}^*) \approx 3\ D(C - Au) + 3\ D(Au - Au).$$

Using $D_e(CAu_4)$ and $D_e(CAu_3^*)$ calculated at the same level of theory and solving the above two equations, we obtain $D(C–Au) \approx 2.8\,eV$ and $D(Au–Au)$ $\approx 0.5\,eV$. These sum up to $\approx 11\,eV$ of total carbon-gold and $\approx 3\,eV$ of total gold–gold interaction in CAu_4. The net aurophilic interaction in this case thus contributes $\approx 21\%$ to the total binding.

From similar calculations for octahedral CAu_6 and its (frozen) CAu_5^* fragment, $D(C–Au) \approx 2.6\,eV$ and $D(Au–Au) \approx 0.2\,eV$, adding up to 6 $D(C–Au) \approx 16$ and 12 $D(Au–Au) \approx 2.2\,eV$ for the total carbon–gold and gold–gold interactions, respectively. The latter value is considerably less relative to that for CAu_4, even though the number of the Au–Au pair interactions is doubled. This leads to $\approx 12\%$ of binding due to aurophilic interaction, about half of the value for CAu_4. Such a reduction can be explained in terms of shorter Au–Au distances and therefore stronger repulsion of the positively charged Au atoms in CAu_6, weakening the net aurophilic interaction. The partial charge on the atoms is implicitly accounted for in their effective interactions.

For the lowest-energy isomers of CAu_n, the negative charge on carbon increases with the number of Au "electron donors", almost linearly up to $n = 4$, then significantly (about twice) more slowly for $n = 5$ and 6, reflecting a reduced ability of the carbon anion to accept more electrons. As a result, the C atom carries $\approx -2e$ in CAu_4, and even more in the hypercoordinated species. Availability of at least two extra electrons could enable formation of extra bonds, such as one and two additional C–Au bonds in CAu_5 and CAu_6, respectively. The charges on gold atoms increase with n slightly, leveling at $\approx 0.5e$ already for $n = 3$. Using these charges and the equilibrium Au–Au distances, a simple sum of Coulomb Au–Au pair repulsions in CAu_6 is about twice that in CAu_4, consistent with the above considerations for the total gold–gold interactions in these systems.

The lowest-energy isomers of CAu_n have moderate (≈ 1 D for $n = 2, 3$), small (≈ 0.2 D for $n = 5$), or no dipole moments (for $n = 4, 6$). Calculated critical electron densities in these isomers are equal (or nearly equal in CAu_5) for all C–Au bonds, and slowly decrease with increasing n. Their values for $n = 2$ to 4 are about half those for the C–H bonds in the corresponding CH_n molecules.

For the lowest-energy isomers of CAu_n, each of vertical spin-excitation energy VE* (effectively the HOMO–LUMO gap), ionisation energy VIE and electron affinity VEA show typical even-odd oscillations in values with increasing cluster size, reflecting spin-pairing for even n (Table 2.1). The VIE values generally decrease with increasing n, while VE* appear to slowly increase and VEA to remain about constant on average, at least in the studied size range.

The CAu_5 cluster offers an interesting example of property evolution with geometry variation and "doping" a small gold cluster by a carbon atom. When the (lowest-energy) planar isomer of Au_5 is transformed into the higher-energy trigonal-pyramidal one, its VE* value drops dramatically, but is then recovered on adding the C atom to its surface (at the Au_3 face). The further transitions to the ATB isomer (with C "sinking" into the cluster) and then to the square-pyramidal isomer do not, however, noticeably affect the VE* value (Table 2.2). At the same time, the VIE and VEA values vary weakly (within 10–15%) with

Table 2.2   Characteristic energies (in eV) of Au_5 and CAu_5 isomers.

System	$D_e{}^a$	$VE^*(S=1/2\to3/2)$	VIE	VEA
Au_5 (planar)	7.54	2.56	6.77	2.60
Au_5 (trigonal bipyr.)	6.90	0.65	6.47	2.23
CAu_5 (capped ATB)	13.9	2.65	6.00	2.31
CAu_5 (ATB)	15.3	2.76	6.05	2.52
CAu_5 (square pyr.)	15.5	2.84	5.95	2.46

$^a Au_5 \to 5\ Au$, $CAu_5 \to C + 5\ Au$

either the shape change or the "doping", the former decreasing and the latter going through a shallow minimum at the 2D to 3D transition in Au_5 and then recovering. Both these alterations are counterintuitive in terms of higher IE and lower EA of C relative to Au.

The open-shell CAu_5 system could be expected to be a reactive species with a low chance of sufficiently long survival in experiments. It turns out that its dimer, $(CAu_5)_2$ can preserve the unique pentacoordination of both carbon atoms while being a stable closed-shell system, consistent with its "magic" total number (18) of valence electrons in atoms. Accordingly, the electron affinity drops dramatically from the monomer to the dimer, from VEA ≈ 2.5 to $\approx 0.6\,eV$. The corresponding variations of VIE and VE* are relatively weaker, the former value increasing and the latter decreasing by $\approx 25\%$. The system can be obtained by joining two CAu_5 units by their square bases in a staggered geometry (Figure 2.2). The binding energy of the monomers is $\approx 4.5\,eV$, and similar to the monomer case, the dimer structure is destabilised by detachment of an axial Au atom.

The above CAu_n $(n>3)$ species can be formally treated as core–shell systems with a single-atom carbon centre surrounded by a one-atom thick layer of gold. Next, we will increase the core size to two atoms.

2.3.1.2 C_2Au_n $(n=1-6)$

From the above analysis of the CAu_n species it appears that the most stable isomers of auro-carbons could be expected to resemble hydrocarbons for up to

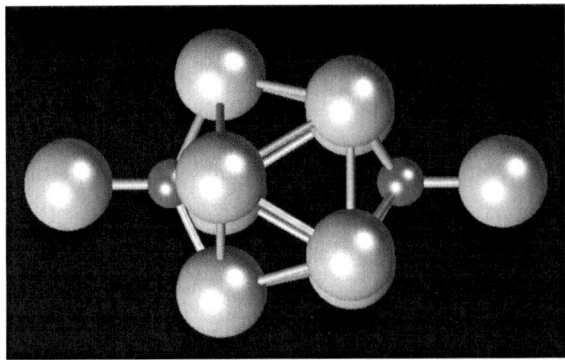

Figure 2.2   Optimised geometry of the $(CAu_5)_2$ dimer.

formal saturation of valence in the C atom, and, for more gold in the system, to have structures possibly different from those of saturated-carbon species with extra Au atoms attached. In this section we check this assumption for some analogous dicarbon-core systems.

The C_2Au and C_2Au_2 systems are linear, similar to the CCH (ethynyl) radical and the acetylene molecule, respectively (Figure 2.3), in accordance with the earlier work.[28] The T-shaped $CCAu_2$ structure (with C_2 perpendicular to Au_2) is a transition state, unlike the stable CCH_2 (vinylidene) radical, likely due to a significant repulsion of large and charged gold atoms close to one another in such a geometry. Such a repulsion is minimised in the linear AuCCAu geometry.

The C_2Au_3 species is T-shaped, with 3rd gold atom attached sideways to the middle of the above AuCCAu system (Figure 2.3), which is different from the

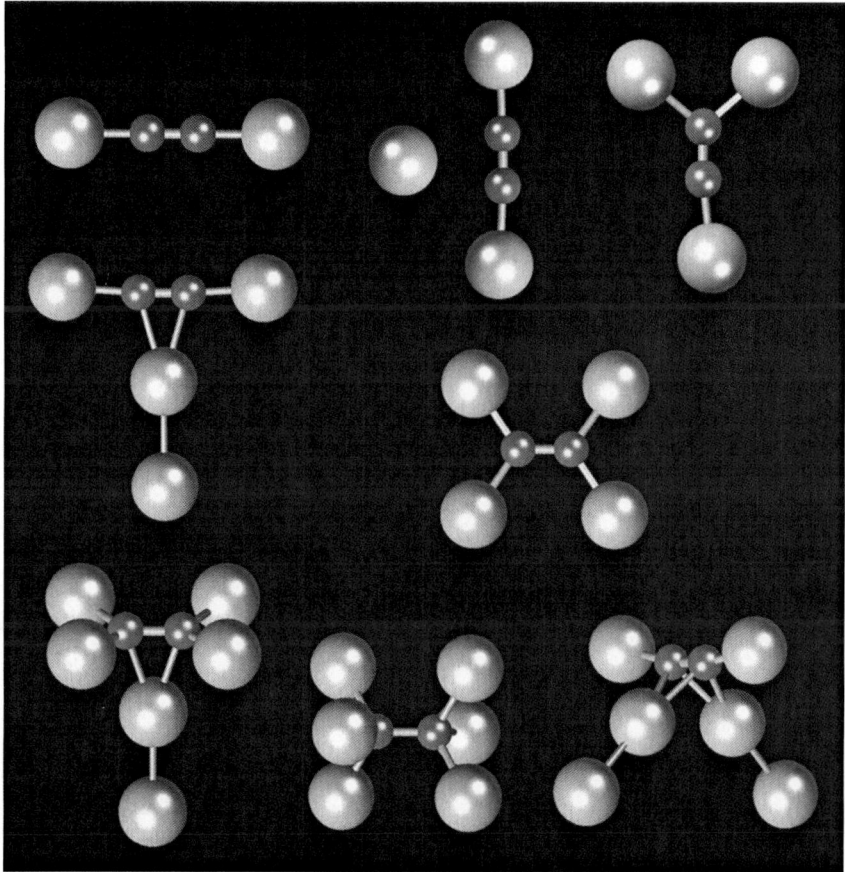

Figure 2.3 Optimised geometries of: C_2Au_2, the T-shaped and vinyl-like isomers of C_2Au_3 (top row); the T-shaped and ethylene-like isomers of C_2Au_4 (middle row); the T-shaped, ethane-like, and right-angled isomers of C_2Au_6 (bottom row).

H_2CCH (vinyl) radical. In fact, the vinyl-like Au_2CCAu structure corresponds to another isomer only $\approx 0.2\,eV$ higher in energy.

Addition of the 4th Au atom can lead to two gold atoms on each end of C_2 in ethylene-like Au_2CCAu_2 isomer (Figure 2.3) identified earlier as well.[28] A rhombic structure with the 4th Au atom attached to dicarbon opposite the 3rd one in CAu_3, which could be expected instead, is a transition state separating equivalent ethylene-like isomers. The slightly (by $\approx 0.3\,eV$) more stable, lowest-energy isomer of C_2Au_4 predicted in the present work is T-shaped, with the gold dimer attached perpendicular to AuCCAu at its middle. Its higher stability is consistent with the above result for C_2Au_3. It is interesting to note a geometric similarity of this structure to that of the weakly bound C_2H_2–H_2 complex.[29] So there appears to be an inversion of the relative stability of the ethylene and T-shaped geometries for the auro-carbon and hydrocarbon counterparts.

For C_2Au_6, two predicted isomers evolve from the above two isomers of C_2Au_4. One, right-angled, has a 2nd gold dimer attached perpendicular the AuCCAu unit at its middle, at an angle of about 90° to the 1st gold dimer (Figure 2.3). There is a transition state separating equivalent such structures, with the 2nd dimer at 180° to the 1st one in a planar cross-shaped geometry. Another isomer, also identified previously,[28] is ethane-like, with two bonded CAu_3 groups in a staggered arrangement. Here, a transition state is similar to that of C_2H_6, with two such groups in the aligned arrangement, *i.e.* with one group rotated about the axis by 30°. The third, most-stable isomer of C_2Au_6 is found to be T-shaped, an alternative extension of T-shaped C_2Au_4, with a single gold dimer attached perpendicular to ethylene-like C_2Au_4 at its middle. These three isomers of C_2Au_6 are within 0.15 eV of one another in energy.

From C_2Au_2 to ethylene-like C_2Au_4, and then to ethane-like C_2Au_6, the equilibrium $R_e(C–C)$ and $R_e(C–Au)$ distances increase from 1.25 Å and ≈ 1.9 Å by ≈ 0.1 Å and ≈ 0.05 Å, respectively, consistent with their variation in the hydrocarbon analogues. Attachment to C_2Au_2 of a gold atom in C_2Au_3 or dimers in T-shaped C_2Au_4 and C_2Au_6 increases $R_e(C–C)$ by ≈ 0.01 Å per atom or ≈ 0.02 Å per dimer, attachment of Au_2 to C_2Au_4 in C_2Au_6 inducing double the latter stretch. The distance between each atom of the C_2 core and the nearest gold atom attached to its middle drops from ≈ 2.5 Å in C_2Au_3 (single Au atom) by ≈ 0.3 Å in the systems with attached Au_2, the dimers remaining almost unperturbed as compared to an isolated gold diatom.

For the lowest-energy isomers the dissociation energies $D_e(C_2Au_n \rightarrow C_2 + n\,Au)$ decrease from ≈ 3.5 to ≈ 2.5 to $\approx 2.3\,eV$ per Au atom for $n = 2$ to 4 to 6, respectively. The last two values are significantly smaller than $3.4 \pm 0.1\,eV$ for CAu_n ($n = 2$–4) due to relaxation of C_2. Detachment of Au from the AuCCAu unit in C_2Au_3 requires $\approx 0.6\,eV$ of energy. The value considerably increases, to $\approx 1.5\,eV$, for detachment of the gold dimer in T-shaped C_2Au_4, and further to $\approx 1.7\,eV$ for the 2nd such dimer in right-angled C_2Au_6. In T-shaped C_2Au_6 the dimer is attached to the Au_2CCAu_2 unit by $\approx 1.8\,eV$.

The above results allow us to conclude that structural analogues of C_2H_n are likely to be at least low-energy isomers of C_2Au_n. However, the auro-carbon lowest-energy isomers can resemble higher-energy, weakly bound hydrocarbon systems still preserving the (unsaturated) hydrocarbon-like unit in their structure but with gold dimers attached perpendicular to its middle. In particular, we could anticipate similar stable structures of larger C_2Au_n, *e.g.* with three, four and perhaps more gold dimers around the AuCCAu unit (for $n = 8$, 10, *etc.*), and possibly with at least two dimers on the opposite sides of the Au_2CCAu_2 unit (for $n = 8$), the relevant further work being in progress. The dicarbon core can thus be surrounded by a gold shell composed of atomic and diatomic units.

To this point, small core–shell auro-carbon systems have been studied, with up to a dicarbon core. In the following section, we will consider similar alumo-carbon species,[30] to compare with another metal component of a different valence.

2.3.2 Small Alumo-Carbon Clusters

2.3.2.1 C_2Al_n (n = 2–6)

The most stable CAl_n isomers have been found previously[15,31] to be similar to those of respective CAu_n for $n = 3$ and 4, except for planar CAl_3 as compared to near-planar CAu_3. A higher-energy isomer of CAl_4 has been predicted to be planar,[15] in qualitative agreement with the above results for CAu_4.

Similar to C_2Au_n, the most stable isomers are acetylene-like AlCCAl for $n = 2$, with another Al atom attached to its middle for $n = 3$, thus confirming previously identified geometries.[32,33] Unlike for the corresponding auro-carbons, however, C_2Al is T-shaped (isosceles-triangular), the linear CCAl system being ≈ 1 eV higher in energy, and there is a ≈ 2 eV higher-energy T-shaped isomer of C_2Al_2, with the aluminium dimer attached perpendicular to C_2 at its middle.

The lowest-energy isomer of C_2Al_4 is found to be rhombic, with two Al atoms attached on the opposite sides of AlCCAl (Figure 2.4), which structure represents a transition state for C_2Au_4. Interaction of dicarbon with the originally rhombic Al_4 thus stretches the aluminium cluster symmetrically into a frame around the carbon core. No ethylene-like isomer is found for this system, unlike for C_2Au_4. In fact, the ethylene-like structure relaxes to a ≈ 1 eV higher-energy nonplanar isomer of C_2Al_4, structurally resembling C_2Al_3 but with two Al atoms added on the same side of AlCCAl and forming a dimer at one end of C. An interesting feature of the rhombic isomer is two planar-tetracoordinated carbon atoms with nonperpendicular C–Al bonds.

Adding two Al atoms at the same or opposite sides of C_2 in C_2Al_4 results in a "side-on" isomer of C_2Al_6, with the Al_2 dimer attached sideways to the still flat but distorted C_2Al_4 base, thus creating an aluminium shell half-enclosing the dicarbon core (Figure 2.4). It is interesting to note that the shell does enclose C_2 completely, by forming a square "belt" around the waist of AlCCAl in the anionic derivative of C_2Al_6. Upon electron detachment from $C_2Al_6^-$, the resulting neutral system recovers its geometry with the dicarbon half-exposed. We could anticipate that adding another aluminium dimer on the exposed side of C_2 could complete the aluminium shell around the carbon core in C_2Al_8.

Isolated Al_6 cluster is a skewed rectangle bipyramid, which geometry would formally match the ethylene-like C_2Al_6 structure with C_2 inside, by analogy

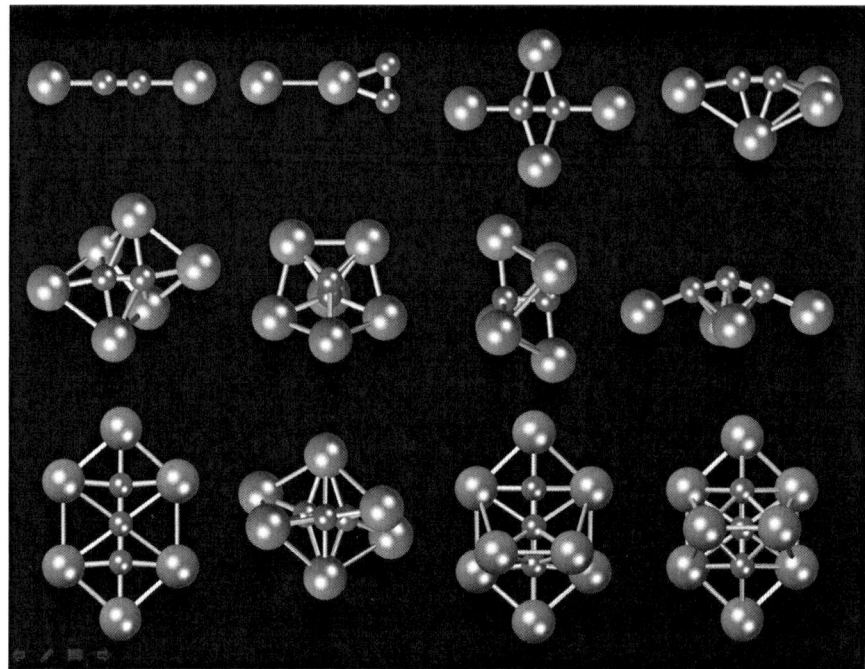

Figure 2.4   Optimised geometries of: the linear and T-shaped isomers of C_2Al_2, planar and nonplanar isomers of C_2Al_4 (top row), the side-on and end-on isomers of C_2Al_6, ethane-like transition state between equivalent end-on isomers, and C_3Al_4 (middle row); the planar and nonplanar (with a C_3-cycle core) isomers of C_3Al_6, and skewed and symmetric isomers of C_3Al_8 (bottom row).

with C_2Al_4. Such a geometry is, however, found to be a transition state between equivalent "end-on" isomers $\approx 0.5\,eV$ lower in energy than the "side-on" one, with the dicarbon protruding from an aluminium "cup". So, this is another example of isomers of auro-carbons correlating to transition states of corresponding alumo-carbons and *vice versa* (as, *e.g.*, for C_2Al_4 and C_2Au_4 above).

We can conclude that while small CAl_n ($n = 2$ to 4) species are structurally similar to respective CAu_n, the C_2Al_n and C_2Au_n systems (except for $n = 2, 3$) have quite different isomers. This could be related to the extra valence electrons in Al as compared to Au, enabling more efficient Al–Al bonding, which possibility is reduced in the C_2Au_n species dominated by C–Au bonding.

For the lowest-energy isomers predicted here, the dissociation energies $D_e(C_2Al_n \rightarrow C_2 + n\,Al)$ per atom drop from $5\,eV$ for $n = 2$ to $\approx 4\,eV$ for $n = 4$ and 6, reflecting a weaker binding of aluminium atoms on the sides of dicarbon as compared to along its axis. The energy $D_e(C_2Al_n \rightarrow C_2 + Al_n)$ for removal of C_2 from the (relaxing) Al_n frame slightly increases with n, being around $9\,eV$. For the systems with even number of Al atoms, relative stabilities can be conveniently discussed also using dissociation energies $D_e(C_2Al_n \rightarrow C_2Al_{n-2} + Al_2)$. Detachment of an aluminium dimer from C_2Al_2, C_2Al_4 and C_2Al_6 requires ≈ 9, ≈ 4 and $\approx 6\,eV$, respectively. The lowest relative stability of

C_2Al_4 with its total of 20 valence electrons in atoms suggests that in systems of atoms with very different electronegativities, such as carbon and aluminium, "magic" numbers of electrons may be unreliable indicators. This is further supported by the highest (also among all system studied) VEA ≈ 1.7 eV and the lowest VE* and VIE of C_2Al_n for $n = 4$. The C_2Al_6 system exhibits the highest VE* ≈ 3.5 eV and VIE ≈ 9.5 eV among all C_nAl_m studied.

The equilibrium distance R_e(C–C) in C_2Al_n steadily increases from 1.25 Å in C_2Al_2 by ≈ 0.1 Å per each additional Al_2, in accordance with more metal atoms donating electrons and thus weakening the C–C bonding. The R_e(C–Al) distances are about same, ≈ 2 Å, for the Al atoms attached axially to C_2, and significantly longer (by ≈ 0.1–0.3 Å) for the atoms attached to its sides.

The charge on the carbon core is -1.5e in C_2Al_2, double that in C_2Al_4, then increases (in absolute value) more weakly in C_2Al_6. The value of the charge on each C atom reaches ≈ -2e, potentially enabling extra bonds for the hyper-coordinated atoms. However, for the carbon formal pentacoordination in the "side-on" isomer and even heptacoordination in the "end-on" isomer of C_2Al_6, no hypervalence is found in terms of the critical points of electron densities (within the atoms-in-molecules, AIM, formalism[34]), the C atoms being tetrabonded. The electron-density distributions indicate a weakly covalent C–Al bonding (with appreciable electron density between atoms) for the axial aluminium atoms and ionic bonding for those on the sides of C_2.

As a next step, alumo-carbons with a triatomic carbon core are considered. One point of interest here is whether or not analogous planar isomers can be found for larger systems. Another aspect is how a less-rigid core may be structurally affected by the considerable atom–atom interactions in the shell.

2.3.2.2 C_3Al_n $(n = 4$–$8)$

Thus, unlike C_2Au_n, not a single C_2Al_n $(n > 2)$ isomer structurally resembles the corresponding C_2H_n. However, low-energy isomers of such alumo-carbons are still obtained when beginning with the hydrocarbon-like geometry, even though the structure alters considerably during optimisation. This is shown in particular for C_2Al_4, both discussed isomers of which are confirmed as the lowest and next-higher-energy structures in a direct extensive scan of the configuration space.[35] Stable systems with a larger tricarbon core can therefore be attempted to be produced also by relaxing their initial geometries similar to those of corresponding C_3H_n.

In particular, beginning with either propyne-like Al_2CCCAl_2 or propadiene-like $AlCCCAl_3$, the same relaxed structure of C_3Al_4 is obtained, with C_3 arching above the bent Al_4 rhombus (Figure 2.4). The longer tricarbon thus does not fit into the same frame as dicarbon in C_2Al_4, hence protruding from and distorting it. This structure can be viewed as a result of interaction between the originally linear C_3 molecule and the originally planar Al_4 cluster, or of addition of a 3rd C atom to the carbon core of rhombic C_2Al_4. It is worth noting that these possibilities thus appear to be, in effect, covered when using hydrocarbon counterparts for initial geometries. An interesting feature is that in the corresponding anion the Al_4 base is shrunk and flat, the carbon core preserving its shape.

Beginning with propene-like $Al_2C=CAl-CAl_3$, the optimised structure of C_3Al_6 is a linear tricarbon in a symmetric hexagonal aluminium frame (Figure 2.4). Such a frame is thus large enough to accommodate the longer core in a planar geometry. Each outer atom of the core exhibits a planar tetra-coordination with bond angles close to $90°$. The initially bent carbon core is thus straightened due to interaction with the aluminium component, exactly opposite to the shape variation for the above C_3Al_4 case.

Another isomer is produced from cyclopropane-like C_3Al_6. The carbon core becomes isosceles-triangular and is surrounded with an aluminium shell of a nonplanar (zigzagged) tetra-atomic frame plus one atom above and one under the core plane, so that the structure resembles Al_6 somewhat distorted by the C_3 cycle inside. This isomer is only $\approx 0.1\,eV$ lower in energy relative to the above planar one, and each carbon atom is formally hexacoordinated. The strain in the core appears to be compensated by the more efficient bonding with the shell and in the shell itself.

Larger C_3Al_n systems can be obtained by adding more Al atoms to either of the above isomers of nearly the same energy. Adding more Al atoms to the planar C_3Al_6 isomer with the exposed carbon core and making new C–Al bonds could be expected to result in lower-energy systems as compared to just increasing the number of the (weaker) Al–Al interactions *via* adding Al atoms to the shell of the other C_3Al_6 isomer.

Two C_3Al_8 isomers are obtained by attaching Al_2 parallel to the face of planar C_3Al_6 and perpendicular to the carbon core (Figure 2.4). In one isomer the core angles away from and in the other one towards the added dimer positioned symmetrically or shifted to one end of the core, respectively. The symmetric isomer is $\approx 0.3\,eV$ lower in energy and preserves an almost flat Al_6 frame. The other, skewed isomer can also be produced by relaxing structurally similar propane-like $Al_3C–CAl_2–CAl_3$ (with C_3 bent accordingly). Both or one, respectively, outer carbon atoms are formally hexacoordinated in these systems.

The equilibrium $R_e(C–C)$ distance in C_3Al_n increases from $n=4$ ($\approx 1.3\,Å$) to $n=6$ marginally (by $\approx 0.01\,Å$) for the planar isomer of C_3Al_6, the transition from a bent (at $\approx 150°$) to a linear core being compensated by more electron donors (Al atoms), or significantly (by $\approx 0.1–0.2\,Å$) for the nonplanar isomer, with the core bending further. Accordingly, for $n=8$ (for more-stable isomer with C_3 bent at $\approx 130°$) the C–C distance increases as compared to planar, or decreases relative to nonplanar C_3Al_6 (with more strongly bent C_3). For the less-stable C_3Al_8 isomer, the core is bent less (by $\approx 150°$), the C–C distance being somewhat shorter.

The dissociation energy $D_e(C_3Al_n \rightarrow C_3 + n\,Al)$ per aluminium atom slightly decreases from $\approx 3.7\,eV$ for $n=4$ to $\approx 3.4\,eV$ for $n=6, 8$. The energy $D_e(C_3Al_n \rightarrow C_3Al_{n-2} + Al_2)$ for detachment of aluminium dimer, however, increases from $\approx 4.5\,eV$ in C_3Al_6 to $\approx 5.5\,eV$ in C_3Al_8. Somewhat counter-intuitively from the structural viewpoint and oppositely to the C_2Al_n case, the energy $D_e(C_3Al_n \rightarrow C_3 + Al_n)$ for removal of the carbon core is smaller for $n = 6$ ($\approx 7\,\mathrm{eV}$) as compared to $n = 4$ ($\approx 8.5\,\mathrm{eV}$), likely due to a larger strain in planar C_3Al_6 as well as a larger structural relaxation of Al_6.

For C_3Al_4 to planar C_3Al_6 to more-stable C_3Al_8 the negative charge on the carbon core increases in absolute value linearly with the number of Al atoms, from $-3.2e$ to $-4.3e$. The core charge is largest, at $-4.5e$, in the nonplanar isomer of C_3Al_6 and the less-stable isomer of C_3Al_8. Except in the nonplanar isomer of C_3Al_6, most of the charge is carried by the outer carbon atoms of the core, up to almost the neutral central atom in planar C_3Al_6. It is worth noting that in the anions extra charge is distributed mainly over the aluminium atoms (positive in the neutral system and thus attracting the extra electron). As a result, the Al atoms in, *e.g.*, $C_3Al_4^-$ become less charged and thus repel each other less, which explains the shrinking Al_4 base.

Dipole moments vary from zero for planar and small for nonplanar C_3Al_6 ($\approx 0.5\,\mathrm{D}$) to significant in C_3Al_4 and both isomer of C_3Al_8 (≈ 1–$2\,\mathrm{D}$). The vertical energies VE*, VIE, and VEA of C_3Al_n all peak (respectively at 3.2, 9.3, 1.5 eV) for $n = 4$ and have minimal values for $n = 6$.

The electron-density distribution indicates ionic bonding between the core and side Al atoms in C_3Al_4, while a weakly covalent bonding in planar C_3Al_6, with the C–Al critical electron densities equal for the axial and side atoms. The outer C atoms appear to make double bonds to the central C atom (not having bonds to the side Al atoms) and thus to formally have five bonds each, which could be facilitated by extra electrons on them (in terms of high negative charges). The AIM critical points indicate six bonds for one of the C atoms in nonplanar C_3Al_6 and for the outer atoms of the carbon core in symmetric C_3Al_8.

The small alumo-carbon clusters with a C_2 core thus exhibit structures very different from those of the corresponding auro-carbons. The fragmented gold shell with separate diatomic units is replaced with a more uniform aluminium shell, the latter tending to a 2D arrangement of atoms around the carbon core in case of its unsaturated bonding.

2.3.3 Larger Auro-Carbon Nanosystems

The above results clearly point to the carbon cores being able to shape the metal shells around them *via* a mixture of covalent and ionic bonding. So originally planar small gold clusters can be wrapped around carbon atom or diatom, as well as a carbon diatom and a triatom can stretch (without restructuring) or considerably restructure the small aluminium clusters (*e.g.* flattening them into 2D shells). In this section we will look at extensions to larger core–shell systems, and at the possibility for a carbon core to stretch a gold shell without a major restructuring of it.

2.3.3.1 $C_5Au_{12,14}$

Both CAu_4 and ethylene-like C_2Au_6 above (Section 2.3.1) exhibit stable structures with constituent methyl-like CAu_3 groups. From CAu_4 to this isomer of C_2Au_6, one gold atom is replaced with such a group, similar to the structure evolution from methane to ethylene. If all four terminal H atoms of methane are replaced by CH_3 groups, neopentane (or tetramethylmethane) is formed. Upon a similar procedure with auro-carbon counterparts, a structurally analogous $C_5Au_{12} = C(CAu_3)_4$ system can be constructed.[36] The so-obtained gold shell around the carbon core correlates to a cuboctahedral isomer of the Au_{13} cluster with the central atom removed. We can also consider another, icosahedral-shaped Au_{13}, and use its different Au_{12} "surface" to accommodate the same C_5 core inside. The icosahedral shell can be seen as composed of four connected gold-atom triangles in a tetrahedral arrangement around its centre, to which four outer atoms of the carbon core can be bonded, again forming CAu_3 groups.

Relaxation of the above two structures only slightly distorts the original geometry of the gold shells, resulting in two isomers of C_5Au_{12}, icosahedral-like (I_h) and cuboctahedral-like (O_h), with the CAu_3 groups and the near-tetrahedral C_5 core preserved (Figure 2.5). The two isomers are nearly degenerate for the smaller basis set used, the O_h isomer becoming ≈ 0.5 eV lower in energy than the I_h one for the single-point calculation (in the frozen geometries) for the larger basis set.

The equilibrium C–C and C–Au distances of ≈ 1.6 and ≈ 2.1 Å are about same for both isomers, being exactly identical or varying within 0.05 Å for the I_h and O_h isomer, respectively. The Au–Au separations of around 3 Å are on average slightly longer for the I_h isomer, varying within 0.3 and 0.5 Å for the respective isomers. The C–Au and Au–Au distances in the CAu_3 groups are nearly the same as in C_2Au_6 calculated at the same level of theory.

The dissociation energy $D_e(C_5Au_{12} \rightarrow C_5 + 12\,Au)$ per gold atom, calculated relative to the lowest-energy (triplet) state of C_5 relaxed from its geometry of the core, is ≈ 1.5 eV. This is considerably lower than ≈ 2.4 eV for CAu_4 calculated at the same level of theory, reflecting a significant core relaxation,

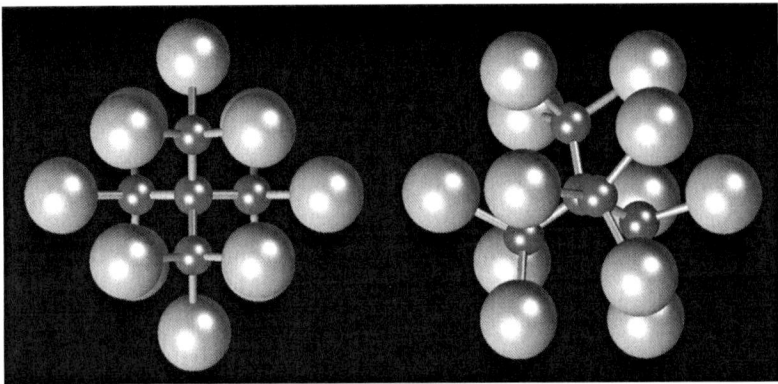

Figure 2.5   Optimised geometries of the O_h and I_h isomers of C_5Au_{12}.

hence its strain inside C_5Au_{12}. By comparison, relative to the frozen core this D_e value is $\approx 2.2\,eV$ per gold atom.

For both isomers of C_5Au_{12}, the (natural) charge distribution is similar and shows a negative carbon core ($\approx -2.9 \pm 0.2\,e$) and equally positive metal shell, qualitatively similar to the smaller systems considered above. In the core, the central C atom carries $\approx -0.3\,e$, about half the charge on each outer C atom.

Unlike the icosahedral Au_{12} shell that collapses if empty inside, the cuboctahedral cage isomer of Au_{12}, composed of three staggered square Au_4 units, is stable and has a closed-shell ground state, the triplet state being only $\approx 0.3\,eV$ higher. It has the same dissociation energy of $\approx 1.5\,eV$ per gold atom as the C_5Au_{12} system. Insertion of the carbon core thus reshapes (inflates) the gold shell mainly by stretching its central Au_4 square unit, while generally preserving the overall structure and stability. Hence, formation of this core–shell system relative to such a cage is approximately energy neutral, the shell inflation being compensated by the core–shell bonding.

Vertical ionisation energy (VIE) and electron affinity (VEA) of C_5Au_{12} are slightly lower (by 10–15%) relative to those of the Au_{12} cage (Table 2.3). The vertical spin-excitation energy $VE^*(S=0\rightarrow1)$, appreciable for the cage cluster, is, however, effectively cancelled in C_5Au_{12}.

We can approximately separate the dependences of the above properties on the gold shell geometry change only and the core–shell interaction *via* charge transfer, by calculating the property values for the shell frozen at the geometry in the total system. The $D_e(Au_{12} \rightarrow 12\ Au)$ is then $\approx 1.2\,eV$ per atom, noticeably smaller due to stretched Au–Au distances and no core–shell Coulomb attraction. The ground state of the empty shell is actually triplet, though the singlet state is only slightly higher in energy (Table 2.3), similar to the situation for C_5Au_{12}, hence VE^* is dominated by the shell shape. The VIE value for the empty shell is decreased relative to that for the gold cage by about a third of the total change, so the core–shell interaction is a major factor here. The VEA value significantly increases (by $\approx 1\,eV$ or 30%) when the cage is inflated, but then decreases back when the core is inserted, to a value $\approx 10\%$ smaller than the original one. This suggests that both the geometry variation and the core–shell interaction affect this property about equally, but in opposite directions, effectively cancelling their contributions. Hence, different

Table 2.3 Characteristic energies (in eV) of the Au_{12} and XAu_{12} systems.

System	$D_e{}^a$	$VE^*(S=0\rightarrow1)$	VIE	VEA
Au_{13} (O_h isomer)	19.2	0.47^b	6.90	3.45
Au_{12} cage (O_h isomer)	18.2	0.74	7.1	3.02
C_5Au_{12} (O_h isomer)	18.6	0.1^b	6.18	2.68
Au_{12} in C_5Au_{12} (O_h), empty frozen shell	14.7	0.19^b	6.78	3.96
$CH_4@Au_{12}$ ($\sim T_d$)	14.8	0.56	7.31	3.57
$H_2@Au_{12}$ (O_h)	16.5	0.45	7.2	3.40

$^a Au_{12} \rightarrow 12\ Au$, $XAu_{12} \rightarrow X + 12\ Au$
$^b S = 1/2 \rightarrow 3/2$ for Au_{13}, $S = 1 \rightarrow 0$ for C_5Au_{12} and Au_{12} shell

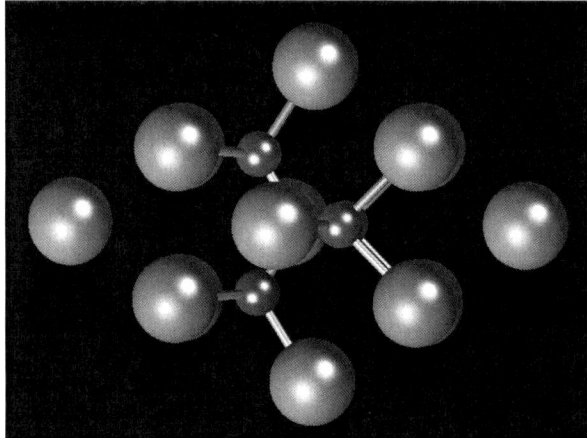

Figure 2.6   Optimised geometry of C_5Au_{14}.

properties exhibit very dissimilar evolutions when going from the Au_{12} cage to the C_5Au_{12} system.

The C_5Au_{12} system is structurally stable to adding more gold atoms. Its O_h isomer generally preserves shape when two Au atoms cap the smaller near-square Au_4 faces on opposite sides along the C_{2v} symmetry axis of the carbon core (Figure 2.6), becoming an elongated C_{2v}-symmetric C_5Au_{14} species. The stable C_5Au_{14} system exhibits a closed-shell singlet state marginally (by ≈ 0.1 eV) lower in energy than the triplet state, exactly opposite to C_5Au_{12} with its triplet ground state nearly degenerate with the singlet state. This could be related to the "magic" total number (34) of valence electrons in the atoms of C_5Au_{14}. Each of the two Au atoms in the formal 2nd shell around C_5 is bonded to C_5Au_{12} by ≈ 1.4 eV, *i.e.* only slightly less than the 1st shell atoms. The ionisation energy of ≈ 6.2 eV and electron affinity of ≈ 2.9 eV are nearly the same as for C_5Au_{12}.

2.3.3.2 $C_{10}Au_{18,21,24}$

Exploring further the structural analogy between the hydrocarbon and auro-carbon systems, the following larger-scale step is attempted. Formal removal of one methyl group of neopentane and dimerisation of the remainder *via* bonding two central C atoms produces a hexamethylethane molecule. A similar procedure for C_5Au_{12} would lead to a C_8Au_{18} counterpart that, however, is unstable, the central C–C bond dissociating upon relaxation.[37] This appears to be due to a strong repulsion between the closely spaced CAu_3 groups on different central carbon atoms. If an additional C_2 unit is introduced between those atoms, the two $C(CAu_3)_3$ components are spaced enough to prevent the dissociation of resulting $C_{10}Au_{18}$ (Figure 2.7), a counterpart of the di-t-butyl-acetylene hydrocarbon. The system thus has a C_3C–CC–CC_3 core surrounded by an

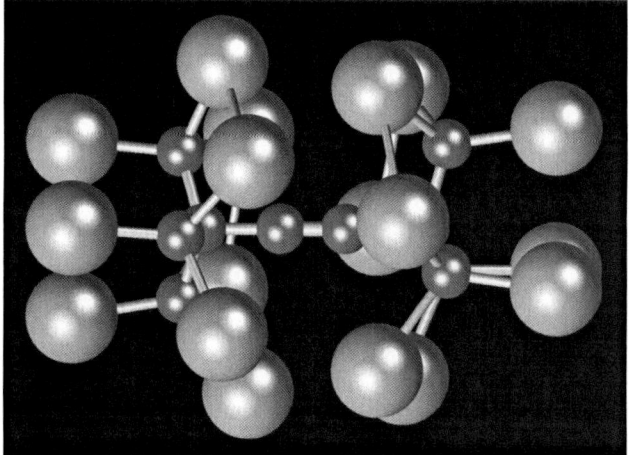

Figure 2.7 Optimised geometry of $C_{10}Au_{18}$.

ellipsoid-like Au_{18} shell composed of two staggered Au_6 rings and two trian-gular Au_3 caps.

Similar to C_5Au_{12}, the singlet and triplet states of $C_{10}Au_{18}$ are nearly degen-erate within 0.05 eV. The ground state is closed-shell, different from the smaller system, likely due to the total of 58 valence electrons in all its atoms, which is another "magic" number, similar to the case of C_5Au_{14} *versus* C_5Au_{12} above.

In accordance with the bonding pattern, the C–C bond lengths increase from the centre (a triple bond), to the periphery of the core (single bonds of the sp–sp^3 and sp^3–sp^3 pairs of hybrids). The Au–Au equilibrium distances between two Au_6 rings are $\approx 3 \pm 0.1$ Å, the rings and caps being farther apart by ≈ 0.5 Å.

The dissociation energy $D_e(C_{10}Au_{18} \rightarrow C_{10} + 18$ Au) per gold atom is cal-culated for a relaxed C_{10} core in its lowest-energy triplet state, the value of ≈ 1.5 eV being the same as for the C_5Au_{12} counterpart. The $D_e(C_{10}Au_{18} \rightarrow C_{10}Au_{17} + Au)$ value of ≈ 1.8 eV is the same for detachment of a single gold atom from either ring or cap, in the latter case the detached atom being replaced by the ring atom, so the product is the same, with a missing ring atom.

Removal of the C_{10} core and relaxation results in a less-symmetric cage isomer of Au_{18} considerably (about twice) contracted along the former core axis and with distorted Au_6 rings, one ring becoming larger. The total dissociation energy of ≈ 1.7 eV per atom is somewhat higher than for $C_{10}Au_{18}$, reflecting the neutrality of and shorter Au–Au distances in the deflated cage. Insertion of the carbon core thus considerably symmetrises and inflates the gold shell *via* mainly stretching it along the core axis, while affecting stability relatively weakly.

The VIE and VEA values of $C_{10}Au_{18}$ are close to those of C_5Au_{12} (within 10%), the former being slightly lower (by ≈ 0.4 eV) and the latter slightly higher (by ≈ 0.2 eV). For the formal step-wise transition from the gold cage to the frozen empty gold shell to the shell with the carbon core inside, these vertical energies of the larger system evolve quite similarly to the case of the smaller system. The only notable difference is the twice as large decrease of VIE (by ≈ 1.2 eV) when the core is inserted into the inflated shell.

The $C_{10}Au_{18}$ auro-carbon preserves its shape when another Au atom is attached to it. The resulting $C_{10}Au_{19}$ system is slightly more stable with the atom added to the side (between the rings) than to the cap (along the axis). The corresponding binding energies for this 2nd-shell atom are 2.3 ± 0.1 eV for the two isomers, which is larger than for C_5Au_{14} above. We can also anticipate a stable $C_{10}Au_{20}$ species with two capping atoms on the opposite sides along its axis.

The barrier for internal rotation of one $C(CAu_3)$ component of $C_{10}Au_{18}$ about the system axis is ≈ 0.3 eV, the associated transition state corresponding to the Au atoms of the two rings being opposite one another (*i.e.* aligned). This geometry can be stabilised by adding atoms into the 2nd gold shell, between the two rings. Attachment of three equally spaced Au atoms produces $C_{10}Au_{21}$, the added atoms occupying positions bridging two rather than capping four atoms in the two rings (Figure 2.8). As a result, the distance between the ring atoms slightly increases to ≈ 3.1 Å. The three atoms are bound by ≈ 2.5 eV per atom relative to relaxed $C_{10}Au_{18}$ (with staggered rings). Subsequent addition of capping Au atoms in the axial positions is expected to lead to stable $C_{10}Au_{22}$ and $C_{10}Au_{23}$ species.

Three more Au atoms added to the 2nd-shell ring around the original system can, however, have stable positions capping four ring atoms in resulting $C_{10}Au_{24}$ (Figure 2.9). The three gold pairs that are both capped and bridged stretch to ≈ 3.3 Å, and the binding energy of the 2nd-shell atoms to $C_{10}Au_{18}$ is then ≈ 2.0 eV per atom. A lower-energy isomer of $C_{10}Au_{24}$ has all six added atoms bridging pairs of opposite atoms in the rings and bound to original $C_{10}Au_{18}$ by ≈ 2.3 eV per atom. One and two capping gold atoms added along the system axis are anticipated to result in stable $C_{10}Au_{25}$ and $C_{10}Au_{26}$.

The charge distribution in $C_{10}Au_{18}$ exhibits a carbon core that is increasingly negative from its (nearly neutral) centre to periphery, and relatively uniformly positive gold shell having most of positive charge in the middle (on the rings). The system thus has a significant quadruple moment whose electric field can significantly polarize the 2nd-shell gold atoms in larger systems, consistent with the above higher binding energies of the added atoms to $C_{10}Au_{18}$. By comparison, in C_5Au_{12} the more centrally symmetric charge distribution can result in a much weaker electric field, with the core and shell fields effectively cancelling each other. Furthermore, the total charge on the C_{10} core is ≈ -4.0 e, significantly larger than on the C_5 core, in accordance with the number of the Au donors in the shell.

So far pure-carbon radical cores are considered, forming bonds to the shell atoms. In the next section, the core is represented by a closed-shell molecule interacting with the metal atom shell noncovalently.

2.3.4   A Gold Cage Structured by a Hydrocarbon Molecular "Dopant"

Another, qualitatively different structural modification of the C_5 radical core is replacement of its outer carbon atoms by hydrogens, leading to a closed-shell

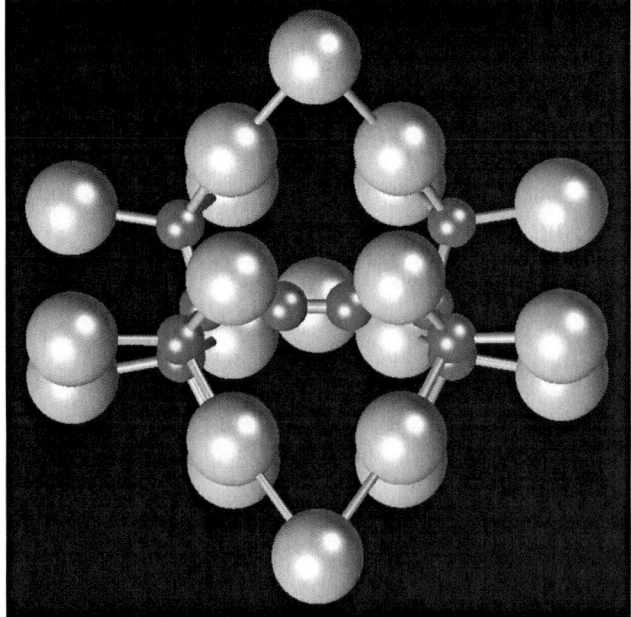

Figure 2.8 Optimised geometry of $C_{10}Au_{21}$.

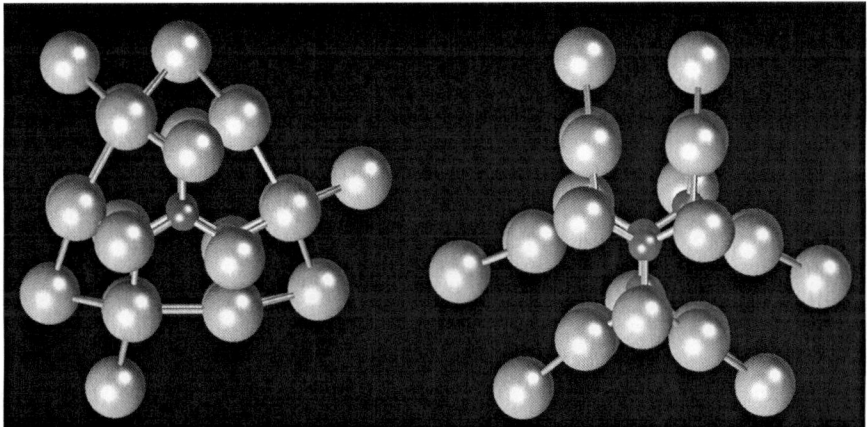

Figure 2.9 Optimised geometries of the isomers of $C_{10}Au_{24}$.

counterpart, tetrahedral methane molecule CH_4. When inserted inside Au_{12} in place of C_5, such a core can either break the gold shell (starting with the O_h isomer of C_5Au_{12}) or transform it considerably (using the I_h isomer as a starting point). In the latter case the molecule is then suspended centrally inside its truncated-tetrahedral-like frame with no bonds formed (Figure 2.10). The

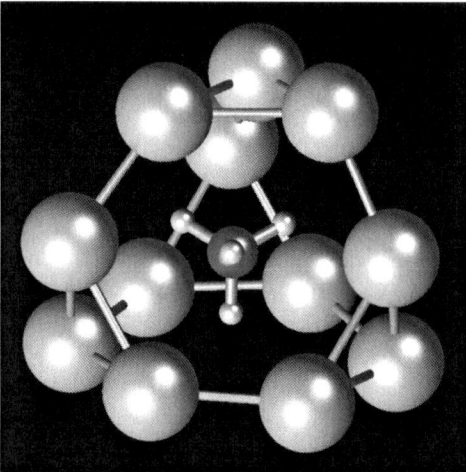

Figure 2.10   Optimised geometry of $CH_4@Au_{12}$.

H-ends of the molecule appear to make large hexagonal holes in the shell composed of four triangular units bonded by their corners.

All equilibrium C–Au and Au–Au distances are $\approx 3.3 \pm 0.2$ and ≈ 2.8–$2.9\,\text{Å}$, respectively, indicating a larger overall size of the gold shell, while on average slightly shorter Au–Au distances as compared to C_5Au_{12}. Each C–H bond points to the centre of Au_6 (nonflat) "facet", each Au_3 "facet" being at the hollow of the methyl side of CH_4, in a staggered arrangement with hydrogen atoms. For the rotated-core geometry with the H ends pointing into the centres of the Au_3 "facets" the energy is $\approx 1.2\,\text{eV}$ higher and upon relaxation the system recovers the original geometry. It is also recovered if the molecule is shifted from the centre of the cage by up to as much as $\approx 1\,\text{Å}$, clearly suggesting the obtained geometry as an isomer of the system.

The gold shell collapses if the molecular core is removed. The energy barrier to such a removal by pulling CH_4 through the hexagonal hole is evaluated to be $\approx 2\,\text{eV}$. The $D_e(CH_4@Au_{12} \rightarrow CH_4 + 12\ \text{Au})$ dissociation energy is $\approx 1.2\,\text{eV}$ per atom, which is somewhat less than for C_5Au_{12}, reflecting no bonding between the core and the shell. For the frozen Au_{12} shell with no core inside, the total dissociation energy is $\approx 0.1\,\text{eV}$ per atom higher, indicating a slightly repulsive core–shell interaction in $CH_4@Au_{12}$, the molecule thus inflating the cage. This is supported by the C–Au distances (for the central C atom) being slightly longer than in C_5Au_{12}.

Unlike the molecular radical (C_5), the similarly shaped molecule (CH_4) thus changes the gold shell structure considerably. This can be associated with noncovalent interaction between the closed-shell core and the shell, different from the core–shell bonding in C_5Au_{12} and thus leading to a different match of the core and shell geometries. In addition, the methane core remains essentially neutral, different from the extensive core–shell charge transfer for the other case.

The closed-shell $CH_4@Au_{12}$ system exhibits a significant singlet–triplet vertical gap VE* (Table 2.3), different from the near-degenerate case of C_5Au_{12}. For the methane-based system the VIE value is $\approx 20\%$ higher and the VEA value $\approx 30\%$ eV higher, consistent and counter intuitive, respectively, with the closed-shell nature of the CH_4 "dopant".

Because of noncovalent interactions between methane and the gold cage, limited test calculations have been carried out at the MP2 level as well. The above geometry of the system is confirmed, and its stability to dissociation into the separate molecule and gold atoms increases by 25% relative to the DFT result.

Hence, CH_4 is a molecule barely fitting into the Au_{12} shell. Moreover, it appears to break the cuboctahedral gold cage (stable by itself) when inserted there. Therefore, next, we consider a smaller (actually the smallest possible) molecule to see if it could fit into such a cage without breaking it.

2.3.5 Hydrogen Trapping in Metal Cluster Nanocages

2.3.5.1 $H_2@Au_{12}$

If H_2 is inserted into the icosahedral-like Au_{12} cluster cage (instead of C_5 in the I_h isomer of C_5Au_{12} above), the cage breaks, unlike for the CH_4 "dopant" above. However, the other, cuboctahedral cage survives the intrusion, and two nearly degenerate isomers of the resulting $H_2@Au_{12}$ system are found, with the molecule oriented approximately along and perpendicular to the cage axis (Figure 2.11). In the latter case, the molecule is aligned with two diagonal atoms of the central square unit of the gold cage. For both isomers, this central square of the cage (directly around the molecule) is stretched by $\approx 0.5\,\text{Å}$ in the Au–Au distance, while the H_2 molecule is essentially intact and neutral.

The ground state of the system is closed-shell, the triplet state being only $\approx 0.1\,\text{eV}$ higher in energy. The $D_e(H_2@Au_{12} \rightarrow H_2 + 12\ \text{Au})$ dissociation energy of $\approx 1.4\,\text{eV}$ per gold atom ($\approx 0.1\,\text{eV}$ per atom smaller than for the relaxed cage) is consistent with some inflation of the cage by the core. The smaller such value found for $CH_4@Au_{12}$ above is in accordance with a larger inflation of the cage by the larger core. The energy of $H_2@Au_{12}$ being higher (by $\approx 1.7\,\text{eV}$) than separate H_2 and Au_{12} cage implies metastability of the core-shell structure.

The VIE and VEA values remain almost unchanged or slightly increase (within 15%) relative to those of the Au_{12} cage (Table 2.3). The VE*(S=0→1) value shows a more pronounced relative decrease (by $\approx 40\%$), remaining small (under 1 eV).

Test MP2 calculations confirm geometries and near-degeneracy of the two isomers of $H_2@Au_{12}$. The system stabilisation relative to separate gold atoms and the molecule is increased by $\approx 40\%$ as compared to the DFT data. This is a larger difference than for the $CH_4@Au_{12}$ above, in accordance with a denser packing of gold atoms in $H_2@Au_{12}$ and an underestimation of the aurophilic interaction by DFT (predicting, for instance, weakly repulsive Au–Au

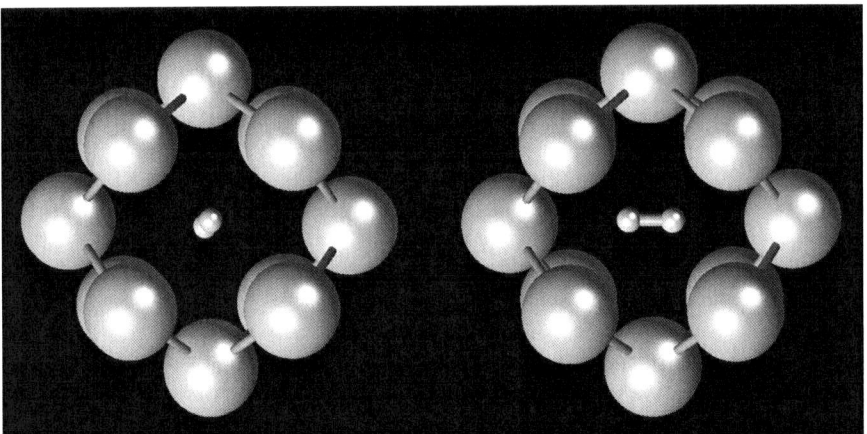

Figure 2.11 Optimised geometries of the isomers of $H_2@Au_{12}$.

interactions in CAu_4 using the same model as in Section 2.3.1.1 above). The system is also confirmed to be metastable relative to $H_2 + Au_{12}$ by $\approx 2\,eV$.

Thus, hydrogen molecules can be accommodated inside a small gold cage, inflating but not distorting it dramatically. For comparison, let us also consider hydrogen inside a cluster cage made of another metal.

2.3.5.2 $H_2@Mg_{8,9,10}$ and $(H_2)_2@Mg_{10}$

Bulk magnesium hydrate, MgH_2, is a promising material for hydrogen storage. It is structurally built of linear HMgH units resembling isolated molecules of such atomic composition. One difficulty with a practical use of this material is the need to break Mg–H bonds to withdraw hydrogen, which requires high temperatures. However, there is another, weakly bound $Mg-H_2$ isomer of the same molecule, of nearly same energy (with two Mg–H bonds replaced by a strong H–H bond). Removal of hydrogen from such a complex can be expected to be easy due to no bond-breaking, or even oppositely, a low temperature may be needed to prevent dissociation of this complex. Structural extensions of the system, capable to trap H_2 more reliably, are Mg_nH_2 clusters. Since a more practical storage requires a higher hydrogen-to-magnesium weight ratio, the smallest clusters are considered.[38]

Appropriate cage structures are known for small Mg_n, including those with $n = 5$ and 7 (trigonal and pentagonal bipyramids), and 9 (tricapped trigonal prism). Attempts to insert hydrogen molecules into Mg_5 and Mg_7 fail, the cages breaking due to being too tight to accommodate the molecule. The Mg_9 cage is the first survivor, with its structure reshaping from three staggered triangles in Mg_9 to two staggered squares capped on one side in $H_2@Mg_9$, all nearest-neighbour Mg–Mg distances being $\approx 3\,Å$. The H_2 molecule is oriented axially and shifted away from the capping atom (Figure 2.12), while being considerably stretched (by $\approx 0.2 \, \text{Å}$) relative to the isolated diatom.

Detachment of capping Mg results in a more symmetric $H_2@Mg_8$ system, with the molecular hydrogen core located centrally inside the magnesium shell of two staggered Mg_4 squares. These squares are slightly larger (by $\approx 0.1 \, \text{Å}$ in side) and further apart than in $H_2@Mg_9$, while H_2 is stretched less, by $\approx 0.1 \, \text{Å}$. The cage geometry is significantly different from that of the isolated Mg_8 cluster (capped pentagonal bipyramid) and is also obtained upon inserting H_2 into its centre and relaxing the system.

Alternatively, attachment of another capping atom to the opposite side of the Mg_9 cage stretches the molecule further and, it turns out, beyond its breaking point. The resulting $H_2@Mg_{10}$ system is represented by two H atoms $\approx 2 \, \text{Å}$ apart (approximately at the centres of the Mg_4 squares), encapsulated in the Mg_{10} shell of two square pyramids joined by their (staggered) bases, all Mg–Mg distances being slightly shorter (by $\approx 0.05 \, \text{Å}$) than in $H_2@Mg_9$. The cage structure differs considerably from that of isolated Mg_{10} (capped trigonal prism

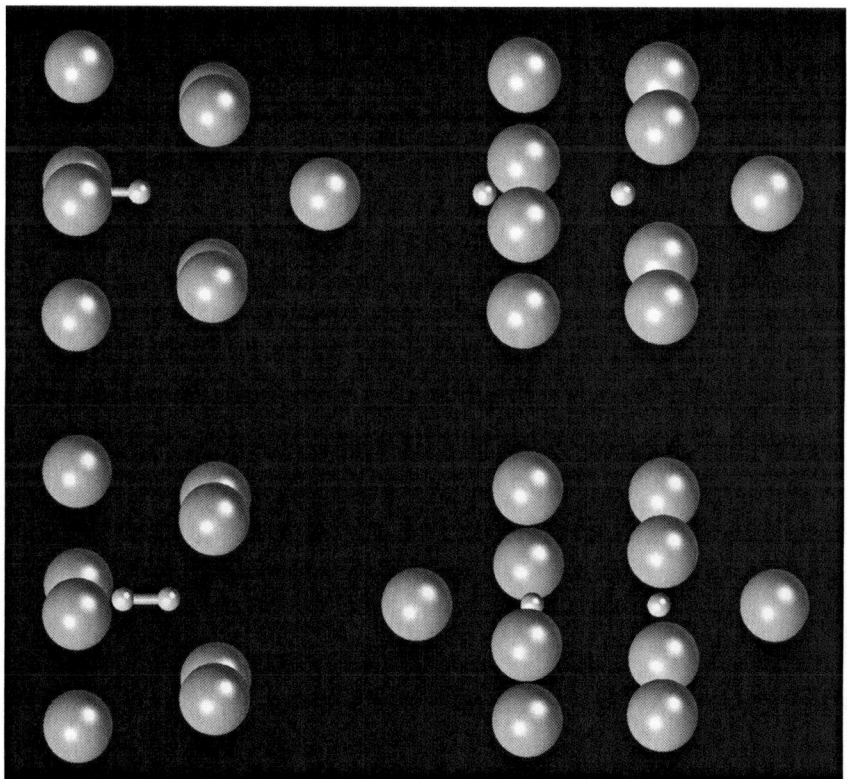

Figure 2.12 Optimised geometries of the isomers of $H_2@Mg_9$ with non- and dissociated H_2 (top row), $H_2@Mg_8$ and $H_2@Mg_{10}$ (bottom row).

that is recovered if H_2 is removed) and is produced also when H_2 is inserted into it.

Uncapping $H_2@Mg_{10}$ does not recover the above $H_2@Mg_9$ structure with nondissociated H_2 but produces another, dissociated-core isomer of the system with two H atoms even slightly further apart (by ≈ 0.1 Å) and shifted axially further away from the other capping atom. The cage slightly shrinks in general as compared to $H_2@Mg_{10}$, and the system is ≈ 0.1 eV lower in energy than the first isomer of $H_2@Mg_9$. The energy barrier between these isomers, *i.e.* for dissociation of hydrogen inside the Mg_9 cage, is ≈ 0.1 eV only. In accordance with the above, this barrier vanishes when $H_2@Mg_9$ is axially capped by another Mg atom.

The evolution of the hydrogen core with the number of atoms in the magnesium shell can be associated with the core–shell charge transfer. With increasing number of Mg atoms, more and more electron density is transferred to H_2, leading to its stretch and finally dissociation. Starting from ≈ -0.3 e on the molecule for $n = 8$, the charge more than doubles (to ≈ -0.8 e on nondissociated H_2) for $n = 9$, and reaches ≈ -1 e on each H atom for $n = 10$, thus in fact leading to a Coulomb explosion of the core, induced by and confined in the shell. Accordingly, the Coulomb attractions contract the increasingly positive cage around the molecule, which explains the above shrinking of the magnesium shell from $n = 8$ to 10. The axial Mg atoms are charged more than the side ones.

The $D_e(H_2@Mg_n \rightarrow H_2 + n \, Mg)$ dissociation energy increases about linearly with n, from ≈ 3 eV for $n = 8$ to ≈ 7 eV for $n = 10$, indicating a detachment energy of ≈ 2 eV per capping atom and considerably smaller values for the Mg_4 squares (≈ 0.4 eV per atom for $n = 8$), consistent with their relative charges. The energy $D_e(H_2@Mg_n \rightarrow H_2 + Mg_n)$ for the molecule withdrawal from the cluster cage is slightly negative, varying from ≈ -0.3 eV to ≈ -0.1 eV to ≈ -0.8 eV for $n = 8$ to 10. The $H_2@Mg_9$ system being thus relatively most stable is formally consistent with the "magic" total number (20) of valence electrons in its atoms. Pressing two H anions closer to each other inside the Mg_{10} shell as compared to the Mg_9 case reduces the system stability. The resulting slight metastability of $H_2@Mg_n$ can be considered as a positive factor in facilitating hydrogen release when needed.

All the above $H_2@Mg_n$ are closed shell systems, similar to $H_2@Au_{12}$. Unlike for the former system, however, the molecule does not dissociate inside the gold cage and remains essentially neutral, similar to the above $CH_4@Au_{12}$ case and different from the case of a magnesium cage. For both metals, the cages with trapped H_2 are metastable, more so for the case of the gold shell exhibiting no significant charge transfer to the core, in accordance with a higher electronegativity of Au as compared to Mg.

An interesting feature of the $H_2@Mg_n$ systems is their slightly negative or near-zero VEA values, varying from ≈ -0.2 eV for $n = 8$ to ≈ 0.1 eV for $n = 10$. This, together with their appreciable VIE and VE* values (≈ 6 to ≈ 7 eV and ≈ 1 to ≈ 3 eV, respectively), would make these species rather nonreactive and thus observable when formed in experiments.

Figure 2.13 Optimised geometry of $(H_2)_2@Mg_{10}$.

Following the above dissociation of H_2 inside Mg_{10}, another cage catalysed and confined reaction can be simulated. Two H_2 molecules are initially placed axially inside the same cage near centres of two Mg_4 squares, and the system is relaxed. Both molecules dissociate and the system goes through a transition state with a newly formed H_2 molecule at the centre and two H atoms symmetrically on its sides, all still lying along the system axis. Then, the central molecule shifts away from the axis and finally merges with one of side H atoms into a symmetric H_3 species with the H–H separation of ≈ 1.1 Å and bent (at $\approx 150°$) "around" the other H atom remaining in the axis ≈ 2 Å away from each atom of H_3 (Figure 2.13). The cage adjusts as well, by stretching the Mg_4 unit around H_3 from a square to a rectangle.

Here, we thus see an example of a magnesium-cage induced reaction that can be written as $(H_2 + H_2 \rightarrow 2H + H_2 \rightarrow H + H_3)@Mg_{10}$. The product system is ≈ 1 eV lower in energy than the above transition state and is metastable by ≈ 1 eV relative to $2H_2 + Mg_{10}$. The reaction is again of a charge-transfer-induced type, with the hydrogen component being negative during the process and represented by $H^{-0.2} + H_3^{-2.2}$ in the end.

2.4 Conclusions

A series of cluster systems of mixed metal–nonmetal composition is considered from the viewpoint of designing stable core–shell systems. The core varies in size from a single atom (C) to a diatom (C_2, H_2) to a larger molecular species (CH_4, $(H_2)_2$, $C_{3,5,10}$), and the shell is constructed of metal atoms (Au, Al, Mg) starting from a few and up to 24.

The gradual increase of the number of gold atoms in small CAu_n clusters indicates a 2D to 3D transition at $n = 4$, much earlier than for pure Au_n (around $n = 13$, ref. 14). The carbon atom is then positioned at the centre of a tetrahedral, square-pyramidal and octahedral shaped layer of gold atoms for $n = 4$ to 6, respectively, for the lowest-energy isomers. Other isomers correspond to different shell shapes, and/or noncentral up to (higher-energy) surface locations of the C atom. Up to $n = 4$, CAu_n are structural analogues of CH_n.

The gold shells around the dicarbon core include diatomic units (in accordance with closed-shell Au_2), at least for the lowest-energy isomers of small C_2Au_n, reflecting interplay between C–Au and Au–Au bonding. The most stable geometries for $n > 2$ resemble those of weakly bound hydrocarbon complexes (hydrogen–acetylene, *etc.*), while the structural counterparts of hydrocarbon molecules, dominated by C–Au bonding, are the lowest-energy isomer for $n = 2$ or the isomers of slightly higher energy for $n \geq 3$.

By comparison, small C_2Al_n (and C_3Al_n) clusters, consistent with higher valence offering more options for Al–Al bonding and with open-shell Al_2, are different and do not exhibit diatomic units as well as (except for $n = 2$) stable hydrocarbon-like structures that in fact may correspond to transition states. In particular, unsaturated alumocarbons (C_2Al_4, C_3Al_6) tend to have lowest- or at least low-energy planar geometries with aluminium 2D shells (frames) around linear carbon cores. The cores can bend and protrude from the frames in the case of their size mismatch (as for C_3Al_4) or in saturated systems (C_3Al_8).

The structural similarity of the low-energy isomers of the small auro-carbon species to analogous hydrocarbons enables starting with appropriate C_nH_m counterparts to build larger core–shell auro-carbons, such as C_5Au_{12} and $C_{10}Au_{18}$. The aurophilic interactions allow further design options employed in extending and realigning the shells in C_5Au_{14} and $C_{10}Au_{21,24}$, with an apparent feasibility of other similar species. In particular, the usually increased stability of systems with "magic" number of valence electrons (like 8 in CAu_4, 34 in C_5Au_{14}, 58 in $C_{10}Au_{18}$) can be another factor to account for in design. Due to an appreciable difference in relative electronegativities of carbon and metals, however, this factor may be of limited validity. For instance, while CAu_4 is the most stable of small CAu_n systems in terms of dissociation and shows a considerable singlet-triplet gap, the larger C_5Au_{14} and $C_{10}Au_{18}$ systems have the closed-shell ground state only marginally lower in energy than the excited open-shell state. In the case of still less electronegative Al, formally "magic" C_2Al_4 is actually least stable among its size-neighbours in either core or shell.

The switch from unsaturated molecules and molecular radicals (such as C_n) to closed-shell molecules like CH_4 and H_2 adds more variety to the core–shell system geometries and related properties due to different (noncovalent) interactions between the core and the shell. In particular, the geometrically C_5-like methane core completely restructures the gold shell as compared to that in C_5Au_{12}, while molecular hydrogen only slightly inflates the cuboctahedral Au_{12} cage and appears to be able to rotate inside it relatively freely. This can allow creation of species with unique geometries and parameters. Due to no core–shell bonding and charge transfer (unlike in C_nAu_m), these systems are metastable (higher in energy) relative to the molecules being outside the shells. Some of the shells (*e.g.* that in $CH_4@Au_{12}$) may collapse with no supporting cores inside, while some cages stable by themselves (like cuboctahedral Au_{12}) may break upon insertion of such a core (*e.g.* CH_4), so there is a composition-specific balance to be found in designing such species.

Some other, sufficiently tight, metal cages, such as Mg_{8-10}, can be significantly reshaped by insertion of even the smallest possible, H_2, molecule. Such systems with shells of less electronegative metal atoms can exhibit a significant core–shell charge transfer. This may result in induced dissociation of the core, and in more complicated chemical reactions between components of a core composed of a few molecules, steered in the confined spaces of and catalysed by the surrounding metal shell, as illustrated by $(2H_2)@Mg_{10}$.

Manifestation of structure–property relationships can be noted in terms of comparing vertical excitation and ionisation energies (VE*, VIE) and electron affinities (VEA) of the empty metal cluster cage and the corresponding core–shell system. For instance, these properties may be affected relatively weakly, as in CAu_5 as compared to Au_5, or vary differently and more strongly, as for other CAu_n, in particular in opposite directions for different n, as for $n = 4$ and 6. Similarly, the properties of a stable empty cage may be altered only slightly by insertion of a small molecule, as for $H_2@Au_{12}$, or more significantly, as in C_5Au_{12}. The properties may also vary in different directions for different molecular cores, such as VEA for $H_2@Au_{12}$ *versus* C_5Au_{12}. Such a variety of possible property modifications makes it possible to selectively design and fine tune desirable parameters of core–shell systems by combining proper cores and shells.

Possible applications of such core–shell systems may include tunable catalysts, molecular (including hydrogen) storage, molecular electronics, nano-structures and nanomaterials, *etc.* In particular, the metastability of species such as H_2 trapped in a metal cage could enable relatively easy release of hydrogen when needed. Moreover, such a metastability can also be considered as energy storage at the molecular/cluster level, providing, in the case of hydrogen, an additional, direct component of accumulated energy.

Acknowledgements

The author is grateful to P. McNelles and G. Kochhar for preliminary calculations on some of the above systems during their 4th-year thesis and Summer research. Acknowledged with gratitude are the financial support of the UOIT and NSERC of Canada, and the technical support of the staff of high-performance computing facilities of the UOIT Faculty of Science and of the SHARCnet distributed academic network of Ontario.

References

1. J. A. Alonso, *Chem. Rev.*, 2000, **100**, 637.
2. E. Roduner, *Chem. Soc. Rev.*, 2006, **35**, 583.

3. M. C. Daniel and D. Astruc, *Chem. Rev.*, 2004, **104**, 293.

4. D. E. Bergeron, A. W. Castleman Jr., T. Morisato and S. N. Khanna, *Science*, 2004, **304**, 84.

5. O. P. Charkin, D. O Charkin, N. M. Klimenko and A. M. Mebel, *Chem. Phys. Lett.*, 2002, **365**, 494.

6. C. Majumder, A. K. Kandalam and P. Jena, *Phys. Rev. B*, 2006, **74**, 205437.

7. V. Kumar, *Comput. Mater. Sci.*, 2006, **36**, 1.

8. M.-M. Rohmer, M. Bénard and J.-M. Poblet, *Chem. Rev.*, 2000, **100**, 495.

9. R. Ferrando, J. Jellinek and R. L. Johnston, *Chem. Rev.*, 2008, **108**, 845.

10. M. P. Johansson, D. Sundholm and J. Vaara, *Angew. Chem. Int. Ed.*, 2004, **43**, 2678.

11. W. Fa and J. Dong, *J. Chem. Phys.*, 2006, **124**, 114310.

12. R. J. C. Batista, M. S. C. Mazzoni, L. O. Ladeira and H. Chacham, *Phys. Rev. B*, 2005, **72**, 085447.

13. C. Chang, A. B. C. Patzer, E. Sedlmayr, D. Sülzle and T. Steinke, *Comput. Mater. Sci.*, 2006, **35**, 387.

14. W. Fa, C. Luo and J. Dong, *Phys. Rev. B*, 2005, **72**, 205428.

15. D. Y. Zubarev and A. I. Boldyrev, *J. Chem. Phys.*, 2005, **122**, 144322.

16. R. Pal, S. Bulusu and X. C. Zeng, *J. Comput. Meth. Sci. Eng.*, 2007, **7**, 185.

17. B. Kiran, H.-J. Zhai, L.-F. Cui and L.-S. Wang, *Angew. Chem. Int. Ed.*, 2004, **43**, 2125.

18. M. Lahav, E. A. Weiss, Q. Xu and G. M. Whitesides, *Nanolett.*, 2006, **6**, 2166.

19. E. J. Bylaska, W. A. de Jong, K. Kowalski, T. P. Straatsma, M. Valiev, D. Wang, E. Aprà, T. L. Windus, S. Hirata, M. T. Hackler, Y. Zhao, P.-D. Fan, R. J. Harrison, M. Dupuis, D. M. A. Smith, J. Nieplocha, V. Tipparaju, M. Krishnan, A. A. Auer, M. Nooijen, E. Brown, G. Cisneros, G. I. Fann, H. Fruchtl, J. Garza, K. Hirao, R. Kendall, J. A. Nichols, K. Tsemekhman, K. Wolinski, J. Anchell, D. Bernholdt, P. Borowski, T. Clark, D. Clerc, H. Dachsel, M. Deegan, K. Dyall, D. Elwood, E. Glendening, M. Gutowski, A. Hess, J. Jaffe, B. Johnson, J. Ju, R. Kobayashi, R. Kutteh, Z. Lin, R. Littlefield, X. Long, B. Meng, T. Nakajima, S. Niu, L. Pollack, M. Rosing, G. Sandrone, M. Stave, H. Taylor, G. Thomas, J. van Lenthe, A. Wong, Z. Zhang, *NWChem, A Computational Chemistry Package for Parallel Computers*, Version 5.0, Pacific Northwest National Laboratory, Richland, Washington 99352-0999 (USA), 2006.

20. A. Ryzhkov and A. Antipin, *ViewMol3D 4.34, a 3D OpenGL viewer for molecular structures.* http://redandr.tripod.com/vm3/

21. *Basis Set Exchange*, v. 1.2.2, Pacific Northwest National Laboratory, Richland, Washington 99352-0999 (USA). https://bse.pnl.gov/bse/portal

22. *NIST Chemistry WebBook* (NIST Standard ref. Database Number 69, June 2005 Release). http://webbook.nist.gov/chemistry/

23. M. Walter and H. Häkkinen, *Phys. Chem. Chem. Phys.*, 2006, **8**, 5407.

24. Q. Sun, Q. Wang, G. Chen and P. Jena, *J. Chem. Phys.*, 2007, **127**, 214706.

25. B. W. Ticknor, B. Bandyopadhyay and M. A. Duncan, *J. Phys. Chem. A*, 2008, **112**, 12355.
26. C. Majumder, *Phys. Rev. B*, 2007, **75**, 235409.
27. W. Zhizhong, Z. Xiange and T. Auchin, *J. Mol. Struct. (Theochem)*, 1998, **453**, 225.
28. P. Zaleski-Ejgierd and P. Pyykko, *Can. J. Chem.*, 2009, **87**, 798.
29. K. Chenoweth and C. E. Dykstra, *Chem. Phys. Lett.*, 2004, **153**, 400.
30. F. Y. Naumkin, *J. Phys. Chem. A*, 2008, **112**, 4660.
31. A. I. Boldyrev, J. Simons, X. Li, W. Chen and L.-S. Wang, *J. Chem. Phys.*, 1999, **110**, 8980.
32. N. A. Cannon, A. I. Boldyrev, X. Li and L.-S. Wang, *J. Chem. Phys.*, 2000, **113**, 2671.
33. X. Li, L.-S. Wang, N. A. Cannon and A. I. Boldyrev, *J. Chem. Phys.*, 2002, **116**, 1330.
34. R. F. W. Bader, *Atoms in Molecules: A Quantum Theory*, Oxford University Press, Oxford, 1990.
35. Y.-B. Wu, H.-G. Lu, S.-D. Li and Z.-X. Wang, *J. Phys. Chem. A*, 2009, **113**, 3395.
36. F. Naumkin, *Phys. Chem. Chem. Phys.*, 2006, **8**, 2539.
37. F. Y. Naumkin, *Chem. Phys. Lett.*, 2008, **466**, 44.
38. P. McNelles and F. Y. Naumkin, *Phys. Chem. Chem. Phys.*, 2009, **11**, 2858.

CHAPTER 3

Self-Assembly of Nanoclusters: An Energy Landscape Perspective

DWAIPAYAN CHAKRABARTI, SZILARD N. FEJER
AND DAVID J. WALES

University Chemical Laboratories, Lensfield Road, Cambridge
CB2 1EW, UK

3.1 Introduction

The emergence of novel functionalities for materials constructed at the nanometer scale makes them attractive for functional materials. Such nanoscale fabrication can be approached *via* two complementary strategies. In a top-down approach, a material is gradually reduced in size, often using radiation or chemical treatment, while a bottom-up approach involves assembly of sub-components. The implicit task in the bottom-up approach is to encode the information of the target structure into the building blocks. However, lack of a sufficiently detailed understanding of the basic rules that govern self-assembly poses a challenge to the rational design of tailor-made building blocks that would self-assemble into specific morphologies. Moreover, the potential building blocks are no longer restricted to atoms and molecules.[1] Thanks to recent advances in synthesis, a spectacular variety of colloidal building blocks differing in shape, pattern, and functionality are now available.[2] While this availability indeed opens up a plethora of routes to functional nanostructures, it also comes with the burden of optimising a large number of control factors

RSC Theoretical and Computational Chemistry Series No. 4
Computational Nanoscience
Edited by Elena Bichoutskaia
© Royal Society of Chemistry 2011
Published by the Royal Society of Chemistry, www.rsc.org

for self-assembly. Computer simulation studies with coarse-grained models that capture the essential physics are a promising means to reduce this burden. Trying to learn from nature in the design of building blocks can also be rewarding.[3] After all, self-assembly is nature's prescription for the creation of complex structures with well-defined functions from simpler building blocks. One of the most striking examples of the remarkable ability of biological matter to robustly self-assemble is the formation of viral capsids, which has a wide range of potential applications in nanotechnology.[4,5]

Colloidal building blocks have enormous potential for programmed self-assembly, since the interactions between them can be tuned.[2,6–12] One of the major challenges, however, is to manipulate the interactions such that the target structure is not only thermodynamically favourable, but is also kinetically accessible.[13] The present account highlights the importance of surveying the underlying potential energy surface in addressing this challenge.

A fundamental difference between nanomaterials and bulk materials lies in the fraction of atoms or molecules in the surface or interface. Characteristic features of this size regime emerge for nanomaterials from the relatively large fraction of building blocks present at the surface. Studies of clusters therefore shed light on a wide variety of phenomena observed at the nanoscale.[14] Here, we provide an overview of our recent efforts on assembling rigid-body building blocks decorated with interaction sites into nanoclusters, emphasising manipulation of the anisotropic interactions.

3.2 Potential Energy Surfaces

The potential energy surface (PES) for any condensed matter system governs the observed structure, thermodynamics, and dynamics.[13] A survey of the PES can therefore provide valuable insight into the likely properties. Our focus is usually on characterising stationary points on the PES, where the gradient vanishes, in particular the local minima and transition states. Here, a geometrical criterion distinguishes transition states from local minima for which all the normal modes have positive curvatures: a transition state has a single imaginary normal mode frequency.[15]

At low temperatures, the most thermodynamically favourable structure of a cluster of N interacting particles is the one that minimises its total energy. The task of finding the global minimum on the PES is, however, nontrivial, especially in the presence of anisotropic interactions. The basin-hopping approach to global optimisation was successfully employed to find the global minima for the clusters described below, which are bound by a variety of anisotropic potentials. This method is based upon hypersurface deformation, where the transformation of the potential energy surface preserves the global minimum, and also the relative energies of the local minima.[16] In practice, the configuration space is explored by taking steps in both the translational and rotational coordinates for rigid bodies. A local geometry optimisation follows each perturbation before the step is accepted or rejected on the basis of a Metropolis (or other suitable) criterion based on the energy of the new local minimum. Since the objective is to escape from the basin of attraction of the current minimum, step sizes are much larger than those typically used in Monte Carlo simulations of thermodynamic properties.

While identifying the global minimum on the PES yields useful structural information, characterisation of the relevant transition states is critical for understanding the dynamics on the experimental timescale. In particular, the kinetic accessibility of the favoured structures must be judged in terms of the energy barriers involved. The discrete path sampling (DPS) approach can be used to obtain this information.[13,17–19] In this method, an initial path between two relevant minima, consisting of a series of intervening transition states and minima on the underlying PES, is determined by double-ended transition state searches, performed with the doubly nudged[20] elastic band[21–23] (DNEB) algorithm. Transition state candidates identified with this approach are then accurately refined using hybrid eigenvector-following techniques.[24,25] The connectivity of a transition state is then defined by the two minima reached by (approximate) steepest-descent paths leaving parallel and antiparallel to the eigenvector with the unique negative eigenvalue; such a minimum-transition state-minimum triplet is termed an elementary rearrangement. The limited-memory Broyden–Fletcher–Goldfarb–Shanno (LBFGS) algorithm of Liu and Nocedal is used for the local minimisations.[26,27] Collections of discrete paths (*i.e.* connected sequences of minima and the intervening transition states on the PES) are generated systematically following the DPS approach from the initial connected path.[19] The objective here is to grow a database of connected stationary points, starting from those in the initial path, by adding all the minima and transition states found during successive connection making attempts for pairs of minima selected from the current database. The discrete path that makes the largest contribution to the two-state rate constant within a steady-state approximation for the intervening minima can be extracted from the DPS database using a network formulation[28] *via* Dijkstra's shortest path algorithm.[29] The original DPS formulation was presented in detail in ref. 17, and more recent developments can be found in refs. 13, 18 and 19.

A pictorial representation of the organisation of the PES, which faithfully represents the energy barriers between the minima, can provide significant insight into the kinetic accessibility of the thermodynamically favoured structures. A two-dimensional representation of the multidimensional PES (the potential energy V is generally a function of $6N$ coordinates for a system of N nonlinear rigid bodies) can be obtained using disconnectivity graphs.[13,30–33] A database of local minima and intervening transition states with their connectivity information is required for the construction of a disconnectivity graph, which is obtained *via* a "superbasin" analysis at a regularly spaced set of energies $V_1 < V_2 < V_3 < \cdots$. The analysis at a given potential energy, V_α, partitions the local minima that lie below V_α into disjoint sets, or "superbasins",[30] such that any two minima within the same set are connected by a pathway where the potential energy never exceeds the threshold V_α. The vertical axis of the graph corresponds to the potential energy increasing upwards, and each distinct superbasin is represented as a point, or node, on the horizontal axis, which can be chosen to space the minima as clearly as possible. The interval between the V_α values is an adjustable parameter, chosen so that the representation of the energy landscape retains the most important topological information. It is also possible to base the vertical axis on enthalpy[34] or free energy.[35–37]

Figure 3.1 illustrates three distinct motifs, which correspond to idealised patterns of organisation in the PES.[31] Every disconnectivity graph is

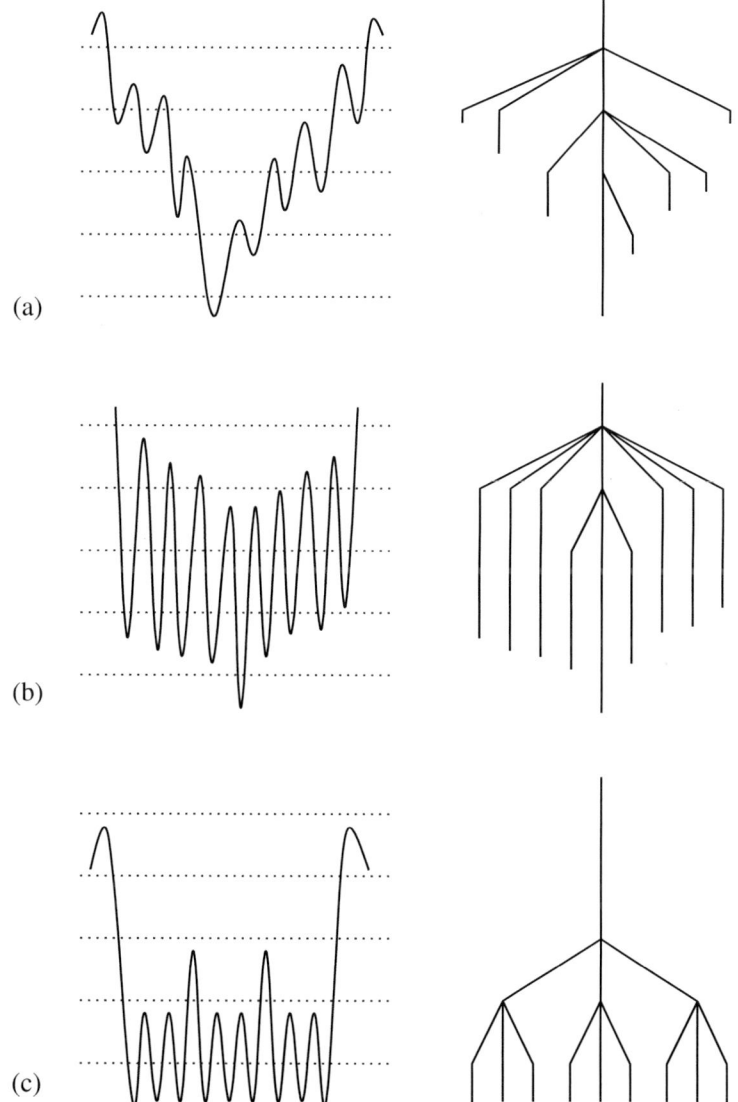

Figure 3.1 One-dimensional potential energy functions (left) and the corresponding disconnectivity graphs (right). The dotted lines indicate the energies at which a superbasin analysis was performed. (a) "palm tree", (b) "willow tree", (c) "banyan tree".[31]

technically a tree graph,[38] because cutting any edge separates the structure into two parts. Hence, the graph corresponding to a well-defined global minimum with relatively small downhill barriers in Figure 3.1(a) may be referred to as a "palm tree".[13,31] Such a pattern supports efficient relaxation to the global minimum over a wide temperature range. For the "willow tree" pattern in Figure 3.1(b), although a well-defined global minimum exists, the downhill barriers are large compared to the potential energy differences between successive minima. Efficient relaxation now requires a relatively high temperature, as for buckminsterfullerene.[31,39,40] Finally, the "banyan tree" pattern in Figure 3.1(c), with several low-lying minima of comparable energies, and two or more distinct energy scales for the barriers between them, has a qualitatively different hierarchical structure, which has been observed for water clusters and models of bulk glass formers.[31,41] Potential energy landscapes with the appearance of a single palm tree or willow tree are the ones that we expect to correspond to efficient "structure seekers"[42] for an appropriate temperature range. These landscapes can also be considered unfrustrated,[43–46] since there are no competitive low-lying minima separated by high barriers.

3.3   Geometry Optimisation of Rigid Bodies

Our approach to surveying the underlying PES for a system of rigid building blocks decorated with interaction sites is critically dependent on the efficiency of geometry optimisation, which in turn depends on the availability of analytic first derivatives of the energy with respect to the rigid-body coordinates. Chasles' theorem, which states that the most general displacement of a rigid body can be described as a translation plus a rotation,[47] underlies our choice for a minimal set of rigid-body coordinates. While the Cartesian representation of the centre-of-mass is the usual choice for the translational coordinates, various possibilities exist for the rotational coordinates.[33,48–52] Euler angles suffer from problems arising from singularities, and the characterisation of transition states and pathways is known to face difficulties.[48] Although the four-parameter unit quaternion provides a singularity-free description of rotation,[53] the unit-norm constraint, which is quadratic in form, makes it less attractive for geometry optimisation. Alternatively, Stone and coworkers used a hierarchy of axis systems.[50,51] A set of rotational coordinates that is distinct from the one describing the orientation of a rigid body was used to obtain the energy derivatives.[50,51] Recently, Wales discussed the use of an angle-axis representation for rotation in the context of site–site isotropic potentials.[33] Such a representation is based on Euler's theorem,[47] which states that a general displacement of a rigid body with one point fixed is a rotation about some axis. The angle-axis representation consists of a unit vector $\hat{\mathbf{p}}$, defining the axis, and an angle θ, describing the magnitude of rotation about that axis. A more concise representation employs an unnormalised vector $\mathbf{p} = [p_1, p_2, p_3] = \theta\hat{\mathbf{p}}$, which we call a rotation vector, whose magnitude, θ, is the angle of rotation, and whose direction is parallel to the axis of rotation. However, the first implementation of the angle-axis scheme for geometry optimisation sacrificed efficiency for flexibility.[33] Moreover, the generalisation to deal with anisotropic potentials was not straightforward within this framework. More recently, Chakrabarti and Wales developed an efficient and robust computational framework for geometry optimisation of rigid bodies, employing the angle-axis representation and using a matrix formulation.[54] This framework, which provided a very flexible user-friendly interface to incorporate any system governed by site–site isotropic or anisotropic potentials, is briefly discussed below.

For the rotation vector $\mathbf{p}$, the corresponding 3×3 rotation matrix $\mathbf{R}$ can be expressed, following Rodrigues' rotation formula,[55] as

$$\mathbf{R} = \mathbf{I} + (1 - \cos\theta)\tilde{\mathbf{p}}\tilde{\mathbf{p}} + \sin\theta\tilde{\mathbf{p}}, \tag{3.1}$$

where $\mathbf{I}$ is a 3×3 identity matrix, and $\tilde{\mathbf{p}}$ is the skew-symmetric matrix obtained from $\hat{\mathbf{p}}$:

$$\tilde{\mathbf{p}} = \frac{1}{\theta}\begin{pmatrix} 0 & -p_3 & p_2 \\ p_3 & 0 & -p_1 \\ -p_2 & p_1 & 0 \end{pmatrix}. \tag{3.2}$$

The product of the skew-symmetric matrix $\tilde{\mathbf{p}}$, obtained from $\mathbf{p}$, and any vector $\mathbf{v}$ returns their crossproduct: $\tilde{\mathbf{p}}\mathbf{v} = \mathbf{p}\times\mathbf{v}$. Eqn (3.1) is the key to our matrix formulation for an angle-axis representation of rotations. If the derivative of a matrix $\mathbf{A}$, $\partial\mathbf{A}/\partial p_k$, is denoted by $\mathbf{A}_k$ ($k = 1, 2, 3$), one has

$$\mathbf{R}_k = \frac{p_k \sin\theta}{\theta}\tilde{\mathbf{p}}^2 + (1 - \cos\theta)(\tilde{\mathbf{p}}_k\tilde{\mathbf{p}} + \tilde{\mathbf{p}}\tilde{\mathbf{p}}_k) + \frac{p_k \cos\theta}{\theta}\tilde{\mathbf{p}} + \sin\theta\tilde{\mathbf{p}}_k, \tag{3.3}$$

where $\theta = (p_1^2 + p_2^2 + p_3^2)^{1/2}$. Here, for example,

$$\tilde{\mathbf{p}}_1 = \frac{1}{\theta}\begin{pmatrix} 0 & p_1 p_3/\theta^2 & -p_1 p_2/\theta^2 \\ -p_1 p_3/\theta^2 & 0 & -(1 - p_1^2/\theta^2) \\ p_1 p_2/\theta^2 & (1 - p_1^2/\theta^2) & 0 \end{pmatrix}. \tag{3.4}$$

In the limit as $\theta \to 0$ the formulation in eqn (3.1) reduces to the identity matrix. Analytical derivatives can then be obtained by considering a Taylor expansion. The result is $\mathbf{R}_k = \tilde{\mathbf{p}}(k)$, where $\tilde{\mathbf{p}}(k)$ is obtained from eqn (3.2) using $\hat{\mathbf{p}}$ with the kth element set to unity.

The second derivatives can be obtained within this framework by successive application of the chain rule. Six additional 3×3 matrices of two types need to be computed, namely $\partial^2 \mathbf{R}/\partial p_k^2$ and $\partial^2 \mathbf{R}/\partial p_l \partial p_k$, which we denote by $\mathbf{R}_{kk}$ and $\mathbf{R}_{kl}$, respectively:

$$
\begin{aligned}
\mathbf{R}_{kk} =& \frac{2p_k \sin\theta}{\theta}(\tilde{\mathbf{p}}_k \tilde{\mathbf{p}} + \tilde{\mathbf{p}}\tilde{\mathbf{p}}_k) + \left(\frac{p_k^2 \cos\theta}{\theta^2} - \frac{p_k^2 \sin\theta}{\theta^3} + \frac{\sin\theta}{\theta}\right)\tilde{\mathbf{p}}^2 \\
&+ (1 - \cos\theta)(2\tilde{\mathbf{p}}_k^2 + \tilde{\mathbf{p}}_{kk}\tilde{\mathbf{p}} + \tilde{\mathbf{p}}\tilde{\mathbf{p}}_{kk}) + \left(-\frac{p_k^2 \sin\theta}{\theta^2} - \frac{p_k^2 \cos\theta}{\theta^3} + \frac{\cos\theta}{\theta}\right)\tilde{\mathbf{p}} \\
&+ \frac{2p_k \cos\theta}{\theta}\tilde{\mathbf{p}}_k + \sin\theta\,\tilde{\mathbf{p}}_{kk},
\end{aligned}
\tag{3.5}
$$

and

$$
\begin{aligned}
\mathbf{R}_{kl} =& \frac{p_k \sin\theta}{\theta}(\tilde{\mathbf{p}}_l \tilde{\mathbf{p}} + \tilde{\mathbf{p}}\tilde{\mathbf{p}}_l) + \left(\frac{p_k p_l \cos\theta}{\theta^2} - \frac{p_k p_l \sin\theta}{\theta^3}\right)\tilde{\mathbf{p}}^2 \\
&+ \frac{p_l \sin\theta}{\theta}(\tilde{\mathbf{p}}_k \tilde{\mathbf{p}} + \tilde{\mathbf{p}}\tilde{\mathbf{p}}_k) + (1 - \cos\theta)(\tilde{\mathbf{p}}_{kl}\tilde{\mathbf{p}} + \tilde{\mathbf{p}}_k \tilde{\mathbf{p}}_l + \tilde{\mathbf{p}}_l \tilde{\mathbf{p}}_k + \tilde{\mathbf{p}}\tilde{\mathbf{p}}_{kl}) \\
&- \left(\frac{p_k p_l \sin\theta}{\theta^2} + \frac{p_k p_l \cos\theta}{\theta^3}\right)\tilde{\mathbf{p}} + \frac{p_k \cos\theta}{\theta}\tilde{\mathbf{p}}_l + \frac{p_l \cos\theta}{\theta}\tilde{\mathbf{p}}_k + \sin\theta\,\tilde{\mathbf{p}}_{kl}.
\end{aligned}
\tag{3.6}
$$

Here, for example,

$$
\tilde{\mathbf{p}}_{11} = \frac{1}{\theta^3}\begin{pmatrix}
0 & p_3(1 - 3p_1^2/\theta^2) & -p_2(1 - 3p_1^2/\theta^2) \\
-p_3(1 - 3p_1^2/p\theta^2) & 0 & p_1(3 - 3p_1^2/\theta^2) \\
p_2(1 - 3p_1^2/\theta^2) & -p_1(3 - 3p_1^2/\theta^2) & 0
\end{pmatrix}, \tag{3.7}
$$

and

$$
\tilde{\mathbf{p}}_{12} = \frac{1}{\theta^3}\begin{pmatrix}
0 & -3p_1 p_2 p_3/\theta^2 & -p_1(1 - 3p_2^2/\theta^2) \\
3p_1 p_2 p_3/\theta^2 & 0 & p_2(1 - 3p_1^2/\theta^2) \\
p_1(1 - 3p_2^2/\theta^2) & -p_2(1 - 3p_1^2/\theta^2) & 0
\end{pmatrix}. \tag{3.8}
$$

The other $\tilde{\mathbf{p}}_{ij}$ matrices can be obtained by substituting the corresponding indices in eqns (3.7) and (3.8). As for the first derivatives, the limit for $\theta \to 0$ can be treated by considering Taylor expansions, giving $\mathbf{R}_{kk} = \tilde{\mathbf{p}}(k)\tilde{\mathbf{p}}(k)$ and $\mathbf{R}_{kl} = \tilde{\mathbf{p}}(k)\tilde{\mathbf{p}}(l)$.

We now outline a prescription for obtaining the first derivatives of the energy with respect to the rigid-body coordinates for site–site isotropic potentials, to highlight the salient features of the angle–axis approach in its matrix formulation. If the coordinates of two rigid bodies are denoted using the superscripts I and J, and the sites within each rigid body by subscripts i and j, then for site–site isotropic potentials the total energy is

$$U = \sum_I \sum_{J<I} \sum_{i \in I} \sum_{j \in J} f_{ij}(r_{ij}),$$

(3.9)

where $r_{ij} = |\mathbf{r}_{ij}| = |\mathbf{r}_i - \mathbf{r}_j|$ and $f_{ij} \equiv U_{ij}^{IJ}$ is the pair potential between sites i and j. If ζ represents one of the six coordinates of rigid body I, then the first derivative of the potential energy is

$$\frac{\partial U}{\partial \zeta} = \sum_{J \neq I} \sum_{i \in I} \sum_{j \in J} f_{ij}'(r_{ij}) \frac{\partial r_{ij}}{\partial \zeta},$$

(3.10)

where $f_{ij}' = df_{ij}(r_{ij})/dr_{ij}$. Here,

$$\frac{\partial r_{ij}}{\partial \mathbf{r}^I} = \hat{\mathbf{r}}_{ij},$$

(3.11)

and

$$\frac{\partial r_{ij}}{\partial p_k^I} = \hat{\mathbf{r}}_{ij} \cdot \frac{\partial \mathbf{r}_{ij}}{\partial p_k^I} = \hat{\mathbf{r}}_{ij} \cdot (\mathbf{R}_k^I \mathbf{r}_i^0),$$

(3.12)

where the following relationship with the vectors in the body-fixed frame (denoted by a superscript 0) was used

$$\mathbf{r}_{ij} = \mathbf{r}^I + \mathbf{R}^I \mathbf{r}_i^0 - \mathbf{r}^J - \mathbf{R}^J \mathbf{r}_j^0.$$

(3.13)

The Hessian matrix for site–site isotropic potentials can also be worked out easily within this framework, which can also deal efficiently with any site–site anisotropic potentials. Further details, including the treatment of anisotropic potentials within this framework, can be found in ref. 54.

The advantage of the above scheme is that all the rigid-body coordinate information is stored in the space-fixed frame and the derivatives of the rotation matrix can be programmed in general using a fixed subroutine, which can be called to deal with site–site isotropic or anisotropic potentials. Terms involving the angle-axis coordinates that appear in calculating the energy and its derivatives can be obtained by the action of the rotation matrix and its derivatives, whose forms are system independent, on vectors or matrices defined in the body-fixed frame [*vide* eqn (3.12)]. The gain in efficiency arises from precomputing these products for a given configuration, thus removing as many operations as possible from the innermost loop over each distinct pair of sites. This framework has been implemented in our global optimisation program, GMIN,[56] and our geometry optimisation program, OPTIM,[57] to incorporate a variety of pair potentials, ranging from site–site isotropic potentials to site–site anisotropic potentials, including single-site anisotropic potentials. Benchmark calculations for minimisation of $(H_2O)_{55}$ clusters described by the TIP4P potential[58] showed that the angle-axis framework in its matrix formulation gained a ten-fold speed-up over the initial implementation, and was only about 10% slower than an Euler-angle scheme, which exploited the specific geometry of the water monomers.[54]

3.4   Illustrative Examples

3.4.1   Nanoscale Chirality

Helical structures are perhaps the most fascinating examples of chiral archi-tectures, and they are common structural motifs in biological macromolecules. It is therefore not surprising that fabrication of helical architectures at the macro- and supramolecular level in synthetic systems has attracted consider-able attention over the years.[59,60] However, helical structures are important in their own right for their diverse potential applications in materials science, especially in optoelectronics, sensors, enantiomeric separation, and asymmetric catalysis.[59,60] These applications rely upon helical order in nanofibres as well as in liquid crystals and crystals.[61–63] Self-organisation of discotic molecules in columnar stacks with helical order has, in particular, been the focus of several crystal or liquid-crystal engineering studies for the exceptional one-dimensional charge carrier mobilities along the columns.[64–66] While a common route to induce helicity in columnar arrangements is inclusion of chiral centres in dis-cotic molecules,[67] noncovalent interactions between achiral discotic molecules, such as π–π stacking, hydrogen bonding, dipolar interactions, and hydrophobic interactions, have also been exploited to fabricate helical columnar structures in specific cases.[60,68,69]

In this section we describe our recent efforts on devising a general strategy for fabrication of nanoscale helical architectures with achiral building blocks, underpinning the physics of emergent chirality. In this task, we first con-sidered assembly of dipolar dumbbells,[70] following a recent experiment that employed colloidal dumbbells with magnetic belts at their waist.[8] In the presence of an applied magnetic field, a linear chain was observed with symmetric dumbbells, while the asymmetry of the dumbbells forced the chain to coil.[8] We modelled the dipolar dumbbells using multiple interaction sites within a rigid-body framework.[70] Each dumbbell involves two spherical lobes, modelled by Lennard-Jones (LJ) sites (labelled 1 and 2), and a point dipole pointing across the axis between the lobes. The total energy of a system of N dumbbells in an electric field $\boldsymbol{E}$ is

$$U^{DB} = \sum_{I=1}^{N-1} \sum_{J=I+1}^{N} \sum_{i\in I}^{1,2} \sum_{j\in J}^{1,2} 4\varepsilon_{ij} \left[\left(\frac{\sigma_{ij}}{r_{ij}}\right)^{12} - \left(\frac{\sigma_{ij}}{r_{ij}}\right)^{6} \right]$$

$$+ \sum_{I=1}^{N-1} \sum_{J=I+1}^{N} \frac{\mu_D^2}{r_{IJ}^3} [(\hat{\boldsymbol{\mu}}_I \cdot \hat{\boldsymbol{\mu}}_J) - 3(\hat{\boldsymbol{\mu}}_I \cdot \hat{\mathbf{r}}_{IJ})(\hat{\boldsymbol{\mu}}_J \cdot \hat{\mathbf{r}}_{IJ})] - \mu_D \sum_{I=1}^{N} \hat{\boldsymbol{\mu}}_I \cdot \boldsymbol{E}. \tag{3.14}$$

Here, $\mathbf{r}_I$ is the position vector for the point-dipole on dumbbell I, $\hat{\boldsymbol{\mu}}_I$ is the unit vector defining the direction of the dipole moment, whose magnitude is μ_D, $\mathbf{r}_{IJ} = \mathbf{r}_I - \mathbf{r}_J$ is the separation vector between dipoles on dumbbells I and J with magnitude r_{IJ}, $\hat{\mathbf{r}}_{IJ} = \mathbf{r}_{IJ}/r_{IJ}$, and r_{IJ} is the separation between LJ sites i and j. The units of energy and length are chosen as the LJ parameters ε_{11} and σ_{11}, respectively. For the LJ interactions we set $\varepsilon_{11} = \varepsilon_{22} = \varepsilon_{12} = 1$ and $\sigma_{11} = 1$. $\sigma_{22} < 1$ was varied to explore the effects of asymmetry with $\sigma_{12} = (\sigma_{11} + \sigma_{22})/2$. With the lobes characterised as spheres with diameters σ_{11} and σ_{22}, an asymmetry parameter was defined as $\alpha = \sigma_{11}/\sigma_{22}$. The direction of the electric field $\mathbf{E} = (0, 0, E)$ was held fixed along the z-axis of the space-fixed frame as its strength, E, was varied. μ_D is then in reduced units of $(4\pi\varepsilon_0\varepsilon_{11}\sigma_{11}^3)^{1/2}$ and E is in $[\varepsilon_{11}/(4\pi\varepsilon_0\sigma_{11}^3)]^{1/2}$, where ε_0 is the permittivity of free space.

Figure 3.2 shows the global minima for clusters of $N = 6$ asymmetric dipolar dumbbells when the asymmetry parameter $\alpha = 2$. A close packing of the smaller spheres allows a slightly distorted hexagonal arrangement of the dipoles [Figure 3.2(a)]. When a strong enough electric field is applied, a single helical strand

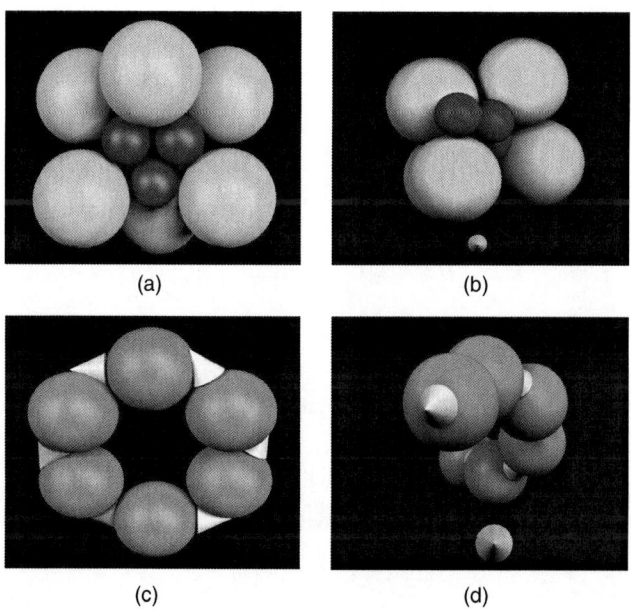

(a)   (b)

(c)   (d)

Figure 3.2 Global minima for clusters of $N = 6$ dipolar dumbbells with the size ratio between the spherical lobes, characterised by the asymmetry parameter $\alpha = 2$. (a) Dipolar dumbbells; (b) dipolar dumbbells in the presence of an applied electric field (top view). The bottom panels (c and d) show the same structures as (a and b) in the same order, but depict for clarity only the position of the point-dipole on the dumbbell axis. Here, dipole moment $\mu_D = 0.7$ and electric field strength $E = 5$. In (b) and (d) the arrow indicates the field direction. Emergence of helicity under the applied field is clearly evident, especially in the reduced representations.

grows along the direction of the field [Figure 3.2(b)]. While the dumbbells tend to align perpendicular to the field because of the dipolar interactions, competition from a second length scale that controls the steric interactions is critical to the emergent supramolecular chirality. Figure 3.3 confirms that asymmetry is indeed a key factor, in agreement with the experiment. When the two competing interactions were manipulated by varying $\sigma_{22} < 1$, μ_D, and E, one or more helical strands emerged as the global minima for small clusters over a range of parameter space.[70] Figure 3.4 illustrates a helical fibre, which was found to be the global minimum for at least 20 asymmetric dipolar dumbbells in the presence of an applied electric field.

Figures 3.2–3.4 show how helical structures can emerge from the interplay between two competing length scales for anisotropic interactions [Figure 3.5(a)]. Such competing length scales seem to be present in DNA, where one characterises the distance between consecutive nucleotides and the other governs the stacking of base pairs. Can we then draw inspiration from nature to design simpler building blocks for self-assembling helical nanostructures? An ellipsoid of revolution is perhaps the simplest building block that provides a realisation of two competing length scales for anisotropic interactions. Oblate ellipsoids, which are often invoked in coarse-grained descriptions of discotic molecules, therefore appear to be promising candidates. Hard ellipsoids of revolution are not suitable as they do not form a columnar phase.[71] We considered two pair potentials for soft ellipsoids of revolution: one suggested by Paramonov and Yaliraki,[72] (PY) and a version of the Gay–Berne potential,[73] modified by Bates and Luckhurst[74] for uniaxial oblate ellipsoids (BLmGBD). The focus here was on parameterisations where the face-to-face configuration of two uniaxial oblate ellipsoids is favoured over the edge-across-edge

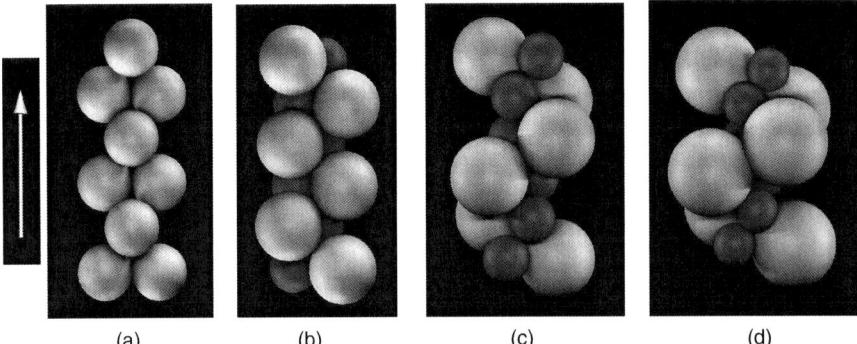

(a)   (b)   (c)   (d)

Figure 3.3   Global minima for clusters of $N = 6$ dipolar dumbbells in the presence of an applied electric field, illustrating the importance of the asymmetry of the dumbbells for this field driven assembly to produce a helical strand. The asymmetry parameter α is increased from 1 to 2 for a dipole moment $\mu_D = 1$ and electric field $E = 5$. (a) $\alpha = 1$; (b) $\alpha = 1.43$; (c) $\alpha = 1.67$; (d) $\alpha = 2$. The arrow points towards the field direction. Helical strands are observed in (c) and (d).

Figure 3.4 A helical strand, which is the global minimum for $N = 20$ dipolar dumb-bells with asymmetry parameter $\alpha = 2.5$, dipole moment $\mu_D = 0.7$ and electric field strength $E = 5$. The electric field is directed vertically and is not shown.

configuration. This bias is conducive to columnar stacking, as for π–π inter-actions in aromatic systems.[67–69]

The PY potential is a generalisation of the LJ potential for ellipsoidal par-ticles, based on the distance of closest approach of two ellipsoids with given orientations, as measured by the elliptic contact function.[72] For identical ellipsoids, the potential involves a set of eight parameters, six of them, $\{a_{1k}\}$ and $\{a_{2k}\}$ ($k = 1, 2, 3$), defining two different shape matrices for the repulsive and attractive parts of the interaction, and the other two, σ_0^{PY} and ε_0^{GB}, defining the length and energy scales, respectively. For the PY model, the parameters were tuned so that the lowest energy configuration for two axially symmetric

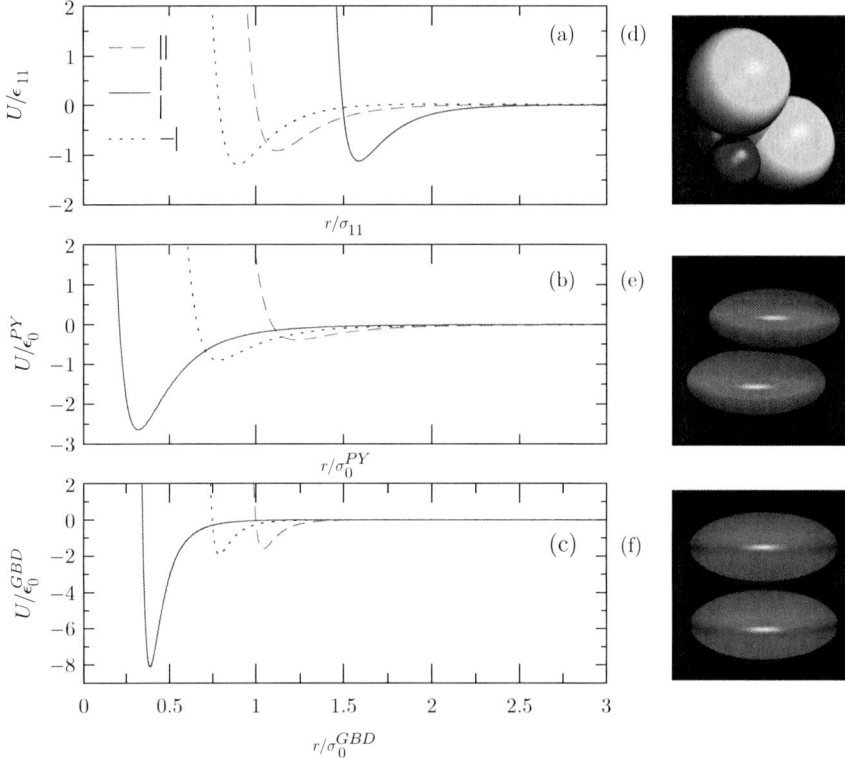

Figure 3.5   The pair potentials as a function of separation between two building blocks at fixed orientations (a–c) and the configurations for the corresponding global potential energy minima (d–f). (a) and (d) correspond to asymmetric dipolar dumbbells in an electric field with $\sigma_{22} = 0.5$, $\mu_D = 0.7$, and $E = 5$ (here r corresponds to the separation between the point dipoles and the field is directed towards the observer). (b) and (e) correspond to ellipsoids bound by the PY potential with parameters $a_{11} = a_{12} = 0.5$, $a_{13} = 0.15$, $a_{21} = a_{22} = 0.45$, $a_{23} = 0.19$, $\sigma_0^{PY} = 1$, and $\varepsilon_0^{PY} = 1$. (c) and (f) correspond to ellipsoids bound by the BLmGBD potential with $\kappa = 0.345$, $\kappa' = 0.2$, $\mu = 2$, $\nu = 1$, $\sigma_0^{GBD} = 1$, and $\varepsilon_0^{GBD} = 1$. The three fixed orientations considered in plotting each of the potentials in (a–c) are the edge-across-edge configuration (dashed), face-to-face configuration (solid), and T-configuration (dotted).

discoids involved an offset geometry [Figure 3.5(b)]. Such an arrangement is not uncommon in discotic molecules.[68] Figure 3.6 shows that for an appropriate parameter set the global minimum for a 13-discoid cluster has a double-helical morphology. However, longer-ranged interactions with increased shape anisotropy were found to support a single helical strand as the global minimum for small clusters, and a representative disconnectivity graph is shown in Figure 3.7.[75] The "palm tree" motif suggests that the helical global minimum is expected to be kinetically accessible over a wide range of temperature.[13,31]

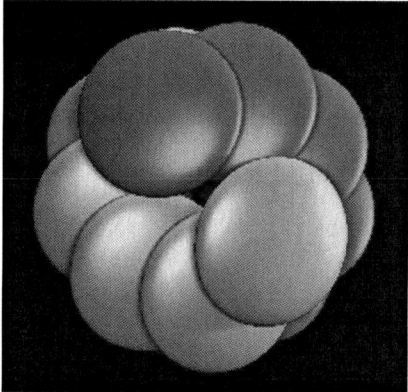

Figure 3.6 Global minimum for a cluster of 13 axially symmetric discoids bound by the Paramonov–Yaliraki potential. A double-helical structure emerges for the following parameter set: $a_{11} = a_{12} = 0.5$, $a_{13} = 0.15$, $a_{21} = a_{22} = 0.45$, $a_{23} = 0.19$, $\sigma_0^{PY} = 1$, and $\varepsilon_0^{PY} = 1$.

It is intriguing that chiral structures also emerged for assemblies of axially symmetric discoids bound by the BLmGBD potential, for which the lowest energy configuration for two discoids does not correspond to an offset geometry [Figure 3.5(c)]. The BLmGBD potential makes use of the orientation-dependent molecular shape parameter, σ, and the energy parameter, ε, to model the interaction between two uniaxial oblate ellipsoids, each having a single site representation.[74] The potential involves four essential parameters, *i.e.* $\{\kappa, \kappa', \mu, \nu\}$, where κ is the aspect ratio of the ellipsoid, $\kappa' = \varepsilon_{ee}/\varepsilon_{ff}$, with ε_{ee} the depth of the minimum of the potential for a pair of ellipsoids aligned parallel in the edge-across-edge configuration, and ε_{ff} the corresponding depth for the face-to-face alignment. The other two parameters control the orientation-dependent depth of the potential. Two additional parameters, σ_0^{GBD} and ε_0^{GBD}, define the length and energy scales, respectively. Figure 3.8 shows the global minima for $N = 49$ axially symmetric discoids bound by the BLmGBD potential, where κ was varied from 0.2 to 0.8 in steps of 0.2 when $\kappa' = 0.2$, $\mu = 2$ and $\nu = 1$. Except for $\kappa = 0.8$, chiral structures were observed, where handedness was clearly established upon symmetry breaking, which in turn was caused by the packing driven by the two competing length scales. The observation that the relative twist of the adjacent stacks varies nonmonotonically with the aspect ratio κ lends firm support to the idea that competing interactions involving two length scales provide a route to the emergence of chiral structures. In all the examples discussed here there is no preference for right- or left-handed structures, since there is no term in the potential to break the corresponding energetic degeneracy.

3.4.2 Self-Assembly of Capsid-Like Shells

The ability of biological matter to robustly self-assemble is remarkable. A fascinating example is provided by the assembly of viral capsids, where a

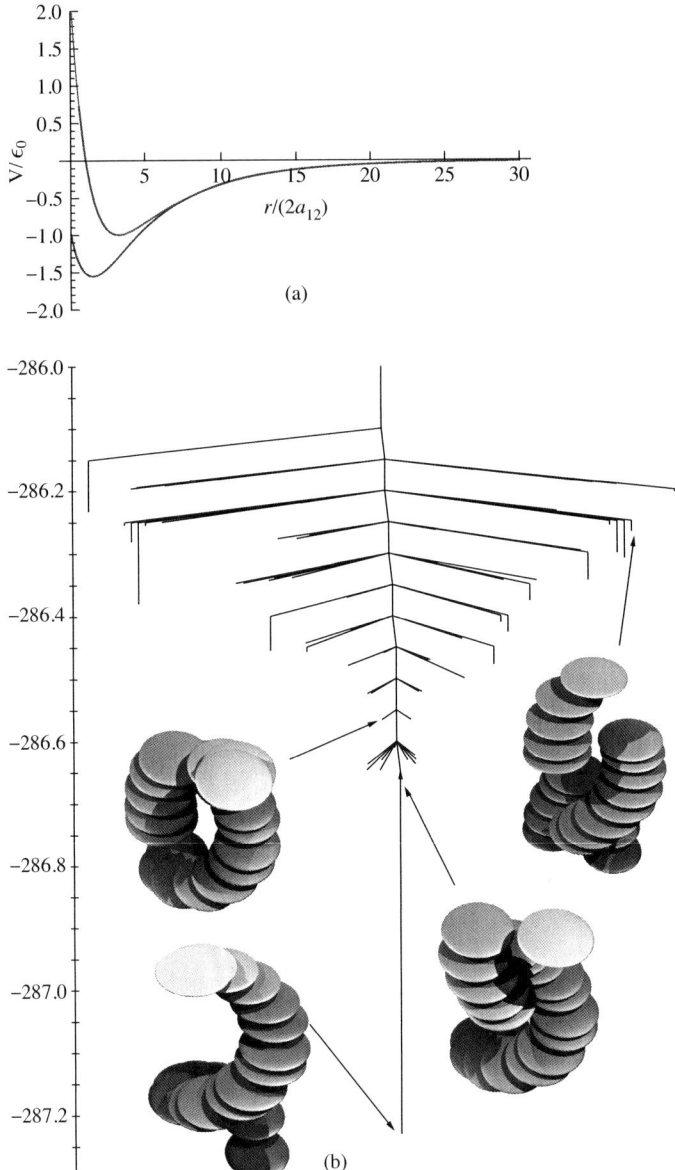

Figure 3.7 Face-to-face (lower curve) and side-by-side (upper curve) interaction profiles for the Paramonov and Yaliraki model as a function of the intercentre distance. Here, $a_{11} = 0.05$, $a_{12} = a_{13} = 0.5$, $a_{21} = 0.4$, $a_{22} = a_{23} = 0.5$, $\sigma_0^{PY} = 18$, and $\varepsilon_0^{PY} = 1$ (b) Disconnectivity graph for the same system, showing the organisation of the landscape for the lowest 100 minima. Structures of four minima are also illustrated.

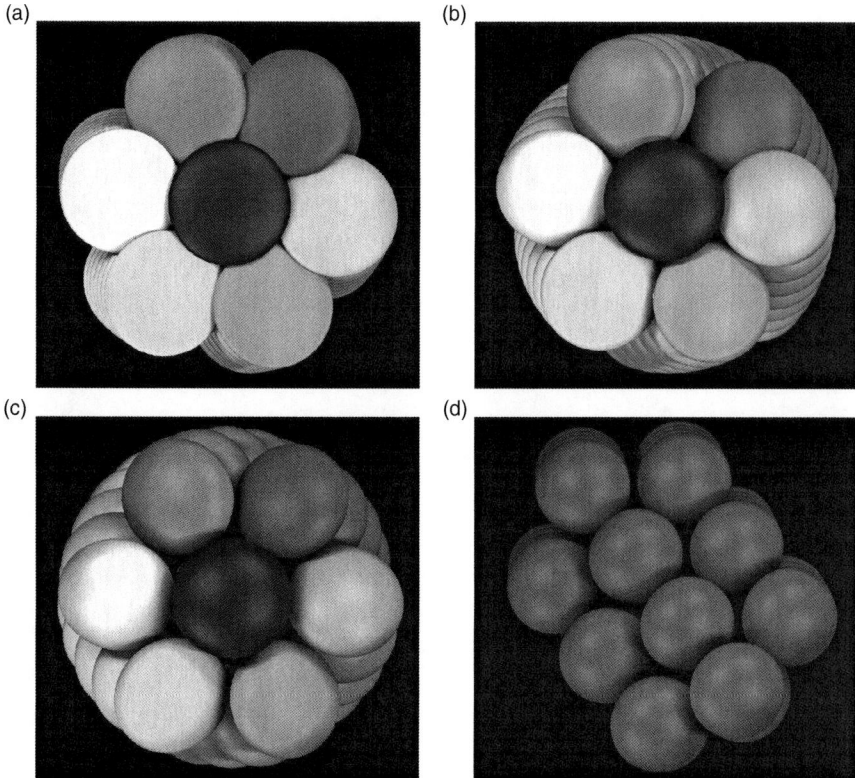

Figure 3.8 Global minima for clusters of $N = 49$ axially symmetric discoids bound by the BLmGBD potential as the aspect ratio of discoids is varied. Here, κ varies for $\kappa' = 0.2$, $\mu = 2$, and $\nu = 1$: (a) $\kappa = 0.2$; (b) $\kappa = 0.4$; (c) $\kappa = 0.6$; (d) $\kappa = 0.8$. The stacks are shaded differently for the chiral structures.

protective coat for the genetic material is formed by a large number of protein subunits. These capsids generally have well-defined structures, and a majority of those formed by simple "spherical" viruses have idealised icosahedral symmetry.[76] In one of the early studies on viral structures, Crick and Watson[77] hypothesised that the subunits must be identical. The stringent criterion of strict equivalence was subsequently relaxed by Caspar and Klug[76] as they put forward the principle of quasiequivalence. The triangulation number T, with values limited to $T = h^2 + hk + k^2 \geq 1$ where h and k are non-negative integers, plays a key role here and explains the formation of icosadeltahedra having $20T$ facets composed of $60T$ subunits, including 12 pentamers and $10(T-1)$ hexamers.

A viral capsid is a familiar example of a nanocage. Understanding capsid assembly can therefore provide valuable insight into nanofabrication of cage-like structures, which has diverse potential applications.[4,5] Here, we focus on the formation of shells with icosahedral symmetry from an appropriate number of rigid building blocks. Each building block, depending on its five-fold or six-fold rotational symmetry, has five or six sites arranged as a regular polygon in the basal plane, and two apex sites, one primary and the other secondary, on opposite sides of this plane, as shown in Figure 3.9.[78] The basal sites of different building blocks interact *via* a Morse potential. While the primary sites interact with each other *via* a repulsive LJ term, the secondary sites do not. The secondary sites, however, interact with the primary sites of the other building blocks *via* an additional LJ term. The total interaction energy between two building blocks is then given by

$$U = \varepsilon_{\text{rep}} \left(\frac{\sigma}{R_{\text{ax}}} \right)^{12} + \varepsilon_{\text{rep}} \sum_{i=1}^{2} \left(\frac{\sigma}{R_i} \right)^{12} + \varepsilon \sum_j \sum_k \left(e^{\rho(1 - R_{jk}/R_e)} - 2 \right) e^{\rho(1 - R_{jk}/R_e)}, \quad (3.15)$$

where R_{ax} is the distance between the two primary apex sites, R_i is the distance between the secondary apex site on ith building block and the primary apex site on the other, and R_{jk} is the distance between sites j and k, with j running over the basal sites on one of the building blocks and k running over the basal sites on the other. ε and R_e, the pair well depth and equilibrium separation, respectively, for each of the Morse contributions, were used as the units of energy and distance. The parameter ρ governs the range of the Morse interaction.[13] This model extends an earlier suggestion,[33] which did not include the secondary apex site. The original model therefore employed pyramidal building blocks, rather than bipyramidal. The LJ range parameter σ was set as

$$\sigma = R_e + r \sqrt{\frac{(\sqrt{5} + 5)}{2}}, \quad (3.16)$$

where r is the basal radius of a building block (Figure 3.9), the Morse parameter $\rho = 3$ as in the original model.[33] If the basal unit has five-fold symmetry, then $r = r_p = 5$. The heights of the apex sites from the base are $h_{1\text{ax}}$ and $h_{2\text{ax}}$.

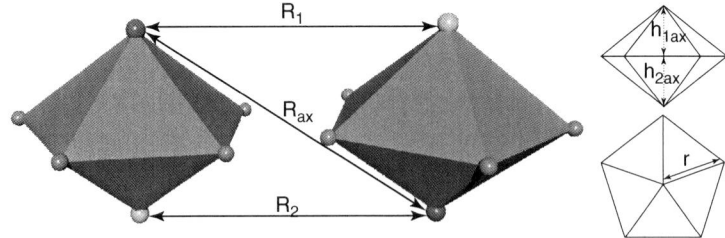

Figure 3.9   (Left) Schematic illustration of the building blocks, showing some of the site-site distances that appear in the pair potential. Primary apex sites are shown in dark grey, the secondary ones in light grey. (Right) Geometrical distances defining a building block.

For a range of values for h, the height of the apex site from the base in the original model, the global minimum of a cluster assembled from 12 pentagonal pyramids was found to have icosahedral symmetry.[33] However, the topology of the low-lying potential energy surface was found to vary with h, suggesting that the kinetic accessibility of this global minimum would be critically dependent on this geometric feature.[33] The disconnectivity graph for $h=r/2$, where the repulsive interaction between adjacent pyramids is minimised in the icosahedral global minimum, exhibited a well defined "palm tree" structure,[33] suggesting efficient relaxation to the icosahedron over a wide range of temperature.[13,31] Although the icosahedron was also found to be the global potential energy minimum for $h=0.35r$ and $0.75r$, the corresponding disconnectivity graphs revealed low-lying minima, which could act as kinetic traps hindering self-assembly to the most stable icosahedral shell.[33] A representative disconnectivity graph for $h=0.75r$ is shown in Figure 3.10(a), where low-lying local minima appear with perturbed icosahedral geometries arising from overlap of one or more pyramid bases. When bipyramidal building blocks were considered instead,[78] the frustration in the landscape was effectively removed for a similar set of parameters: $r=5$, $h_{1ax}=h_{2ax}=0.75r$, $\varepsilon_{rep}=0.28$ [Figure 3.10(b)]. In this case, the disconnectivity graph contains a characteristic "palm-tree" motif, with the icosahedral structure retained as the global minimum.

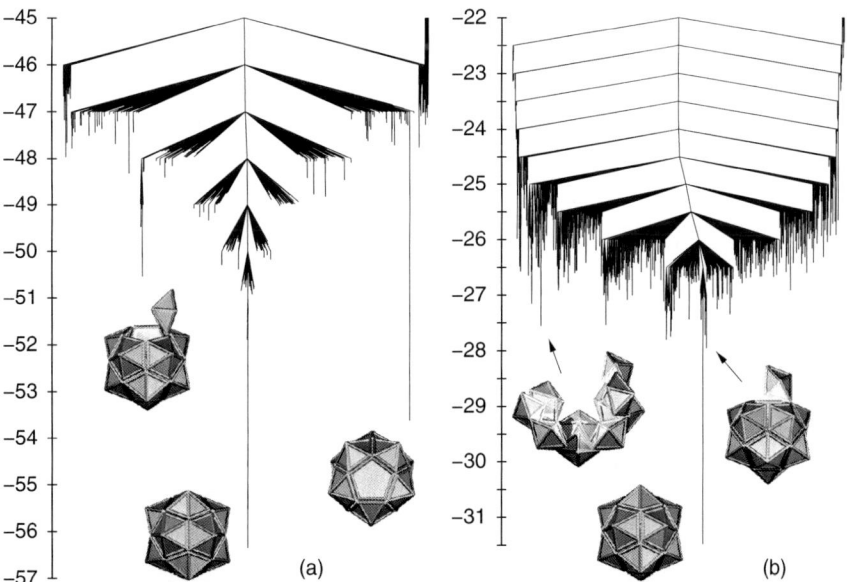

Figure 3.10   Disconnectivity graphs for the $T=1$ system: (a) original model with pyramidal building blocks, (b) extended model with bipyramidal building blocks. Note that the energy scale is different, due to the change in the net repulsive interactions in the two models.

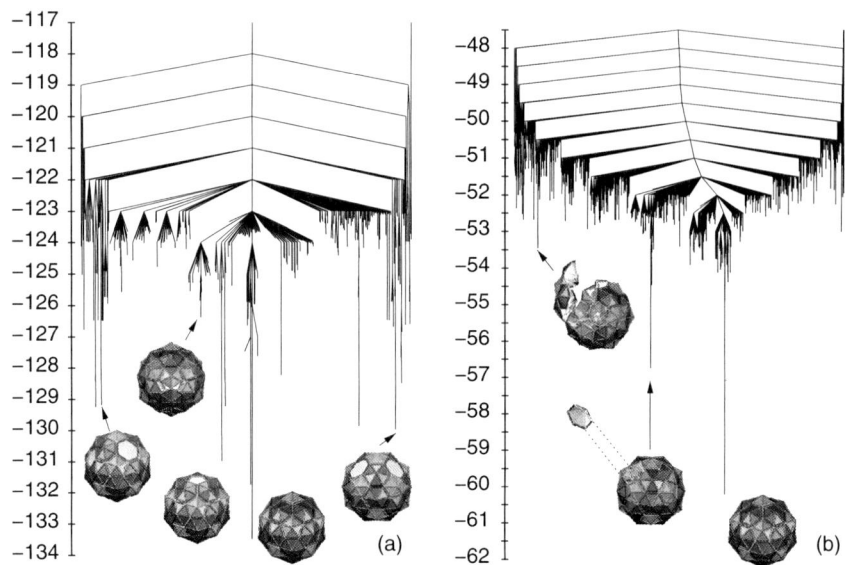

Figure 3.11   Disconnectivity graphs for the $T=3$ system: (a) original model, $r_p = 5$, $h_{1ax,p} = 0.5r_p$, $h_{1ax,h} = 0.5r_h$, $\varepsilon_{rep,hh} = \varepsilon_{rep,pp} = \varepsilon_{rep,ph} = 0.5$; (b) extended model with an extra apex site. The parameters that differ from the ones used in the original model are $h_{1ax,p} = h_{1ax,h} = 0.5r_p$, $\varepsilon_{rep} = 0.4$, $h_{2ax,p} = h_{2ax,h} = 5$.

The $T=3$ shell is obtained from 12 pentagonal and 20 hexagonal units. The base edges of the pentamers and hexamers were taken to be equal to ensure efficient geometric packing. This choice resulted in a ratio of the basal radii $r_h/r_p = 2\sin 36°$, where r_h and r_p are the radii of the hexamers and pentamers, respectively. The range parameter for the hexamer–pentamer basal site interaction was defined by the arithmetic mean of the individual pentamer–pentamer and hexamer–hexamer range parameters, *i.e.* $\sigma_{ph} = (\sigma_{pp} + \sigma_{hh})/2$. The disconnectivity graphs shown in Figure 3.11 confirm that bipyramidal building blocks are better structure seekers than the pyramidal blocks originally considered.[78]

3.5   Conclusion

The success of a bottom-up approach to nanofabrication *via* self-assembly is critically dependent on the choice of building blocks. A rapid growth in the experimental manipulation of colloidal building blocks through the last decade has made a rich variety of building blocks available.[2,6–12] However, *a priori* design of building blocks for programmed self-assembly requires a thorough understanding of the driving forces. Many computational studies have recently been devoted towards achieving this goal.[79–84]

Our recent efforts in designing rigid building blocks for efficient self-assembly into functional nanostructures rely upon a survey of the underlying potential energy surface. To this end, a robust and efficient computational framework for geometry optimisation of rigid bodies was developed. Our illustrative examples demonstrate how helical nanostructures can emerge from the interplay between two competing length scales for anisotropic interactions between the building blocks, and have again confirmed that the "palm tree" topology indeed leads to efficient self-assembly, as shown in recent simulations.[85]

Appendix A. Paramonov–Yaliraki Potential

The Paramonov–Yaliraki (PY) potential[72] is based on the distance of closest approach of two ellipsoids, as measured by the elliptic contact function (ECF):

$$F(\mathbf{A}, \mathbf{B}) = \max_{\lambda} \mathscr{S}(\mathbf{x}(\lambda), \lambda) = \max_{\lambda} \mathscr{S}(\lambda), \tag{3.17}$$

where the function $\mathscr{S}(\mathbf{x}, \lambda)$ is a linear combination of quadratic forms $\mathscr{A}(\mathbf{x})$ and $\mathscr{B}(\mathbf{x})$ for the ellipsoids i and j, centred at $\mathbf{r}_a$ and $\mathbf{r}_b$, respectively:

$$\mathscr{S}(\mathbf{x}, \lambda) = \lambda \mathscr{A}(\mathbf{x}) + (1 - \lambda)\mathscr{B}(\mathbf{x}), \tag{3.18}$$

where

$$\mathscr{A}(\mathbf{x}) = (\mathbf{x} - \mathbf{r}_a)^T \mathscr{A}(\mathbf{x} - \mathbf{r}_a),$$
$$\mathscr{B}(\mathbf{x}) = (\mathbf{x} - \mathbf{r}_b)^T \mathscr{B}(\mathbf{x} - \mathbf{r}_b),$$

and $0 < \lambda < 1$. The matrices $\mathbf{A}$ and $\mathbf{B}$ are defined as

$$\mathbf{A} = \sum_{k=1}^{3} a_k^{-2} \hat{\mathbf{u}}_k \otimes \hat{\mathbf{u}}_k, \quad \mathbf{B} = \sum_{k=1}^{3} b_k^{-2} \hat{\mathbf{v}}_k \otimes \hat{\mathbf{v}}_k, \tag{3.19}$$

where $\{\hat{\mathbf{u}}_k\}$ and $\{\hat{\mathbf{v}}_k\}$ are sets of orthonormal unit vectors specifying the axes of ellipsoids i and j, respectively, "$\otimes$" denotes a dyadic product, and $\{a_k\}$ and $\{b_k\}$ are the lengths of the semiaxes of the two ellipsoids. The PY potential is given by

$$U_{ij}^{\mathrm{PY}} = 4\varepsilon_0^{\mathrm{PY}} \left[\left(\frac{\sigma_0^{\mathrm{PY}}}{r_{ij} - r_{ij} F_1(\mathbf{A}_1, \mathbf{B}_1)^{-1/2} + \sigma_0^{\mathrm{PY}}} \right)^{12} - \left(\frac{\sigma_0^{\mathrm{PY}}}{r_{ij} - r_{ij} F_2(\mathbf{A}_2, \mathbf{B}_2)^{-1/2} + \sigma_0^{\mathrm{PY}}} \right)^{6} \right],$$

where $F_1(\mathbf{A}_1, \mathbf{B}_1)$ and $F_2(\mathbf{A}_2, \mathbf{B}_2)$ are the "repulsive" and "attractive" elliptic contact functions, constructed using different shape matrices:

$$\mathbf{A}_1 = \sum_{k=1}^{3} a_{1k}^{-2} \hat{\mathbf{u}}_k \otimes \hat{\mathbf{u}}_k, \quad \mathbf{B}_1 = \sum_{k=1}^{3} b_{1k}^{-2} \hat{\mathbf{v}}_k \otimes \hat{\mathbf{v}}_k,$$

and

$$\mathbf{A}_2 = \sum_{k=1}^{3} a_{2k}^{-2}\hat{\mathbf{u}}_k \otimes \hat{\mathbf{u}}_k, \quad \mathbf{B}_2 = \sum_{k=1}^{3} b_{2k}^{-2}\hat{\mathbf{v}}_k \otimes \hat{\mathbf{v}}_k,$$

where $\{a_{1k}\}$ and $\{b_{1k}\}$ are the lengths of the semiaxes of the repulsive ellipsoidal core for ellipsoids A and B, respectively, while $\{a_{2k}\}$ and $\{b_{2k}\}$ denote the lengths of the semiaxes for the attractive ellipsoidal core.

Appendix B. Modified Gay–Berne Potential

Bates and Luckhurst modified the Gay–Berne (GB) pair potential[73] to model oblate uniaxial ellipsoids.[74] This potential (BLmGBD) makes use of the orientation-dependent molecular shape parameter, σ, and the energy parameter, ε, to model the interaction between two uniaxial oblate ellipsoids. Each ellipsoid i has a single site representation, defined by the position, $\mathbf{r}_i$, of its centre of mass and a unit vector, $\hat{\mathbf{e}}_i$, along the short axis. The BLmGBD pair potential is given by

$$U_{ij}^{\text{GBD}} = 4\varepsilon(\hat{\mathbf{r}}_{ij}, \hat{\mathbf{e}}_i, \hat{\mathbf{e}}_j)(\rho_{ij}^{-12} - \rho_{ij}^{-6}), \tag{3.20}$$

where $\rho_{ij} = (r_{ij} - \sigma(\hat{\mathbf{r}}_{ij}, \hat{\mathbf{e}}_i, \hat{\mathbf{e}}_j) + \sigma_{ff})/\sigma_{ff}$. Here, σ_{ff} defines the thickness, r_{ij} is the distance between the centres of mass of ellipsoids i and j, and $\hat{\mathbf{r}}_{ij} = \mathbf{r}_{ij}/r_{ij}$ is a unit vector along the intermolecular vector, $\mathbf{r}_{ij}$. The potential involves four essential parameters, *i.e.* $\{\kappa, \kappa', \mu, \nu\}$ and two additional parameters σ_0^{GBD} and $\varepsilon_0^{\text{GBD}}$ as described before. σ_0^{GBD} also defines the cross-sectional diameter, and hence the aspect ratio $\kappa = \sigma_{ff}/\sigma_0^{\text{GBD}}$. σ and ε both depend on $\hat{\mathbf{e}}_i$ and $\hat{\mathbf{e}}_j$ as well as on $\hat{\mathbf{r}}_{ij}$:

$$\sigma(\hat{\mathbf{r}}_{ij}, \hat{\mathbf{e}}_i, \hat{\mathbf{e}}_j) = \sigma_0^{\text{GBD}}\left[1 - \frac{\chi}{2}\left\{\frac{(\hat{\mathbf{e}}_i \cdot \hat{\mathbf{r}}_{ij} + \hat{\mathbf{e}}_j \cdot \hat{\mathbf{r}}_{ij})^2}{1 + \chi(\hat{\mathbf{e}}_i \cdot \hat{\mathbf{e}}_j)} + \frac{(\hat{\mathbf{e}}_i \cdot \hat{\mathbf{r}}_{ij} - \hat{\mathbf{e}}_j \cdot \hat{\mathbf{r}}_{ij})^2}{1 - \chi(\hat{\mathbf{e}}_i \cdot \hat{\mathbf{e}}_j)}\right\}\right]^{-1/2}, \tag{3.21}$$

with $\chi = (\kappa^2 - 1)/(\kappa^2 + 1)$ and

$$\varepsilon(\hat{\mathbf{r}}_{ij}, \hat{\mathbf{e}}_i, \hat{\mathbf{e}}_j) = \varepsilon_0^{\text{GBD}}[\varepsilon_1(\hat{\mathbf{e}}_i, \hat{\mathbf{e}}_j)]^{\nu}[\varepsilon_2(\hat{\mathbf{r}}_{ij}, \hat{\mathbf{e}}_i, \hat{\mathbf{e}}_j)]^{\mu}, \tag{3.22}$$

where the exponents μ and ν are adjustable,

$$\varepsilon_1(\hat{\mathbf{e}}_i, \hat{\mathbf{e}}_j) = [1 - \chi^2(\hat{\mathbf{e}}_i \cdot \hat{\mathbf{e}}_j)^2]^{-1/2}, \tag{3.23}$$

and

$$\varepsilon_2(\hat{\mathbf{r}}_{ij}, \hat{\mathbf{e}}_i, \hat{\mathbf{e}}_j) = 1 - \frac{\chi'}{2}\left[\frac{(\hat{\mathbf{e}}_i \cdot \hat{\mathbf{r}}_{ij} + \hat{\mathbf{e}}_j \cdot \hat{\mathbf{r}}_{ij})^2}{1 + \chi'(\hat{\mathbf{e}}_i \cdot \hat{\mathbf{e}}_j)} + \frac{(\hat{\mathbf{e}}_i \cdot \hat{\mathbf{r}}_{ij} - \hat{\mathbf{e}}_j \cdot \hat{\mathbf{r}}_{ij})^2}{1 - \chi'(\hat{\mathbf{e}}_i \cdot \hat{\mathbf{e}}_j)}\right], \tag{3.24}$$

with $\chi' = [(\kappa')^{1/\mu} - 1)]/[(\kappa')^{1/\mu} + 1)]$.

Acknowledgement

DC gratefully acknowledges support from the Oppenheimer Fund.

References

1. G. M. Whitesides and M. Boncheva, *Proc. Natl. Acad. Sci. USA*, 2002, **99**, 4769.
2. S. C. Glotzer and M. J. Solomon, *Nature Mater.*, 2007, **6**, 557.
3. S. Zhang, *Nat. Biotechnol.*, 2003, **21**, 1171.
4. T. Douglas and M. Young, *Science*, 2006, **312**, 873.
5. M. Fischlechner and E. Donath, *Angew. Chem. Int. Ed.*, 2007, **46**, 3184.
6. J. A. Fan, C. Wu, K. Bao, J. Bao, R. Bardhan, N. J. Halas, V. N. Manoharan, P. Nordlander, G. Shvets and F. Capasso, *Science*, 2010, **328**, 1135.
7. S. Sacanna, W. T. M. Irvine, P. M. Chaikin and D. J. Pine, *Nature*, 2010, **464**, 575.
8. D. Zerrouki, J. Boudri, D. Pine, P. Chaikin and J. Bibette, *Nature*, 2008, **455**, 380.
9. J.-W. Kim, R. J. Larsen and D. A. Weitz, *J. Am. Chem. Soc.*, 2006, **128**, 14374.
10. Y.-S. Cho, G.-R. Yi, J.-M. Lim, S.-H. Kim, V. N. Manoharan, D. J. Pine and S.-M. Yang, *J. Am. Chem. Soc.*, 2005, **127**, 15968.
11. G. Zhang, D. Wang and H. M. öhwald, *Angew. Chem. Int. Ed.*, 2005, **44**, 7767.
12. A. Yethiraj and A. van Blaaderen, *Nature*, 2003, **421**, 513.
13. D. J. Wales, *Energy Landscapes*, Cambridge University Press, Cambridge, 2003.
14. F. Baletto and R. Ferrando, *Rev. Mod. Phys.*, 2005, **77**, 371.
15. J. N. Murrell and K. J. Laidler, *Trans. Faraday Soc.*, 1968, **64**, 371.
16. D. J. Wales and J. P. K. Doye, *J. Phys. Chem. A*, 1997, **111**, 5111.
17. D. J. Wales, *Mol. Phys.*, 2002, **100**, 3285.
18. D. J. Wales, *Mol. Phys.*, 2004, **102**, 891.
19. J. M. Carr and D. J. Wales, *J. Phys. Chem. B*, 2008, **112**, 8760.
20. S. A. Trygubenko and D. J. Wales, *J. Chem. Phys.*, 2004, **120**, 2082.
21. G. Henkelman and H. Jónsson, *J. Chem. Phys.*, 1999, **111**, 7010.
22. G. Henkelman and H. Jónsson, *J. Chem. Phys.*, 2000, **113**, 9978.
23. G. Henkelman, B. P. Uberuaga and H. Jónsson, *J. Chem. Phys.*, 2000, **113**, 9901.
24. L. J. Munro and D. J. Wales, *Phys. Rev. B*, 1999, **59**, 3969.
25. Y. Kumeda, D. J. Wales and L. J. Munro, *Chem. Phys. Lett.*, 2001, **341**, 185.
26. J. Nocedal, *Math. Comput.*, 1980, **35**, 773.
27. D. Liu and J. Nocedal, *Mathematical Programming*, 1989, **45**, 503.
28. D. A. Evans and D. J. Wales, *J. Chem. Phys.*, 2004, **121**, 1080.
29. E. W. Dijkstra, *Numerische Mathematik*, 1959, **1**, 269.
30. O. M. Becker and M. Karplus, *J. Chem. Phys.*, 1997, **106**, 1495.

31. D. J. Wales, M. A. Miller and T. R. Walsh, *Nature*, 1998, **394**, 758.
32. M. A. Miller and D. J. Wales, *J. Chem. Phys.*, 1999, **111**, 6610.
33. D. J. Wales, *Philos. Trans. R. Soc. A*, 2005, **363**, 357.
34. T. F. Middleton and D. J. Wales, *J. Chem. Phys.*, 2003, **118**, 4583.
35. S. V. Krivov and M. Karplus, *J. Chem. Phys.*, 2002, **117**, 10894.
36. D. A. Evans and D. J. Wales, *J. Chem. Phys.*, 2003, **118**, 3891.
37. D. J. Wales and T. Bogdan, *J. Phys. Chem. B*, 2006, **110**, 20765.
38. G. Chartrand, *Introductory Graph Theory*, Dover, New York, 1985.
39. T. R. Walsh and D. J. Wales, *J. Chem. Phys.*, 1998, **109**, 6691.
40. Y. Kumeda and D. J. Wales, *Chem. Phys. Lett.*, 2003, **374**, 125.
41. V. K. de Souza and D. J. Wales, *J. Chem. Phys.*, 2008, **129**, 164527.
42. K. D. Ball, R. S. Berry, R. E. Kunz, F.-Y. Li, A. Proykova and D. J. Wales, *Science*, 1996, **271**, 963.
43. J. D. Bryngelson and P. G. Wolynes, *Proc. Natl. Acad. Sci. USA*, 1987, **84**, 7524.
44. H. Frauenfelder, S. G. Sligar and P. G. Wolynes, *Science*, 1991, **254**, 1598.
45. P. E. Leopold, M. Montal and J. N. Onuchic, *Proc. Natl. Acad. Sci. USA*, 1992, **89**, 8721.
46. P. G. Wolynes, J. N. Onuchic and D. Thirumalai, *Science*, 1995, **267**, 1619.
47. H. Goldstein, *Classical Mechanics*, Addison-Wesley, Reading, MA, 1980.
48. D. J. Wales and I. Ohmine, *J. Chem. Phys.*, 1993, **98**, 7257.
49. S. L. Price, A. J. Stone and M. Alderton, *Mol. Phys.*, 1984, **52**, 987.
50. P. L. A. Popelier and A. J. Stone, *Mol. Phys.*, 1995, **82**, 411.
51. P. L. A. Popelier, A. J. Stone and D. J. Wales, *Faraday Discuss.*, 1994, **97**, 243.
52. M. P. Allen and G. Germano, *Mol. Phys.*, 2006, **104**, 3225.
53. S. L. Altmann, *Rotations, Quaternions, and Double Groups*, Clarendon Press, Oxford, 2003.
54. D. Chakrabarti and D. J. Wales, *Phys. Chem. Chem. Phys.*, 2009, **11**, 1970.
55. S. Belongie, From MathWorld – A Wolfram Web Resource, created by Eric W. Weisstein. http://mathworld.wolfram.com/RodriguesRotation-Formula.html.
56. D. J. Wales and T. V. Bogdan, GMIN: A program for finding global minima and calculating thermodynamic properties from basin-sampling; http://www-wales.ch.cam.ac.uk/GMIN.
57. D. J. Wales, OPTIM: A program for optimising geometries and calculating reaction pathways; http://www-wales.ch.cam.ac.uk/OPTIM.
58. W. L. Jorgensen, J. Chandrasekhar, J. D. Madura, R. W. Impey and M. L. Klein, *J. Chem. Phys.*, 1983, **79**, 926.
59. E. Yashima, K. Maeda, H. Iida, Y. Furusho and K. Nagai, *Chem. Rev.*, 2009, **109**, 6102.
60. D. Pijper and B. L. Feringa, *Soft Matter*, 2008, **4**, 1349.
61. C. C. Lee, C. Grenier, E. W. Meijer and A. P. H. J. Schenning, *Chem. Soc. Rev.*, 2009, **38**, 671.
62. J. W. Goodby, *J. Mater. Chem.*, 1991, **1**, 307.
63. O. R. Evans and W. Lin, *Acc. Chem. Res.*, 2002, **35**, 511.

64. D. Adam, P. Schuhmacher, J. Simmerer, L. Häussling, K. Siemensmeyer, K. H. Etzbachi, H. Ringsdorf and D. Haarer, *Nature*, 1994, **371**, 141.
65. L. SchmidtMende, A. Fechtenkötter, K. Müllen, E. Moons, R. H. Friend and J. D. MacKenzie, *Science*, 2001, **293**, 1119.
66. V. Percec, M. Glodde, T. K. Bera, Y. Miura, I. S. I, K. D. Singer, V. S. Balagurusamy, P. A. Heiney, I. Schnell, A. R. A, H. W. Spiess, S. D. Hudson and H. Duan, *Nature*, 2002, **419**, 384.
67. H. Engelkamp, S. Middelbeek and R. J. M. Nolte, *Science*, 1999 **284**, 785.
68. I. Azumaya, D. Uchida, K. Takako, A. Yokoyama, A. Tanatani, H. Takayanagi and T. Yokozawa, *Angew. Chem. Int. Ed.*, 2004, **43**, 1360.
69. Z. Wang, V. Enkelmann, F. Negri and K. Müllen, *Angew. Chem. Int. Ed.*, 2004, **43**, 1972.
70. D. Chakrabarti, S. N. Fejer and D. J. Wales, *Proc. Natl. Acad. Sci. USA*, 2009, **106**, 20164.
71. D. Frenkel and B. M. Mulder, *Mol. Phys.*, 1985, **55**, 1171.
72. L. Paramonov and S. N. Yaliraki, *J. Chem. Phys.*, 2005, **123**, 194111.
73. J. G. Gay and B. J. Berne, *J. Chem. Phys.*, 1981, **74**, 3316.
74. M. A. Bates and G. R. Luckhurst, *J. Chem. Phys.*, 1996, **104**, 6696.
75. S. N. Fejer and D. J. Wales, *Phys. Rev. Lett.*, 2007, **99**, 086106.
76. D. L. Caspar and A. Klug, *Cold Spring Harbor Symp. Quant. Biol.*, 1962, **27**, 1.
77. F. H. C. Crick and J. D. Watson, *Nature*, 1956, **177**, 473.
78. S. N. Fejer, T. R. James, J. Hernández-Rojas and D. J. Wales, *Phys. Chem. Chem. Phys.*, 2009, **11**, 2098.
79. E. Bianchi, J. Largo, P. Tartaglia, E. Zaccarelli and F. Sciortino, *Phys. Rev. Lett.*, 2006, **97**, 168301.
80. G. Foffi and F. Sciortino, *J. Phys. Chem. B*, 2007, **111**, 9702.
81. E. Jankowski and S. C. Glotzer, *J. Chem. Phys.*, 2009, **131**, 104104.
82. T. Chen, Z. Zhang and S. C. Glotzer, *Proc. Natl. Acad. Sci. USA*, 2007, **104**, 717.
83. X. Zhang, Z. Zhang and S. C. Glotzer, *J. Phys. Chem. C*, 2007, **11**, 4132.
84. S. Torquato, *Soft Matter*, 2009, **5**, 1157.
85. I. G. Johnston, A. A. Louis and J. P. K. Doye, *J. Phys.: Condens. Matter*, 2010, **22**, 104101.

Phase Transition under Confinement

JAYANT K. SINGH,[1] HUGH DOCHERTY[2] AND
PETER T. CUMMINGS[2,3]

[1] Department of Chemical Engineering, Indian Institute of Technology
Kanpur, UP 208016, India; [2] Department of Chemical and Biomolecular
Engineering, Vanderbilt University, Nashville, TN 37235, USA; [3] Center for
Nanophase Materials Sciences, Oak Ridge National Laboratory, Oak Ridge,
TN 37831, USA

4.1   Introduction

The confinement of fluid in nanopores is a common occurrence in sensors and
devices. As such, understanding the phase equilibria, and having knowledge of
structural and transport properties, of confined fluids is important in the development of new technologies for manufacturing and in the modification of current
methods. Interestingly, the phase behaviours of fluid in porous material are dramatically different from those of the bulk fluid, because of the competition of fluid–
fluid and fluid–wall interaction energies. Moreover, the geometry of the adsorbate
can cause the adsorbent to behave as a two-dimensional fluid in nanotubes or as a
one-dimensional fluid, as for gases stuck in the corners of rectangular pores.[1,2] These
geometrical constraints and the presence of external forces are the primary source for
a range of different phase transitions, such as layering, prewetting and capillary
condensation and a shift in critical properties and melting/freezing properties.[3]

   Bearing this in mind, this chapter deals with the molecular simulation of fluids
confined in nanopores. Our aim is twofold: first, to give a review of recent

RSC Theoretical and Computational Chemistry Series No. 4
Computational Nanoscience
Edited by Elena Bichoutskaia
© Royal Society of Chemistry 2011
Published by the Royal Society of Chemistry, www.rsc.org

molecular simulation techniques for phase transitions under confinement, with relevant case studies and secondly, to demonstrate the possibilities and limitations of the simulation methods discussed here, showing recent results for some systems and properties selected from the fields of research of the authors. The chapter is divided into two sections, one dealing with vapour–liquid transitions, the other focusing on fluid-solid transitions. Within each section, we provide a short review of appropriate methodologies together with examples of the studies that utilised them. In this manner, we provide details regarding the thermodynamic and structural properties of nanoconfined systems and highlight important considerations in the application of molecular simulation techniques. We begin in the following section by providing a very brief and general introduction to molecular simulation for confined fluids aimed at those not familiar with these methods. A more complete description of these techniques may be found in the excellent books of Frenkel and Smit[4] and Allen and Tildesley.[5]

4.2 Molecular Simulation for Confined Fluids

The investigation of phase transitions and dynamical properties of fluid under confinement using molecular simulation is primarily done using two methods, *viz.*, Monte Carlo simulation and molecular dynamics. These methods invariably use the statistical mechanics formalism, which connects thermophysical properties such as pressure, energy and other macroscopic properties to the collection of microstates of systems of molecules, all having in common extensive properties. A *microstate* of a system of molecules is a complete specification of all positions and momenta of all molecules. A Collection of microstates is also called an ensemble.[5,6] Most commonly, we encounter the total energy, the total volume, and/or the total number of molecules (of one or more species, if a mixture) as extensive properties. Examples of ensembles are:

- microcanonical ensemble (fixed number of particles, volume and energy);
- canonical ensemble (fixed number of particles, volume and temperature);
- isothermal–isobaric ensemble (fixed number of particles, pressure and temperature);
- grand-canonical ensemble (fixed chemical potential, volume and temperature).

A Monte Carlo simulation generates configurations of molecules using an equilibrium probability distribution function for a statistical mechanical ensemble. For example, for the canonical ensemble, the probability of obtaining a configuration of N molecules $\{r_1, r_2, ..., r_N\} \equiv r^N$, *i.e.* the probability density (π) is defined as:

$$\pi(\mathbf{r}^N) = \frac{\exp[-U(\mathbf{r}^N)/k_B T]}{\int \exp[-U(\mathbf{r}^N)/k_B T] d\mathbf{r}^N} = \frac{\exp[-U(\mathbf{r}^N)/k_B T]}{Z}, \qquad (4.1)$$

where, U is the total potential energy of the configuration, Z is the configurational part of the partition function,[7] k_B is the Boltzmann's constant and T is the temperature.

The most elementary procedure in Monte Carlo simulations involves generating a trial configuration by making a perturbation to the configuration of molecules by selecting a molecule randomly and displacing it a small amount from its present position. Subsequently, the ratio of probabilities of the trial and original configurations is computed and the trial configuration would be accepted according to the Metropolis algorithm:[8]

$$p_{acc} = \min[1, \exp(-\beta\Delta U)], \tag{4.2}$$

where ΔU is the change in potential energy between the trial and original configuration and $\beta = 1/k_B T$.

In the case of $p_{acc} = 1$, we accept the trial otherwise, a random number is generated, *rand*, from a uniform distribution in the interval [0,1]. The trial configuration is then accepted if $P_{acc} > rand$. If the trial is accepted, the new configuration is taken as the next state in what is known as a Markov chain;[5] otherwise, the original configuration is taken as the next state in the chain. Averages are collected over many configurations, as shown below:

$$\langle A \rangle = \frac{\sum\limits_{i=1}^{N\text{sample}} A(\mathbf{r}^N)\exp[-\beta U(\mathbf{r}^N)]}{\sum\limits_{i=1}^{N\text{sample}} \exp[-\beta U(\mathbf{r}^N)]}. \tag{4.3}$$

While Monte Carlo simulations rely on configurational sampling of the phase space, molecular dynamics is a deterministic approach and its formalism is based on classical mechanics. In particular, molecular dynamic considers a system of molecules and integrates the governing equation of motion through time, *i.e.* successive configurations of the system are generated by integrating Newton's laws of motion. While these methods are commonly applied successfully to the prediction of a wide range of bulk fluid properties, their usage for confined fluids is mainly limited to nonreactive systems due to limitations of appropriate force fields or molecular models. Nevertheless, these methods are most useful for gaining a qualitative understanding of the properties of confined fluids.

4.3 Vapour–Liquid Transition under Confinement

A number of Monte-Carlo methods exist for the study of vapour–liquid (or capillary condensation) phase transitions in confined systems. In the case of vapour–liquid phase transition, grand canonical Monte Carlo (GCMC) simulation and Gibbs-ensemble Monte Carlo (GEMC) simulation are widely used.[6] Other methods that have been applied less frequently to these problems are semigrand Monte Carlo (SGMC), histogram-reweighting and histogram-biasing methods, and molecular dynamics.[6] Below, we give a brief review of the GCMC simulation technique and its application to confined fluid-phase equilibria.

4.3.1 Grand Canonical Monte Carlo

Simulations in the grand-canonical ensemble require a constant chemical potential μ, volume V, and temperature T. The microstate probability density distribution for this ensemble is given by,

$$\pi_s = \frac{1}{\Xi} \frac{V^{N_s}}{\Lambda^{3N_s} N_s!} \exp[-\beta(U_s - \mu N_s)], \tag{4.4}$$

where V is the volume, $\beta = 1/k_B T$ is the inverse temperature, k_B is the Boltzmann's constant, Λ is the de Broglie wavelength, Ξ is the grand partition function, U_s is the interaction energies of particles, N_s, in the microstate, s. The macrostate probability, $\Pi(N)$, is calculated by summing all microstate probabilities at a constant number of molecules, N. The mathematical formula can be expressed as,

$$\Pi(N) = \sum_{N_s = N} \pi_s. \tag{4.5}$$

Vapour–liquid equilibrium behaviour under confinement requires that the simulation should be able to sample all important ranges of particle numbers. Simulation at equilibrium chemical potential yields a bimodal distribution in number probability distribution, with peaks corresponding to the stable phases. However, in practice a chemical potential corresponding to saturation conditions is usually not known *a priori*. Fortunately, we can make use of probabilities of visiting microstates during the standard Metropolis method.[8] Let us consider a move that attempts to visit a state t with M number of molecules from a state s with N number of molecules. The acceptance probability is defined as,

$$a(s \rightarrow t) = \min[1, \pi_t/\pi_s]. \tag{4.6}$$

In the usual Monte Carlo approach, the above acceptance probabilities are not recorded and system information is lost that could be exploited to estimated saturation conditions and other properties using the transition matrix Monte Carlo (TMMC) approach.[9] The TMMC method is a simple book-keeping scheme to obtain the macrostate probability. In this scheme, the acceptance probability is recorded in a collection matrix C, for each MC move, as shown in eqns (4.8) and (4.9), regardless of whether the move is being accepted or not.

$$C(N \rightarrow M) = C(N \rightarrow M) + a(s \rightarrow t), \tag{4.7}$$

and

$$C(N \rightarrow N) = C(N \rightarrow N) + 1 - a(s \rightarrow t). \tag{4.8}$$

The macrostate transition probability can then be obtained from the matrix C at any time using the following expression,

$$P(N \rightarrow M) = \frac{C(N \rightarrow M)}{\sum\limits_{O} C(N \rightarrow O)}. \tag{4.9}$$

The detailed balance expression[5,6] can be utilised to obtain the macrostate probabilities:

$$\Pi(N)P(N \rightarrow M) = \Pi(M)P(M \rightarrow N). \tag{4.10}$$

Configuration space is sampled in GCMC simulation by incorporating various MC moves. For a pure simple fluid simulation (for binary mixtures see Shen and Errington[10] for more details), only deletion, addition, and displacement moves are employed, considering one molecule at a time. The possible state change will be from one of the following choices: $N \rightarrow N$, $N \rightarrow N - 1$, $N \rightarrow N + 1$, hence the transition probability matrix is tridiagonal. In such conditions, a sequential approach would be an efficient and simple way to obtain the macrostate probabilities,

$$\ln \Pi(N + 1) = \ln \Pi(N) - \ln \left[\frac{P(N + 1 \rightarrow N)}{P(N \rightarrow N + 1)} \right]. \tag{4.11}$$

As discussed earlier, GCMC simulations are conducted at a specified value of the chemical potential, which is not necessarily close to the saturation value. In practice, the coexistence chemical potential is usually not known *a priori*. To determine the phase-coexistence value of the chemical potential, the histogram-reweighting method of Ferrenberg and Swendson[11] is generally used. This method enables one to shift the probability distribution obtained from a simulation at chemical potential μ_o to a probability distribution corresponding to a chemical potential μ using the relation,

$$\ln \Pi(N; \mu) = \ln \Pi(N; \mu_o) + \beta(\mu - \mu_o)N. \tag{4.12}$$

The above relation is usually used to find the chemical potential that produces a probability distribution $\Pi_c(N)$ where the areas under the vapour and liquid regions are equal. Saturated densities are related to the first moments of the vapour and liquid peaks of the coexistence probability distribution.

4.3.2 Crossover from 3D-Like to 2D-Like under Confinement

The above method (GCMC + TMMC also known as GC-TMMC) can be used to understand the crossover behaviour of critical properties of confined fluids from 3D to 2D. Various experimental results[12–16] suggest that vapour-liquid critical temperatures are suppressed under confinement due to the reduction in the coordination number. This decrease in the critical temperature increases as pore size decreases. However, due to wide pore-size distribution and irregular pore geometry, it has not been possible to establish a quantitative relation between the critical point shift and pore size. Singh et al.[17] have recently illustrated the crossover behaviour, from 3D-like to 2D-like, of a square-well fluid confined in a slit pore as a function of reducing pore size. In their work, fluid–fluid interaction and wall–fluid interaction is represented by the following square-well potentials:

$$u_{ff}(r) = \begin{cases} \infty, & 0 < r < \sigma_{ff} \\ -\varepsilon_{ff}, & \sigma_{ff} \leq r < \lambda_{ff}\sigma_{ff}, \\ 0, & \lambda_{ff}\sigma_{ff} \leq r \end{cases}$$

(4.13)

$$u_{wf}(z) = \begin{cases} \infty, & 0 < z < \sigma_{ff}/2 \\ -\varepsilon_{wf}, & \sigma_{ff}/2 \leq z < \lambda_{wf}\sigma_{ff}, \\ 0, & \lambda_{wf}\sigma_{ff} \leq z \end{cases}$$

where r is the interparticle separation distance, z is separation distance of the particle from the surface, $\lambda_{ff}\sigma_{ff}$ is the fluid–fluid potential well diameter, ε_{ff} is the depth of the fluid–fluid potential well, σ_{ff} is the diameter of the fluid–fluid hard core, $\lambda_{wf}\sigma_{ff}$ is the fluid-wall potential well diameter and ε_{wf} is the depth of the fluid–wall potential.

Critical parameters are estimated using coexistence data obtained *via* the previously described GC-TMMC and a least square fit scaling law and rectilinear approach.[18] The vapour–liquid envelope for confined SW fluids for varying slit-pore size is shown in Figure 4.1.

Fisher and Nakanishi,[19] with the help of scaling arguments, showed that the decrease in the critical temperature in larger pores should obey the relation, $(T_{cb} - T_{cp})/T_{cb} = kH^{-1/\nu}$, where ν is the critical exponent for the correlation length, T_{cb} is the critical temperature of the bulk, T_{cp} is the critical temperature in pore and k is a proportionality constant. To account for the strongly adsorbed layer on pore walls, it is necessary to replace the true pore width H, by

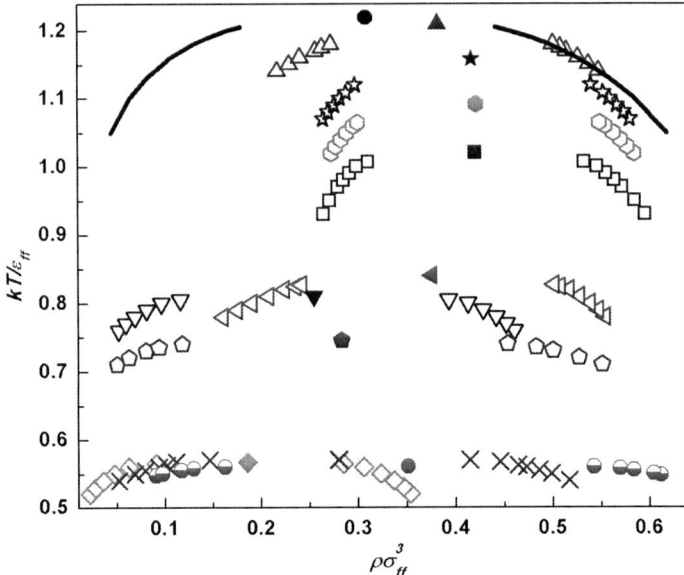

Figure 4.1 Temperature–density vapour–liquid coexistence curve for square-well fluid in a slit pore of width, H, varying from 40 to $1\sigma_{ff}$ for $\varepsilon_{wf}/\varepsilon_{ff} = 4$. Solid curves represent the bulk coexistence densities with filled black circles as the bulk critical points. Open and filled symbols represent the coexistence densities and critical points respectively. From top, symbols triangle, star, hexagon, square, left triangle, inverted triangle, pentagon, diamond, cross and half-filled circle represent $H/\sigma_{ff} = 40$, 12, 8, 6, 4, 3, 2, 1.9, 1.25 and 1, respectively.

a modified pore width, H_{eff}, which accounts for the adsorbed layers that form prior to capillary condensation. Using the Ising 3D (bulk) correlation length critical exponent[20] $v(3D) = 0.63$, and Ising 2D correlation length critical exponent[21,22] $v = 1$, one can analyse the crossover behaviour of critical properties of confined fluids from 3D to 2D in pores as shown by Singh *et al.*[17] Figure 4.2 presents a case study and demonstrates how the crossover changes with surface attractions.

Specifically, it shows that crossover behaviour is seen to substantially change with the surface field. In the above case, the slit width around which crossover occurs jumps from 8 to 16 molecular diameters when the surface-fluid interaction is increased three fold. Crossover behaviour is expected to change with the shape of the pore. For example, for hard pores, crossover from 3D to 2D in a cylindrical pore for square-well fluids occurs at 28 molecular diameters (diameter of the cylindrical tube[23]) compared to in a slit pore where it occurs at 5 molecular diameters.

In this section, we have explained some of the interesting behaviour and structural changes of confinement induced vapour–liquid phase transitions that differentiate these systems from their bulk counterparts. Further to this, we have explained how to apply simulation techniques to gain useful information

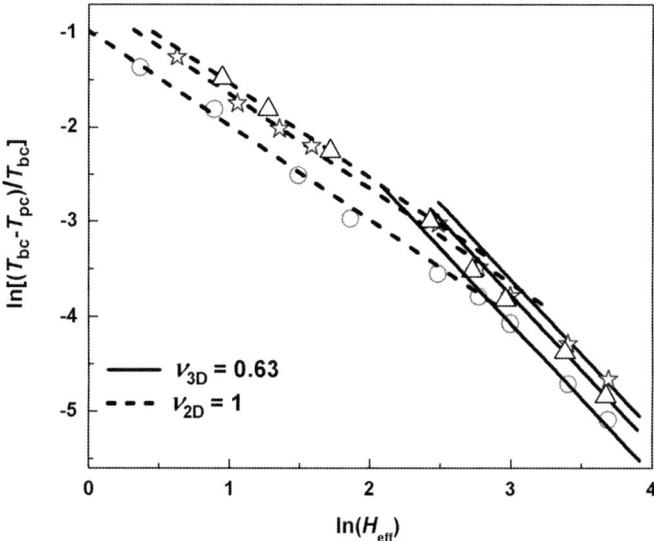

Figure 4.2 Crossover from 3D to 2D of square-well fluid for three different surface attractions, $\varepsilon_{wf}/\varepsilon_{ff} = 2$, 4 and 6 are shown by symbols circle, triangle and star. ν is the correlation length critical exponent. Solid and dotted lines represent the 3D and 2D regimes, respectively.

regarding confinement effects. Specifically, the above discussion is of fundamental concern pertaining to the crossover from 3D to 2D behaviour. The successful practical application of confined fluid-phase behaviour requires understanding the adsorption mechanism of guest molecules in porous medium.[24] The GC-TMMC technique can provide an effective means to study such systems as shown by Chen and Sholl.[25] Now, we turn to another important aspect; the subject of solid-liquid transitions induced by confinement under the context of nanotribology.

4.4 Fluid–Solid Transition under Confinement

Ask any high-school physics class what occurs when two surfaces in contact are moved relative to each other and they will probably answer something to the effect that kinetic energy is converted into thermal energy due to friction. While this may be desirable when attempting to slow a moving vehicle, in the majority of situations friction is an undesirable phenomenon that may have catastrophic implications. When dealing with nanoscale devices, the issue of friction is further complicated and exacerbated since, in contrast to macroscopic systems, kinetic friction at the nanoscale is also a function of surface roughness and, crucially, contact area. It is somewhat ironic that the enormous surface-volume area, which gives nanoscale systems many of their amazing and unique properties, means that nanodevices, such as nanoelectromechanical systems (NEMS), are particularly susceptible to wear and damage due to friction affecting their implementation and durability. If we are to be able to gain the technological advances promised by the advent of NEMS, clearly the issue of friction needs to be resolved.

Returning to our hypothetical high-school students, it seems likely that their reply to the question of "how can the problem of friction be reduced" would be "lubricate the surfaces" and of course they would be correct. In macroscopic devices, lubrication is typically achieved by placing the appropriate lubricant (most frequently, but not always, a liquid) between the surfaces whose relative motion is to be facilitated. Unfortunately, when dealing with nanoscale devices, things are, once again, not quite so straightforward. Put simply, unexpected things happen at the nanoscale. Of particular interest here is the role of nanoconfinement on phase behaviour.

In terms of lubrication, the transition of most concern is the liquid to solid (*i.e.* freezing) which raises the question as to whether confinement to the nanoscale can induce such a change (*i.e.* from fluid to solid)? Unfortunately, the answer to this has been somewhat of a grey area for the last two decades. In order to understand why this is the case, it is first necessary to understand the history of how nanoconfined fluids are studied experimentally.

In the late 1960s, Tabor and Winterton[26] developed an experimental mechanism capable of measuring the force between surfaces (*in vacuo*) separated by less than 10 nm. The purpose of their work was to demonstrate experimentally the theoretical prediction of a change in the nature of van der Waals forces at separations on the order of several nanometers. To measure force at such small distances they identified three basic requirements of the equipment.

1. The confining surfaces must be smooth enough to avoid steric hindrance in the approach.
2. An accurate method of measuring the separation between the surfaces.
3. A sensitive means of measuring the attractive force.

The first of these was accomplished by the use of mica, which may be cleaved along natural planes resulting in a step-free molecularly smooth surface with a relatively large surface area. This property of mica has ensured that it remains amongst the most commonly used materials in the study of nanoconfined phenomena. In the 1970s, Isrealachvili and Adams[27] adapted this technique to the study of nanoconfined fluids, noting that it is possible to *indirectly* measure changes in the confined fluid *via* changes to the confining surfaces. In essence, the method consists of gradually moving two mica surfaces towards each other whilst immersed in the fluid to be confined.

Initially, and still primary, was the measurement of the attractive/repulsive forces between the surfaces, leading to such an experimental setup being known as a surface-force apparatus (SFA). A point of interest, and at the root of the debate described later, is that, to date, all theories as to what occurs when a fluid is nanoconfined have been based on the measurement of forces on the surfaces, rather than direct measurement of the confined material.

The 1980s saw the widespread application of SFA experiments to fluids ranging from linear alkanes, such as dodecane,[28] to cyclical and globular molecules such as cyclohexane[29] and octamethylcyclotetrasiloxane (OMCTS).[30] Interestingly, although encompassing disparate molecular geometries, all the systems studied exhibited qualitatively similar behaviour. Specifically, when confined to the order of several molecular diameters or less, their behaviour is consistent with that of a crystalline solid. Even though it is not possible to experimentally observe directly what occurs under this extreme level of confinement, the theory of solidification under extreme nanoconfinement (3 molecular layers or less) obtained widespread acceptance. Since this solid-like behaviour was observed at conditions at which the bulk system is fluidic (typically atmospheric temperature and pressure) this phenomena is often called confinement-induced freezing.

In contrast to the solid-like behaviour observed for extreme nanoconfine-ment, when the separation n (measured in molecular diameters) of the mica surfaces is sufficiently large, the confined fluid exhibits bulk-like liquid beha-viour. Thus, it should be clear that there must exist a phase transition as a function of separation. On the basis of this observation, there were a number of investigations in the 1990s aiming at understanding this transition from liquid-like to solid-like behaviour. In particular, experiments probed the existence of a sharp transition from disordered to ordered below a critical number of layers n_c. Unfortunately, in this case, the experimentalists studying these systems did not manage to come to a consensus.[31] Rather, there emerged two opposing theories based on the differing indirect experimental observations. The first of these theories is based upon experimental evidence of an abrupt and reversible, many-orders-of-magnitude increase in viscosity, and the onset of stick-slip behaviour, as a function of separation. As such, the theory posits that the transition is first order in nature. In contrast to this, the second theory is based on evidence that suggests the transition is continuous in nature and thus is equivalent to vitrification.[32,33] Clearly, these disparate theories could be reconciled by observing directly the structural changes that occur when a fluid is nanoconfined.

Unfortunately, while direct observation of the confined fluid is an obvious solution, achieving such resolution (at the molecular level) for a confined sys-tem is far from trivial and has yet to be achieved (although attempts are being made).[34] However, thought experiments of possible causes for the differences in experimental observations may be made and then tested by repeating the experiments noting any changes in the behaviour of the confining surfaces. And in fact this is what has occurred. Interestingly, rather than leading to a broad consensus on the effects of nanoconfinement, each of the proposed reasons has been strongly criticised and refuted by the other side, leading to a vigorous decades-long debate and further polarising opinion in an already contentious area of research. A probable cause of the animosity in this debate is that, due to the inability to observe directly structural changes, the proposed causes have often been based on perceived weaknesses in opposing experiments. What is needed is a tool with molecular resolution, which brings us to the role of computational nanoscience.

With its inherent nanoscale resolution, molecular simulation is an indispensible tool in the study of confinement induced fluid–solid phase transitions and a range of techniques exist capable of providing both a direct description of the effects of nanoconfinement as well as insight into the underlying physics of such systems. Here, we divide these techniques into methods of simulation and methods of analysis (*i.e.* determining the nature of any structural changes) and will deal with them sequentially starting with simulation methods.

4.4.1   Simulation of Confinement-Induced Fluid–Solid Phase Transitions

The first point to consider in the design of these simulations is the level of detail required to provide a sufficiently accurate description of the phenomena one is interested in studying. A review of the literature on these systems reveals that three common levels of model accuracy exist. Here, we briefly describe each, noting previous applications and drawing the reader's attention to important considerations when deciding which level of accuracy is most appropriate.

4.4.1.1   Models

4.4.1.1.1   Primitive Models.   At the lowest level of detail, the confining surfaces are considered to be structureless, while the confined molecules are modeled as spheres or chains of spheres depending on the confined fluid to be studied. The molecules and surface may be purely repulsive or attractive interactions might also be considered, however, the surfaces are typically repulsive. An example of work using this level of molecular description is that of Ayappa and coworkers, who have published a number of articles[35,36] in which they consider nanoconfined Lennard-Jones spheres. Their focus is on understanding the structure of the confined solid-like phases and considers both the effect of surface separation on a generic Lennard-Jones potential as well as, in a later work, a potential chosen to mimic octamethylcyclotetrasiloxane (OMCTS). Using the generic potential,[36] they observe that opposed to a single solid-like phase, there exists a sequence of solid phases as a function of surface separation (*i.e.* level of confinement). Using a parameter set representative of OMCTS,[35] they demonstrate not only the same phases as for the generic potential, but find evidence that even if the layers in contact with the surface are disordered, the central layers in a confined system may be ordered. Other examples include the work of Fortini and Dijkstra[37] who calculated the complete phase diagram of solid–liquid and various crystal structures for a system of hard spheres between hard plates. The work of Fortini and Dijkstra is discussed in more detail later in this chapter with regard to the structural analysis of confined systems.

The use of these relatively simple systems facilitates the calculation of more complex properties, such as free energy, which, coupled with their relatively low computational cost, allows for a more complete investigation, compared to the more detailed methods described below. Thus, such methods are particularly suited to getting a quick and rough qualitative description, providing information about the major factors affecting fluid–solid transitions. However, they are not able to provide a reliable or accurate approximation to real systems and, as such, miss much of the subtle details of what occurs in these systems. Finally, these primitive models play an additional, and very important, role as the basis for statistical equations of state aimed at modeling confined systems.

4.4.1.1.2 United-Atom-Level Approach. At the next level of accuracy, the confining surfaces are modeled as structured and the confined molecules are described at a united atom (UA) level. While the surfaces are structured, the structure is typically simplified, most commonly face-centred-cubic (FCC). The particles forming the surface may be allowed to vibrate around their lattice points, however, this is not necessary and the same confined structure will be obtained if the particles are held stationary. This level of molecular model is very common and has led to the identification of a number of important factors determining whether or not a fluid–solid transition occurs, as well as the nature of the formed solid. Here, we provide examples of two important studies utilising united-atom molecular models, noting their specific contributions.

We begin by considering the pioneering work of Cui *et al.*[38–40] who performed molecular dynamics simulations of dodecane confined between mica sheets. In their work, Cui *et al.* described the mica sheets using an FCC lattice of Lennard-Jones spheres and used the Siepmann, Karaborni and Smit (SKS) alkane model for the dodecane molecules. In addition to providing insight into confinement induced freezing, the success of the work of Cui *et al.*[38–40] resulted in their choice of models becoming standard in the simulation of nanoconfined alkanes. Despite that, it is worth noting that the SKS model for alkanes may be replaced by any number of UA level models (*e.g.* OPLS-UA) and the same system behaviour obtained.

Cui *et al.*[38–40] performed constant normal pressure molecular dynamics simulations and made a number of interesting discoveries; however, here we consider only two. The first of these is that they successfully showed that dodecane undergoes a phase transition from a disordered fluid state to a layered and herringboned-ordered solid-like state as a function of confinement (*i.e.* surface separation). The second insight from their work is the role of wall–fluid interaction strength on confinement induced freezing. Specifically, they showed that in order for a fluid–solid phase transition to occur, the wall–fluid interaction strength must be sufficiently stronger than the fluid–fluid interaction. Interestingly, this is in agreement with the global theory of phase transitions under nanoconfinement developed by Radhakishnan *et al.*[41,42] It is thus clear that care must be taken to ensure the use of an appropriate wall–fluid potential if the correct behaviour is to be observed. Interestingly, if the wall–wall interaction strength is chosen in order to match the experimental surface energy of mica, as was done for the work of Cui *et al.*[38–40] Lorentz-Berthelot mixing rules result in a wall–fluid interaction that is approximately 4.5 times the fluid–fluid interaction strength for dodecane.

Other important information obtained using UA level models may be found in the work of Jabbarzadeh and coworkers,[43,44] who have presented a series of articles in which they investigate how changing the surface structure of the confining sheets affects the structure of nanoconfined dodecane. Making use of the same models and parameters as Cui *et al.*[38–40] Jabbarzadeh *et al.*[43,44] not only confirm the formation of a herringbone-ordered structure for FCC surfaces, but show that when the FCC is replaced with BCC surfaces, the herringbone structure still forms, but rotated by 45°, ensuring that the alignment of the confined structure with the surface is retained. Further to this, they consider the effect of an amorphous surface structure, noting that as the confined surfaces become increasingly amorphous, the induced layered herringbone structure becomes less well defined. This is particularly interesting since, based on the experimental evidence that fluids of different geometry behave similarly, it is often thought that any transition is purely a function of confinement, *i.e.* surface structure does not play a role.

Although the united-atom (UA) approach described above has provided valuable insight into the effects of nanoconfinement, it could be argued that approximations in the models and simulations may introduce artificialities that affect the results. Specifically, concerns exist that the relatively small system size, together with the shape and periodic nature of the simulation box, stabilised the herringbone-ordered structure. In addition to this, the relatively simplistic models, particularly those describing the mica surfaces, might not adequately represent the experimental system. With regard to the role of system size, it is worth noting that in their recent work Jabbarzadeh *et al.*[43,44] used mica surfaces approximately 4 times the surface area of those used in the work of Cui *et al.*[38–40] (51.84 nm^2 *vs.* 12.96 nm^2) and that, in our own work, we have used even larger sheets (80–400 nm^2). In both cases, the system exhibited the same behaviour (*i.e.* the formation of a herringbone-ordered structure). Thus, it seems that system size does not play a determining role in earlier simulation studies. We recently addressed concerns over the use of simplistic models using atomistically detailed simulations,[45,46] as explained next.

4.4.1.1.3 All-Atom Approach. While we describe the highest level as one of the three common levels of accuracy, we note that the use of such accurate models is quite recent, primarily due to computational limitations. However, it seems likely they will continue to become increasingly popular in the next few years now that computational resources have made them feasible. These models consist of trying to describe the confined systems, both the surface and the fluid, in a fully atomistic manner including the incorporation of explicit partial charges where appropriate. As an example we will use our recent work[45,46] simulating systems of dodecane and cyclohexane nano-confined between mica sheets in which we make use of the atomistic mica model of Heinz *et al.*[47] coupled with the OPLS all-atom model for the alkanes. There are many advantages to the use of these models; however, perhaps the most important is that by using a fully flexible model of mica and the confined fluid, together with explicit coulombic interactions, our results may be considered to be truly indicative of what occurs in the real system, opposed to merely being valid for a model mica–alkane. Following this logic, we have performed isothermal–isobaric simulations for systems of nanoconfined dodecane as well as cyclohexane.

In the case of dodecane, we observe behaviour that is qualitatively similar to that seen in the work of Cui *et al.*[38–40] and Jabbarzadeh *et al.*[43,44,48–50] *i.e.* the formation of a layered and herringbone-ordered solid. However, there are a few differences that are worth pointing out. First, while united atom approach using FCC (or BCC) surfaces results in patches of parallel dodecane molecules aligned perpendicularly, the patches for the atomistic models are aligned at approximately 60° (or 120°). Secondly, we see the formation of a stable layered structure up to 9 layers thick, in comparison to the 7 layers observed by Cui *et al.* The reasons for these differences may be realised by noting the insights provided by the work of Cui *et al.*[38–40] and Jabbarzadeh *et al.*[43,44,48–50] That is to say that the increased number of stable layers is mostly likely a result of a stronger wall–fluid interaction, whilst the 60° alignment is consistent with the hexagonal surface structure of the confining surfaces. This demonstrates that while a good qualitative description may be obtained with a united-atom approach, in order to capture fully the more subtle details of confinement induced phenomena, fully atomistic simulations are required.

We move now to consider the role of the geometry of the confined fluid. This is motivated by the experimental observation that fluids ranging from linear to cyclic exhibit the same behaviour upon nanoconfinement. As such, confinement-induced phase transitions are thought to not depend on the fluid being commensurate with the confining structure. Here, we investigate whether the shape of the confined molecule affects the formation of a layered and ordered structure by simulating the nanoconfinement of cyclohexane that, in addition to differing in geometry to dodecane, was involved in early SFA experiments.[51]

The results of our simulations show that, in common with dodecane, cyclohexane exhibits an abrupt transition to a layered and ordered structure as a function of confinement. However, it only forms a stable structure up to 5 layers thick, compared to the 9 layers observed for dodecane. This value compares well to Perkin *et al.*'s experimental value of 6 stable layers.[52] In terms of the form of the structure (see Figure 4.3), we note that while, as expected, the structure is different from that of dodecane, it is still hexagonal in nature; this highlights the importance of the surface structure of the confining material and suggests that the common assumption that this is not the case needs to be reassessed.

In summary, the use of atomistically detailed molecular models provides a level of fidelity and, thus, confidence that cannot be matched by the other

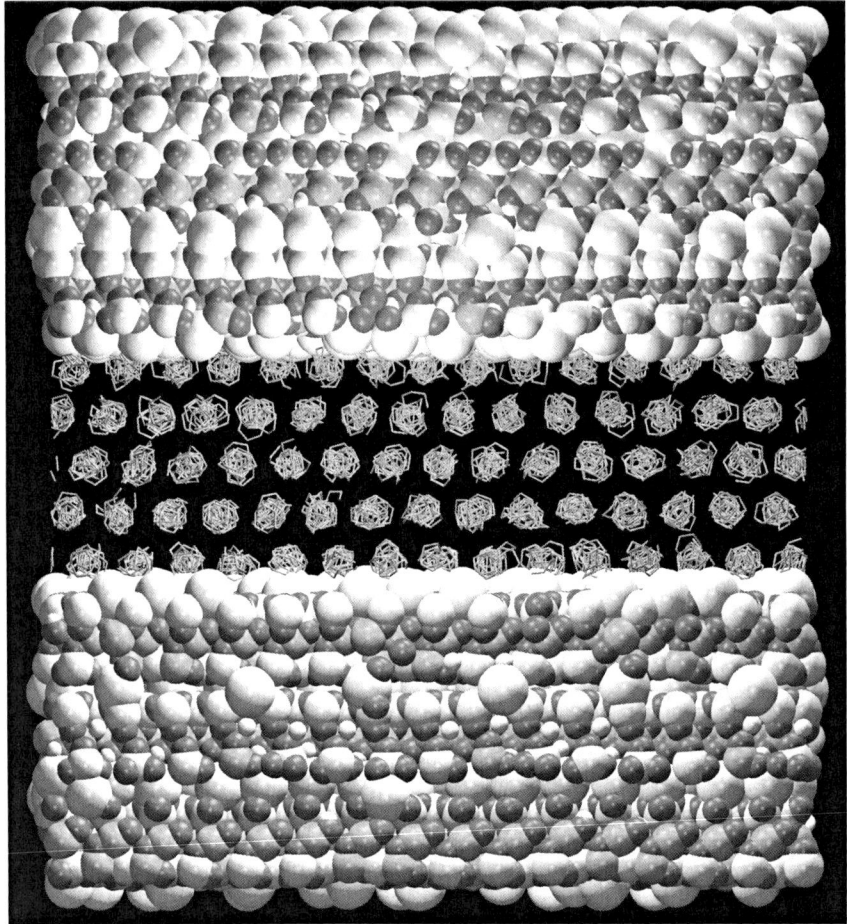

Figure 4.3   Side view of an equilibrated all-atom simulation of cyclohexane confined between atomistically detailed mica surfaces. Note the formation of a 5-layered hexagonal solid-like structure.

models described here. Thus, if one wishes to state with confidence that one's simulations describe a particular experimental system, there is no substitute. However, the use of such models is not without its cost that, in this case, is the dramatically larger computational time required.

4.4.1.2   Choice of Ensemble

Another important decision in the simulation of confined system (and of course in any simulation) is the choice of ensemble in which to perform the simulation. Here, we outline some choices that have been successfully applied to these systems.

4.4.1.2.1 Canonical Ensemble. The canonical ensemble is perhaps the most suitable choice for two types of studies. The first is that performed by Dijkstra *et al.*,[53] *i.e.* when trying to determine the confined structure as a function of density. There are a few difficulties to be aware of when performing such a study. Primarily, when the confining surfaces are not allowed to move, jamming, particularly of longer-chain molecules, may be problematic requiring, at best, unfeasibly long simulation times. The second type of simulation for which we recommend the use of a canonical ensemble is the study of phase coexistence under confinement. Although this is technique is perhaps most easily applied to the study of vapour–liquid equilibria, it should be possible to simulate a system consisting of a confined solid and fluid phase in equilibrium.

4.4.1.2.2 Isothermal Isobaric. Given that the majority of experimental studies take place under atmospheric conditions, a natural choice for simulation studies is the use of the isothermal–isobaric ensemble. This ensemble was the method of choice for both Cui *et al.*[38–40] and Jabbarzadeh *et al.*[43,44,48–50] in their studies of nanoconfined dodecane and, as such, we will use dodecane between FCC mica surfaces as an example to further clarify the use of this ensemble.

The setup of the system requires placing *N* dodecane molecules between FCC surfaces (See Figure 4.4). The system may be three-dimensionally periodic, or

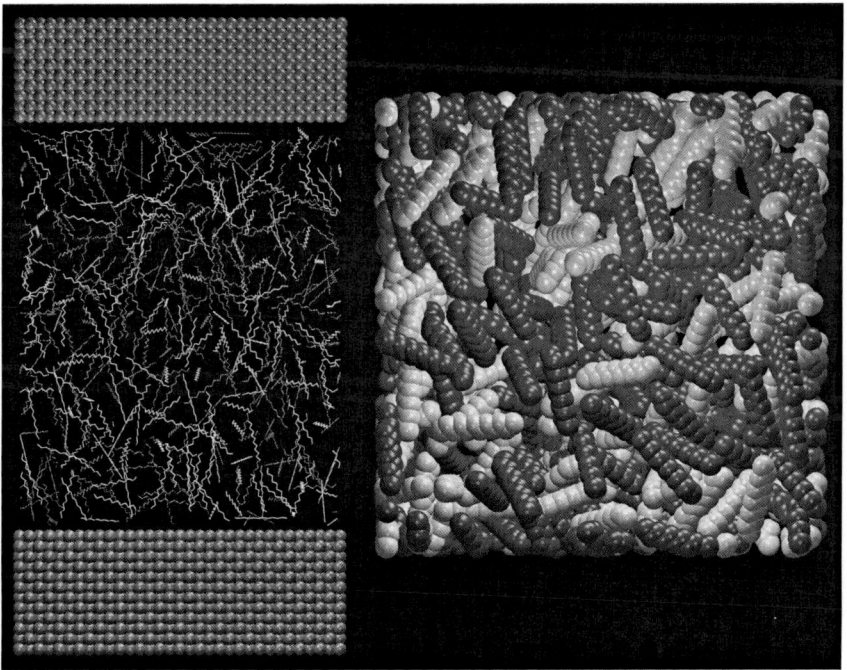

Figure 4.4 Side and top view of a typical starting configuration for an isothermal–isobaric simulation of dodecane confined between FCC surfaces.

may have what is known as a slab geometry (*i.e.* not replicated in the dimension perpendicular to the confining surfaces). At this stage, a straightforward isothermal–isobaric ensemble simulation with the pressure-controlled anisotropically in all three dimensions (only possible for flexible surfaces) may be performed or, commonly, the pressure may be controlled just in the dimension normal to the confining surfaces (suitable for rigid surfaces). In either case, the observed effects of nanoconfinement will be the same. What occurs will depend not only on the imposed temperature and pressure, but also on the number of fluid molecules confined. Specifically, if the number of dodecane molecules is less than the number required to form n_c layers, the fluid will undergo a transition to a layered and ordered solid-like structure. Otherwise, it will remain fluidic in nature.

There are a few caveats to this. The dodecane molecules will tend to form an integer number of layers, forming what is known as a herringbone structure (Figure 4.5) however, the transition to this structure is much more rapid (less simulation required) and the final structure contains fewer flaws if the number of dodecane molecules is chosen to correspond as near as possible to the number required to form n "perfect" layers. How to know how many molecules in a perfect layer? One option is to assume a reasonable level of packing and use this to guess/predict. Alternatively, the layers immediately adjacent to the confining surfaces tend to form quite rapidly (due to the strong wall–fluid interaction) and are pretty much independent of the number of layers formed in the ordered system. Thus, running a relatively short simulation allows for the formation of a not perfect but fairly well-packed first layer from which the number of molecules per layer may be "counted." Of course, these two methods may be combined.

Figure 4.5   Side and top view of an equilibrated configuration for an isothermal–isobaric simulation of dodecane confined between FCC surfaces. Note the formation of 5 distinct layers with an intralayer herringbone structure (top view).

The isothermal–isobaric ensemble appears to provide a reasonable balance between fidelity to experimental studies and computational cost, with its biggest criticism being that the periodic nature of the simulation cell may introduce an artificial stability in the confined structure as well as affecting the nature of its ordering (*e.g.* promote a herringbone structure). With regards to the latter point, it is worth noting that simulations performed using grand canonical molecular dynamics, described below, result in the same herringbone structure, suggesting this concern is probably unfounded.

4.4.1.2.3 Isothermal–Isobaric Grand Canonical Molecular Dynamics (NPT-GCMD).
Another alternative is the grand canonical molecular dynamics technique of Gao *et al.*[54] This method may be thought of as taking the isothermal–isobaric simulation cell described above (Figure 4.4) and, placing a large amount of bulk fluid on two opposite sides of the confining sheets such that molecules are free to enter (Figure 4.6), and leave, the slit from the bulk fluid (*i.e.* the confined fluid and surrounding bulk are in equilibrium). The system is three-dimensionally periodic, with the box length in the lateral direction used to maintain the pressure. This method has two key advantages. First, the confining surfaces remain stationary (*i.e.* the pore width is constant), which eliminates confinement rate as a source of error. Secondly, since molecules are free to enter and leave the pore, the number of confined molecules is not artificially fixed. Typically, the results of this technique are consistent with isothermal–isobaric simulations. For example, for the united-atom n-dodecane/FCC LJ system a herringbone ordered structure with the same structure and orientation relative to the confining surfaces is obtained. The drawback to this method is that it requires a significantly larger number of fluid molecules, resulting in increased computational cost. Given this, the technique is probably best used sparingly as a check of the computationally cheaper isothermal–isobaric method.

4.4.1.3 Analysis

Having dealt with methods to perform simulations of confined fluid–solid transitions, we turn to the determination of whether a transition has occurred as well as measures of the structure of the confined system. The global order of the system indicates whether a system is fluid-like or solid-like and is characterised by the bond-order parameters introduced by Steinhardt *et al.*[54] Here, a bond is not a chemical bond but a vector joining two neighbouring atoms

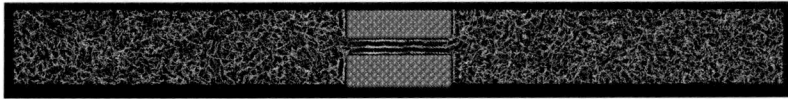

Figure 4.6 Side view of an equilibrated NPT-GCMD simulation of united-atom dodecane confined between FCC Lennard-Jones mica surfaces with a fixed surface separation of 3 dodecane.

whose distance is less than a cutoff radius. In order to calculate bond-order parameters, a spherical harmonic associated with every bond r_{ij} in the system needs to be determined. The quantity is defined as,

$$Q_{lm} = Y_{lm}(\theta(\mathbf{r}_{ij}), \phi(\mathbf{r}_{ij})), \qquad (4.14)$$

where Y_{lm} is the spherical harmonic and θ and ϕ are the polar and azimuthal angles of the bond with respect to a fixed coordinate frame that can be arbitrarily chosen. The global bond-order parameter Q_l which is rotationally invariant is then defined as follow,

$$Q_l \equiv \left(\frac{4\pi}{2l+1} \sum_{m=-l}^{l} |\bar{Q}_{lm}|^2 \right)^{1/2}, \qquad (4.15)$$

where $\bar{Q}_{lm}$ is the average of Q_{lm} s over all bonds in the system. The presence of any order under confinement can be identified by the use of Q_6 as the order parameter. Q_6 is very sensitive to the global structure, which increases significantly when order appears and vanishes for the isotropic fluid.

To illustrate the above, we take an example from our recent work,[55] in which we considered the simple case of a square-well fluid confined in a hard-sphere cylindrical pore and investigated the fluid–solid transition and structural changes under confinement. Using NPT-based molecular dynamics simulation, P_{zz}, the spreading pressure (defined as pressure along the axis of the cylindrical tube) is calculated at different densities. Separate NVT-based Monte Carlo simulations were also conducted to verify the results. Figure 4.7 presents a typical spreading pressure *vs.* density isotherm. Phase equilibria data is estimated from the P_{zz} equality criteria (which is not a strict condition). Results from both types of simulations are in agreement. The discontinuity in the equation of state, which is observed by both methods, is indicative of a fluid–solid transition. This comparison indicates that both methods are accurate enough to determine the equation of state of both fluid and solid branches. However, for determining the fluid–solid coexistence regime, MD is better than NPT MC, as MD has a much smaller relaxation time, which enables better accuracy in determining the coexistence density regime and coexistence pressure, while NPT MC generally cannot pinpoint these properties with such accuracy. However, it is noted that precise calculation of phase equilibria is not feasible from this approach, for which a separate free-energy calculation is necessary, which is described briefly later.

Even though the equation of state, as shown in Figure 4.7, illustrates a coexistence region, it is appropriate to look at the structure of states near the coexistence region to verify the liquid- or solid-state behaviour. As stated earlier, precise solid–liquid equilibria calculation under confinement from pressure-density isotherms is not possible; hence, investigation of the structural order parameter is necessary as a check. Figure 4.8 illustrates a bulk bond-order parameter Q_6 as a function of density in a pore of size $D = 2.2\,\sigma_{\text{ff}}$. Q_6 for

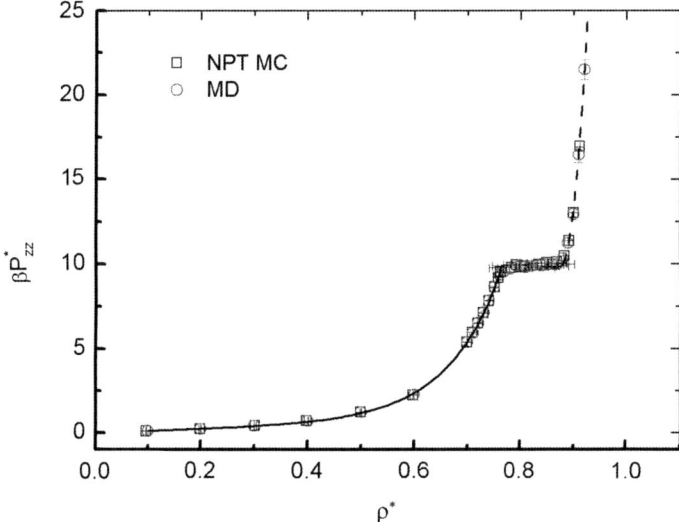

Figure 4.7 Comparison of the equation of state (reduced axial pressure *vs.* reduced numerical density) of SW molecules of $\lambda_{ff}=1.5$ confined in cylindrical hard pores with diameter, $D/\sigma_{ff}=2.2$ obtained by isobaric–isothermal Monte Carlo (NPT MC) and molecular dynamic (MD) simulations. Here, squares indicate NPT MC results, and circles the MD result. The solid line indicates an analytical fit of the result at the fluid branch, and the dashed line is the second-order polynomial fit to the solid branch. Error bars are the standard deviation of 5 independent runs.

the above system is found to be nonzero for dilute fluid due to a confinement effect. The system with a random set of N_b bonds, has a Q_l value of $1/\mathrm{sqrt}(N_b)$, but not zero[56] as within the tight confinement, since the direction of bonds joining each pair of atoms is not stochastic but has limited orientations. An interesting phenomenon has been observed with the increase of the density of the fluid. Q_6 increases to reach a value around 0.15 at a reduced density around 0.75. Similar behaviour is found for the 2D bond-orientation parameter, which is described by eqn (4.16)

$$\psi_6 = \left| \frac{1}{M} \sum_{i=1}^{M} \frac{1}{N_i} \sum_{j=1}^{N_i} \exp(i6\theta_{ij}) \right|, \qquad (4.16)$$

where M is the number of particles in the 2D layer (in the above case it would be a cylindrical layer), N_i is the number of nearest neighbours of particles i. The value of ψ_6 is equal to 1 for a perfect hexagonal plane, but far from 1 for a disordered phase.

We should not take agreement between the global bond order and local order to be the general behaviour for confined solids. As shown by Coasne and coworkers[57] the bulk bond-order parameter is not a suitable parameter to study

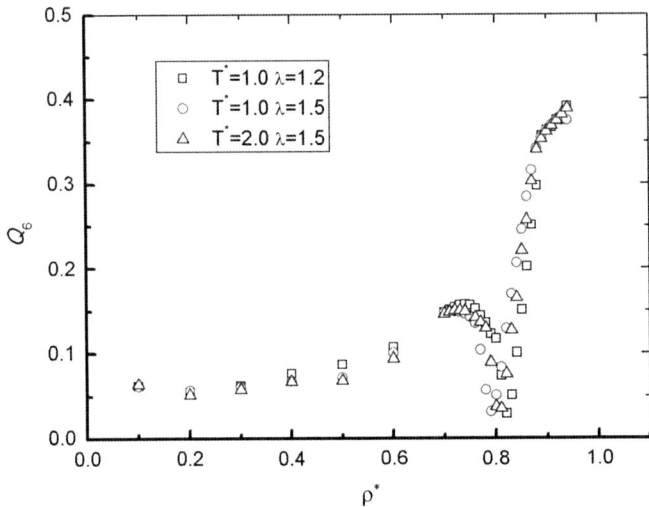

Figure 4.8 The bond-order parameter Q_6 as the function of density, $\rho^* = \rho\sigma_{\text{ff}}^3$, for the SW molecules confined in a cylindrical hard pore of $D = 2.2\sigma_{\text{ff}}$ at different temperature and potential ranges.

the structural parameters as it is sensitive to the structure of the surface. For a crystal phase, the $\bar{Q}_{lm}$ components add up coherently. However, for example, for disorder porous materials, Q_6 would provide erroneous information about the structure as a large degree of disorder of the porous materials would prevent coherency of the orientational order of the confined phase. Hence, a more appropriate parameter for the study is a 2D hexagonal order parameter (inplane 2D orientational order parameter). Other than order-parameter calculations, radial distribution functions can also provide necessary structural information for confined fluids. For example, Figure 4.9 illustrates the axial distribution function $g(z)$, for the aforementioned confined square-well fluid case. Corresponding molecular snapshots are shown in Figure 4.10.

At a reduced density of $\rho = 0.30$ ($\sigma_{\text{ff}} = 1$ and the asterisk is dropped in the rest of the chapter for the sake of convenience), $g(z)$ shows a peak at $z = 1.0$ and a valley at $z = 1.2$, which happen to be the hard-core diameter and the extent of the potential well, respectively, and it quickly decays to 1.0 as z increases, indicating the system does not have a long-range order at this density. For $\rho = 0.75$ and 0.80, peaks appear at intervals of approximately 0.33 with the largest peak occurring at $z = 1.0$ before dropping smoothly to $g(r) = 1.0$. In contrast to $\rho = 0.30$, $g(r)$ at $z = 0$ is nearly zero. Interestingly, the position of the peaks matches with the axial distance of the peaks of tetrahedrons. This supports the evidence from bond-order parameters, which suggest there is a partially ordered phase of spiral packing of tetrahedra around this density range. For densities of 0.85 (within coexistence densities), 0.89 (around melting point), and 0.92 (solid phase), a sharp peak appears at $z = 0$ and is followed by a series of peaks periodically spaced at intervals of about 0.87, with the largest peak

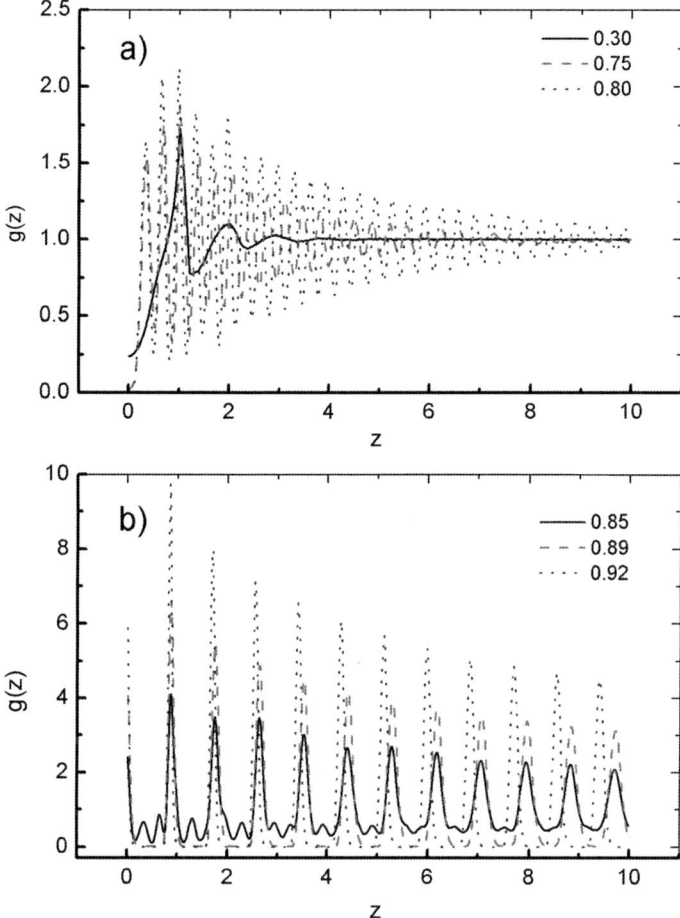

Figure 4.9 Axial pair distribution function $g(z)$ for SW molecules confined in a cylindrical pore of $D = 2.2\sigma_{ff}$ at $kT/\varepsilon_{ff} = 1.0$ and $\lambda_{ff} = 1.2$, at different average densities, $\rho^* = 0.30$ (dilute fluid), 0.75 (dense fluid), 0.80 (around freezing density), 0.85 (within fluid–solid coexistence), 0.89 (around the melting density), and 0.92 (solid). Note that z is normalised by σ_{ff}.

being the first of these. This phenomenon indicates that the tetrahedral packing has transformed to triangular packing. Note that for the density within coexistence, there exist two small peaks between the sharp peak at $z = 0$ and $z = 0.87$, which indicates that both tetrahedral and triangular packings coexist in the system. Similar behaviour is observed for the case of diameter, $D, = 2.5$, however, the twisted helical structure disappears at $\rho = 0.95$, where the structure is transformed to a square shape consisting of four particles sharing the same z against another nearest square layer by rotating by an angle of $\pi/4$ (see Figure 4.10). In addition to the inplane positional correlation function, the inplane bond-orientational pair correlation function, $G_6(r)$, which measures for

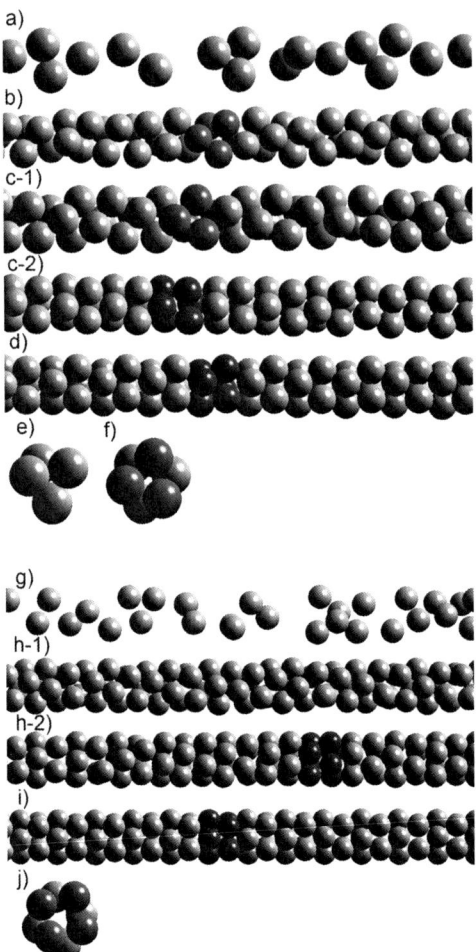

Figure 4.10 Snapshots for SW molecules confined in narrow cylindrical pores of $D = 2.2$ (*a* to *d*), and $D = 2.5$ (*g* to *i*) at different average densities: *a*, $\rho = 0.3$ (dilute fluid), *b*, $\rho = 0.75$ (dense fluid, spiral packing of tetrahedron), *c*, $\rho = 0.85$ (fluid–solid coexistence), *d*, $\rho = 0.92$ (solid, triangular packing), *g*, $\rho = 0.3$ (dilute fluid), *h*, $\rho = 0.85$ (fluid–solid coexistence) and *i*, $\rho = 0.95$ (solid, square packing). The tetrahedral, triangular, and square packings are shown in *e*, *f*, and *j*, respectively. Two different structures are found in the corresponding coexistence densities in $D = 2.2$ (*c*-1 and *c*-2) and $D = 2.5$ (*h*-1 and *h*-2).

each 2D plane the correlation between the local bond-order parameter, ψ_6 (r) at two positions separated by a distance *r*:

$$G_6(r) = \langle \psi_6^*(0)\psi_6(r) \rangle. \tag{4.17}$$

Similar to the positional pair correlation function, the G_6 function would possess a short-range orientation order for liquid phases and long-range orientation order for crystal layers. Hence, analysis of the inplane 2D pair correlation function and orientation correlation function jointly can provide the transition between liquid and solid. Nevertheless, precise calculation of coexistence liquid–solid phases can only be done by evaluating the chemical potential of the individual phases or, equality, of the grand potential $\Omega = F - \mu N$.[37]

Finally for this section, we note that, in addition to inducing a fluid–solid transition, confinement can change stable crystal structures and may lead to a variety of different and new structures, which are not seen in bulk solids, as shown by Fortini and Dijkstra[37] for hard spheres confined between two hard plates. The authors determine for the first time the complete phase diagram of solid–liquid and various crystal structure of a confined hard-sphere system by calculating the free energy using a thermodynamic integration technique. The coexisting solid phase consists of crystalline layers with either triangular (T) or square (S) symmetry (similar to that observed for square-well fluids in hard cylindrical pores). A sequence of crystal structures, from $n\text{T} \rightarrow (n+1)\text{S} \rightarrow (n+1)\text{T} \cdots$, is observed by the authors with increasing plate separation. Here, n is the number of crystal layers. This behaviour is in agreement with experiments on colloids.[58]

4.5 Summary

We finish by noting that, in addition to providing interesting insight into the nanoscale causes of macroscopic phenomena, the study of nanoconfined systems is of vital importance to the fledgling development of nanoscale devices. Direct experimental observation of what occurs in these systems is difficult and expensive, when it is possible at all. This may be due to the speed at which a particular phenomenon occurs (*e.g.* wetting) or simply due to the lack of a tool with sufficient resolution to observe what is occuring (*e.g.* confinement induced freezing). This weakness provides an exciting opportunity to apply any of a wide range of increasingly powerful molecular simulation techniques to the study of nanoconfined systems. With their inherent molecular resolution such methods are a synergetic complement to current experimental studies, providing direct insight into the underlying mechanisms of nanoconfinement-induced behaviour.

Acknowledgements

We are grateful for support of our work by the Department of Science and Technology, India. Part of the work presented in this chapter was supported by the National Science Foundation through grant CHE-0626259. In addition, it used resources of the National Energy Research Scientific Computing Center, which is supported by the office of Science of the U.S. Dept. of Energy under Contract No. DE-AC02-05CH11231, and resources of the National Center for

Computational Sciences at Oak Ridge National Laboratory, which is supported by the Office of Science of the U.S. Department of Energy under Contract No. DE-AC05-00OR22725.

References

1. M. J. Bojan and W. A. Steele, *Carbon*, 1998, **36**, 1417.
2. G. M. Davies and N. A. Seaton, *Carbon*, 1998, **36**, 1473.
3. L. D. Gelb, K. E. Gubbins, R. Radhakrishnan and M. S. Bartkowiak, *Rep. Prog. Phys.*, 1999, **62**, 1573–1659.
4. D. Frenkel and B. Smit, *Understanding Molecular Simulation: From Algorithms to Applications*, Academic Press, London, 1996.
5. M. P. Allen and D. J. Tildesley, *Computer Simulation of Liquids*, Clarendon Press, Oxford, 1987.
6. D. Frenkel and B. Smit, *Understanding Molecular Simulation: From Algorithms to Applications*, 2nd edn, Academic Press, 2002.
7. D. A. McQuarrie, *Statistical Mechanics*, Harper & Row, New York, 1976.
8. N. Metropolis, A. W. Rosenbluth, M. N. Rosenbluth, A. H. Teller and E. Teller, *J. Chem. Phys.*, 1953, **21**, 1087–1092.
9. J. R. Errington, *J. Chem. Phys.*, 2003, **118**, 9915.
10. V. K. Shen and J. R. Errington, *J. Chem. Phys.*, 2005, **122**, 064508.
11. A. M. Ferrenberg and R. H. Swendsen, *Phys. Rev. Lett.*, 1988, **61**, 2635–2638.
12. G. V. Burgess, D. H. Everett and S. Nuttall, *Pure Appl. Chem.*, 1989, **61**, 1845.
13. A. de Keizer, T. Michalski and G. Findenegg, *Pure Appl. Chem.*, 1991, **63**, 1495.
14. W. D. Machin, *Langmuir*, 1994, **10**, 1235.
15. A. P. Y. Wong, S. B. Kim, W. I. Goldburg and M. H. W. Chan, *Phys. Rev. Lett.*, 1993, **70**, 954.
16. A. P. Y. Wong and M. H. W. Chan, *Phys. Rev. Lett.*, 1990, **65**, 2567.
17. S. K. Singh, A. Saha and J. K. Singh, *J. Phys. Chem. B.*, 2010, **114**, 4283–4292.
18. J. K. Singh, J. Adhikari and S. K. Kwak, *Fluid. Phase. Equil.*, 2006, **248**, 1–6.
19. M. Fisher and H. Nakanishi, *J. Chem. Phys.*, 1981, **75**, 5857.
20. R Guida and Z.-J. J, *J. Phys. A: Math. Gen.*, 1998, **31**, 8103.
21. R. Evans, Marini Bettolo Marconi U and T. P, *J. Chem. Soc. Faraday Trans.*, 1986, **82**, 1763.
22. R. Evans, Marconi Marini Bettolo U and T. P, *J. Chem. Phys.*, 1986, **84**, 2376.
23. S. K. Singh and J. K. Singh, *Fluid Phase Equil.*, 2010, **300**, 182.
24. A. H. Fuchs and A. K. Cheetham, *J. Phys. Chem. B.*, 2001, **105**, 7375.
25. H. B. Chen and D. S. Sholl, *Langmuir*, 2006, **22**, 709.
26. D. Tabor and R. H. S. Winterton, *Proc. R. Soc. London Ser. A*, 1969, **312**, 435.

27. J. N. Israelachvili and G. E. Adam, *Nature*, 1976, **262**, 773.
28. H. K. Christenson, D. W. R. Guen, R. G. Horn and J. N. Israelachvili, *J. Chem. Phys.*, 1987, **87**, 1834.
29. M. L. Gee, P. M. Meguiggan, J. N. Israelachvili and A. M. Homola, *J. Chem. Phys.*, 1990, **93**, 1895.
30. J. N. Israelachvili, M. L. Gee, P. Mcguiggan, A. M. Homola and P. Thompson, *J. Phys.: Condens. Matter.*, 1990, **2**, SA89.
31. A. L. Demirel and S. Granick, *Phys. Rev. Lett.*, 1996, **77**, 2261.
32. S. Granick, *Science*, 1991, **253**, 1374.
33. H. Hu, G. A. Carson and S. Granick, *Phys. Rev. Lett.*, 1991, **66**, 2758.
34. S. C. Bae, J. S. Wong, M. Kim, S. Jiang, L. Hong and S. Granick, *Philos. Trans. R. Soc. London. Ser. A*, 2008, **366**, 1433.
35. K. G. Ayappa and C. Ghatak, *J. Phys. Chem. B.*, 2007, **111**, 14299.
36. K. G. Ayappa and C. Ghatak, *J. Chem. Phys.*, 2002, **117**, 5373.
37. A. Fortini and M. Dijkstra, *J. Phys.: Condens. Matter.*, 2006, **18**, L371.
38. S. T. Cui, P. T. Cummings and H. D. Cochran, *J. Chem. Phys.*, 2001, **114**, 7189.
39. S. T. Cui, P. T. Cummings and H. D. Cochran, *Fluid. Phase. Equil.*, 2001, **183**, 381.
40. S. T. Cui, C. McCabe, P. T. Cummings and H. D. Cochran, *J. Chem. Phys.*, 2003, **118**, 8941.
41. R. Radhakrishnan, K. E. Gubbins and M. Sliwinska-Barkowiak, *J. Chem. Phys.*, 2000, **112**, 11048.
42. R. Radhakrishnan, K. E. Gubbins and M. Sliwinska-Barkowiak, *J. Chem. Phys.*, 2002, **116**, 1147.
43. A. Jabbarzadeh, P. Harrowell and R. I. Tanner, *Phys. Rev. Lett.*, 2005, **94**, 126103.
44. A. Jabbarzadeh, P. Harrowell and R. I. Tanner, *Phys. Rev. Lett.*, 2006, **96**, 206102.
45. H. Docherty and P. T. Cummings, *Soft Matter*, 2010, **6**, 1640.
46. P. T. Cummings, H. Docherty, C. R. Iacovella and J. K. Singh, *AICHE J.*, 2010, **56**, 842.
47. H. Heinz, H. Koerner, K. L. Anderson, R. A. Vaia and B. L. Farmer, *Chem. Matter*, 2005, **17**, 5678.
48. A. Jabbarzadeh, P. Harrowell and R. I. Tanner, *Tribol. Int.*, 2007, **40**, 1574.
49. A. Jabbarzadeh, P. Harrowell and R. I. Tanner, *J. Phys. Chem. B.*, 2007, **111**, 11354.
50. A. Jabbarzadeh, P. Harrowell and R. I. Tanner, *J. Chem. Phys.*, 2006, **125**, 034703.
51. J. Klein, *Nature*, 1980, **288**, 248.
52. S. Perkin, L. Chai, N. Kampf, U. Raviv, W. Briscoe, I. Dunlop, S. Titmuss, M. Seo, E. Kumacheva and J. Klein, *Langmuir*, 2006, **22**, 6142.
53. A. Fortini, A. Hynninen and M. Dijkstra, *J. Chem. Phys.*, 2006, **125**, 094502.
54. P. J. Steinhardt, D. R. Nelson and M. Ronchetti, *Phys. Rev. B.*, 1983, **28**, 784.

55. H. C. Huang, W. W. Chen, J. K. Singh and S. K. Kwak, *J. Chem. Phys.*, 2010, **132**, 224504.
56. M. D. Rintoul and S. Torquato, *J. Chem. Phys.*, 1996, **105**, 9258.
57. B. Coasne, S. K. Jain, L. Naamar and K. E. Gubbins, *Phys. Rev. B*, 2007, **76**, 085416.
58. A. B. Fontecha, H. J. Scope, H. Konig, T. Palberg, R. Messina and H. Lowen, *J. Phys.: Condens. Matter.*, 2005, **17**, S2779.

CHAPTER 5

Simulating Thermomechanical Phenomena of Nanoscale Systems

P. ALEX GREANEY AND JEFFREY C. GROSSMAN

Department of Materials Science and Engineering, 13-5049 MIT,
7 Massachusetts Avenue, Cambridge, MA 02139-4307, USA

5.1 Introduction

The manner in which we use and control heat is of great importance both technologically, and within our daily lives. Roughly 90% of the energy that we use today has been involved in the generation to and conversion from heat (*e.g.*, coal or gas power plants, combustion engines for transportation, and nuclear energy, to name only a few). The efficiency of these processes is limited by the control we have over the storage and transport of heat within materials. Most practical thermal-energy systems that are involved in transport, storage, and/or conversion operate far from their limits.

Since thermal transport is a materials phenomenon, it is useful to compare our understanding of it with other transport properties such as the conduction of electrons. Since the discovery of electricity, research on charge transport in materials has pushed the extremes of electrical conductivity, which now spans over 20 orders of magnitude. In 1947 the transistor was invented, which permitted external tuning of electrical conductivity in the solid state—and forming the foundation for information processing and beginning the digital age. The societal impact of research in electrical conduction has been and will continue

RSC Theoretical and Computational Chemistry Series No. 4
Computational Nanoscience
Edited by Elena Bichoutskaia
© Royal Society of Chemistry 2011
Published by the Royal Society of Chemistry, www.rsc.org

to be enormous. In contrast, heat transport in condensed matter has received much less attention. For example, the thermal conductivity of current solid materials spans only 5 orders of magnitude at room temperature and its external tunability is limited. With the exception of the Carnot limit to heat engines, we really do not know what limits, if any, exist regarding the fundamental processes of thermal energy transport and conversion, and storage. There are currently no thermally superconducting materials, nor are any existing materials perfect thermal insulators.

An equivalent paradigm shift in thermal transport would be the use of frequency-dependent thermal properties; that is, exploiting properties of materials that depend on the thermal energy being concentrated at specific frequencies. Such opportunities may exist in new thermal phenomena that occur in nanoscale materials. The aim of this chapter is to provide an overview of the methods for simulating thermal phenomena at the nanoscale. After giving a brief overview of some of the experimental progress that is driving this field we survey the current computational methods highlighting their strengths, and the major challenges that still face the simulation of heat. This is followed by two detailed case studies in which selected methods have been used to study nanoscale thermomechanical behaviour, showing how these tools can be used, and what insight they can give. The examples given are in the study of dissipation in CNT resonators, and frequency-dependent heat transport across weak nanoscale interfaces. The chapter finishes with a discussion of the major challenges still facing simulation of heat, and the outlook to new applications of thermal properties.

5.2   Examples of Recent Experimental Progress

There have been recently a number of exciting advances in our ability to control heat flow that have primarily been achieved with the use of nanoscale systems. Accompanying the advances in engineering thermal properties have been breakthroughs in or ability to measure thermal transport and to characterize thermal behaviour at this small scale. Together, these experimental developments show that we can extend the current boundaries of thermal conductance, and they call on scientists to develop more accurate and predictive methods for computing thermal conductance and for simulating thermomechanical phenomena in nanoscale systems. A brief survey of the experimental advances in nanoscale thermal properties that is motivating the need for new computational approaches is given here.

5.2.1   Increasing Thermal Conductivity

As long ago as 1941[1] it was recognised that the mean free path of longitudinal acoustic phonons diverges as their frequency approaches zero.[i] In the absence of internal scattering mechanisms this would mean that a material's thermal

---

[i] An infinitely long-wavelength mode is a quasistatic compression that must have an infinite mean free path in order to satisfy Newton's laws.

conductivity would *increase* with the size of the material. In bulk materials there are a plethora of defects, interfaces and other scattering mechanisms that limit the mean free path, and make the thermal conductivity size invariant in practice. Single-walled carbon nanotubes and graphene sheets are composed of a single layer of atoms and are not bulk materials (one cannot distinguish between a surface and an internal atom in a CNT). With no interfacial scattering the only mechanisms for restricting the mean free path of long-wavelength modes are anharmonic phonon interactions and isotopic effects. The result of this is that single-walled CNTs can exhibit very large thermal conductivity along their length, and that the tubes must be very long (up to 10 μm) before κ stops increasing with tube length.[2–5] CNTs hold much hope for engineering new materials with very efficient thermal conduction for uses in thermal management; however, much of the heat transported along a CNT is conveyed by flexural and torsional modes. These important heat-carrying modes only move unimpeded when the CNT is free standing; the modes are quickly scattered if the CNT is in contact with a substrate (or another CNT) reducing the thermal conduction by as much as two orders of magnitude.[6,7] Nevertheless, applications have been proposed for using CNTs as heat guides. A very recent and innovative application has been to use CNTs as a heat guide to set up rapidly propagating thermal reaction waves in a reacting medium.[8] Bundles of CNTs are coated with an exothermically reactive mixture, and when the reaction is initiated rapid heat conduction along the tube speeds up the propagation speed of the reaction front. Most interestingly the passage of the reaction front has been found to be accompanied by a large electrical thermopower pulse in the tubes that is proportional to the speed of reaction propagation.

The use of CNTs in other (nonreactive) thermal management applications has also been demonstrated. Vertically aligned "forests" of CNTs have been proposed for efficient thermal management of computer chips, where removal of heat can be the limiting factor in device performance, and where the power densities that must be removed can be extremely high.[9] Vertically aligned forest structures utilize the full thermal transport potential of the CNTs, limited only by contact resistance at the CNT ends. Other researchers have used embedded networks of CNTs to engineer highly anisotropic thermal conductivity in composite materials and to tailor the cooling properties of oils with the addition of a suspension of CNTs.[10] In these cases the heat is carried across a sparse network of CNTs. In order to travel long distances the heat must be passed between CNTs at the nexus points where they overlap within the network, and this process has its own resistance, which is found to be large. Transport in these networks—both electrical and thermal—exhibits a percolation transition in which if the network is too sparse there are insufficient nodal connections linking the tubes and they will not create a contiguous pathway through the material. Remarkably, the percolation transition for conduction of electricity occurs at a lower network density than it does for the conduction of heat. This is surprising as the networks are composed of CNTs with a distribution of different diameters and chiralities with roughly only one third of them being metallic; this implies that there is also a wide distribution of thermal contact resistance, with many junctions being electrically conductive but highly resistive to heat.

5.2.2   Reducing Thermal Conductivity

In crystalline solids a fundamental lower limit of thermal conductivity can be conceived by imagining that the mean free path of a phonon cannot be any shorter than the intrinsic lattice spacing (similar conceptual limits can be derived for amorphous solids). These limits are somewhat artificial, however, being only a limit in which one model of heat flow no longer makes sense. Thus, while these limits can provide useful guidelines when designing thermal systems, there are no rigid reasons why thermal resistance cannot surpass these limits in some cases. This has in fact been observed experimentally for roughened Si nanowires[11] that have been shown to transport heat below the perceived lower limit of conductance.

Beyond the fundamental question on limits of thermal transport it is often desirable to increase the thermal resistance of a given material without greatly altering other properties. An important technological example of this are thermoelectric materials, for which it is desirable to minimize lattice thermal conduction without altering the electrical properties of the material. In this case, one can take advantage of the fact that the mean free paths of charge carriers (holes and electrons) are much smaller than the mean free path of lattice vibrations (phonon wave packets), and thus one can nanostructure the material on a length scale that will confound traveling-wave phonons, but that is invisible to charge carriers. Experimentally, this was first approached by growing multilayered materials with controlled layer thicknesses. This is a top-down approach that allows one to finely control the structure at the angstrom level in one direction. Another top-down approach is to create a nanoscale grain structure in the material by deformation or powder processing. This approach has been successful for further improving the thermopower of established thermoelectric semiconductor materials.[12] Examples of other more exotic approaches to suppressing thermal conduction include roughened nanowires, nanoporous materials, and metamaterials that possess a phonon bandgap. Recently, films containing nanoscale tubular pores running through the film thickness have been found to suppress thermal conductivity by several orders of magnitude.[13,14]

5.2.3   Characterisation

In addition to advances in the ability to fabricate structures at the nanoscale there has been significant progress in the ways that one can measure thermal properties. Length-dependent thermal conductance has been measured on single CNT, and Si nanowires, lying across a set of heating stages.[4,15–17] Using this approach Chang *et al.* demonstrated thermal rectification along carbon and boron nitride nanotubes that have been asymmetrically decorated with platinum carbide.

An alternative noncontact method of measuring heat transport properties in nanoscale structures is by time-domain thermoreflectance. This is a pump-probe technique in which a short (pico or nanosecond) laser pulse locally heats the sample (raising the temperature by only a few degrees). The heating changes the optical reflectance of the sample, and a probe laser beam is used to measure this change in reflectance as the heat dissipates. Cahill and coworkers[18,19] have found that by making the region of material that is heated smaller than the mean free path of some of the phonons these phonons leave the hot region ballistically, allowing one to measure the contribution of different phonon frequencies to thermal transport.

5.2.4 Nanoscale Phenomena

In addition to the greater range of thermal conductance that can be achieved with nanoscale materials, the nanometer-sized systems provide new thermal phenomena that are unique to this length scale. Making one or two dimensions of a system small (while leaving the others macro-or mesoscopically large) results in objects that are in effect two-, or one-dimensional. These objects still live in a three-dimensional space (mathematically they are said to have a codimension greater than one). Low-dimensionality systems possess new low-frequency phonon mode shapes in which the object deforms into the unconstrained dimension. Films and surfaces gain surface Lamb waves, beams possess flexural and torsional modes, and tubes and fullerenes have radial breathing modes and cyclops modes.[20]

The symmetries of these low-dimensional modes can give rise to subtle thermal effects. Interatomic potentials are asymmetric, near equilibrium close to harmonic, as we move further from equilibrium the potential softens as the bond is stretched and stiffens when compressed. The most important anharmonic term in the representation of an atomic bond is the cubic term, $d^3 E_{ij}/dR_{ij}^3$. Yet the symmetry of a nanostructured material combined with the symmetry of the vibrational mode can result in modes that are symmetrical, and where the fourth-order anharmonic term dominates. To see this, consider a nanotube. Imagine sitting on the surface of a zigzag nanotube watching the motion of atoms as the tube undergoes a torsional oscillation. We see that for each torsional cycle an atomic bond stretches twice; the mode is symmetrical and the cubic term in the bond potential $d^3 E_{ij}/dR_{ij}^3$ results in a quartic term $d^4 E_{tor}/da_{tor}^4$ dominating the anharmonicity of the modes. Following the motion of the radial breathing mode (RBM), on the other hand, we see that it is asymmetric, and the cubic term $d^3 E_{RMB}/da_{RBM}^4$ dominates. Thus, through the increased freedom of a codimension greater than one, it is possible to obtain vibrational modes with a mixture of differing symmetry of anharmonicity, although the structure is constructed with only asymmetric bonding potentials. Another simple example of this is a tensioned one-dimensional chain of atoms connected by harmonic potentials. The "string-like" modes of the chain are very strongly anharmonic because as the string oscillates the chain length is increased, increasing the overall tension in the chain. This effect can result in a negative coefficient of thermal expansion along the chain length, and is observed in carbon nanotubes that contract along their axis when they get hotter.[21,22]

Reducing the dimensionality of an object can cause other significant changes in properties. Diffusion in systems with two or fewer dimensions is space filling—that is given an infinite amount of time a random walker will visit every lattice site—this is not the case for three-dimensional spaces. A mathematical connection between diffusion and heat conduction results that heat flow in low-dimensional objects does not obey Fourier's law. For the case of one-dimensional systems the thermal conductivity of the system is found to scale with the length of the system, $\kappa \sim L^{\alpha}$, with α being reported between 0.3,[23] and 1/3.[24] This has important implications for modeling thermal transport in low dimensions: in simulation one often extrapolates the scaling of their results with system size in order to estimate macroscale properties or to check convergence. This result implies that there is *no* length scale convergence in one-dimensional systems.

5.2.5 Quantum Phenomena

The examples given thus far are purely classical arising only due to restricted dimensionality; however, advances in characterisation and fabrication at the nanoscale have allowed researchers to measure several purely quantum effects of atom motion. Roukes and coworkers, by cooling down a free-standing SiN membrane supported by four phonon waveguides, were able to freeze out all but the lowest-frequency vibrational modes in the waveguides and were able to measure ballistic conduction through just the four lowest-frequency modes (one longitudinal, two flexural, and one torsional) in each waveguide.[25] They found the quantum of thermal conductance, g_o, to match the Landauer equation $g_o = \pi^2 k_B^2 T/(3h)$, approaching the quantum limit of measurement.

A different quantum measurement effect that is close to the grasp of current experimental methods is the direct observation of zero-point motion and the Heisenberg uncertainty in a mesoscopically large mechanical system. Gigahertz nanoscale resonators can be frozen to their ground state at milikelvin temperatures, but one must also be able to sensitively detect the displacements associated with this motion which are on the subangstrom scale. Beyond establishing that quantum-mechanical effects are observable in mechanical systems containing many billions of atoms, interesting applications for quantum computing are found if the system is coupled to a quantum-mechanical two-level system.[26]

5.2.6 Far-from-Equilibrium Behaviour

Traditionally, the way that we have controlled and transport heat in solids has been close to equilibrium—a usage that is reflected in the established theories of heat flow such as the Boltzmann transport theory (BTT) in which heat carriers transport thermal energy between regions that are in local thermal equilibrium. In recent years, however, there have been examples of nanoscale devices (or phenomena) that take advantage of thermal behaviour far away from equilibrium, that is, where thermal occupation of vibrational modes does not correspond to the Bose–Einstein distribution (the local temperature is not well defined) but is instead athermally distributed amongst the system's vibrational modes. Operating away from equilibrium allows one to take advantage of the mechanical properties of a system at a particular set of frequencies and thus enables the exploitation of a wider range of thermal behaviour than can be achieved at equilibrium.

An athermal phonon population (APP) can arise in suspended carbon nanotube (CNT) resonators through either frequency-specific Joule heating of optical phonons (at K and Γ),[27–30] or by direct driving[31,32] (or cooling[33]) of low-frequency flexural modes. In the former case this causes the electrical conduction to saturate at high voltage, and in the latter the resulting APP can dramatically reduce the resonator's quality factor.[34,35] APPs can also arise when heat is conducted across CNT interfaces.[10,36] Taking advantage of the APP leads to strategies for engineering interfacial thermal conductivity,[37,38] and may be important in the recently discovered thermal power waves in CNTs.[8] Biological systems can also display nonequilibrium thermal-energy distributions. Enzymatic reactions can result in a large heating of localised modes in the protein. This heat must be dissipated efficiently without denaturing the enzyme, and is transferred through a restricted set of localised vibrational modes without heating the enzyme as a whole[ii].[39,40] A similar "energy funneling" phenomenon is observed in virus capsids[41] in which laser heating of high-frequency modes is funneled into a handful of low-frequency mechanical modes—an effect that can may be exploited for selectively destroying harmful viruses.[42] In the second of the case studies given later in this chapter we discuss APPs that arise from frequency-selective transmission of heat in two weakly interacting objects, and we show how this can be exploited for applications in chemical sensing.

5.3 Survey of Simulation Methods

There are two approaches to computing thermal properties of nanoscale systems. If the theory behind the property of interest is well understood one may compute the property directly, perhaps using simulation or first-principles calculations to obtain the values of input parameters. An example of this approach is the use of density-functional theory (DFT) to compute a material's phonon density of states and then using that information to compute the specific heat as a function of temperature. This approach works well for calculating intrinsic properties of a system that are not dependent on size or geometry. An alternative approach that is valuable for system-specific

---

[ii] A similar concept is important for barrierless termolecular reactions.

properties is to use atomistic simulation to model the behaviour directly. This approach can be thought of as performing an experiment within the computer (*in silico*) instead of in the laboratory. An advantage of direct simulation is that as one has the full atomistic trajectory of the system in time it is often possible to dissect the data to determine features that are contributing strongly to the phenomenon under study, and to gain insight into how one can alter the system to change it. Both approaches—direct calculation, and computer experiment—are important and complementary as will become evident in the rest of this chapter.

One of the most important thermal properties for engineering applications is the thermal conductivity of a material, κ, which is defined by the phenomenological Fourier's law that relates the steady-state[iii] thermal flux, J_Q, to the local temperature gradient:

$$J_Q = -\kappa \cdot \nabla T. \tag{5.1}$$

Here, we review several well-established methods for predicting the lattice contribution to κ at the nanoscale. Calculation of κ provides good examples of both direct computation approaches and *in silico* experiment, and in this context we discuss the general advantages and challenges of these two approaches (beyond just computing κ). At the end of this section we go beyond the mean thermal conductance to examine far-from-equilibrium thermal behaviour, and the methods can be used as the basis of studying other thermomechanical phenomena in nanoscale systems, as will be shown in the case studies that follow this section.

5.3.1   Direct Computation of κ

In crystalline solids, heat is carried by moving scattering lattice vibrations, and one can therefore write the thermal flux using the Boltzmann transport equation (BTE):[iv]

$$J_Q = \sum_{\text{branches}} \int_0^\infty \omega \hbar \beta v(\omega) \lambda(\omega) \rho_v(\omega) \frac{\partial N(\omega, T)}{\partial T} \nabla T. \tag{5.2}$$

Here, the sum is performed over all phonon branches and polarisations, and $v\,(\omega)$, $\lambda\,(\omega)$, and $\rho_v\,(\omega)$ are, respectively, the group velocity, mean free path, and volumetric density of states for phonons with frequency ω. β is a

---

[iii] We restrict discussion to the steady-state properties. Inclusion of time dependence to Fourier's law ignores the fact that heat travels with a finite velocity and thus for describing transient thermal solutions one must use the relativistic heat conduction equation, or the Cattaneo equation, which is beyond the scope of this chapter.

[iv] Note that the BTE is a very general equation for describing the motion of particles in a fluid that is away from equilibrium. Only in 2010 has the general form of the equation been rigorously proven for systems close to equilibrium. For simplicity, we jump straight to the form appropriate for the motion of a phonon gas, and further simplify by assuming isotropy so that κ becomes a scalar quantity rather than a tensor.

dimensionless geometric factor and $N(\omega, T)$ is the number of phonons occupying a mode with frequency ω at temperature T. Strictly, one must solve for $N(\omega, T)$ self-consistently; however, close to equilibrium this can be approximated by the Bose–Einstein distribution.

The BTE is very general and may be extended to include the ballistic contributions to transport that can occur in very small systems. Computing κ becomes a matter of computing β, v, λ, and ρ_v, and the accuracy of one's prediction relies on how accurately each of these terms are computed. The terms v and ρ_v can be computed from first-principles methods. The physics of scattering in the system is encapsulated in the mean free path term and includes: 3- and 4-phonon anharmonic scattering processes; electron–phonon interactions; and scattering from surfaces, interfaces, and impurities. The detail with which one includes these factors will determine the ultimate accuracy of the prediction.

One important advantage of directly calculating κ from the BTE is that the quantum mechanical nature of heat is treated correctly—both in terms of the selection rules for phonon creation and annihilation, and the occupation of the vibrational spectrum. The difficulty is in quantifying all the detailed geometry-specific phonon transition rates.

In a system with a high degree of disorder, propagating phonon states are not well defined and the phonon gas model of transport is no longer appropriate. Allen and Feldman[43] and others have developed a model description relevant to such a system. They classify the quanta of vibrational energy (*vibrons*) in an amorphous system into: *propagons*, that reside in low-frequency plane-wave-like modes; *diffusons*, that reside in nonlocalised but non-propagating (stationary) modes; and *locons*, that occupy high-frequency localised modes that are above the mobility edge. The majority of heat conducted through these systems is mediated by phonon transitions between diffusion modes.

5.3.2 Computing κ by Direct Simulation

A material's intrinsic resistance to conducting heat is due to phonon scattering that stems from anharmonicity in the interatomic interactions. As enumerating and computing all the scattering processes is difficult, a potentially more appealing approach is simply to accurately compute the anharmonicity of the atomic interactions and then use this within molecular dynamics (MD) simulations to measure the thermal conductivity *in silico*. The simulation will numerically account for all of the phonon-scattering processes without having to identify them *a priori*.

The most intuitively straightforward method for computing κ in this manner is to impose a thermal gradient in a system and simulate it using MD. One can straightforwardly set up a molecular dynamics simulation in which two (distant) slabs of atoms within a solid are thermostatted to different temperatures. By measuring the work done by the thermostats at each side of this thermocouple one finds the heat current that is transported down the thermal gradient in the intervening material between the slabs. However, while this approach most resembles a real experiment, it is fraught with several challenges, some that are common to all MD simulations, and some (such as the size of the simulation cell and the use of thermostats) that can be overcome with more carefully designed simulations.

In the computational setup described above, the temperature in the intervening material does not vary smoothly between the two thermostatted regions but experiences a discontinuous step at the boundary with the thermostatted regions—making the temperature gradient hard to control computationally. More significant, (especially in terms of computational affordability) is that the computational cell must be large enough to accommodate the mean free path of the heat-carrying phonons; and these can be several hundreds of nanometers. While this is large from the point of view of computational tractability, it is often still very small compared to the temperature difference, resulting in a large ΔT, which can throw into question the validity of the linear response approximation.

An alternative approach is to switch the roles of cause and effect, so that rather than impose a temperature gradient one imposes a fixed thermal flux by adding (non-net-translational) kinetic energy at a constant rate to the atoms in one region, and removing heat at the same rate in the cold slab. This procedure was first developed by Müller-Plathe[44] and is often referred to as reverse nonequilibrium molecular dynamics (RNEMD). The approach is applicable to many transport phenomena and is also used for computing the viscosity of fluids.[45] The approach still suffers from requiring a very large computational cell but the system converges to the steady state more rapidly than the direct nonequilibrium MD approach described above.

It is possible to move away from the nonequilibrium approaches altogether, and instead to simulate a system at equilibrium in the microcanonical ensemble (NVE) and make use of the fluctuation-dissipation theorem that relates the linear response properties of a system out of equilibrium to the dissipation of thermal fluctuations within the system at equilibrium. The Green–Kubo method[46] relates the thermal conductance κ, to the autocorrelation function of the fluctuations in the heat flux J_Q, at equilibrium:

$$\kappa = \frac{1}{T^2 k_B V} \int\limits_0^\infty dt\, C(t), \tag{5.3}$$

with k_B the Boltzmann constant, V the volume of the cell and the correlation function $C(t)$ defined by:

$$C(t) = \int d\tau J_Q(\tau) J_Q(\tau + t) = \langle J_Q(\tau); J_Q(\tau + t)\rangle, \tag{5.4}$$

$$J_Q(t) = \frac{dR}{dt}, \tag{5.5}$$

$$R = \sum_i r_i h_i. \tag{5.6}$$

Here, the term R can be considered the centre of energy, as the sum of the positions of all atoms weighted by the sum of their potential and kinetic energies.[v] One advantage of the Green–Kubo method (in addition to its computational simplicity) is that as the heat-flux correlation function decays much more rapidly than the heat-carrying phonon's mean lifetime, the system size that must be simulated can be many times smaller than is required for nonequilibrium MD methods.[47]

For all of their intuitive appeal, using MD simulation to *in silico* measure κ suffers from several important and fundamental challenges, the most important of which is that the simulations are *classical*—that is, the trajectories of the atoms are integrated according to Newtonian mechanics without any quantum-mechanical phenomena included. This is usually justified by stating that for most elements at room temperature the atoms are sufficiently heavy that their de Broglie wavelength is negligible and the atoms can be treated classically. A notable exception is hydrogen, whose small mass does make quantum-mechanical effects significant and makes simulation of water particularly troublesome.[77] While this reasoning is true it overlooks two other quantum effects, first, that the permitted energy of vibrational modes is quantised with occupation $E_n = \omega\hbar(n + \frac{1}{2})$, and that the quanta of energy (that we refer to as phonons, not the modes that they occupy) are bosons and so fill the available states according to the Bose–Einstein distribution.

This has *two* consequences: First, not all modes have a uniform amount of energy—low-frequency modes will hold more energy than higher-frequency modes according to relation $\langle E_i(T) \rangle = \omega_i\hbar/exp(\omega_i\hbar\beta) - 1$, where, $\beta = 1/k_BT$, is the reciprocal temperature. At temperatures below the Debye temperature, T_D, the high-frequency modes with be unoccupied and will possess only their zero-point energy. This stands in direct contrast to a classical system in which all the modes may have a continuum of energy, and where on average *all* the modes will have the same energy regardless of the temperature. For this reason molecular dynamics simulations are usually performed at temperatures above T_D, where at least the participation of all vibrational modes is a reasonable assumption (although the filling of the modes is still incorrect). Alternatively, simulations may be performed at low temperatures if the phenomena of interest only involves low-frequency modes, and if it can be shown that the activity of high-frequency modes does not influence the behaviour of these low-frequency modes.

The second consequence of the quantum harmonic oscillator is that when heat is transferred from one mode to other modes (*i.e.* when phonons are

---

[v] The potential energy of an atom can be poorly defined for many-body potentials; however, it is usual to approximate the local bonding energy to be shared between participating atoms in a pairwise fashion.

created and annihilated) the participating phonons must follow strict selection criteria such that the sum of the frequencies of the participating modes before and after the transition event are equal. No such restrictions are present in classical systems. The importance of this for MD simulations is that one cannot assume that the processes observed are directly transferable to real quantum-mechanical systems. Goddard and coworkers have shown that in the case of diamond at temperatures above T_D, the quantum correction to the classical Green–Kubo calculated κ is small.[47]

Besides the fundamental quantum-mechanical deficiencies of MD simulations there are also some practical considerations. Thermodynamic properties are an ensemble average over all the possible states of the system. While in principle MD simulations are ergodic and will correctly sample phase space the time it takes to do so is very long, much longer than the duration that can be feasibly achieved in a single simulation. In a single MD simulation the systems trajectory only samples a narrow region of phase space, if we are to properly average over phase space we must run many simulations starting with very different initial conditions in order to attain a representative sampling on the ensemble.

5.3.3 Gaining Insight from Simulations

Despite the many limitations and deep-rooted challenges for simulating thermal phenomena with MD methods, it is still a widely used tool that can be powerful for gaining physical insight if it is used judiciously. The main reason for this is that one knows the positions and velocities of every atom in the system during a MD simulation (knowledge that is out of reach from a real experiment). Armed with this raw data it is often possible to tease out subtle mechanisms of heat transport, scattering, and dissipation. This can be particularly useful when designing structures that suppress or enhance the mobility of phonons within a particular band of frequencies. Extending the analogy of the computer experiment these methods are like the characterisation tools of the real experimenter: they do not interfere with the physics used to simulate the system, but are used to interpret the results of simulation. The last part of this review of computational methods focuses on these numerical "characterisation" methods for distilling insight from raw MD simulation data.

Considerable insight can be gained simply by watching an animation of atomic motion in an MD simulation. One gains a mental picture of what the atoms are doing and one can quickly spot anomalous behaviour. As a method this is not very rigorous, it is a challenge to be sure that the animation is representative of the statistical ensemble, and it is difficult to disseminate movies in the traditional publishing format. However, the human brain is very good at spotting visual patterns (sometimes too good, finding patterns that are not there): watching a movie of simulated atom motion can quickly help one to identify bugs, or spot whether seemingly interesting results are merely artifacts of the system setup. It can also help identify interesting and physically meaningful behaviour on which more sophisticated and rigorous numerical characterisation methods can be trained.

A more rigorous method of "watching" the atoms was employed with great success by Phillpot, *et al.*[48] These researchers launched phonon wave packets at high-angle grain boundaries in Si and taking snapshots of the average kinetic energy of the atoms within slices of the computational cell, they followed the envelope of vibration as it collided with the interface, and were able to identify the mixture of scattered and transmitted wave packets. The method was used to determined which vibrational frequencies are transmitted most efficiently across the grain boundaries.

Much information can be learned from the vibrational spectrum of a nanoscale object, which as was mentioned above, can differ greatly from bulk systems. One consolation of the classical limitations of MD is that we can compute the vibrational spectrum (more or less) for free from any *equilibrium* MD simulation that we perform by Fourier transforming the autocorrelation function of the total kinetic energy. This can be done for any temperature, and while the spectrum must be weighted by the Bose–Einstein distribution to obtain the correct intensities the MD will capture the shift in mode frequencies due to filling of anharmonic modes.

The zero-temperature phonon spectrum for a nanoscale system can also be computed directly by diagonalising the Hessian matrix or the dynamical matrix (depending on whether the system is molecular or extended). Both of these methods have the advantage that in addition to the frequencies of the modes, one also obtains their eigenvectors. For very large systems with many millions of atoms, finding *all* the eigenvectors becomes computationally expensive; however, one can search sequentially for the lowest-frequency modes by recasting the matrix diagonalisation as an energy-minimisation problem, a method developed by Sankey *et al.*[49] for computing atomistically the vibrational modes of virus capsids.

Knowing the eigenvectors for the vibrational modes permits one to search through them, classifying the modes based on their symmetry or their spatial character, and to identify which modes are dominating the thermal phenomenon under study. For example, if one has a complex system of weakly interacting molecular units, such as a double-walled CNT, it permits one to identify modes that are confined to each individual CNT and those modes that are shared between the tubes. Obviously the shared modes will be most important for transmitting heat between the inner and outer walls of the CNT.

There are other ways of classifying modes based on their spatial extent. Allen and Feldman *et al.*[50] have classified the modes of amorphous Si, into propagating, diffusive, and localised modes. The lowest-frequency modes extend across the full extent of the system and can be thought of as propagating wave-like deformation modes. The wavelengths are much longer than the scale of the structural heterogeneity and the mode feels the average mechanical properties of the material—the limit being the infinite-wavelength mode corresponding to a homogeneous eigenstrain. At intermediate frequencies the modes become weakly localised, in that the wavelength of the modes becomes comparable to their mean free path. Above this so called Ioffe–Regal limit[51] the modes are diffusive, and the majority of heat is transported by energy exchange through these modes. The highest-frequency modes are short and localised around structural anomalies. These transport little heat and are said to be above the mobility edge.

Various approaches for measuring the degree to which a mode is localised have been developed by different researchers. Galli *et al.*[14,52] have computed the "participation ratio", p of the eigenmodes in a number of nanoscale systems with one or two reduced dimensions. The participation ratio of mode i is a measure of the fraction of atoms in the system that are participating in the displacement of this mode and is defined as:

$$p_i = \frac{1}{N} \left(\sum_{j=1}^{N} (\boldsymbol{\varepsilon}_{ij} \cdot \boldsymbol{\varepsilon}_{ij})^2 \right)^{-1}, \tag{5.7}$$

where N is the total number of atoms, and $\boldsymbol{\varepsilon}_{ij}$ is the displacement vector of the jth atom due to mode i with $\sum_{j=1}^{N} \boldsymbol{\varepsilon}_{ij} \cdot \boldsymbol{\varepsilon}_{ij} = 1$. Fully delocalised modes have p ranging between 1 and 0.5 with localised modes having smaller p with the limit that a mode localised to a single atom has $p = 1/N$. Yu and Leitner[40,53] have sorted the vibrational modes of a large protein molecule by their exponential localisation length, ξ by best fitting $\exp(|\boldsymbol{r} - \boldsymbol{r}_o|/\xi)$ to the decay of the mode's amplitude away from the mode's centre ($\boldsymbol{r}_o$). Using this approach, an inverse correlated was found between localisation length and mode frequency.

Armed with the knowledge of some or all of the vibrational modes of a system one can project these modes onto the atomic displacements and velocities at any time during an MD simulation to obtain the instantaneous amplitude, $a(t)$, and velocities, $\dot{a}(t)$, of the modes.

$$a_i(t) = \sum_{j=1}^{N} \boldsymbol{\varepsilon}_{ij} \cdot (\boldsymbol{r}_j(t) - \boldsymbol{r}_j^o), \tag{5.8}$$

$$\dot{a}_i(t) = \sum_{j=1}^{N} \boldsymbol{\varepsilon}_{ij} \cdot \boldsymbol{v}_j(t). \tag{5.9}$$

From computing the velocity of individual modes in this way during an equilibrium MD simulation one can calculate the mode's lifetime, τ_i, by integrating the normalised autocorrelation function of the mode velocity.[6,54] Integrating gives the time it takes for the velocity autocorrelation function to decay, which is the *average* time after which the oscillation has lost coherence with itself. Galli and coworkers have shown the power of this approach by multiplying the computed lifetimes by the phonon group velocity, v_g (the gradient of the phonon dispersion) to calculate the phonon mean free paths, $\lambda = v_g \tau$. The mean free path appears in the BTE (eqn (5.2)), and thus computing this from simulation permits one to attribute each mode's contribution to the total heat transport.

Mode projection also yields insightful information from nonequilibrium MD simulations where knowledge of the modes not only allows one to analyse the results of a simulation but also excite a simulation away from equilibrium in a carefully chosen way. Adding an instantaneous velocity and/or displacement to the system along a mode's eigenvector excites a single vibrational mode. After excitation, by simulating in the microcanonical ensemble one can follow the system as it relaxes back to equilibrium. This scheme has been widely used for uncovering the detailed mechanisms of energy transfer in a diverse array of systems, including: phonon scattering at grain boundaries,[48,55] energy relaxation within protein molecules[41,56] and damping in CNT resonators.[34,35] More detailed examples of the use of the phonon projection are given in the two case studies later in this chapter. Here, we give a pedagogical discussion of frequency-dependent energy transfer. The aim is to show the types of information that can be gleaned from MD simulations using mode projection, and some of the areas in which care must be taken.

As a test system, let us consider the case of two identical carbon (10, 0) carbon nanotubes. The tubes are lying adjacent and parallel to each other, and they are weakly bound together by the van der Waals interactions between them. The tubes are (10, 0) tubes with a 4.2-nm periodic repeat distance, and they are in a vacuum. Full details of the simulation setup can be found in ref 36. After optimising the structure of an *isolated* single tube, its eigenmodes are computed using the frozen phonon method. Note that this only gives stationary solutions, not traveling modes—it is equivalent to diagonalising the dynamical matrix at Γ for the whole computational cell. The second identical tube is now introduced and the system is again optimised—the two tubes are attracted to each other and flatten very slightly where they "touch" to maximize this attraction. The normal modes of the isolated CNT are not the normal modes of this new composite system; nevertheless, it is instructive to interpret the transfer of energy in terms of the modes of the isolated CNT. We now excite just one mode in one CNT by displacing it. In this example, we choose the radial breathing mode RBM of the left-hand CNT that we call tube A.[vi] The system of the excited and relaxed tube is simulated in the microcanonical ensemble. Mode projection is used to compute the amplitude and velocity of each mode measured relative to the ground state of the relaxed double-tube structure at every timestep (r_j^o in eqn (5.8) is the atomic positions in the optimised double-tube system). Before the projection is performed care must be taken to unwrap any atoms that have crossed the periodic boundaries, and if the tube has randomly rotated it must be mapped back into the orientation in which the Hessian matrix was computed. The set of mode energies at a time t are represented as a smoothed spectrum by summing a set of Lorentzian functions centred at the

---

[vi] Besides the mode that has been externally excited, the tubes are at absolute zero. This is an unusual and hypothetical situation that is seemingly problematic for classical dynamics. However, as all modes have *zero* energy using classical mechanics it is justified any time $t = 0$. Moreover, as we are only following the first stages of the relaxation of this energy before all become classically occupied, the use of MD remains justified for the short timescale simulated here. The simulations were repeated at finite temperatures and the same behaviour was observed.

mode frequencies and weighted by the mode energies. This approach to obtaining the vibrational spectrum is considerably more computationally expensive than the velocity correlation method, and less straightforward to implement; however, it gives us the excited spectrum at every time step, rather than averaged over some simulation time. Plotting the instantaneous spectrum against time gives the surface plots shown in Figure 5.1.

Examining the plots in Figure 5.1 reveals a surprising amount of information. First, we can follow the first cascade of energy as the excitation of the RBM in the "hot" CNT relaxes towards equilibrium. Four distinct energy-transfer events are distinguishable: (a) First, energy is transferred resonantly from the RBM in the hot tube to the RBM in cold tube. The transfer is frequency-selective and the frequency of the heat is preserved after being exchanged between tubes; (b) Next, the energy in the cold tube's RBM is scattered anharmonically into a mode with exactly half the frequency of the RBM. It can be shown by examining the eigenmodes that this frequency halving transition arises because the RBM is *asymmetrically* anharmonic whereas the half-frequency mode is not;[36] (c) The half-frequency mode in the cold tube transfers the excitation resonantly back to the equivalent mode in the hot tube, where (d) some of the energy is anharmonically scattered back into the initially excited RBM.

In addition to following the cascade of energy transitions, the projection data reveals much information about the nature of the participating modes. While the projection scheme computes the exact instantaneous kinetic energy of each

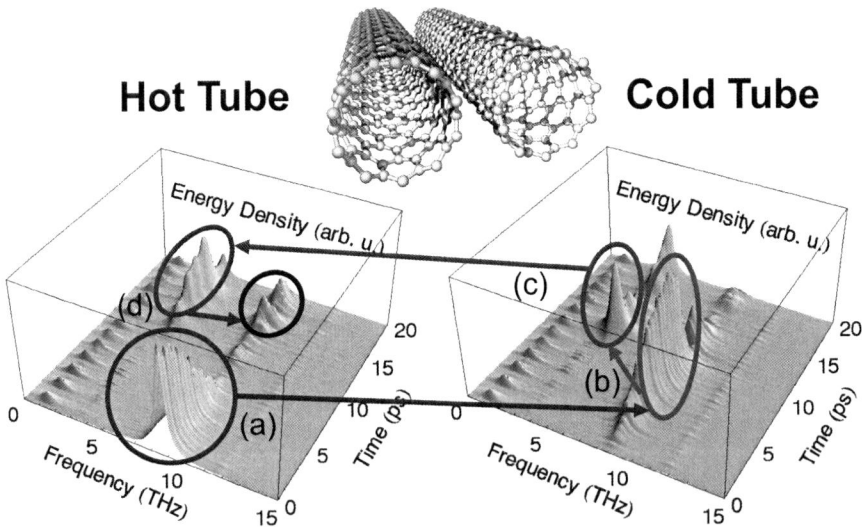

Figure 5.1 Plots showing the relaxation of the excited RBM in one of a pair of parallel carbon nanotubes. Four distinct energy transfer events are distinguishable: (a) hot tube resonantly exchanges energy from RBM into RBM in cold tube, (b) energy is scattered anharmonically within cold tube from RBM into a modes with half the frequency, (c) resonant exchange of energy from cold tube to hot between half frequency modes, and (d) anharmonic scattering within hot tube from half frequency mode to RBM.

mode, the potential energy is only estimated using the harmonic approximation. This inaccuracy can be used to our advantage. Comparing the over-(and/or under) estimation of the potential energy with the kinetic energy of a mode provides important information regarding the degree of anharmonicity in the mode, whether the anharmonic terms stiffen or soften the mode, and whether anharmonicity is asymmetric (as is the case for the radial breathing mode of a CNT). This information can be compared with and corroborated by computing the frequency shift of the mode relative to the frozen-phonon limit (computed from the power spectrum of the mode's velocity autocorrelation function). Comparing the amplitudes of a mode's potential and kinetic energy can also be used to identify nonresonant oscillations. The row of small peaks below 1 THz in the spectra of both the hot and cold tubes are an example of this. It is found that this oscillation has very little kinetic energy and is mostly comprised of a quasi-static elastic deformation of the low-frequency flattening modes of the tube. The excitation is artificial, and is not caused by these modes oscillating but by a much slower oscillation between the two tubes as a whole. The tubes "chatter" together under the influence of the van der Waals interaction. As the tubes come close together they flatten slightly (like a bouncing tennis ball) and it is the deformation of this flattening that gives rise to the serrated row of peaks.

Thus far, we have discussed the use of mode projection as a characterisation tool for interpreting the results of MD simulation; however, as mode projection gives information about the instantaneous state of the system there is no reason why this information cannot be fed back into the simulation and used to direct it. For example, knowing the eigenmodes of a system and using projection to determine their occupancy allows one to envision using external driving forces to regulate the energy in specially selected modes. This could be used for: continually driving a system away from equilibrium (as is done experimentally in driven nanomechanical resonators); enforcing a Bose–Einstein occupation of the classical modes; or even in schemes for efficiently searching phase space for rare events. While such procedures are not yet widely used they are being actively developed. Praprotnik *et al.* have shown that molecular dynamics can be performed efficiently in phonon space just as easily as in Cartesian space,[57] while Parinello *et al.*[58] have developed a Lengevin thermostatting algorithm that uses history-dependent (that is, correlated) noise to drive particular frequency modes without the need for mode projection. The continued development of these types of algorithms means that MD—for all of its classical shortcomings—will continue to be an important and powerful tool for studying nanoscale thermal behaviour both at equilibrium, and far from it.

5.4 Example: Heat Flow in a Nanoscale Material; Intrinsic Dissipation in CNT Resonators

Carbon nanotubes possess a number of properties that make them attractive for use as resonating members in many nanoscale devices. The use of CNT resonators has already been demonstrated as a radio tuner,[59] an entire radio receiver,[60] and a radio transmitter.[61] In addition, CNT resonators have been used to measure minuscule masses,[31] even down to the mass of a single Au atom,[32] and to approach the quantum limits of vibration.[62] In addition to these demonstrated applications the potential uses for CNT resonators are much broader, being applicable to any nanoscale device that requires controlled vibration, such as gyroscopes, mechanical processing of signals, and simple mechanical time keeping. The reason why CNTs are so suitable are multifold: CNTs' extraordinarily high stiffness combined with their low density enables them to attain very high natural frequencies, and frequency sensitivity. That CNTs are quasi-one-dimensional or string-like provides well-defined strategies for tuning their frequency—for example one may alter their length or put them under tension. Finally, CNTs can be both driven, and sensed, electronically by a number of different methods, making it possible to integrate CNTs into more complex nanoelectromechanical systems (NEMS) in which the resonator is only one part of the device.

To date, the biggest impediment to the use of CNTs as resonators is their very poor quality factor, Q, which when measured at ambient temperatures (and under conditions of constant driving) falls in the range of 8–300. (Q is defined to be 2π times the inverse fraction of oscillator energy lost per cycle, and may be thought of as roughly the number of oscillations it takes for the energy to be reduced by 99.8%.) This result has been resistant to improvement, having been observed both under vacuum and at ambient pressure; in both cantilevered and doubly clamped geometries with many different clamping methods; and through many different measurement techniques.[31,63–66] Only recently by cooling to milikelvin temperatures have Qs in excess of 10^5 been attained.[62] The universally poor ambient temperature results suggest that an intrinsic damping mechanism may be dominant. Macro- or mesoscopic theories of intrinsic damping from sources such as switching of defect states, thermoelastic damping, and phonon drag relate dissipative behaviour to the thermal energy in the system, that is, the background temperature T_{bg}. These theories have been successfully used to describe dissipation in some nanoscale systems, particularly those where phonons are diffusive and phonon lifetimes are shorter than the period of the resonator. Roukes *et al.*[67,68] and others[69] have suggested that the intrinsic thermoelastic damping mechanism is capable of producing very low Q factors in CNTs due to their very small surface-to-volume ratio—although other researchers disagree on the importance of this mechanism.[70] Previous computational work by Jiang *et al.*, of an open-ended, cantilevered CNT found $Q = 1500$ at 293 K, with the unexpected temperature dependence: $Q \sim T^{-0.36}$, although the authors did not explicitly identify an intrinsic damping mechanism.

The question of whether a poor quality factor is intrinsic is one into which *in silico* experiments can lend considerable insight. In the computer, one can simulate a carefully chosen idealised test system in which all extrinsic sources of dissipation have been removed. Simulating such a system it is possible to see: First, if intrinsic damping alone can account for the poor Q factors; and second, to identify the mechanisms of intrinsic damping with the goal of finding ways to mitigate them. We recently performed such a study that leads to the discovery of a new and surprising dissipation mechanism—one that can yield Mpemba-like behaviour in the cooling of an excited mode. The work is described here in order to serve as one example of how computational methods can be used to gain insight into nanoscale thermal phenomena.

The messy system of a doubly clamped suspended CNT resonator was represented in the idealised test system as a short section of a periodically repeated (along its axis) single-walled CNT, isolated in a vacuum and free from defects and mass impurities. This setup removes dissipation sources from defect migration, clamping friction, and gas damping. The effect of periodic boundary conditions is to add a clamping of sorts by restricting the number and wavelength of the CNT's flexural modes. The frequencies of flexural modes in the unclamped tube are higher than the modes of the same wavelength in the doubly clamped counterpart as there is no centre-of-mass motion. Instead of simulating the resonator under conditions of constant driving—as is the case for the experimentally measured Q—the resonator was simulated in the microcanonical ensemble as an initially excited flexural mode was allowed to the ring-down.

A typical simulation proceeded as follows: (1) The structure of a (10,0) CNT (simulated using the AIREBO potential for carbon–carbon interactions[71]) was relaxed, and the periodic repeat distance optimised.[vii] (2) the stiffness matrix for the system was computed and then diagonalised to yield the tube's eigenmodes, and their frequencies. (3) The tube was heated to a desired background temperature, T_{bg}, and allowed to equilibrate. (4) An instantaneous velocity was added to the system along *one* particular eigendirection (usually that of the second flexural mode) such that the total *average* temperature of the system is raised by the amount T_{ex}. (5) Ring-down was simulated (NVE) during which the vibrational energy distribution in all of the modes of the CNT is tracked using the mode-projection algorithm described above. The excitations of the flexural mode were large; however, despite the energetic excitations it was found that the mode remained reasonably harmonic, containing at most a 5% anharmonic contribution to the potential energy.

This study relies extensively on *in silico* experiments to investigate dissipation mechanisms. It is therefore imperative to ensure that the computational methods used are meaningful. It should be noted that the simulations were performed using classical molecular dynamics at temperatures well *below* the Debye temperature for the CNT. This is justified *a posteriori* by the finding that it is low-frequency modes that are participating in the dissipation mechanism. Simulating at low temperatures allows observation of the dissipation with little obfuscating thermal noise, and therefore is preferable if it can be physically justified. Leaving aside the issue of mode occupancy, a more fundamental problem is the use of classical mechanics to simulate the

---

[vii] The length of the tube was typically 8.4 nm, which it should be noted has an aspect ratio considerably lower than a typical NEMS resonator, and additionally possesses no residual axial tension. The short tube length was chosen to reduce the computational cost of the mode-projection scheme.

dissipation of an energy that is quantised. Yet, as a more suitable method that includes quantised dynamics is lacking, classical dynamics is used with the understanding that the physical interpretation of the results is limited. Similarly, interatomic forces were computed using empirical potentials that were formulated to capture the energetics of carbon and hydrogen bonding across a range of bonding coordinations. There is no reason to expect this to correctly represent the anharmonic character of the carbon–carbon bond in a CNT. Thus, it is necessary to verify the robustness of the reported simulation results to changes in interatomic potential. The results of the simulations *are* found to be sensitive to changes in the interatomic potential; however, the overall qualitative behaviour is not. This may be in part because the dissipative behaviour is due to the shapes of the low-frequency modes, which are largely dictated by the tubular geometry rather than the details of the potential. As the qualitative trends do not depend on how the simulations are performed one is justified to draw general and meaningful conclusions.

The ring-down curves for the second flexural mode of a 8.4-nm long (10,0) CNT, with a background temperature of $T_{bg} = 5$ K, and with increasing levels of initial excitation is shown in Figure 5.2(a). It can be seen that the attenuation of the oscillation follows a sigmoidal path, with larger initial excitations being completely damped in a shorter time than softer excitations. Figure 5.2(b) shows the ring-down profile for a 150 K excitation in tubes with increasing thermal background. As the background temperature rises the region of fastest attenuation is moved to earlier times.

To interpret the attenuation profiles in Figures 5.2(a) and (b) it is instructive to consider the attenuation of a simple damped harmonic oscillator with temperature-dependent dissipation. The equation of motion for such an oscillator is given by

$$\ddot{u} = -\omega^2 u - B(T_{bg})\dot{u}, \tag{5.10}$$

where u, ω, B, are respectively the oscillator's displacement, (undamped) frequency, and mass-weighted drag coefficient, with the overdot indicating the derivative with respect to time. If the period of oscillation is short in comparison to the attenuation time ($\omega \gg B/2$), then we may ignore the oscillatory behaviour, noting instead that the rate at which energy is lost to drag is proportional to twice the kinetic energy, and thus the total energy in the oscillator decays according to

$$\dot{E}(t) = -B(T_{bg})E(t). \tag{5.11}$$

As a first approximation, the drag term is assumed to take the form of a first-order Taylor expansion: $B(T_{bg}) = B_o + B'\Delta T_{bg}$, with B' positive. The increase in background temperature is simply the energy lost from the oscillator divided by C, the specific heat of the background, so that $\Delta T_{bg} = (E_o - E(t))/C$. Solving eqn (5.11) gives the attenuation profile,

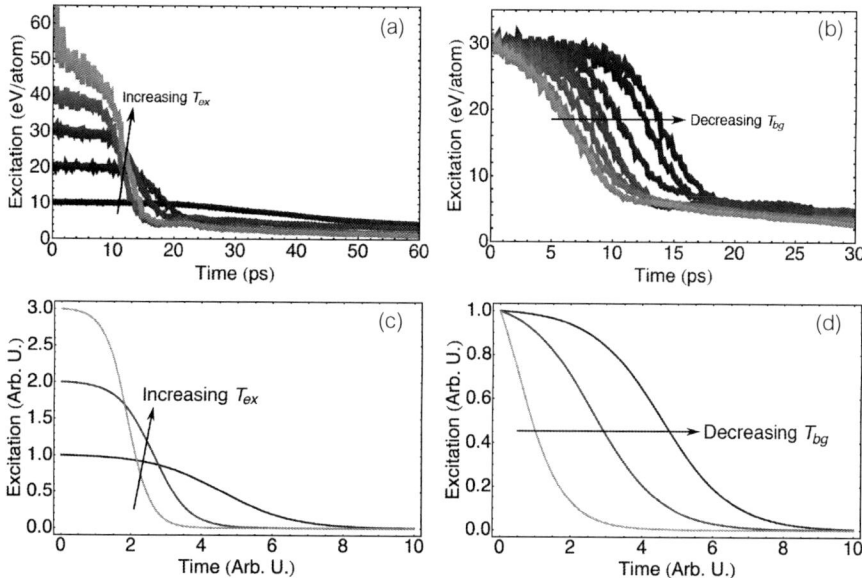

Figure 5.2 Plots (a) & (b) show the MD simulated ring-down profiles for the second flexural modes in 8.4 nm long (10,0) CNTs. In all cases the data is the average of 10 separate simulations with differing initial conditions; the data is plotted with a broad line thickness chosen such that it encloses the deviation of the averaged data. Plot (a) shows tubes with initial $T_{bg} = 5$ K and $T_{ex} = 50, 100, 150, 200,$ and 300 K. Plot (b) show the attenuation profile in tubes with initial $T_{ex} = 150$ K, and $T_{bg} = 5, 10, 50, 100, 150, 200, 400,$ and 400 K. Plots (c) & (d) show trends in simulation profile of a closed, damped harmonic oscillator (eqn (5.12)) when independently increasing T_{ex}, and T_{bg}, respectively.

$$E(t) = E_0 \frac{B_f}{B_f - B_i(1 - e^{B_f t})}, \qquad (5.12)$$

(plotted in Figures 5.2(c) and (d)) where $B_i = B_0$ is the initial damping coefficient, and $B_f = B_0 + B'E_0/C$ is the damping coefficient when the system is fully relaxed. This simple model displays many of the features of the molecular dynamics simulations in Figures 5.2(a) and (b), including an inflection in the attenuation profile (if the damping coefficient more than doubles as the system relaxes), as well as the trends that arise from independently increasing T_{ex} or T_{bg}. Using a simplified "toy" model such as this is often very instructive for interpreting simulation results. In this case the model that might have initially seemed as surprising sigmoidal attenuation of the resonator's ringing is entirely consistent with the computational setup. However, the model is also *overly* simple and there are features in the simulation results that the model cannot capture; still, by using the model as a starting point one can also learn where to look for new and interesting behaviour.

5.4.1 Mpemba-Like Behaviour

A consequence of simulating in the microcanonical ensemble is that the CNT is a closed system; energy dissipated from the excited flexural mode accumulates in the rest of the vibrational modes of the tube, raising T_{bg}. Increasing T_{ex} independently from T_{bg} changes the average temperature of the system and thus one would expect a different attenuation profile. More remarkable is the cooling of the excited mode in systems in which the same total energy, $T_t = T_{bg} + T_{ex}$, but differing *initial* partitioning of this energy between the flexural mode and the background (shown in Figure 5.3(a). Starting a simulation further from equilibrium (that is with larger initial T_{ex}/T_{bg}) causes the system to reach equilibrium in a shorter time! We refer to this astounding, and counterintuitive behaviour, as "Mpemba-like" in analogy with the Mpemba effect[72] in which it is observed that when hot water and cold water are put into a freezer the hot water freezes first.[viii]

While the dissipation mechanisms active during the cooling of water are very different from the damping within a CNT we will see that in both cases the system is able to reach equilibrium faster because both systems contain a hidden variable, and in both cases the cooling results in changes in the hidden variable that do not follow a unique pathway.

Figure 5.3(b) shows the ring-down time for increasing initial T_{ex} in systems for which $T_{ex} + T_{bg} = 300$ K. Two measures of the ring-down time are plotted: the time taken for excitation to be damped to a fixed lower threshold; and the interval over which the excitation is diminished by a set fraction—both measures decrease the further one starts from equilibrium. As with the Mpemba effect, relaxing faster to equilibrium the further one starts from it can only occur if the cooling pathway is not unique, but instead depends on the system's initial conditions—as illustrated by the inset plot in Figure 5.3(a).

From the ring-down profile of the simulated CNT resonator one can compute the Q factor at any time t as $Q(t) = -\omega E(t)/E(t)$, where ω is the

---

[viii]The Mpemba effect was and still is somewhat controversial. There are a number of differing explanations for the phenomenon that range from the practical to the fundamental. An example of the former is the hot water in the ice cube tray melts a layer of surface ice on the freezer shelf thus making better thermal contact. Just one example of a fundamental explanation of the Mpemba effect is that the cooling of the water is mediated by convection currents that have inertia and momentum. Once established in a hot liquid—with a large driving force they are maintained as the liquid cools. Cooling then depends on the average temperature T (which has no memory) and the convective flow C (which *is* dependent on the liquid's thermal history). The convective flow C provides the history-dependent hidden variable. The Mpemba effect is also controversial because of the manner in which it became popularised—through the observations of a Tanzanian school by George Mpemba (although the effect had been commented on much earlier by other natural philosophers, including Aristotle and Francis Bacon[73])—and how it seems to directly challenge *intuitive* scientific understanding. In fact the Mpemba effect makes no challenge to established science or the scientific method, it only holds a mirror to the way that we do science *in prac tice* and the barriers that we have for updating our personal intuitive scientific understanding when we are presented with more evidence. A good discussion of the Mpemba effect, and its history is given by Jeng.[73]

frequency of the excited mode.[ix] Figure 5.3(c) shows the change in the Q factor accompanying the attenuation of a CNT. The Q drops from close to 1900 by more than 95% to 37 and then is seen to recover towards 1000 after 20 ps. The origin of this huge suppression of the Q and its subsequent recovery can be understood by using the projection algorithm to track how the energy populates the background modes once it has been dissipated from the flexural mode. The evolution of the background population is shown in Figure 5.3(d). It can clearly be seen that the energy goes first into low-frequency modes—close to the frequency of the flexural mode—before dispersing across the full vibrational spectrum of the CNT. This nonuniform filling of the background modes is due to a small set of key "gateway modes" that act as strongly nonlinear channels for dissipation. These gateway modes and the role that they play is examined in more detail later in this section. However, without knowing how the gateway modes work it can be seen that the nonequilibrium distribution of energy in the background modes that they give rise to provides a hidden variable that is the necessary ingredient in for Mpemba-like behaviour.

The very simple model in eqn (5.12) does not exhibit the Mpemba-like behaviour that is observed in CNT's simulations. Reducing the initial T_{ex}/T_{bg} ratio in the model has the effect of starting the attenuation from further along the same universal ring-down pathway, and thus increasing the initial distance from equilibrium always results in longer cooling time. Additionally, in this model the Q-factor decreases monotonically in time and does not show the recovery that is seen in Figure 5.3(c).

Armed with the insight from the nonuniform filling of the thermal background gained from the mode projection (Figure 5.3(d)) one can construct a slightly more sophisticated model of the CNT damping process. Rather than assuming that the energy is dissipated into a single reservoir of background modes we can subdivide the background into two. One subreservoir, called the low-frequency background, is the set of modes that interact strongly (and nonlinearly) with the excited flexural mode. The remainder of the modes that interact less strongly make up the other thermal bath that is referred to as the high-frequency background. The strongly interacting group is referred to as the low-frequency background because in Figure 5.3(d) it can be seen that the modes that receive dissipated energy first have in general lower frequencies—although it is important to note that the frequency of the mode is of no importance for this model. The relaxation of this three-body model is now governed by three heat dissipation rates: The power dissipated from the excited mode into the low-frequency background, $p_{ex \to l}$, power dissipated from the flexural mode into the high-frequency background, $p_{ex \to h}$, and the rate of heat transfer between the low- and high-frequency backgrounds $p_{l \to h}$. To correctly couple the three thermal reservoirs in a manner that reaches the correct thermodynamic equilibrium a phenomenological form of the dissipation eqn (5.11) is modified such that

---

[ix] In practice, this is done by fitting a smoothing spline to the data in order to minimize the fluctuations in $\dot{E}(t)$.

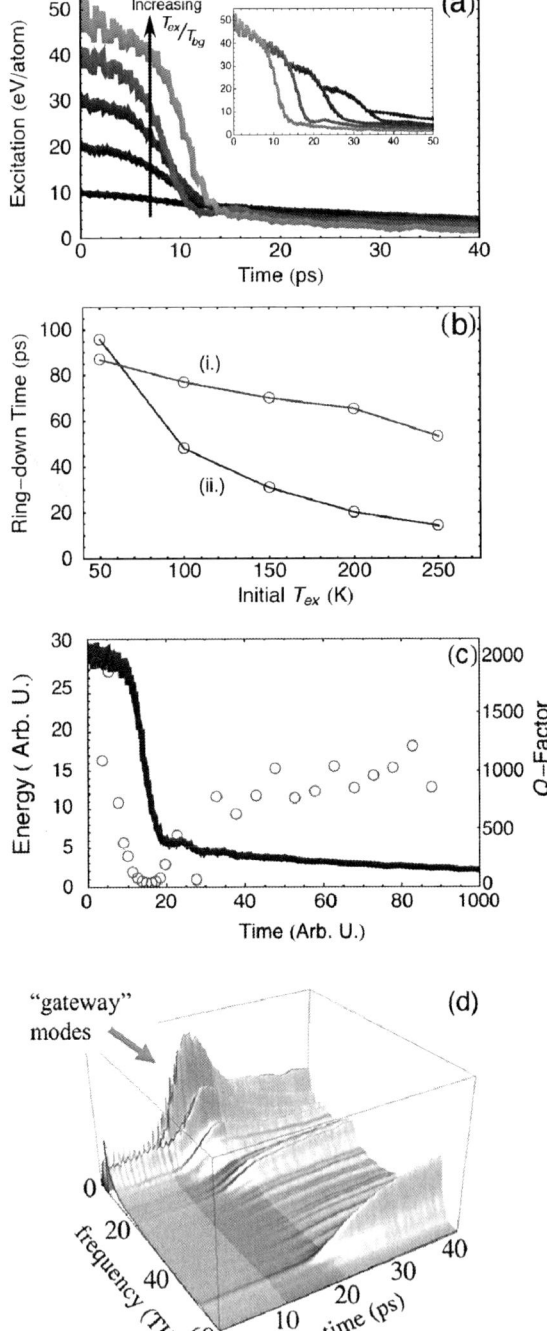

"gateway" modes

$$p_{\alpha \to \beta}(t) = -B_{\alpha\beta}(T_\beta)\left(\frac{T_\alpha}{C_\alpha} - \frac{T_\beta}{C_\beta}\right)C_{\text{tot}}, \tag{5.13}$$

where C_α and C_{tot} are specific heat of the individual reservoir α, and the total specific heat of all the reservoirs combined. The quantity $T_\alpha C_{\text{tot}}/C_\alpha$ then represents the temperature (that is the locally time averaged kinetic energy) of a mode within reservoir α. Note that this is different from T_α that was defined to represent the heat in the sets of modes in terms of its contribution to the temperature of the system as a whole.

This model in eqn (5.13) is solved numerically using as an initial condition that the temperature of the two background sets are the same. It is found that the extra degree of freedom in the system afforded by the two background reservoirs is sufficient to reproduce all of the features of the simulated CNT ring-down data—including the observation of Mpemba-like behaviour, and the recovery in Q. Figure 5.4(a) shows a relaxation profile of the model fit to the CNT attenuation profile from Figure 5.3(c), for which the 3 linear dissipative terms, one nonlinear dissipative term, and the number of modes in the low-frequency background were used as fitting parameters. Figure 5.4(b) uses the same model parameters as in (a) with differing $T_{\text{ex}}/T_{\text{bg}}$ ratios showing Mpemba-like behaviour.

The model plotted in Figures 5.4(a) and (b) is intended to be illustrative rather than predictive. It shows that there are two key ingredients that are needed in order to observe the Mpemba-like behaviour seen in the CNTs: nonlinear dissipation (caused by heating), and an internal degree of freedom (caused by heterogeneous heating of the background modes). For the fit in Figure 5.4(a) six fitting parameters were used, which is a lot, and one must not read too much meaning into them. There is, however, one parameter from which some insight can be gained: the number of modes in the low-frequency background. The tail in the ring-down profile occurs when the excited flexural mode comes into local

---

Figure 5.3   Plot (a) shows MD simulated total ring-down profiles (computed as for Fig. 5.2(a) and (b)) for the simulations in which the total energy $T_{\text{bg}} + T_{\text{ex}} = 300$ K but with different initial partitioning ratios, $T_{\text{ex}}/T_{\text{bg}}$. As with the Mpemba effect the mode cools faster if it starts hotter! The inset plot shows the cooling curves shifted in time so that the more weakly excited simulations commence on the cooling path for more strongly excited simulations. It can be clearly seen that there is no universal cooling trajectory. Total ring-down times plotted in (b), measured as the time taken to decay to an excitation of 1.5 eV/atom (i.), and time to lose 88% of initial energy (ii.). Plot (c) shows ring-down (solid line) overlaid with the Q-factor (circles) for initial $T_{\text{bg}} = 5$ K, $T_{\text{ex}} = 150$ K. The surface plot (b) shows how the dissipated energy from this simulation is distributed over the spectrum of the CNT's background vibrational modes. The region of rapid damping is marked by (ii.). It can clearly be seen that the dissipated energy does not reach the high-frequency background modes until the end of the period of anomalous dissipation.

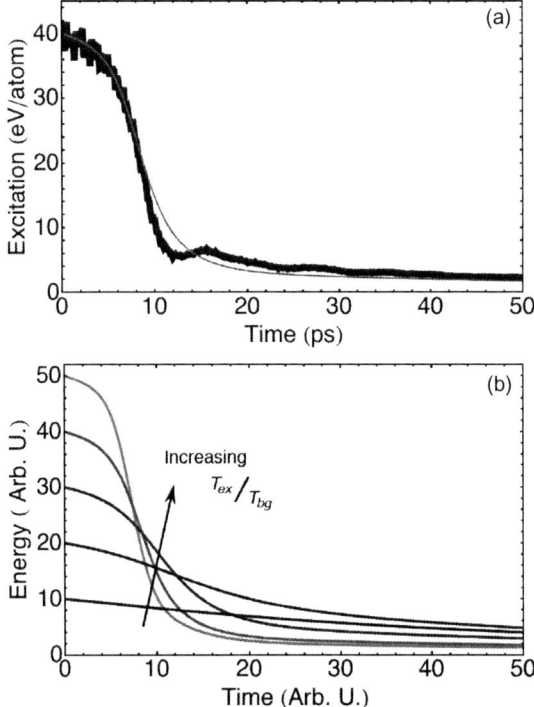

Figure 5.4 Attenuation profile predicted by the coupled dissipation model (eqn (5.13)). Plot (a) show the model (thin line) with parameters crudely fit to simulation data (thick line) for a CNT with initial $T_{bg} = 100$ K, and $T_{ex} = 200$ K. Plot (b) shows the model prediction (with the same fitting parameters) for the simulation data plotted in Figure 5.3(a). It is clear that the fit is far from perfect, however the major features remain. Most importantly, the model reproduces the Mpemba-like behaviour.

equilibrium with the low-frequency background, and the initial height of this tail in turn depends on the number of modes in this low frequency set—that is, the number of modes that interact strongly with the excited flexural mode. To obtain a good fit to the CNT simulation results this number must be between about 3 and 10 that agrees well with the finding from the mode projection that there are a small number of gateway modes that trigger rapid dissipation.

The Mpemba effect in the freezing of water is counterintuitive because one assumes that as hot water has cooled down it passes thought the same state as water that is initially cool. This is not the case. The rate of dissipation of heat from the cooling water depends not on the average temperature of the water but on a number of history-dependent hidden variables such as the temperature gradient and convective circulation. While the detailed origins of the Mpemba effect in water are not fully agreed upon it is clear that the effect is possible because of internal degrees of freedom within the system that are not described by the average temperature of the system. In this work it is demonstrated that an abstraction of the same phenomena is possible in other systems such as CNT resonators. It has been shown that: (1) the dissipation results in the formation a history-dependent athermal filling of vibrational modes that is not described by the average temperature, and (2) the dissipative state of the system is extremely sensitive to this athermal phonon population.

5.4.2 Gateway Modes

Having established the importance of gateway modes for Mpemba-like cooling it is worth examining the role that the gateway modes play. Again this can be done by taking advantage of the experiments being performed *in silico*.

Using the mode projection it is possible to identify which of the low-frequency modes were receiving the dissipated energy and within this subset of low-frequency modes it is found that there are two "gateway modes" that act as strongly nonlinear channels for dissipation. For the 8.4 nm (10,0) CNT these gateway modes are: a third flexural mode that is coplanar with the excited mode; and the fundamental bending mode, out-of-plane with the excitation. The energy in both of these modes, and the remainder of the energy in the background are shown in Figure 5.5(a). Once the gateway modes accumulate a little bit of energy they open efficient channels for rapid dissipation from the excited flexural mode. In the simulation it is possible to externally add a small excitation to a gateway mode. Exciting either of the gateway modes by as little as 1.5 K triggers the immediate onset of strong dissipation and suppression of the Q factor. Similarly, exciting the other low-frequency recipient modes had no effect.

That the gateway modes are the first and third flexural modes of the periodic CNT arouses suspicion that the observed gateway behaviour in the MD simulations is simply an artifact of the limited system size. It is necessary then to conduct two tests of the size dependence. The most trivial test is to repeat the study for CNTs with increasing periodic repeat distance to establish that gateway-mode dissipation is unique to neither (10,0) nor 4.2 nm periodically repeated tubes. A second, and more insightful, test of the size dependence on the gateway modes is to simulate a longer CNT in which the excited flexural mode still exists, but which is incommensurate with the wavelength of the two gateway modes.

Ring-down of the equivalent 1.5 THz flexural modes of 8.4 nm tubes was simulated in tubes 1.5, 2 and 3 times as long.[x] The attenuation of these modes is plotted in Figure 5.5(c). Both of the gateway modes in the 8.4-nm tube are incommensurate with the 12.5-nm tube and they do not exist. In this intermediate-sized tube the system finds a different but less-effective gateway mode (a flexural mode with wavelength 6.3 nm) to dissipate the energy. The longer 16.7-nm tube possesses all the vibrational modes of the 8.4-nm tube in addition to a further 2400 modes. However, despite the additional modes the gateway

---

[x] That is, modes with the same frequency as the third, fourth and sixth flexural modes of these tubes, respectively.

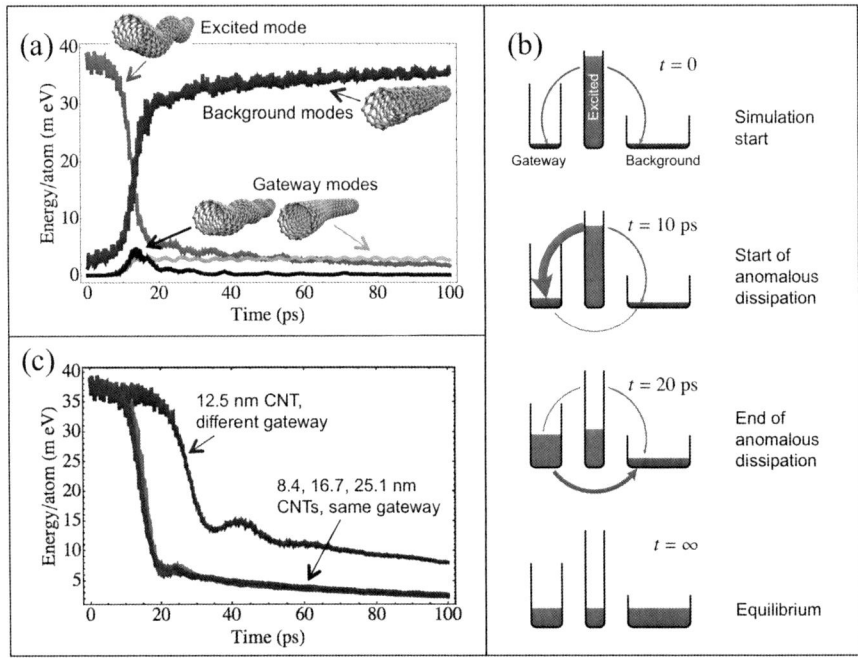

Figure 5.5   Plot (a) shows the energy in the excited flexural mode, the high and low frequency gateway modes, and the remainder of the background modes, during ring-down. Graphic (b) shows a schematic representation of the dissipation pathway and the role of the gateway modes in the CNT. The energy in the excited mode, the gateway modes, and the remaining background is represented by the filling of three receptacles, with the thickness of the arrows indicating the rate of energy transfer between receptacles. Plot (c) shows the ring-down of the 1.5 THz flexural mode in tubes with 8.4, 12.5, 16.7, and 25.1 nm periodic repeat distances.

modes of the 8.4-nm tube are the preferred path for dissipation. The longest 25.1-nm tube possesses all of the modes of the 12.5-nm and 16.7-nm tubes but the gateway modes of the 8.4-nm tube become active first. This hints that there could be many modes that can perform the gateway role; however, those that are most effective at it do so first, eliminating the need for other modes to act as dissipation gateways. Importantly, these show that the action of gateway modes is not an artifact of the computational cell size but does depend on which gateway modes are permitted by the periodic boundaries of the computational cell.

It is now possible to pull together the understanding of the Mpemba-like behaviour with the insight into the gateway modes into a cartoon model of the dissipation process in the CNT, shown schematically in Figure 5.5(b). Computer "experiments" were performed of the intrinsic dissipation in CNTs. For reasons of computational expedience, dissipation was observed during ring-down rather than under the experimentally measured conditions of constant

driving, however this approach revealed the formation of an athermal population of background phonons and a concomitant change in quality factor.

The consequence of these observations for a real NEMS device could be of
great practical importance. A real CNT resonator under vacuum can only lose
heat through its clamped ends. It is therefore very likely that under conditions
of constant driving the occupancy of phonon modes will be far from the
equilibrium distribution, and that this will in some way impact the qualify
factor of the tube. The implication is two-fold: (1) In order to properly
understand the *Q* factor of driven systems one must consider the whole phonon
population, (2) by understanding the full phonon population distribution we
can gain strategies for externally changing it and thus can engineer the dissipative properties of the system. Moreover, if it is found that an athermal
phonon affects the dissipative properties of other systems besides CNTs these
issues will also be relevant in other NEMS devices. If this is the case then the
gateway mechanism provides one with several avenues for improving *Q* factor.
Resonator lengths could be selected carefully so as to exclude important
gateway modes, or tubes could be decorated with other nanoparticles in order
to shift the frequency of gateway modes. Alternatively, gateway modes could be
identified and externally cooled (such as by laser cooling[33] or quantum back
action[74]). The phenomenon of triggering rapid damping externally by *adding*
more energy could find important applications such as in the ability to minimize unwanted vibrations in CNTs. It could be used for actively blocking heat
flow along CNTs and providing a way to mechanically filter thermal transport.

5.5 Example: Heat Flow Between Nanoscale Materials; Exploiting Frequency-Selective Thermal Transport for Chemical Sensing

At the beginning of this chapter it was discussed that exploiting far-from-
equilibrium conditions in nanoscale systems can permit us to take advantage of
heat that is athermally distributed across frequency—that is, using the properties of a select frequency band. In the survey of the computational "characterisation tools" the example was given of frequency-selective thermal
transport between weakly coupled objects. In this case study, it is shown how
this concept could be applied in a new approach to chemical sensing that could
be both highly sensitive *and* label-free. Simulation in combination with
numerical characterisation tools are used to provide a proof-of-concept, and
then to develop guidelines to help experimental researched in the development
of the method in practice.

Current strategies for sensing hazardous airborne chemicals, such as pollu-
tants or chemical-warfare agents, must compromise between focused sensitivity
and breadth of vigilance. Mechanically based chemical methods provide one
avenue to ultrasensitive detection. These methods detect binding of analyte to a
resonator by the accompanying change in surface stress or resonant frequency.
The sensitivity of this method increases inversely to the lengths scale of the

resonator, with nanoscale resonators able to detect an analyte in parts per trillion. The specificity of the method, however, is provided by the functionalised coating—and thus the flexibility of the approach to a rapidly evolving watch-list of chemical threats depends on how rapidly one can develop newly targeted chemical functonalisation. In contrast, spectroscopic methods for chemical sensing are "label-free", that is, they require no preconditioning in order to identify a given analyte. Unfortunately, they can only achieve satisfactory sensitivity by concentrating and segregating the analyte with a chromatography step.

As a route to chemical sensing that provides both sensitivity and label-free selectivity it has been proposed to exploit frequency-selecting thermal transport that can occur across weak nanoscale interfaces to probe analyte molecules in a way that is both mechanical and spectroscopic in nature. The method, referred to as nanomechanical spectroscopy (NRS),[75] uses an array of tuned nanomechanical resonators that become excited in the presence of a hot analyte with a particular vibrational frequency. (The graphics (a) and (b) in Figure 5.6 illustrate the NRS concept and its comparison to optical spectroscopy.) Taken together, the array of resonators can be likened to a stringed musical instrument; the vibrational spectrum of each analyte strikes a unique chord that can be used to identify it. Sensing the analyte is then reduced to resolving the relative excitations of the nanoresonator array. This task of "listening" to the strings is itself nontrivial; it requires being able to measure an excitation in a single vibrational mode of an ultrahigh-frequency resonators, with nanosecond resolution.

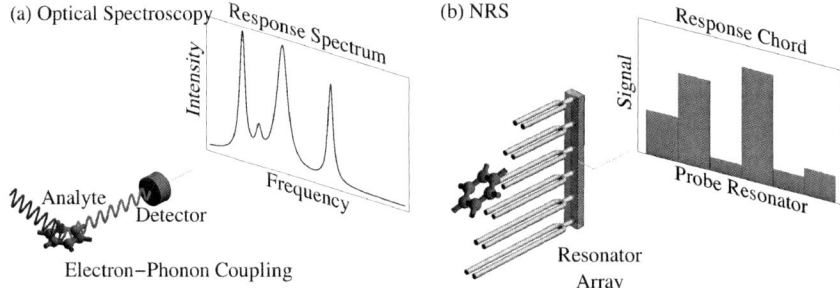

Figure 5.6   A schematic comparison of NRS with traditional optical spectroscopy. Panel (a) depicts in very general terms an optical spectroscopy method, in which the vibrational modes of an analyte are interrogated with a laser. Electron–phonon coupling transduces the illuminating radiation and the measured spectrum that results is used to identify the analyte. Panel (b) depicts the proposed NRS method in which a heated analyte interacts directly with a series of nanomechanical probe resonators. If a probe is in resonance with a vibrational mode in the analyte, then vibrational energy is exchanged; thus, the analyte excites a unique chord in the array of probes that can be used to identify it.

In order to demonstrate the NRS principle we turn again to an idealised test system that can be easily simulated with MD and that can be analysed unambiguously. The simplest possible analyte is used, namely an H_2 dimer. This has only *one* vibrational mode—the bond stretch—and the frequency of this mode is tuned by making the H atoms artificially heavy. For the probe resonator a (10,0) single-walled (and periodically repeated) CNT is used. This allows us to efficiently bring all of the characterisation apparatus described above to bear on this problem. To start, the dimer is positioned randomly at the CNT surface and given a large ($\sim$1100 K) excitation of its bond stretch (as shown in Figure 5.7). The CNT is initially at absolute zero. (Classical simulation of this system setup is justified by the same arguments used for the case of two parallel CNTs above). The system is simulated in the microcanonical ensemble. The analyte (dimer) and probe (CNT) are weakly coupled *via* the van der Waals forces, and this mediates the transfer of energy from the dimer to the tube. The weak binding also permits a slow bouncing oscillation of the dimer on the tube surface, causing the dimer to migrate randomly along the tube. Figure 5.8 shows the evolution of the spectrum of vibrations excited in a CNT that is in contact with a dimer tuned to oscillate at 10 THz.

It can clearly be seen from Figure 5.8 that for a 10-THz analyte most of the vibrational energy that is transferred to the CNT occupies modes with frequencies closest to 10 THz, *i.e.* modes in resonance with the dimer. This demonstrates that information about the vibrational frequency of the analyte is communicated to the probe, and remains localised in frequency for an experimentally measurable duration. The dissipation of energy in the excited mode in the tube is small, as the tube was initially at 0 K. Simulations in which the tube has an initial background temperature show the same qualitative behaviour although with increased dissipation from the resonantly excited

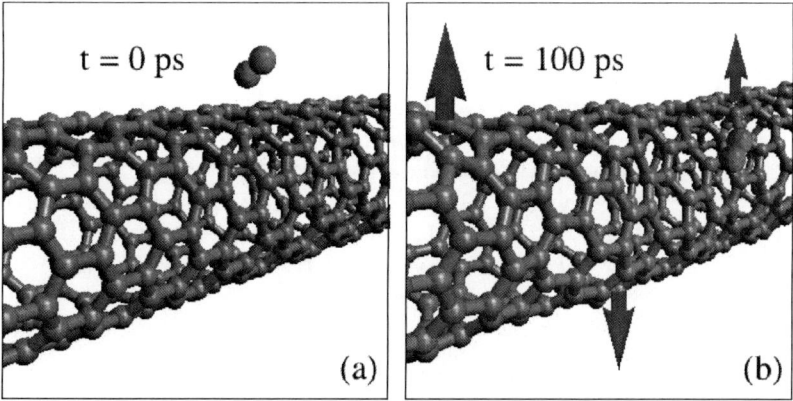

Figure 5.7 Typical simulation setup. Panels (a) and (b) show respectively the initial and final configurations of the system. In panel (b) the internal displacements within the CNT have been artificially amplified to show that it is a flexural mode in the tube that has been primarily activated.

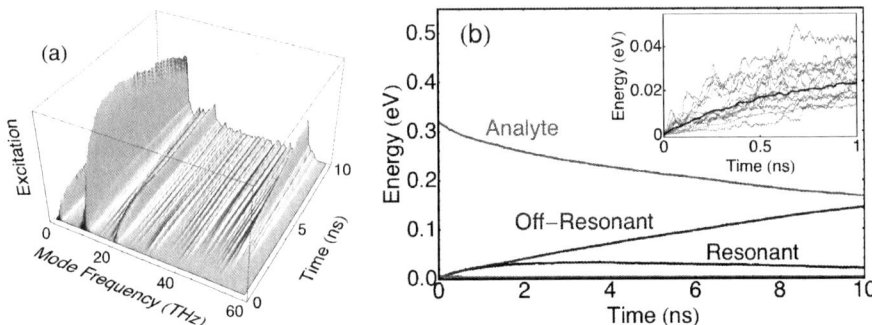

Figure 5.8 Mean excitation of CNT due to interaction with a single dimer vibrating at 10 THz. The plots are the average of 29 simulations. Plot (a) shows the full excitation spectrum in the CNT. A clear resonant peak is visible at 10 THz. Plot (b) shows the partitioning of the thermal energy between the stretch mode of the analyte, and the resonant modes and background models in the CNT. The inset plot shows the first nanosecond of resonant signal for the first 15 individual simulations (thin lines) with the mean plotted in bold. It can be seen that the trajectories are stochastic, with energy repeatedly exchanging back and forth between the dimer and CNT.

mode. An interesting feature in Figure 5.8(a) is the small excitation of modes with twice the frequency of the dimer. This arises from the strong anharmonicity in the van der Waals interaction rectifying (frequency doubling) the force that the dimer exerts on the tube.

Figure 5.8(b) shows the partitioning of the total thermal energy in the system between the dimer's stretch mode, and the resonant and off-resonant modes in the CNT. It can be seen that in the 10 ns simulated less than half of the energy in the dimer's stretch has transferred to the CNT, and very little energy is transmitted into the dimer's rotational and translational modes (due to the atomic mass of the dimer being relatively large). The set of resonant modes in the CNT was chosen to be those within 0.4 THz of the dimer frequency—a window just wide enough to enclose the full resonant ridge along 10 THz in Figure 5.8(a). The plotted "resonant signal" is the summed energy of the modes in this window—effectively a narrow-bandpass filtering of the full vibrational spectrum. The inset plot shows individual heating trajectories of the resonant modes over the first nanosecond for 15 of the simulations. It can be seen that there is a wide distribution in the trajectories but also that each individual trajectory follows a rapidly fluctuating stochastic pathway in which heat is both added and removed randomly, with only a gradual accumulation of energy.

Averaging a portion of the spectrum in Figure 5.8(a) at early times between 90–100 ps gives a measure of the response of the CNT resonance probe to a 10-THz analyte vibration that can be compared to the response due to analytes of different frequencies. By repeating the simulations with dimers of different frequencies one obtains a map of the response of the probe to analytes spanning its entire frequency range, the first 20 THz of which is shown in Figure 5.9. The fact that the CNT response surface is excited primarily along the diagonal in this plot shows that the excitation of the probe remains resonant—that is, sharply localised at the dimer frequency—for all analyte frequencies. This result represents a critical proof of concept important for a functional NRS: it ensures that there is a one-to-one mapping between the frequency of the measured probe signal and the vibration in the analyte that caused it. There are no strong off-resonant signals that could cause a false positive. Moreover, the presence of the analyte does not strongly alter the frequency of the modes of the CNT— which would blur the frequency specificity of the probe. Unlike mass-sensing approaches that rely on the analyte binding causing a frequency shift in the resonator, here the analyte is only very weakly bound to the probe through the van der Waals interaction, which vitiates the effects of mass loading.

Having established that the principle of the NRS method works one can examine the simulations more closely to glean information that will help optimize the NRS device performance. First, the simulations give a picture of the analyte moving randomly on the surface of the CNT rapidly exchanging energy back and forth with the CNT. We can use this picture as the basis for a simple mathematical model of stochastic energy transfer. This model predicts that one can maximize the rate of energy transfer and narrow the resonant frequency band if one can engineer the strength of interaction between the dimer and CNT and the frequency of their bouncing. Secondly, in Figure 5.8(b) it can be seen that both the resonant and off-resonant modes of the CNT become excited. An obvious question is: are the background modes of the CNT excited only due to dissipation from the resonant modes or are they also excited directly by the analyte (and if so is there anything that we can do to prevent it?). Here, we can take advantage of the noise due to the constant random shuttling of energy. It is found that the noise in the analyte and the noise in the background modes have very different frequency components. Moreover, it is found that the noise in the resonant mode is closely anticorrelated to both the analyte and background, but there is no correlation between the analyte's noise and the background modes. In other words, the resonant channel for energy transfer is extremely efficient and almost all the energy transferred into the CNT is mediated by resonant exchange—there is little improvement to be made here. Finally it can be seen from the plot in Figure 5.9 that the resonant signal measured in the CNT is stronger for lower frequencies. The CNT is effectively self-filtering for the low-frequency signal. This property can be used advantageously and can help set some guidelines for the realisation of a working NRS device. The most sensitive sampling frequency of each probe resonator is its lowest-frequency mode; thus one should design the spectroscope so that one can monitor the occupation of this mode in each of the probes in the array. This one guideline vastly simplifies the process of selecting nanoscale objects to act as probe resonators, by choosing families of objects with controllable geometries that change the fundamental frequency. Several examples of this are fullerenes where the radius of the fullerene determines the frequency of its breathing and flattening modes or the fundamental mode of a suspended or

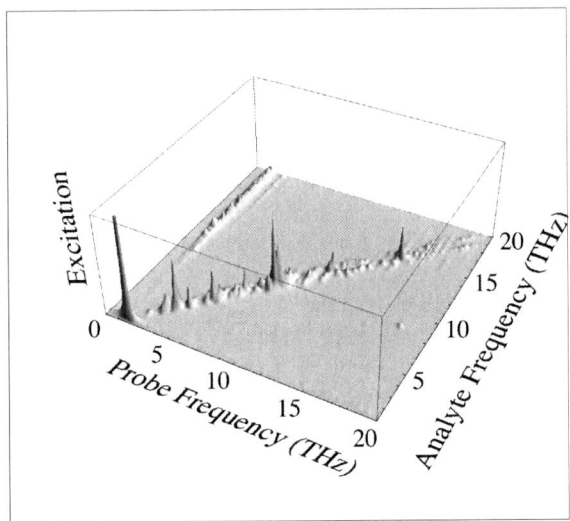

Figure 5.9 Scan of the excitation response of a CNT nanomechanical resonance probe to molecular vibrations with frequencies over the entire frequency range of the CNT. Only the first 20 THz of the scan are plotted as the response of the probe at higher frequencies was negligible. The response spectrum of the probe for each analyte frequency is an average of the excited spectrum between 90 to 100 ps of the simulation.

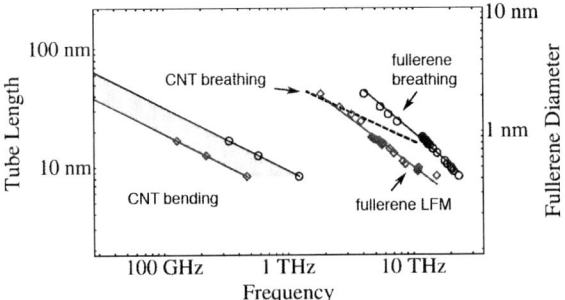

Figure 5.10 Mode frequency for families of nanoscale objects. The data plotted at high frequencies show the frequency of the lowest frequency mode and the breathing mode for fullerenes as a function of their diameter. The dashed line shows the frequency of the radial breathing mode for CNTs as a function of their diameter (with data plotted using Eklund and Dresselhaus fitting formula $\omega\,[\mathrm{cm}^{-1}] = 244/d^2\,[\mathrm{nm}]^{76}$). The shaded region plotted at lower frequencies shows the frequencies of the lowest frequency flexural mode of a single walled CNTs as a function of length for a range of stable chiralities and diameters. The mode frequencies for both the fullerenes and CNTs were calculated using the REBO potential.

cantilevered nanowire or nanotube, determined by length and radius. By way of example, the lowest-frequency modes for fullerenes and the flexural modes of CNTs are plotted as a function of their tuning dimension in Figure 5.10. It can be seen that together these modes span nearly 3 decades in frequency space and thus, if used as nanomechanical resonance probes for NRS, could potentially permit detection of a wide range of molecular vibrations.

In this case study we have shown that lots of subtle information can be learned from MD simulations of even the simplest systems. However, we must beware not to push the interpretation too far. We have shown that the resonant effect is significant, but in the simulation the transfer of energy is *continuous* not by discrete jumps as is the case for a quantum system. In this sense we have made progress to the extent possible for a *classical* MD approach, but there is much that can be learned if we could correctly simulate quantised energy transfer. There is much work still to be done in this field, and much to be understood by doing so.

5.6 Conclusions and Outlook

In this chapter we have explored different approaches to simulating thermal behavior in nanoscale systems, and the relationship between simulation and direct computation. The chapter started with a survey of experimentally demonstrated thermal properties that can be found in nanoscale systems. Making use of nanoscale structured systems has the potential to radically alter the way that we both use, and think about heat. The fundamentals of thermodynamics were solved during the industrial revolution in the era of the steam engine but heat remains as technologically important now as it was then. Moving to the nanoscale allows us to control heat in unprecedented ways, permitting engineering of phonon band structure, and manipulation of structure at the length scale of the phonon mean free path. Achieving the full potential of heat at the nanoscale requires not only devising novel ways of exploiting heat but also advancing our fundamental understanding of thermal energy in nanoscale systems, and developing the computational tools required to study it. As we hope this chapter has demonstrated, this is an exciting field, in which there is much work to be done.

Acknowledgements

This work was supported by the Defense Threat Reduction Agency—Joint Science and Technology Office for Chemical and Biological Defense (Grant HDTRA1-09-1-0006).

References

1. I. Pomeranchuk, *Physical Review*, 1941, **60**, 820.
2. N. Mingo and D. A. Broido, *Nano Letters*, 1005, **7**, 1221–1225.

3. S. Berber, Y.-K. Kwon and D. Tománek, *Physical Review Letters*, 2000, **84**, 4613–4616.

4. C. W. Chang, D. Okawa, H. Garcia, A. Majumdar and A. Zettl, *Physical Review Letters*, 2008, **101**, 075903.

5. C. Yu, L. Shi, Z. Yao, D. Li and A. Majumdar, *Nano Letters*, 2005, **5**.

6. D. Donadio and G. Galli, *Physical Review Letters*, 2007, **99**, 255502.

7. Philip Kim, L. Shi, A. Majumdar and P. L. McEuen, *Physica B*, 2002, **323**, 67–70.

8. W. Choi *et al.*, *Nature Mater.*, 2010, **9**, 423–429.

9. T. Tong *et al.*, *IEEE Transactions on Components and Packing Technologies*, 2007, **30**, 92–100.

10. S.T. Huxtable *et al. Nature Materials*, 2003, **2**, 731.

11. W. R. Kima and A. Majumdara, *Nanotoday*, 2007, **2**, 40–47.

12. B. Poudel *et al.*, *Science*, 2008, **320**, 634–638.

13. J.-H. Lee, G. A. Galli and J. C. Grossman, *Nano Letters*, 1008, **8**, 3750–3754.

14. He, Y., Donadio, D., Lee, J.-H., Grossman, J. C., Galli, G. Unpublished.

15. L. Shi *et al.*, *Journal of Heat Transfer*, 2003, **125**, 881–888.

16. L. Shi *et al.*, *Journal of Applied Physics*, 2009, **105**, 104306.

17. D. Li *et al.*, *Applied Physics Letters*, 2003, **83**, 2934–2937.

18. D. G. Cahill, K. Goodson and A. Majumdar, *Journal of Heat Transfer*, 2002, **124**, 223–241.

19. Y. K. Koh and D. G. Cahill, *Physical Review B*, 2007, **76**, 75207.

20. Gao, G., Çağin, T. and III, W. A. G. Nanotechnology 9, 184–191 (1998).

21. Y.-K. Kwon, S. Berber and D. Tomanek, *Physical Review Letters*, Jan 2004, **92**(1), 015901.

22. G. Cao, X. Chen and J. W. Kysar, *Journal of the Mechanics and Physics of Solids*, 2006, **54**(6), 1206–1236.

23. S. Lepri, R. Livi and A. Politi, *Physical Review Letters*, Mar 1997, **78**(10), 1896–1899.

24. O. Narayan and S. Ramaswamy, *Physical Review Letters*, Oct 2002, **89**(20), 200601.

25. K. Schwab, E. Henriksen, J. Worlock and M. Roukes, *Nature*, 2000, **404**, 974–977.

26. K. C. Schwab and M. L. Roukes, *Physics Today*, 2005, **58**(7), 36–42.

27. Z. Yao, C. L. Kane and C. Dekker, *Physical Review Letters*, Mar 2000, **84**(13), 2941–2944.

28. M. Steiner *et al.*, *Nature Nanotechnology*, 2009, **4**, 320–323.

29. M. Lazzeri and F. Mauri, *Physical Review B.*, Apr 2006, **73**(16), 165419.

30. N. Vandecasteele, M. Lazzeri and F. Mauri, *Physical Review Letters*, May 2009, **102**(19), 196801.

31. B. Lassagne, D. Garcia-Sanchez, A. Aguasca and A. Bachtold, *Nano Letters*, 2008, **8**, 3735–3738.

32. K. Jensen, K. Kim and A. Zettl, *Nature Nanotechnology*, 2008, **3**.
33. A. Schliesser, P. Del'Haye, N. Hooshi, K. J. Vahala and T. J. Kippenberg, *Physical Review Letters*, 2006, **97**, 243905.
34. P. A. Greaney, G. Lani, G. Cicero and J. C. Grossman, *Nano Letters*, 2009, **9**, 3699–3703.
35. Greaney, P. A., Lani, G., Cicero, G. and Grossman, J. C. Metallurgical Transactions. under review.
36. P. A. Greaney and J. C. Grossman, *Physical Review Letters*, 2007, **98**, 125503.
37. H. Zhong and J. R. Lukes, *Physical Review B.*, Sep 2006, **74**(12), 125403.
38. Z. Xu and M. J. Buehler, *ACS Nano*, 2009, **3**, 2767.
39. K. Moritsugu, O. Miyashita and A. Kidera, *Physical Review Letters*, 2000, **85**, 3970.
40. X. Yu and D. M. Leitner, *The Journal of Physical Chemistry B.*, 2003, **107**, 1698–1707.
41. E. C. Dykeman and O. F. Sankey, *Journal of Physics: Condensed Matter*, 2009, **21**, 505102.
42. K.-T. Tsen *et al.*, *Journal of Biomedical Optics*, 2009, **14**, 064042–064051.
43. P. B. Allen and J. L. Feldman, *Physical Review B*, Nov 1993, **48**(17), 12581–12588.
44. F. Müller-Plathe, *The Journal of Chemical Physics*, 1997, **106**(14), 6082–6085.
45. F. Müller-Plathe, *Phys. Rev. E*, May 1999, **59**(5), 4894–4898.
46. R. Kubo, M. Yokota and S. Nakajima, *Journal of the Physical Society of Japan*, 1957, **12**(11), 1203–1211.
47. Che, J., Çagin, T., Deng, W. and III, W. A. G. The Journal of Chemical Physics 113(16), 6888–6900 (2000).
48. S. R. Phillpot, P. K. Schelling and P. Keblinski, *Journal of Materials Science*, 2004, **40**, 3143–3148.
49. E. C. Dykeman and O. F. Sankey, *Physics Review Letters*, 2008, **100**, 028101.
50. J. L. Feldman, M. D. Kluge, P. B. Allen and F. Wooten, *Physical Review B*, Nov 1993, **48**(17), 12589–12602.
51. S. N. Taraskin and S. R. Elliott, *Physical Review B*, May 2000, **61**(18), 12031–12037.
52. D. Donadio and G. Galli, *Nano Letters*, PMID: 20163124 2010, **10**(3), 847–851.
53. D. M. Leitner, *Annual Review of Physical Chemistry*, 2008, **59**(1), 233–259.
54. A. J. H. McGaughey and M. Kaviany, *Physical Review B*, Mar 2004, **69**(9), 094303.
55. A. Bodapati, P. K. Schelling, S. R. Phillpot and P. Keblinski, *Physical Review B*, 2006, **74**(24), 245207.
56. K. Moritsugu, O. Miyashita and A. Kidera, *Physical Review Letters*, Oct 2000, **85**(18), 3970–3973.

57. M. Praprotnik and D. Janežiè, *Journal of Chemical Information and Modeling*, 2005, **45**(6), 1571–1579.
58. G. Bussi, D. Donadio and M. Parrinello, *The Journal of Chemical Physics*, 2007, **126**(1), 014101.
59. C. Rutherglen and P. Burke, *Nano Letters*, 2007, **7**, 3296–3299.
60. K. Jensen, J. Weldon, H. Garcia and A. Zettl, *Nano Letters*, 2007, **7**(11), 3508–3511.
61. J. Weldon, K. Jensen and A. Zettl, *Physical Status Solidi (B)*, 2008, **245**, 2323–2325.
62. G.A. Steele *et al.*, *Science*, 2009, **325**(5944), 1103–1107.
63. V. Sazonova *et al.*, *Nature*, 2004, **431**, 284–287.
64. D. Garcia-Sanches *et al.*, *Physical Review Letters*, 2007, **99**, 085501.
65. H. B. Peng, C. W. Chang, S. Aloni, T. D. Yuzvinski and A. Zettl, *Physical Review Letters*, 2006, **97**, 087203.
66. B. Witkamp, M. Poot and H. S. J. van der Zant, *Nano Letters*, 2006, **6**, 2904–2908.
67. R. Lifshitz and M. L. Roukes, *Physical Review B*, 2000, **61**, 5600–5608.
68. K. L. Ekinci and M. L. Roukes, *Review of Scientific Instruments*, 2005, **76**(6), 061101.
69. K. Jensen, H. B. Peng, A. Zettl, in: *Nanoscience and Nanotechnology*, (IEEE, ed.), (2006). Cat Number 06EX411C.
70. C. Seoánez, F. Guinea and A. H. C. Neto, *Physical Review B*, 2007, **76**, 125427.
71. S. J. Stuart, A. B. Tutein and J. A. Harrison, *Journal of Chemical Physics*, 2000, **112**, 6472–86.
72. E. B. Mpemba and D. G. Osborne, *Physics Education*, 1969, 172–175.
73. M. Jeng, *American Journal of Physics*, 2006, **74**, 514–522.
74. M. D. LaHaye, O. Buu, B. Camarota and K. C. Schwab, *Science*, 2004, **304**, 74–77.
75. P. A. Greaney and J. C. Grossman, *Nano Letters*, PMID: 18707177 2008, **8**(9), 2648–2652.
76. M. S. Dresselhaus and P. C. Eklund, *Advances in Physics*, 2000, **49**(6), 705–814.
77. J. A. Morrone and R. Car, *Phys. Rev. Lett.*, 2008, **101**, 017801.

CHAPTER 6

Computational Electrodynamics Methods

NADINE HARRIS,[1] LOGAN K. AUSMAN,[1]
JEFFREY M. McMAHON,[1,2] DAVID J. MASIELLO[3]
AND GEORGE C. SCHATZ[1]

[1] Northwestern University, Department of Chemistry, International Institute for Nanotechnology, 2145 Sheridan Road, Evanston, Illinois, 60208-3113, USA; [2] Argonne National Laboratory, Center for Nanoscale Materials, 9700 S. Cass Avenue, Argonne, Illinois, 60439, USA; [3] University of Washington, Department of Chemistry, Box 351700, Seattle, Washington, 98195-1700, USA

6.1   Introduction

Noble-metal nanoparticles of varying geometries have stimulated considerable research interest owing to their unique optical properties. Nanospheres with diameters in the size range of a few to a few hundred nanometers have large absorption cross-sections that surpass their geometric cross-sections. This is the result of localised surface plasmon resonances (LSPRs), which are collective excitations of the conduction electrons of the nanosphere that are resonantly excited by coupling to the electric field of the incident light.[1,2]

It is now well established that the optical response of an individual nano-particle can be tuned by changing the dielectric of the surrounding medium, or even more dramatically by changing the nanoparticle structure itself.[3] This has prompted great interest in the synthesis of a wide variety of structures other than spheres, including shells,[4–7] rods[4–9] and triangular prisms,[4–9] to name a few, where the main aim is to create monodisperse nanoparticles having a LSPR at the

RSC Theoretical and Computational Chemistry Series No. 4
Computational Nanoscience
Edited by Elena Bichoutskaia
© Royal Society of Chemistry 2011
Published by the Royal Society of Chemistry, www.rsc.org

desired wavelength. Electrodynamic modelling has played an important role in this research, as it is possible to calculate the extinction and scattering properties of nanoparticles of essentially any size and shape with great precision by solving Maxwell's equations for the interaction of light with them. Other optical properties can also be modelled, including plasmon-enhanced fluorescence and absorption, which are of interest in medical diagnostics and solar-energy applications. In addition, the electromagnetic contribution to the enhancement factor that is operative in surface-enhanced Raman spectroscopy (SERS) can be determined by evaluation of the plasmon-enhanced electric fields evaluated at the positions of the molecules, which are usually at a nanoparticle's surface.

This chapter is dedicated to the description of several computational methods employed to model the optical responses of individual nanoparticles. The choice of technique is partly determined by the shape of the nanoparticle, as spherically symmetric ones can be described using analytical solutions based on Mie theory, while for others grid-based numerical techniques are often used, such as the finite-difference time-domain (FDTD) method and the discrete-dipole-approximation (DDA) method. Computational considerations and accuracy also play in this decision, and for some applications there are other methods that are preferred, such as the Whitney-form finite-element method, or the dyadic Green's function approach.

The Mie theory, FDTD and DDA methods are well known, and are particularly well suited for the properties we consider in this chapter, so a special focus herein will be on the use of these in applications that "go beyond" what is done with standard "off-the-shelf" codes. In the case of Mie theory, we describe how to use this theory to estimate the dipole-reradiation (DR) contribution to the SERS enhancement factor, which is a more rigorous expression for describing electromagnetic effects than has been done in the past with plane-wave scattering methods. For FDTD, we show how to incorporate a spatially nonlocal dielectric response into the theory so that it is possible to describe very small particles where conventional electrodynamics is no longer appropriate. Finally, for DDA we present a method for coupling electronic-structure theory for a molecule near a nanoparticle with the DDA description of the electrodynamics, thereby making it possible to study the influence of the nanoparticle on the optical properties of the molecule, including absorption and Raman scattering.

Section 6.2 describes the mathematical basis of Mie theory and then the extension of it to include DR effects into SERS. Section 6.3 provides a brief introduction to FDTD, with Section 6.4 detailing the inclusion of nonlocal effects. Finally, Section 6.5 describes DDA and Section 6.6 presents the use of it in a many-body Green's-function formalism to determine how molecular absorption is perturbed by the presence of a metal nanoparticle.

6.2 Classical Mie Theory and the Inclusion of Dipole Reradiation in SERS

Mie theory is the classical electrodynamics of light, represented as a plane wave, scattering from a spherical particle. Because of spherical symmetry, it is possible to separate variables in Maxwell's equations and thereby to generate analytical solutions that are expressed in terms of Bessel functions, Legendre polynomials and other functions. Mie theory is thoroughly described in many places.[1,10–11] The development below provides a brief introduction so that subsequent material concerned with Raman-intensity modelling can be understood.

6.2.1 Vector Spherical Harmonics

In Mie theory, the electric and magnetic fields associated with light scattering from a spherical particle are expanded in partial waves and then a system of linear equations is solved to enforce the appropriate boundary conditions. That is, the electric and magnetic fields are expressed in the form:

$$\mathbf{E} = \sum_i c_i^e \mathbf{M}_i^e, \tag{6.1}$$

$$\mathbf{H} = \sum_i c_i^h \mathbf{M}_i^h, \tag{6.2}$$

where c_i are coefficients and $\mathbf{M}_i$ are vector-basis functions. These basis functions are solutions of the vector Helmholtz equation:

$$\nabla^2 \mathbf{M}(\mathbf{r}) + k^2 \mathbf{M}(\mathbf{r}) = 0. \tag{6.3}$$

Since we are interested in solving the problem of electromagnetic scattering due to a spherical target, the solutions to this equation in spherical-polar coordinates will form the basis vector functions that we use for our expansion. In electrodynamics the magnitude of the wavevector (also called the wave number), k, is defined as $k = 2\pi\sqrt{\varepsilon\mu}/\lambda$. Here, $\mathbf{M}$ is a divergenceless vector that has the property[1]:

$$\mathbf{M}(\mathbf{r}) = \nabla \times \mathbf{r}u(\mathbf{r}), \tag{6.4}$$

and the solutions of $u(\mathbf{r})$ in the spherical-polar coordinate system have the form:

$$u_{lm}(\mathbf{r}) = z_l(kr)P_l^m(\cos\theta)e^{im\phi}; \quad \begin{matrix} l = 0, 1, 2, \dots \\ m = -l, -l+1, \dots, l-1, l \end{matrix}. \tag{6.5}$$

The radial solution, $z_l(kr)$, is a spherical Bessel function. The spherical Bessel functions of the first and second kinds are given as $j_l(kr)$ and $y_l(kr)$, respectively. The third important function for the purposes described here is the Bessel function of the third kind, also called the Hankel function, which is a linear combination of $j_l(kr)$ and $y_l(kr)$ such that:

$$h_l^{(1)}(kr) = j_l(kr) + iy_l(kr). \tag{6.6}$$

The two angular solutions are $P_l^m(\cos\theta)$, which is the associated Legendre function and $e^{im\phi}$. Together these are related to the scalar spherical harmonics by:

$$P_l^m(\cos\theta)e^{im\phi} = \sqrt{\frac{4\pi}{2l+1}\frac{(l+m)!}{(l-m)!}}\,Y_{lm}(\theta,\phi). \tag{6.7}$$

Since $\mathbf{M}_{lm}(\mathbf{r})$ does not contain a $\hat{\mathbf{r}}$ component, it does not span all space. In order to form a complete basis, it is necessary to construct another vector function independent from $\mathbf{M}_{lm}(\mathbf{r})$ that also satisfies the vector wave equation. This vector function is given as[1]:

$$\nabla \times \mathbf{M}_{lm}(\mathbf{r}) = k\mathbf{N}_{lm}(\mathbf{r}). \tag{6.8}$$

Using the vector spherical harmonics as basis functions, the incident electric field can be written:

$$\mathbf{E}_i(\mathbf{r},\omega) = \sum_{l=1}^{\infty}\sum_{m=-l}^{l}\left[P_{lm}\mathbf{N}_{lm}^{(1)}(k_0\mathbf{r}) + q_{lm}\mathbf{M}_{lm}^{(1)}(k_0\mathbf{r})\right], \tag{6.9}$$

where p_{lm} and q_{lm} are expansion coefficients. The subscript "0" on the wave number indicates that the field is in a medium with a dielectric function, ε_0, and a permeability, μ_0. The superscript "1" on the vector spherical harmonic functions indicates that the Bessel function of the first kind, $j_l(k_0r)$, is used. The coefficients are general but can be determined for the specific type of field such as plane-polarised or dipole.

The corresponding electric field inside the sphere is:

$$\mathbf{E}_1(\mathbf{r},\omega) = \sum_{l=1}^{\infty}\sum_{m=-l}^{l}\left[c_{lm}\mathbf{N}_{lm}^{(1)}(k_1\mathbf{r}) + d_{lm}\mathbf{M}_{lm}^{(1)}(k_1\mathbf{r})\right]. \tag{6.10}$$

Again, the radial variation of the vector spherical harmonics is due to the function $j_l(k_1r)$ in order to ensure that the field is finite at the origin. Here, the subscript "1" on the wave number indicates that the dielectric function, ε_1, and permeability, μ_1, are those of the material that makes up the sphere.

In a similar manner we can write an expression for the scattered field, but we now must concern ourselves with the behaviour of the field as it extends outward towards infinity instead of its behaviour at the origin. In this case, we replace the Bessel function j_l with $h_1^{(1)}(k_0r)$, and the scattered electric field is:

$$\mathbf{E}_s(\mathbf{r},\omega) = \sum_{l=1}^{\infty}\sum_{m=-l}^{l}\left[a_{lm}\mathbf{N}_{lm}^{(3)}(k_0\mathbf{r}) + b_{lm}\mathbf{M}_{lm}^{(3)}(k_0\mathbf{r})\right]. \tag{6.11}$$

The magnetic fields corresponding to all of these electric fields are obtained using:

$$\mathbf{H}(\mathbf{r}, \omega) = \frac{-i}{k}\sqrt{\frac{\varepsilon}{\mu}}\nabla \times \mathbf{E}(\mathbf{r}, \omega). \tag{6.12}$$

Due to eqn (6.8), the magnetic field expansions involve the switching of the vector spherical harmonic functions in eqns (6.9)–(6.11) along with multiplication by k.

Using eqns (6.9)–(6.11), along with the associated magnetic fields, and given the coefficients of the incident field, p_{lm} and q_{lm}, the four unknown expansion coefficients, a_{lm}, b_{lm}, c_{lm}, and d_{lm} can be found for a sphere of radius $r = a$ by solving a system of linear equations that enforce electromagnetic boundary conditions on the sphere surface and that also have the appropriate asymptotic form for a plane wave scattering from the sphere.[1] The resulting scattering coefficients, a_{lm} and b_{lm}, are:

$$a_{lm} = -\frac{\varepsilon_1 j_l(k_1 a)[k_0 a j_l(k_0 a)]' - \varepsilon_0 j_l(k_0 a)[k_1 a j_l(k_1 a)]'}{\varepsilon_1 j_l(k_1 a)[k_0 a h_l^{(1)}(k_0 a)]' - \varepsilon_0 h_l^{(1)}(k_0 a)k_1 a j_l(k_1 a)]'}p_{lm} \tag{6.13}$$

$$= \alpha_l p_{lm}$$

$$b_{lm} = -\frac{\mu_1 j_l(k_1 a)[k_0 a j_l(k_0 a)]' - \mu_0 j_l(k_0 a)[k_1 a j_l(k_1 a)]'}{\mu_1 j_l(k_1 a)[k_0 a h_l^{(1)}(k_0 a)]' - \mu_0 h_l^{(1)}(k_0 a)k_1 a j_l(k_1 a)]'}q_{lm}. \tag{6.14}$$

$$= \beta_l q_{lm}$$

We note that in most cases the magnetic permeability μ_1 is close to the vacuum value μ_0 and the result is that these are taken to be unity.

6.2.2 Plane-Polarised Electric-Field Expansion in Vector Spherical Harmonics

The expression for an electromagnetic plane wave of frequency ω propagating in the positive z-direction, with the electric field vector polarised in the x-direction is given as:

$$\mathbf{E}_i(\mathbf{r}; \omega) = E_0 e^{ikr\cos\theta}\hat{\mathbf{x}}, \tag{6.15}$$

where $\hat{\mathbf{x}}$ is a unit vector in the x-direction. Using this we can now obtain the incident coefficients, p_{lm} and q_{lm}. In order to calculate q_{lm}, we substitute eqn (6.15) into eqn (6.9), multiply both sides by $\mathbf{M}_{lm}^{(1)}(k_0\mathbf{r})^*$, and then rearrange to get[1]:

$$q_{lm} = \frac{\iiint \mathbf{M}_{lm}^{(1)}(k_0\mathbf{r})^* \cdot E_0 e^{ikr\cos\theta}\hat{\mathbf{x}}r^2 \sin\theta dr d\theta d\phi}{\iiint |\mathbf{M}_{lm}^{(1)}(k_0\mathbf{r})|^2 r^2 \sin\theta dr d\theta d\phi}. \tag{6.16}$$

The resulting coefficients that are only nonzero for $m = \pm1$ and:

$$q_{l1} = \frac{i^{l+1}}{2} \frac{2l+1}{l(l+1)} E_0 \tag{6.17}$$

$$q_{l(-1)} = \frac{i^{l+1}}{2}(2l+1)E_0. \tag{6.18}$$

The derivation of p_{lm} is similar only $\mathbf{M}_{lm}^{(1)}(k_0\mathbf{r})$ is replaced by $\mathbf{N}_{lm}^{(1)}(k_0\mathbf{r})$. As with the q_{lm}, the values of p_{lm} are only nonzero for $m = \pm1$ and are:

$$p_{l1} = \frac{i^{l+1}}{2} \frac{2l+1}{l(l+1)} E_0 \tag{6.19}$$

$$p_{l(-1)} = -\frac{i^{l+1}}{2}(2l+1)E_0. \tag{6.20}$$

For a plane wave with amplitude E_0 we have the scattering cross-section, C_{sca}, which is:

$$C_{\text{sca}} = \frac{4\pi}{k_0^2|E_0|^2} \sum_{l=1}^{\infty} \frac{l(l+1)}{2l+1} \left\{ l(l+1)\left(a_{l1}^2 + b_{l1}^2\right) \right. \tag{6.21}$$
$$\left. + \frac{1}{l(l+1)}\left(a_{l(-1)}^2 + b_{l(-1)}^2\right) \right\}.$$

Dividing eqn (6.21) by a sphere with cross-sectional area, πa^2, we obtain the scattering efficiency, Q_{sca}:

$$Q_{\text{sca}} = \frac{C_{\text{sca}}}{\pi a^2}. \tag{6.22}$$

The extinction cross-section C_{ext} can be obtained in a similar fashion and is:

$$C_{\text{ext}} = -\frac{4\pi}{k_0^2|E_0|^2} \sum_{l=1}^{\infty} \frac{l(l+1)}{2l+1} \text{Re}\left\{ l(l+1)(a_{l1}p_{l1}^* + b_{l1}q_{l1}^*) \right. \tag{6.23}$$
$$\left. + \frac{1}{l(l+1)}\left(a_{l(-1)}p_{l(-1)}^* + b_{l(-1)}q_{l(-1)}^*\right) \right\}$$

and the extinction efficiency, Q_{ext}, is:

$$Q_{\text{ext}} = \frac{C_{\text{ext}}}{\pi a^2}. \tag{6.24}$$

6.2.3 Dipole Electric-Field Expansion in Vector Spherical Harmonics

The incorporation of a dipole emitter within the framework of Mie theory, as is needed to describe fluorescence or SERS, requires that a dipole field be expressed in the form of vector spherical harmonics. In order to do this we need to find p_{lm} and q_{lm} as given in eqn (6.9) for a dipole field. Given these it is then possible to solve the boundary-value problem for optical scattering in a similar fashion to the plane-wave case. We begin by noting that the electric field of an oscillating induced dipole, $\mathbf{p}$, can be obtained by using the relation:

$$
\begin{aligned}
\mathbf{E}_{dip}(\mathbf{r};\omega) &= \frac{4\pi i \omega}{c^2} \int \mathbf{G}_0(\mathbf{r},\mathbf{r}';\omega) \cdot \mathbf{J}(\mathbf{r}';\omega) d^3 r' \\
&= \frac{4\pi k^2}{\varepsilon\mu} \int \mathbf{G}_0(\mathbf{r},\mathbf{r}';\omega) \cdot \mathbf{p}\delta(\mathbf{r}' - \mathbf{r}_0) d^3 r', \\
&= \frac{4\pi k^2}{\varepsilon\mu} \mathbf{G}_0(\mathbf{r},\mathbf{r}_0;\omega) \cdot \mathbf{p}
\end{aligned}
\tag{6.25}
$$

where in the above $\mathbf{G}_0$ is the free-space tensor Green's function. What remains to be found is an expression for $\mathbf{G}_0$ in a vector spherical harmonic basis. One way that this can be accomplished is *via* an eigenfunction expansion, also termed the Ohm–Rayleigh method, as shown by Tai.[12] The result is the closed form of $\mathbf{G}_0$, which is:

$$
\begin{aligned}
\mathbf{G}_0(\mathbf{r},\mathbf{r}_0;\omega) = &\frac{ik}{4\pi} \sum_{l=1}^{\infty} \sum_{m=-l}^{l} (-1)^m \frac{2l+1}{l(l+1)} \\
&\cdot \begin{cases} \left[\mathbf{N}^{(1)}_{l(-m)}(k\mathbf{r}_0)\mathbf{N}^{(3)}_{lm}(k\mathbf{r}) + \mathbf{M}^{(1)}_{l(-m)}(k\mathbf{r}_0)\mathbf{M}^{(3)}_{lm}(k\mathbf{r}) \right], r > r_0 \\ \left[\mathbf{N}^{(3)}_{l(-m)}(k\mathbf{r}_0)\mathbf{N}^{(1)}_{lm}(k\mathbf{r}) + \mathbf{M}^{(3)}_{l(-m)}(k\mathbf{r}_0)\mathbf{M}^{(1)}_{lm}(k\mathbf{r}) \right], r < r_0 \end{cases}
\end{aligned}
\tag{6.26}
$$

With this result and use of eqns (6.25) and (6.9) we obtain the following expressions for p_{lm} and q_{lm}. When $r > r_0$:

$$
p_{lm} = (-1)^m \frac{ik^3}{\varepsilon\mu} \frac{2l+1}{l(l+1)} \mathbf{N}^{(1)}_{l(-m)}(k\mathbf{r}_0) \cdot \mathbf{p}
\tag{6.27}
$$

$$
q_{lm} = (-1)^m \frac{ik^3}{\varepsilon\mu} \frac{2l+1}{l(l+1)} \mathbf{M}^{(1)}_{l(-m)}(k\mathbf{r}_0) \cdot \mathbf{p}
\tag{6.28}
$$

and when $r < r_0$:

$$
p_{lm} = (-1)^m \frac{ik^3}{\varepsilon\mu} \frac{2l+1}{l(l+1)} \mathbf{N}^{(3)}_{l(-m)}(k\mathbf{r}_0) \cdot \mathbf{p}
\tag{6.29}
$$

$$q_{lm} = (-1)^m \frac{ik^3}{\varepsilon\mu} \frac{2l+1}{l(l+1)} \mathbf{M}^{(3)}_{l(-m)}(k\mathbf{r}_0) \cdot \mathbf{p}. \qquad (6.30)$$

One of the main differences between the field coefficients for a dipole electric field and those of a plane-polarised electric field is that for a dipole electric field the coefficients may be nonzero for any value of m, while in plane-polarised fields the values of m may only be ± 1.

6.2.4 Electromagnetic Mechanism of Surface-Enhanced Raman Scattering

As an example of the use of Mie theory, we consider the evaluation of the electromagnetic enhancement factor for SERS. SERS has recently seen renewed interest due to advances in nanoparticle substrate preparation and functionalisation for SERS analytes[13,14] and improved technologies for Raman scattering. Since the initial discovery of SERS in the 1970s[15–17] there have been two proposed mechanisms for the scattering enhancement: the electromagnetic mechanism (EM) due to plasmon excitation and the chemical mechanism (CM)[21–25] that arises from charge-transfer interactions between molecule and substrate. It is generally assumed that the electromagnetic enhancement mechanism provides the majority of the SERS enhancement, but what is less well known is the rigorously correct expression for this enhancement factor.[18]

The EM enhancement arises because the electric field of the electromagnetic wave that is incident on a molecule is enhanced when the molecule is on the surface of a nanostructure compared to the same molecule in solution or vacuum. Usually, this mechanism is correlated with LSPR excitation in the nanoparticle.[11,16–18] and usually the EM contribution to the enhancement factor is taken to be $|\mathbf{E}_{loc}(\omega)|^2|\mathbf{E}_{loc}(\omega_s)|^2$, where $\mathbf{E}_{loc}$ is the electric field at the position of the molecule, and ω and ω_s are the incident and Stokes shifted frequencies, respectively. This expression defined what we will call the plane-wave (PW) approximation to the EM, as it assumes that the incident and scattered field enhancements are both derived from the scattering of plane waves from the metal nanostructure. However, Kerker used concepts derived from Mie theory[10] to show that although the incident photon enhancement factor is given by $|\mathbf{E}_{loc}(\omega)|^2$, the emitted photon enhancement factor should be derived from the Stokes frequency component of the emission by the dipole induced in the absorbed molecule. This emitted radiation interacts with the nanostructure, leading to plasmon enhancement in the scattered field. This phenomenon is referred to as dipole reradiation (DR) and it has been recently shown[19] that the DR/PW comparison can show factor of 2–3 differences for large particles.

The determination of the SERS enhancement factor within the context of Mie theory utilises the theories developed above by performing a plane-wave scattering calculation followed by a calculation to determine the DR contribution to the EM enhancement. We begin with a molecule located at $\mathbf{r}_0$ in the presence of a sphere. The sphere scatters an incident plane-polarised EM field and the resulting electric field that the molecule experiences is:

$$\mathbf{E}_{loc}(\mathbf{r}_0; \omega) = \mathbf{E}_i(\mathbf{r}_0; \omega) + \mathbf{E}_s(\mathbf{r}_0; \omega), \tag{6.31}$$

where the subscript "s" refers to the scattered field. This is then operated on by the Raman polarisability of the molecule to produce an oscillating dipole moment[20]:

$$\mathbf{p} = \boldsymbol{\alpha}_R \cdot \mathbf{E}_{loc}(\mathbf{r}_0; \omega). \tag{6.32}$$

This is the same $\mathbf{p}$ that was employed before. The molecule then re-emits radiation in the form of a dipole electric field at the Stokes-shifted frequency ω_s, and this field is scattered by the sphere to produce a secondary scattered electric field. The result is then measured at $\mathbf{r}$ in the far-field so that our SERS electric field is:

$$\mathbf{E}_{SERS}(\mathbf{r}; \omega_s) = \mathbf{E}_{dip}(\mathbf{r}; \omega_s) + \mathbf{E}_{s2}(\mathbf{r}; \omega_s). \tag{6.33}$$

In the absence of a sphere the Raman scattering is driven by the incident plane-polarised electric field, and the resulting Raman scattered field is:

$$\mathbf{E}_{RS}(\mathbf{r}; \omega_s) = \mathbf{E}_{dip,0}(\mathbf{r}; \omega_s). \tag{6.34}$$

The SERS enhancement factor is obtained by noting that the general form for a radiating field with an arbitrary order of multipoles is of the form[20]:

$$\mathbf{E} \approx \mathbf{F} \frac{e^{ikr}}{r}, \tag{6.35}$$

where $\mathbf{F}$ is the vector amplitude of the field. Therefore, we define the DR SERS enhancement factor as:

$$G(\mathbf{r}; \omega, \omega_s) = \frac{|\mathbf{F}_{SERS}(\mathbf{r}; \omega, \omega_s)|^2}{|\mathbf{F}_{RS}(\mathbf{r}; \omega, \omega_s)|^2}. \tag{6.36}$$

6.2.5 Results: Isolated Sphere Dipole Reradiation (DR) Effects

Here, we examine the SERS enhancement factor associated with a molecule interacting with a single Ag sphere. We pay particular attention to the relationship between the PW approximation and the result obtained by rigorously including DR. We assume that we are in the limit of zero Stokes shift so that the

PW is approximately $|\mathbf{E}_{loc}|^4$. While the enhancement factors due to single spheres are not large, these results will help highlight issues that arise when DR is included. In what follows we use the dielectric data for Ag as reported by Johnson and Christy.[21] We also mention that the incident electric field is assumed to be plane polarised in the $\hat{\mathbf{x}}$-direction and propagating in the $\hat{\mathbf{z}}$-direction, and the medium is taken to be vacuum. The molecule location is denoted by the coordinates r_0, θ_0 and ϕ_0 while the detector location is denoted by the coordinates r_s, θ_s and φ_s and depictions of these are shown in Figure 6.1.

The molecule is considered to be a classical point dipole with an isotropic response. While there are likely to be many interesting effects due to different responses, the choice of an isotropic α_R is done for simplicity and we note that other responses could be considered. Additionally a quantum molecular response can also be included by incorporating the many-body theory of SERS,[22] which will be described in Section 6.6.

In Figure 6.2 we present EM enhancement factors for a 100-nm diameter isolated sphere.

The molecule is located along the x-axis (parallel to $\mathbf{E}_i$) 0.5 nm from the surface of the sphere. We choose two different detector locations: (1) a configuration where $\varphi_s = 90°$ and (2) an off-axis direction where $\varphi_s = 45°$. In both cases $\theta_s = 90°$. These directions are somewhat arbitrarily chosen but are typical of directions used for SERS measurements.

An examination of Figure 6.2 shows that the maximum values of the PW enhancement factor (dotted line) and the $\varphi_s = 90°$ DR enhancement factor (dashed line) occur at $\lambda = 408$ nm and are $\sim 2.78 \times 10^3$ and $\sim 2.74 \times 10^3$, respectively, corresponding to an error of 1.5%. When the observation point is moved to $\varphi_s = 45°$ (dash-dot line) the maximum G value decreases to 1.52×10^3.

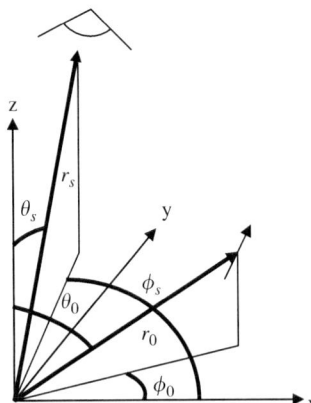

Figure 6.1   Scheme that shows the relevant geometries associated with the molecule locations (r_0, θ_0, ϕ_0) and detector locations (r_s, θ_s, φ_s). Note that the wavevector is along the z-axis and the polarisation is along x. Reprinted with permission from: L. K. Ausman and G. C. Schatz, *J. Chem. Phys.* 2009, **131**, 084706. Copyright 2009, American Institute of Physics.

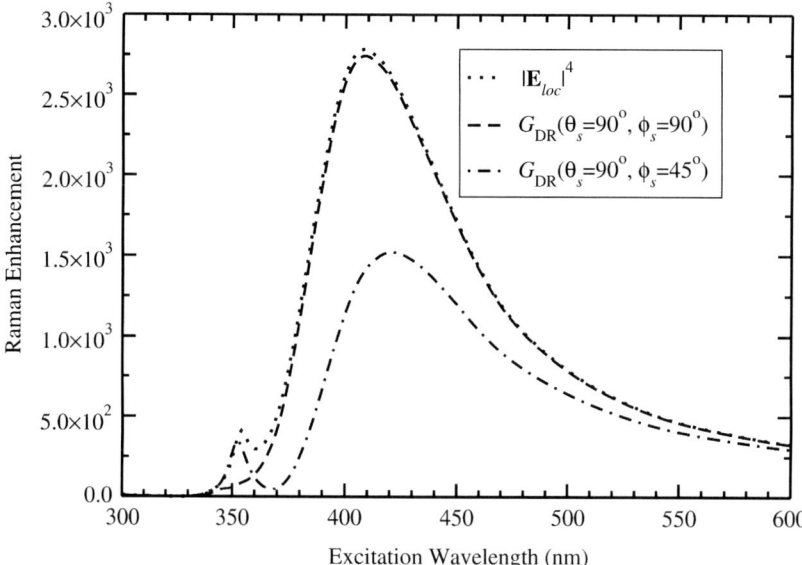

Figure 6.2 SERS excitation profile for single 100-nm diameter Ag sphere with a molecule located 0.5 nm from the sphere surface along the axis parallel to the incident *E*-field polarisation. The dotted line is the PW approximation result, the dashed line is the $\varphi_s = 90°$ DR result, and the dash-dotted line is the DR result for a detector at $\varphi_s = 45°$. In both detector locations $\theta_s = 90°$.

Also, for this detector location we can see that there are significant differences between PW and DR. In addition to the decrease in the enhancement for this configuration, the location of the maximum enhancement redshifts from 408 nm to 420 nm. The reason for this good agreement when $\varphi_s = 90°$ is that the induced oscillating dipole in the molecule due to the locally enhanced field from the particle, and the induced dipole in the particle due to the molecule, are parallel to $\mathbf{E}_i$. Therefore, the resulting far-field amplitudes of the fields are all dipolar in character and inphase with each other at a point normal to the direction of the induced dipole moments. Figure 6.3(a) shows this interference, which in this case is constructive. The discrepancy between DR and PW results for $\varphi_s = 45°$ arises because the dipole emitted by the molecule and by the particle are not quite inphase for angles φ_s away from 90°. Note that phase interference alone would typically lead to smaller DR intensities than PW as purely dipolar emission is maximised at $\varphi_s = 90°$ and has the most constructive interference.

Another interesting feature of Figure 6.2 occurs near 354 nm where the error becomes much larger for $\varphi_s = 90°$, as $|\mathbf{E}_{loc}|^4$ shows a quadrupole resonance peak that is not excited by DR for this geometry. These results are all consistent with the phase interference model pictured in Figure 6.3(a) together with the increased importance of quadrupole excitation (Figure 6.3(b)) for the 100 nm

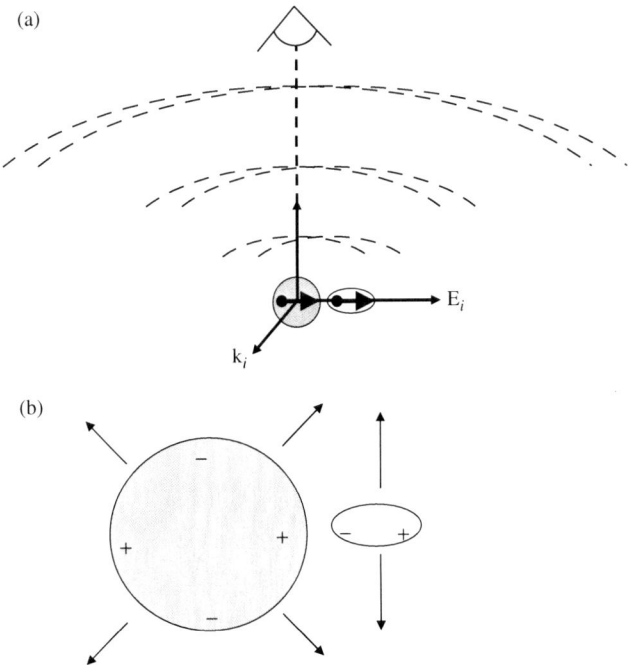

Figure 6.3   (a) A schematic showing the induced dipole in the molecule, and the dipole induced in the particle by the molecule. The interference of emission between these dipoles determines the dependence of the far-field emission on detection angles. (b) The molecular dipole, now represented in terms of positive and negative charges, along with the quadrupole charges induced in the particle. (The particle dipole charges are omitted.) The oscillating quadrupole charges preferentially emit at 45° and 135°, while the dipoles emit at 90°, so the superposition of these emissions can have stronger emission at 45° than would dipole-alone emission. Reprinted with permission from: L. K. Ausman and G. C. Schatz, *J. Chem. Phys.* 2009, **131**, 084706. Copyright 2009, American Institute of Physics.

sphere. This error decreases when quadrupole excitation in the particle can give enhanced emission for angles closer to $\varphi_s = 45°$, as schematically depicted in Figure 6.3(b).

Calculation of SERS enhancements due to a single Ag sphere in earlier work[31,33,36] showed that the PW approximation is nearly exact for a small sphere when the molecule is located along the polarisation axis of the incident field and for an observation point that is normal to this direction. However, different detector locations can result in dramatic errors due to phase-interference effects.

6.3   FDTD: Finite-Difference Time-Domain Method

While analytical solutions are available to calculate the optical properties for metal spheres, shells and small ellipsoids (Mie theory), there are no generally applicable analytical solutions for arbitrarily shaped nanoparticles, such as cubes, rods or triangular prisms. It is therefore necessary to resort to numerical methods, for example FDTD and DDA. Both of these methods use cubic grids to represent the structure(s). Furthermore, DDA is an approximate frequency-domain method, while FDTD is an exact time-domain one. The DDA method will be described in Section 6.5. In this section, a brief introduction to FDTD is presented so as to give important background material for our subsequent discussion of nonlocal effects in Section 6.4. For a more detailed description of FDTD the reader is referred to refs. 23 and 24.

The aim of FDTD is to solve Faraday's law (eqn (6.37)) and Ampere's law (eqn (6.38)) in differential form over a grid-based region.[25]

$$\varepsilon \frac{\partial}{\partial t} \vec{E} = \nabla \times \vec{H} - \vec{J} \tag{6.37}$$

$$\mu \frac{\partial}{\partial t} \vec{H} = -\nabla \times \vec{E}. \tag{6.38}$$

The FDTD method is based upon the Yee algorithm, which is the most popular discretisation and propagation scheme.[23,26] In essence, the Yee algorithm shifts $\vec{E}$ and $\vec{H}$ relative to each other by half-grid points and a leap-frog scheme is used to take discrete time steps. Other finite-difference algorithms are also available, including the Lax–Wendroff method. In this algorithm, $\vec{E}$ and $\vec{H}$ are updated at the same time, as opposed to the leap-frog scheme discussed above. A benefit of the Lax–Wendroff method[27] is that one-sided differences, which are unstable when using the leap-frog scheme, can be used to truncate the region.[25]

If a 2D system is considered, the electric field has only x- and y-components, while the magnetic field has only a z-component. Denoting the time step as τ allows the following equations to be used:

$$E_x^{n+\frac{1}{2}} = E_x^{n-\frac{1}{2}} + \frac{\tau}{\varepsilon} \left[\frac{\partial H_z^n}{\partial y} - J_x^n \right] \tag{6.39}$$

$$E_y^{n+\frac{1}{2}} = E_y^{n-\frac{1}{2}} + \frac{\tau}{\varepsilon} \left[-\frac{\partial H_z^n}{\partial x} - J_y^n \right] \tag{6.40}$$

$$H_z^{n+1} = H_z^n - \frac{\tau}{\mu} \left[\frac{\partial E_y^{n+\frac{1}{2}}}{\partial x} - \frac{\partial E_x^{n+\frac{1}{2}}}{\partial y} \right]. \tag{6.41}$$

As previously mentioned, the FDTD method is time based, however, frequency-domain spectra can be calculated using Fourier transforms.

The FDTD algorithm can also be used to calculate the optical response of 3D nanostructures, however, these calculations are computationally more demanding than 2D ones. Nevertheless, the implementation of parallel processing has afforded 3D calculations in which overall grid dimensions larger than one micrometre have been considered.[28] General formulas for 3D FDTD have been given in several places.[23,24,29]

One limiting factor of the FDTD method, as with all discretisation techniques, is what is known as the "stair-casing" error. This is caused by the use of a discrete grid to approximate a nanostructure. If the grid spacing is too large, then the stair-casing error becomes evident in the calculated optical spectra. To overcome this, smaller grid spacings are required to resolve fine features of nanostructures, which can be computationally expensive for large-sized structures.

6.4   FDTD: Inclusion of Nonlocal Effects in the Finite-Difference Time-Domain Method

Interest in metallic structures with features on the order of ten nanometers or less has significantly increased as experimental techniques for their fabrication have become possible.[30] Even if the features involve many hundreds of atoms or more, so that a continuum level of description is adequate, their optical responses cannot be correctly modelled using the FDTD technique described in Section 6.3 due to the importance of spatially nonlocal dielectric effects.[31]

In the small-particle size limit, the dielectric constant $\varepsilon(\vec{k}, \omega)$ is a function of both the wavevector $\vec{k}$ and angular frequency ω (which are independent). In this case, the constitutive relationship between the electric displacement field ($\vec{D}$) and the electric field ($\vec{E}$) is:

$$\vec{D}(\vec{k}, \omega) = \varepsilon(\vec{k}, \omega)\vec{E}(\vec{k}, \omega). \tag{6.42}$$

The $\vec{k}$-dependence in $\varepsilon(\vec{k}, \omega)$ leads to a spatially nonlocal relationship between $\vec{D}(\vec{x}, \omega)$ and $\vec{E}(\vec{x}, \omega)$ when eqn (6.42) is Fourier transformed to the spatial domain.[32] Physically, this implies that since $\vec{D}(\vec{x}, \omega)$ is related to the electric polarisation, the polarisation at one point in space depends on both the local $\vec{E}(\vec{x}, \omega)$ and the $\vec{E}(\vec{x}, \omega)$ in its neighbourhood.

Spatially nonlocal dielectric effects have long been known to affect the optical responses of structures with features less than approximately 10 nm. For example, anomalous absorption is experimentally observed in thin metal films[33,34] and theoretically it is found that the inclusion of a $\vec{k}$-dependence in $\varepsilon(\vec{k}, \omega)$ accounts for this through the excitation of longitudinal (or volume) plasmons,[35] so-called because they are longitudinal to $\vec{k}$ and are contained within the volume of the structure (unlike LSPRs, which propagate along a metal/dielectric interface). Recently, such a $\vec{k}$-dependence has been found necessary to describe the blueshifting of LSPRs in small Au nanoparticles[36] relative to local predictions.[31]

While theoretical methods to describe nonlocal effects have been available for some time for basic geometries, including spherical structures[37,38] or aggregates thereof,[39-42] cylinders[39] and metal films,[35] only recently has a general method to calculate the optical responses of arbitrarily shaped structures been developed[31,40] and applied.[41,42] In this section, this method is overviewed had the optical responses of Au nanowires are compared in the local and nonlocal limits.

6.4.1 Theoretical Approach

Before Maxwell's equations can be solved an explicit form for $\varepsilon(\vec{k}, \omega)$ in the constitutive relationship between $\vec{D}(\vec{k}, \omega)$ and $\vec{E}(\vec{k}, \omega)$ (eqn (6.42)) must be specified. Since the explicit examples below are for Au structures, $\varepsilon(\vec{k}, \omega)$ can be described in the continuum limit by three separate components:

$$\varepsilon(\vec{k}, \omega) = \varepsilon_\infty + \varepsilon_{inter}(\omega) + \varepsilon_{intra}(\vec{k}, \omega) \tag{6.43}$$

the value as $\omega \to \infty$, ε_∞, a contribution from d-band to sp-band electronic transitions, $\varepsilon_{inter}(\omega)$, and a contribution due to the excitation of the conduction electrons, $\varepsilon_{intra}(\vec{k}, \omega)$. The notation in eqn (6.43) highlights the $\vec{k}$ and ω dependencies. $\varepsilon_{inter}(\omega)$ is physically described by a multipole Lorentz oscillator model[1]:

$$\varepsilon_{inter}(\omega) = \sum_n \frac{\Delta\varepsilon_{Ln}\omega_{Ln}^2}{\omega_{Ln}^2 - \omega(\omega + i2\delta_{Ln})}, \tag{6.44}$$

where n is an index labelling the individual d-band to sp-band electron transitions that occur at ω_{Ln}, $\Delta\varepsilon_{Ln}$ is the shift in relative dielectric constant at the transition, and δ_{Ln} is the electron dephasing rate. Because there are two interband transitions in Au at optical frequencies (near 3 and 4 eV[21]) $n = 2$ is used for the calculations in this section.

$\varepsilon_{intra}(\vec{k}, \omega)$ is responsible for both the plasmonic optical response of a metal structure and nonlocal effects. Based on quantum-mechanical considerations in a DFT formalism, $\varepsilon_{intra}(\vec{k}, \omega)$ can be physically described by the hydrodynamic Drude model (which reduces to the local Drude expression for electron motion if $\vec{k} \to 0$)[43]:

$$\varepsilon_{intra}(\vec{k}, \omega) = -\frac{\omega_D^2}{\omega(\omega + i\gamma) - \beta^2\vec{k}^2}, \tag{6.45}$$

where ω_D is the plasma frequency, γ is the collision frequency, and β^2 depends on a variety of factors, including the dimensionality of the system, energies considered, and quantum-mechanical effects. In this section, $\beta^2 = v_F^2/2$ where v_F is the Fermi velocity ($1.39 \cdot 10^6$ m/s for Au).[31]

By fitting eqn (6.43) to the experimentally determined data of bulk Au[21] using physical considerations (fitting terms over the appropriate energy ranges, using the limit of $\vec{k} \rightarrow 0$, *etc.*), parameters of $\varepsilon_\infty = 3.559$, $\omega_D = 8.812$ eV, $\gamma = 0.0752$ eV, $\Delta\varepsilon_{L1} = 2.912$, $\omega_{L1} = 4.693$ eV, $\delta_{L1} = 1.541$ eV, $\Delta\varepsilon_{L2} = 1.272$, $\omega_{L2} = 3.112$ eV, $\delta_{L2} = 0.525$ eV are found.[31,40]

Inserting eqns (6.42) and (6.43) (using eqns (6.44) and (6.45)) into the Maxwell–Ampère law in $\vec{k}$-space for a time-harmonic field, $-i\omega\vec{D}(\vec{k}, \omega) = i\vec{k} \times \vec{H}(\vec{k}, \omega)$, gives:

$$-i\omega\varepsilon_0\varepsilon_\infty \vec{E}(\vec{k}, \omega) + \sum_n \vec{J}_{Ln}(\vec{k}, \omega) + \vec{J}_{HD}(\vec{k}, \omega) = i\vec{k} \times \vec{H}(\vec{k}, \omega), \qquad (6.46)$$

where the $\vec{J}_{Ln}(\vec{k}, \omega)$ are polarisation currents associated with eqn (6.44),

$$\vec{J}_{Ln}(\vec{k}, \omega) = -i\omega\varepsilon_0 \frac{\Delta\varepsilon_{Ln}\omega_{Ln}^2}{\omega_{Ln}^2 - \omega(\omega + i2\delta_{Ln})} \vec{E}(\vec{k}, \omega) \qquad (6.47)$$

and $\vec{J}_{HD}(\vec{k}, \omega)$ is a (nonlocal) polarisation current associated with eqn (6.45),

$$\vec{J}_{HD}(\vec{k}, \omega) = i\omega\varepsilon_0 \frac{\omega_D^2}{\omega(\omega + i\gamma) - \beta^2\vec{k}^2} \vec{E}(\vec{k}, \omega). \qquad (6.48)$$

Equations of motion for the currents in eqns (6.47) and (6.48) can be obtained by multiplying through each equation by the appropriate denominator and inverse Fourier transforming ($-i\vec{k} \rightarrow \nabla$ and $-i\omega \rightarrow \partial/\partial t$):

$$\frac{\partial^2}{\partial t^2}\vec{J}_{Ln}(\vec{x}, t) + 2\delta_{Ln}\frac{\partial}{\partial t}\vec{J}_{Ln}(\vec{x}, t) + \omega_{Ln}^2\vec{J}_{Ln}(\vec{x}, t) = \varepsilon_0\Delta\varepsilon_{Ln}\omega_{Ln}^2\frac{\partial}{\partial t}\vec{E}(\vec{x}, t) \qquad (6.49)$$

$$\frac{\partial^2}{\partial t^2}\vec{J}_{HD}(\vec{x}, t) + \gamma\frac{\partial}{\partial t}\vec{J}_{HD}(\vec{x}, t) - \beta^2\nabla^2\vec{J}_{HD}(\vec{x}, t) = \varepsilon_0\omega_D^2\frac{\partial}{\partial t}\vec{E}(\vec{x}, t). \qquad (6.50)$$

Using FDTD techniques from Section 6.3, eqns (6.49) and (6.50) can easily be solved self-consistently with Faraday's Law (eqn (6.37)) and the inverse Fourier-transformed form of eqn (6.46):

$$\varepsilon_0\varepsilon_\infty\frac{\partial}{\partial t}\vec{E}(\vec{x}, \omega) + \sum_n \vec{J}_{Ln}(t) + \vec{J}_{HD}(\vec{x}, \omega) = \nabla \times \vec{H}(\vec{x}, \omega), \qquad (6.51)$$

along with the requirement that Gauss' laws, $\nabla \cdot \vec{D} = \rho$ and $\nabla \cdot \vec{B} = 0$, where ρ is the charge density, remain satisfied.[40] By not updating eqns (6.49) and (6.50) outside of the nonlocal materials, arbitrarily shaped structures can be numerically simulated. The full set of nonlocal finite-difference equations, including their derivation, can be found in ref. 40.

6.4.2   Optical Responses of Au Nanowires in the Local and Nonlocal Limits

In this subsection, the optical responses of Au nanowires illuminated with light polarised with the $\vec{E}$ field in the plane are studied using the method described above. As an initial example, the optical responses of cylindrical nanowires are calculated in the local and nonlocal limits. Extinction cross-sections for nanowires with radii of $r = 2$ to 8 nm are shown in Figure 6.4.

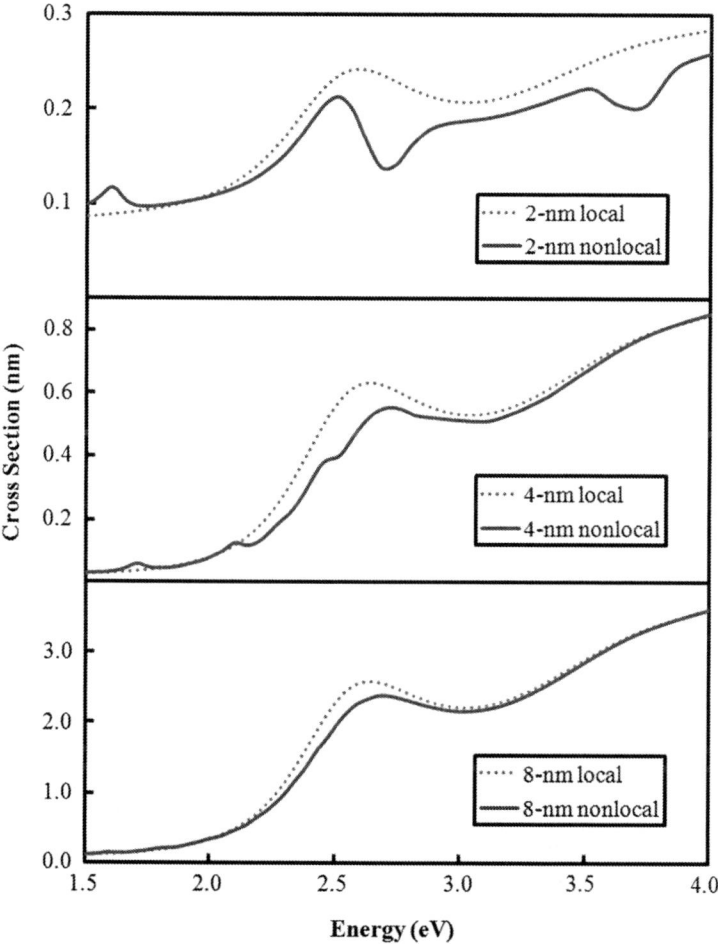

Figure 6.4   Optical responses of Au cylindrical nanowires with radii of (top) 2, (middle) 4, and (bottom) 8 nm. Full curve is nonlocal theory; broken curve is local theory. J. M. McMahon, S. K. Gray, and G. C. Schatz, *Phys. Rev. Lett.*, 2009, **103**, 097403. Copyright (2009) by the American Physical Society.

There are a couple of interesting effects that occur in the nonlocal results. First, for the $r = 2$ nm nanowire, anomalous absorption peaks are seen when nonlocal effects are included (these are also seen in the $r = 4$ nm results, but to a lesser extent). These arise from the optical excitation of longitudinal plasmons that cannot be described by local electrodynamics. These can be confirmed by looking at intensity profiles of $|\vec{D}(\vec{x}, \omega)|^2$ at the anomalous absorption energies; Figure 6.5.

Inside the nanowire, discrete standing-wave longitudinal plasmons are seen with wavelengths of:

$$\lambda_L = \frac{2d}{m}, \tag{6.52}$$

where $d = 2r$ and $m = 1, 3, 5, \ldots$ – *i.e.* odd numbers of half-wavelengths fit inside d. Note that only the $m = 3, 5$, and 7 "modes" are explicitly shown in Figure 6.5. These results are in sharp contrast to the local result of a relatively uniform profile of $|\vec{D}(\vec{x}, \omega)|^2$, but explain the appearance of discrete absorption peaks in Figure 6.4.

In addition to anomalous absorption, nonlocal effects are found to blueshift the dipolar LSPR resonances, relative to local calculations. For example, at $r = 4$ nm there is a 0.014 eV blueshift. This can be understood by looking at the form of eqn (6.45). The interplay between ω and $\vec{k} = (2\pi/\lambda_L)$ causes a nonlocal LSPR to appear at a higher energy compared to the local limit (*e.g.*, the $\vec{k} = 0$ Drude model).

The importance of nonlocal effects is quickly lost with increasing r, since the discrete nature of the longitudinal plasmon modes becomes less important; as $r \to \infty$, m must go to ∞ for the mode to remain at an optical energy, resulting in $\lambda_L \to 0$. For example, in Figure 6.4, when $r = 4$ nm there is only minor anomalous absorption and LSPR blueshifting, and when $r = 8$ nm it is hardly distinguishable (but nonetheless still noticeable, due to many closely spaced modes).

Nonlocal effects have been found to be particularly strong in structures with apex features.[31,40] As an example, the optical responses of Au (equilateral) triangular nanowires are now considered. The optical responses of nanowires with side lengths of $l = 5$ to 40 nm are shown in Figure 6.6.

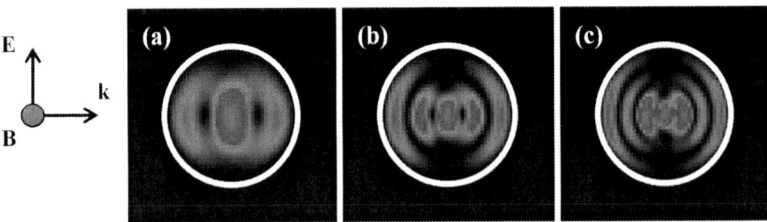

Figure 6.5 Longitudinal plasmons inside a $r = 2$ nm Au cylindrical nanowire at energies of (a) 1.60, (b) 2.71, and (c) 3.69 eV. The polarisation and direction of incident light are indicated; the nanowire is outlined in white. J. M. McMahon, S. K. Gray, and G. C. Schatz, *Phys. Rev. Lett.*, 2009, **103**, 097403. Copyright (2009) by the American Physical Society.

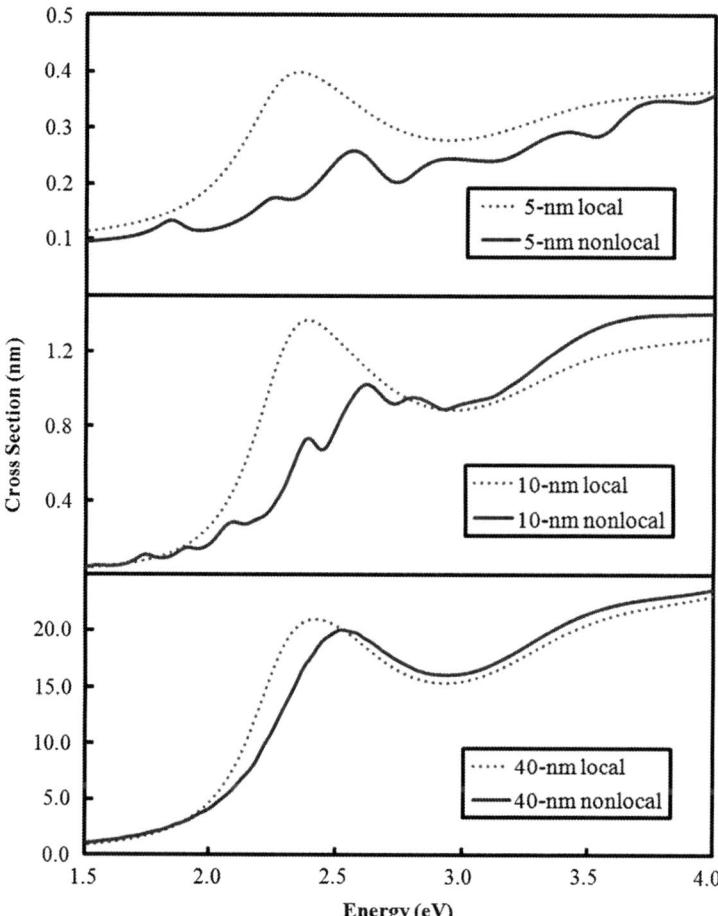

Figure 6.6 Optical responses of Au triangular nanowires with side lengths of (top) 5, (middle) 10, and (bottom) 40 nm. Full curve is nonlocal theory; broken curve is local theory. J. M. McMahon, S. K. Gray, and G. C. Schatz, *Phys. Rev. Lett.*, 2009, **103**, 097403. Copyright (2009) by the American Physical Society.

Similar to the cylindrical nanowires above, nonlocal effects lead to both anomalous absorption and LSPR blueshifting, however, the effects are much stronger in these structures. For example, when $l = 5$ nm, the anomalous absorption is so strong that the dipolar LSPR is hardly distinguishable. Additionally, these effects persist for sizes much larger than the analogous cylindrical nanowires. For instance, even at $l = 40$ nm there is a 0.1 eV LSPR blueshift.

It is interesting to again look at profiles of $|\vec{D}(\vec{x}, \omega)|^2$. Figure 6.7 shows such profiles for the $l = 5$ nm nanowire at some of the anomalous absorption energies.

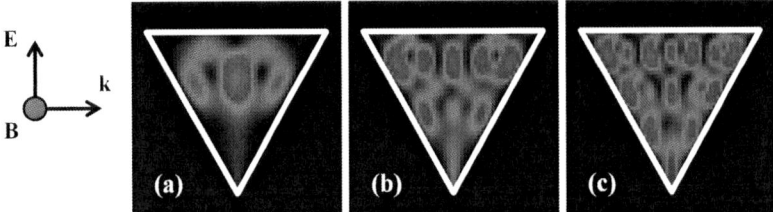

Figure 6.7   Longitudinal plasmon modes inside a $l = 5$ nm Au triangular nanowire at
energies of (a) 1.88, (b) 2.82, and (c) 3.71 eV. The polarisation and
direction of incident light are indicated; the nanowire is outlined in white.
J. M. McMahon, S. K. Gray, and G. C. Schatz, *Phys. Rev. Lett.*, 2009,
103, 097403. Copyright (2009) by the American Physical Society.

As expected, longitudinal plasmons are excited inside the structures with
wavelengths given by eqn (6.52), where d in this case is the distance between the
nanowire sides along the longitudinal direction. Comparison of Figures 6.5 and
6.7 reveals an interesting difference between the cylindrical and triangular
structures. In the latter, the longitudinal plasmons are found to be sustained at
discrete vertical positions, at each point where eqn (6.52) is satisfied (which
increases with m). The ability of such structures to strongly sustain longitudinal
plasmons at multiple positions (particularly near the apex where d can be arbitrarily small and low-order longitudinal plasmon modes can always be sustained)
is one reason why nonlocal effects are so strong.[31] Another reason is that large-
magnitude $\vec{k}$ components can be generated by the apices. By generalising the
results above to other apex structures, it can be suggested that longitudinal
plasmon modes can always be sustained (at least near the apex), which can cause
nonlocal effects to remain important for arbitrarily large sizes.[31]

Before leaving this section, it should be noted that the specific applications
herein focused on far-field optical responses. However, near-field properties, such
as electromagnetic field enhancements, can also be significantly affected by nonlocal effects. For a more complete discussion of such effects, see refs. 40, 42 and 44.

6.5   DDA: The Discrete Dipole Approximation Method

The discrete dipole approximation was first developed in 1973 by Purcell and
Pennypacker[45] as a numerical technique to describe scattering of electromagnetic radiation from grains of dust in the interstellar medium. This
technique is relatively popular based largely on its implementation by Draine,
et al.[46] into the freely available software package DDSCAT.[47,48]

The DDA method has advantages and disadvantages compared to FDTD. A
disadvantage is that it is not an exact method; we show below that convergence
of the method to the exact solution of Maxwell's equation is only guaranteed
for a homogeneous, infinite system. Comparisons of extinction cross-sections
for spherical metal nanoparticles with Mie theory results suggest that convergence to within 10% is not unusual, however, comparable convergence for
other particle shapes is not guaranteed (later we provide an example). In

addition the convergence of near-field properties such as surface electric fields is usually found to be poorer than for far-field properties. Of course, DDA suffers from stair-casing errors as mentioned previously for FDTD. A positive aspect of DDA is that the gridding need only cover the region where the metal is located (no grid points are needed to describe empty space around the particle), so fewer grid points are used than in FDTD. This makes DDA calculations somewhat faster for some three-dimensional studies. However, the scaling with size of the system is better for FDTD, so a properly parallelised FDTD code can usually treat an overall system size that is larger than DDA.

6.5.1 Mathematical Formalism of DDA

The DDA technique represents the target, which can be arbitrarily shaped, by an array of N dipoles with a cubic grid, an example of which is shown in Figure 6.8. Here, the target is a 20-nm edge length cube. It then calculates the polarisability, α_i of each dipole based upon the dielectric function and position within the target r_i. Once the dipole polarisabilities have been calculated it is then possible to calculate the target absorption C_{abs} and scattering, C_{sca} cross-sections, which is described below.[46,49–51]

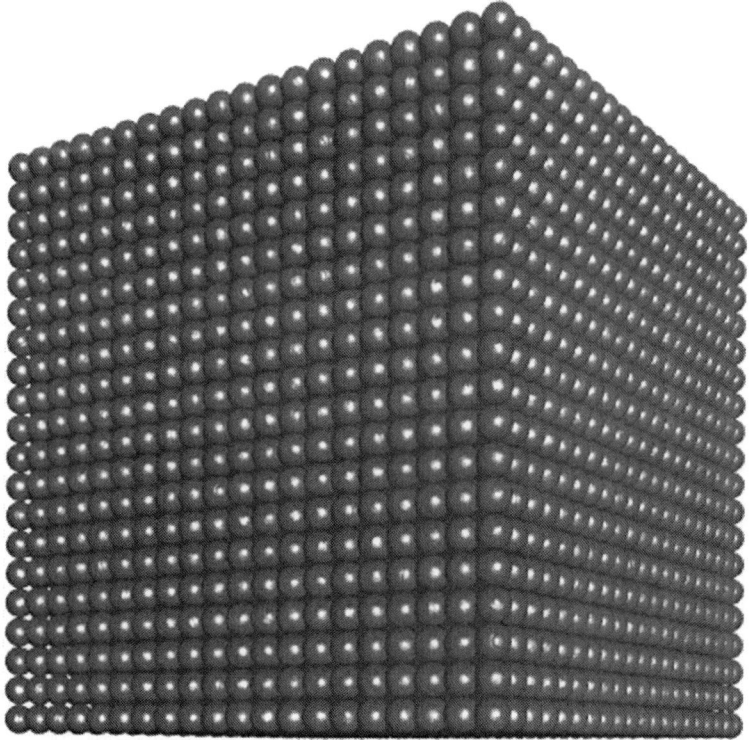

Figure 6.8   20-nm edge length cube represented by a cubic grid of dipoles. Each dipole has a diameter of 1 nm. Total number of dipoles in the target structure is 8000.

The index $i = 1, 2, \ldots N$ represents each dipole within a cubic lattice then, upon application of a plane wave field, the induced dipole moment, P_i is given by[52]:

$$P_i = \alpha_i \cdot E_{\text{loc},i}, \tag{6.53}$$

where α_i is the dipole polarisability tensor and the electric field $E_{\text{loc},i}$ is the sum of the incident electric field $E_{\text{inc}}(r_i, t)$ and the radiated electric field $E_{\text{dip},i}$ from the surrounding $N - 1$ dipoles[52]

$$E_{\text{loc},i} = E_{\text{inc}}(r_i, t) + E_{\text{dip},i}, \tag{6.54}$$

with

$$E_{\text{inc}}(r_i, t) = E_0 e^{i(kr_i - \omega t)}, \tag{6.55}$$

where E_0 is the amplitude of the electric field, k is the wavevector, r_i is the position vector, ω is the angular frequency and t is time.

The local electric field $E_{\text{dip},j}$ is the radiated electric field from dipole j at position r_j as given by[46]

$$E_{\text{dip},i} = \sum_{j \neq i} A_{ij} \cdot P_j. \tag{6.56}$$

This implies

$$E_{\text{loc},i} = E_0 e^{i(kr_i - \omega t)} - \sum_{j \neq i} A_{ij} \cdot P_j. \tag{6.57}$$

Each element A_{ij} is defined by the following equation for an electric field from a radiating or oscillating dipole[46,52]

$$A_{ij} \cdot P_j = \frac{e^{ikr_{ij}}}{r_{ij}^3} \left\{ k^2 r_{ij} \times (r_{ij} \times P_j) + \frac{(1 - ikr_{ij})}{r_{ij}^2} [r_{ij}^2 P_j - 3 r_{ij}(r_{ij} \cdot P_j)] \right\}, \tag{6.58}$$

where $i \neq j$.

Substituting eqn (6.57) into eqn (6.53) and simplifying to find $E_{\text{inc}(r_i, t)}$ gives a set of $3N$ linear equations[46]

$$E_{\text{inc}}(r_i, t) = \frac{P_i}{\alpha_i} + \sum_{j \neq i} A_{ij} \cdot P_j. \tag{6.59}$$

The Parallel Iterative Methods (PIM)[47] package within DDSCAT uses iteration of the Preconditioned BiConjugate Gradient with Stabilisation (PBCGST) algorithm to solve eqn (6.59) for P_i. Once P_i is known, the cross-sections C_{abs} and C_{sca} can be calculated from eqns (6.60)[52] and (6.61),[49] respectively.

$$C_{abs} = \frac{4\pi k}{|E_{inc}|^2} \cdot \sum_{i=1}^{N} \left\{ \text{Im}[P_i \cdot (\alpha_i^{-1})^* P_i^*] - \frac{2}{3} k^3 |P_i|^2 \right\} \tag{6.60}$$

$$C_{ext} = \frac{4\pi k}{|E_{inc}|^2} \cdot \sum_{i=1}^{N} \text{Im}(E_{inc,i}^* \cdot P_i). \tag{6.61}$$

In the first implementation of DDA the Clausius–Mossotti relation (CMR)[50] was used to calculate the dipole polarisabilities based upon the complex refractive index m such that

$$\alpha_{CMR} = \frac{3d^3}{4\pi} \left(\frac{m^2 - 1}{m^2 + 2} \right), \tag{6.62}$$

where d is the interdipole spacing. This approach is exact in the quasistatic regime, however, it is not exact at finite wavelengths because eqn (6.62) does not consider radiative reaction.

In 1993 Draine[53] introduced the lattice-dispersion relation (LDR) that required that the polarisability of dipoles on an infinite lattice give the same dispersion relation as a continuum medium.[50] In the LDR approach, the radiative reaction correction occurs naturally and the polarisability is given by

$$\alpha_{LDR} = \frac{\alpha_{CMR}}{1 + \left(\frac{\alpha_{CMR}}{d^3} \right) [(b_1 + m^2 b_2 + m^2 b_3 S)(kd)^2 - \left(\frac{2}{3} \right) i(kd)^3]}, \tag{6.63}$$

where $b_1 = -1.8915316$, $b_2 = 0.1648469$, $b_3 = -1.7700004$ and $S = \sum_j (a_j e_j)^2$ with a and e being the unit and polarisation vectors, respectively, and the directions $j = 1, 2, 3$ corresponding to the dipole lattice axes.

The LDR guarantees convergence of the DDA for a homogeneous infinite material, however, its applicability to light interacting with nanoparticles is based on experience. Going to smaller dipole spacings makes the corrections to eqn (6.63) smaller, but there are limits as to how small a grid spacing can be used. When using DDSCAT[47] the authors recommend that the interdipole spacing d be small compared to any structural lengths in the target and the wavelength λ of the incident light. The second criterion is satisfied with a high degree of accuracy if $|m|kd < 0.5$. The difficulty here is that as the number of dipoles within the target increases so does the computational time (roughly as $N \ln N$). So a compromise must be met between calculation accuracy and computational expense.

In addition to this, for dipoles near the particle surfaces, the local field must be corrected. Local field corrections arise even in the quasistatic regime due to unbalanced dipole interactions for dipoles near particle surfaces, leading to an effective polarisability that depends on position.[54] In 2004 Collinge[50] developed the surface-corrected-lattice-dispersion relationship (SCLDR) that is a modified version of the LDR that accounts for both finite wavelengths and local field effects. However, the SCLDR can only be applied to targets with an analytical solution in the electrostatic limit,[54] which prevents the use of SCLDR on arbitrarily shaped nanostructures.

6.5.2 Optical Response of Au Cube: Comparing DDA and FDTD Convergence

In this subsection the extinction cross-section of a 30-nm edge length Au cube is calculated using both the DDA and 3D FDTD methods. In each case dipole spacings or grid spacings of 0.5, 1, 3 and 5 nm are considered. The aim here is to provide an example of convergence of the extinction cross-section spectra as the dipole spacing is reduced. In both cases the dielectric constant of Au in vacuum from Johnson and Christy was used.[21] The faces of the cube extend along the *x*-, *y*- and *z*-axes and the incident wavevector (along the *x*-axis) is incident upon the *y*-planar face of the cube. Figures 6.9 and 6.10 show the extinction spectra as a function of wavelength calculated using DDA and 3D FDTD methods, respectively.

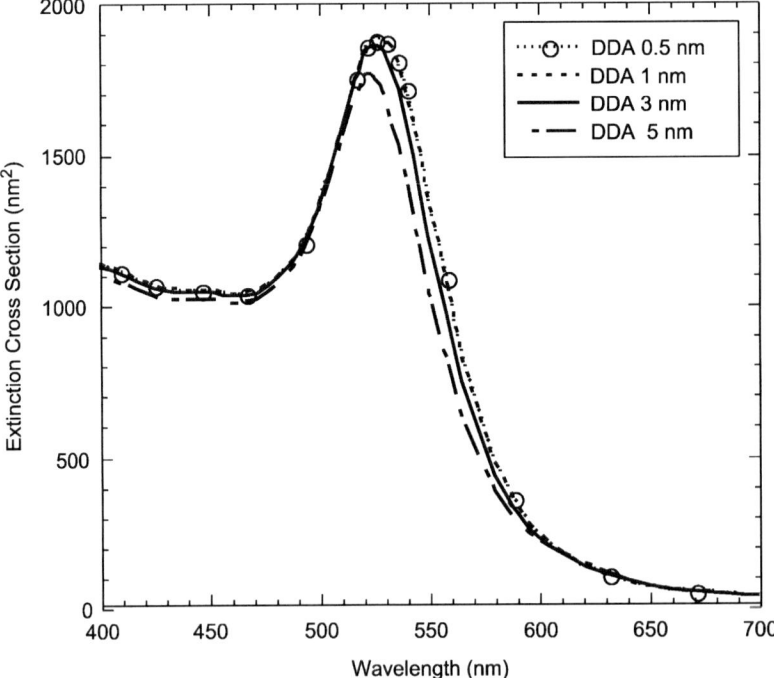

Figure 6.9 Extinction cross-section spectra for a 30-nm edge length Au cube with interdipole spacings of 0.5, 1, 3 and 5 nm. Calculations were performed using the DDA technique (DDSCAT).

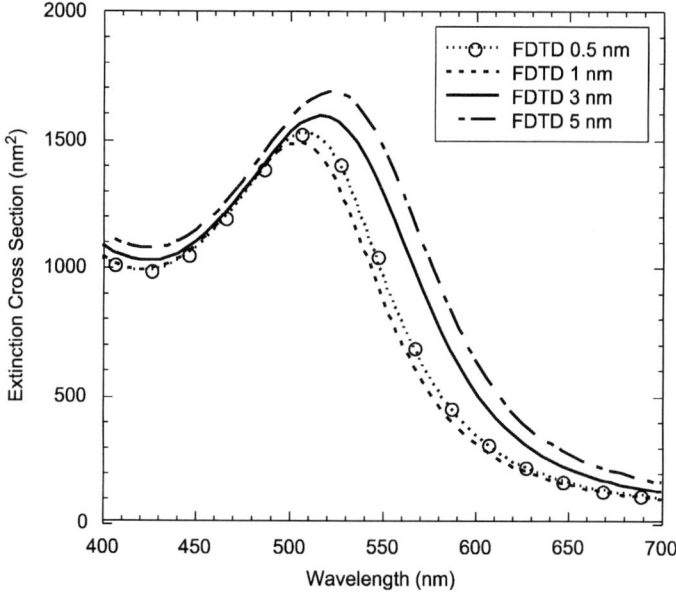

Figure 6.10 Extinction cross-sections for a 30-nm edge length Au cube with grid spacings of 0.5, 1, 3 and 5 nm. Calculations were performed using the 3D FDTD technique.

Consider the DDA results in Figure 6.9, it can be seen that the spectrum for the 5-nm dipole spacing is not converged because only 219 dipoles were present in the cube structure. However, convergence is achieved for a dipole spacing of 1 nm (528 nm peak wavelength), which corresponds to 27 000 dipoles within the structure. Decreasing the dipole spacing further to 0.5 nm increases the number of dipoles within the cube 8 times (216 000), yet this does not alter the peak extinction wavelength (528 nm) or cross-sectional magnitude. Similar convergence behaviour occurs in the 3D FDTD calculations (Figure 6.10). For the larger grid spacings (3 and 5 nm) convergence is again poor, however, the 1-nm and 0.5-nm grid results are reasonably similar. Note also that there are some differences between the wavelength of the plasmon maximum, with FDTD being 20 nm to the blue and the peak cross-section is also different ($\sim$1500 *versus* $\sim$1800 nm^2). These are the sort of errors that can arise with DDA, however, a subtle point is that in the FDTD method the dielectric constant is described using a model rather than the actual measured data. The determination of the near-field response requires finer dipole or grid spacings and careful attention to convergence.

6.6 Modelling LSPR of Metal Nanoparticles Coated with Resonant Molecules

In this section we discuss our recent work[55] where DDA calculations for a metal nanoparticle are combined with electronic-structure calculations for molecules near the nanoparticles so as to determine the modifications to optical spectra of both the particle and the molecule that arise from the molecule–metal interactions. This work is based on a new many-body Green's-function formalism that is capable of describing a variety of plasmon-enhanced linear spectroscopies in which the DDA method is used to determine local fields that interact with a molecule that is treated *via* linear response quantum mechanics.

LSPR spectroscopy (either extinction or scattering of light by nanoparticles) is highly sensitive to small changes in the refractive index of the medium surrounding metal nanostructures.[56–59] Thus, making this a potential tool for the development of chemical and biological sensors.[60,61] Usually, it is applied to the measurement of the plasmon wavelength of the metal particle under conditions where the nanoparticles being observed dominate the spectrum, as the adsorbed molecules in a monolayer concentration is too low to permit detection of the molecules. However, there has been on-going interest in the case where molecules present on the nanoparticle surface can also have sufficient absorption that they are also observable.[62,63] In the example we study below, we consider a model system that consists of pyridine molecules on a spheroidal particle surface, and we choose both the electronic structure model and the metal particle shape so that the plasmon and molecular resonances can be tuned relative to one another.

6.6.1   Many-Body Green's-Function Formalism

Our formalism[22] is based upon a quantum many-body Green's-function description of the molecule's electronic dynamics within the Born–Oppenheimer approximation, where its interactions with the classical electromagnetic field as well as the classical plasmons of a nearby metal particle are built in perturbatively. Details of our method and its numerical implementation can be found in refs. 22 and 55 and will only be briefly summarised here.

Generalising the model of Gersten and Nitzan[64] where the molecule and metal are each treated as classical dipoles, we discretize the metal system into a collection of polarisable points,[48] driven by the external excitation field as well as other electric-dipole fields. For the particle, each point dipole is assigned a metallic polarisability derived from the lattice dispersion relation and empirical dielectric data.[65] Together with this discrete-dipole approximation (DDA) description of the metal we add additional dipoles representing the molecular system, each with polarisabilities computed from a separate quantum-mechanical calculation within a modified version of the Q-Chem software package.[66] A self-energy is evaluated that reflects the influence of the metal particle on the electronic states of the molecule. This includes image effects between the molecule and metal that lead to damping and shifting of the electronic states of the molecule due to the metal particle.[22,55]

6.6.2   Results: Pyridine Interacting with an Ag
Spheroidal Particle

We use our formalism,[22,55] to study the wavelength shift of the combined pyridine–spheroidal silver particle extinction in the monolayer coverage limit of the molecule in comparison to that of the bare metal. The plasmon resonance position is well known to be highly sensitive to its local dielectric environment and displays rich dispersive behaviour in the optical regime where the environment has resonances that overlap spectrally with those of the plasmon. Figure 6.11 shows the averaged extinction cross-sections (upper) and extinction coefficients (lower) of the coupled pyridine–silver nanospheroid system as a function of excitation energy and plasmon-resonance position. Our calculations reveal that the pyridine polarisability and extinction (upper, grey) are sensitive to nanoparticle shape, plasmon resonance position and width, relative molecule–metal separation and orientation, and field-polarisation direction. For comparison, we also compute the pyridine-extinction cross-section based upon an empirical and constant parametrisation of the molecule–metal interaction (upper, black) that is consistent with theory work from Jensen *et al.*[67] Progression from a pyridine-coated prolate spheroid (left) with radii (50,100,50) nm, to sphere (middle) with radii (50,50,50) nm, to oblate spheroid (right) with radii (50,1,50) nm allows for the *y*-axis plasmon resonance to be tuned from red-detuned to nearly resonant with the lowest-lying molecular resonance of pyridine; the molecule–metal separation is 1.6 nm in all calculations. Extinction coefficients of the coupled system in the monolayer-coverage limit (lower, grey) and bare metal (lower, black) are shown with the excitation field polarised along the *y*-axis as well as polarised along the *z*-axis in the oblate case for the bare metal only (lower right, dotted black). The latter reveals how a resonant molecular monolayer can mediate the coupling between *y*-polarised light and the previously dark *z*-polarised plasmon of the oblate spheroid.

6.7 Conclusion

This chapter has focused on a number of commonly used analytical and numerical electrodynamics methods that can be used to model the optical properties of plasmonic nanostructures, with emphasis on nonconventional applications of these methods to problems that have recently been of interest in the surface spectroscopy field. We first considered the analytical approach of Mie theory where a comparison between our SERS model, which incorporates dipole reradiation, and the more traditional plane-wave approximation was performed. Use of the dipole reradiation expression to calculate the electromagnetic enhancement for a molecule located on a single Ag nanosphere showed excellent agreement with the plane-wave approximation when the nanosphere was small compared to the wavelength of the incident light and providing the detector location was orthogonal to both the incident-field polarisation and propagation direction. However, when the detector location was not in this specific location or the molecule was situated other than along the direction of the incident field polarisation, where the interaction between molecule and particle emission is important, then the plane-wave approximation was shown to be poor. This can be understood in terms of interference effects between light directly emitted by the molecule, and that which scatters off the particle.

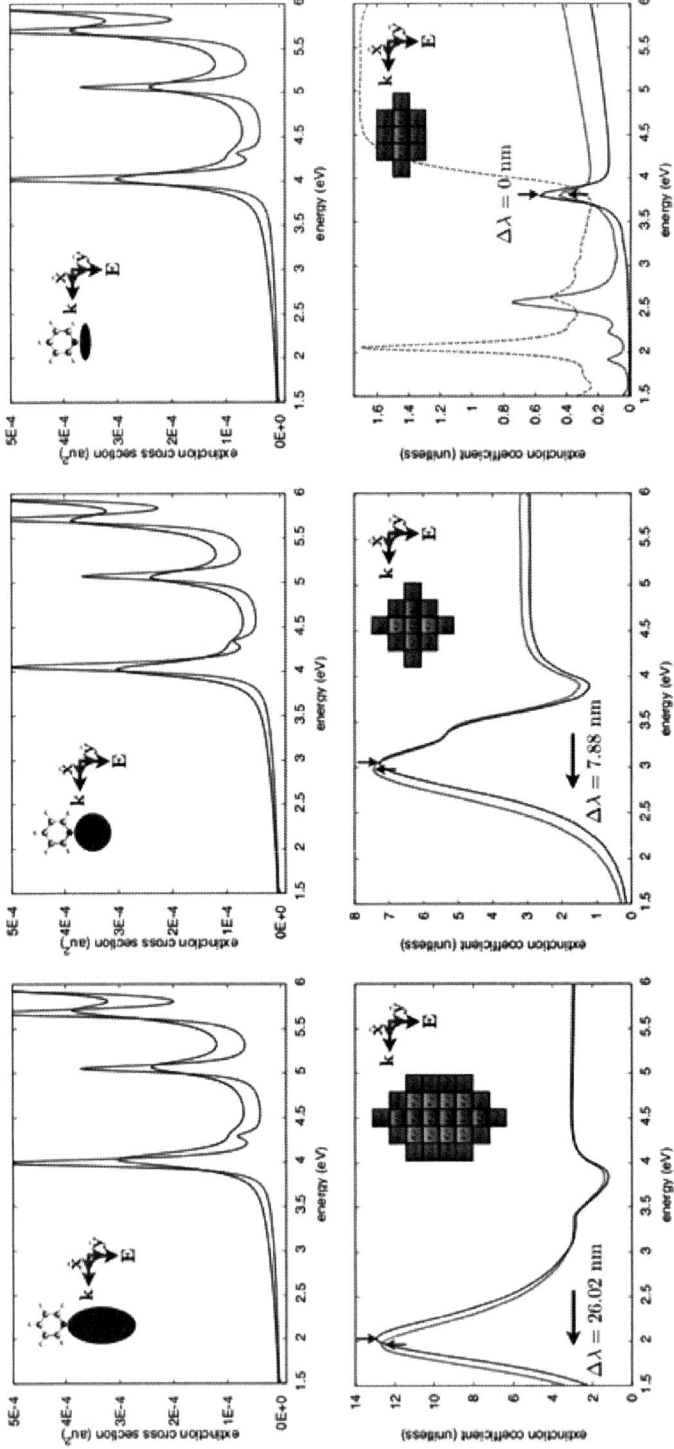

A description of the FDTD method was briefly introduced and then an implementation of Maxwell's equations was overviewed that can model the spatially nonlocal dielectric response of arbitrarily shaped structures. The formulation was based on a derived equation of motion for the hydrodynamic Drude model that can be solved self-consistently with Maxwell's equations using FDTD techniques. The optical responses of cylindrical and triangular Au nanowires were presented that demonstrated the importance of including such nonlocal effects when describing metal–light interactions at the nanometer length scale, especially now that experimental investigation at this scale is becoming possible.

The mathematical formalism of a second numerical technique, the DDA method, was introduced and we provided comparisons between FDTD and DDA results for a cubic nanoparticle that provide insights concerning the nature of the convergence of the two calculations as grid spacing is reduced. In addition, we described a hybrid quantum-mechanics/classical electrodynamics method that provides capabilities to describe the changes in optical properties of both the molecule and particle due to their mutual interactions. The linear response and scattering of a pyridine–Ag spheroidal system was calculated as a function of excitation energy and aspect ratio. Varying the aspect ratio allowed us to tune the position of the LSPR of the spheroid from nearly resonant to red-detuned from the molecule's lowest-lying electronic excitation leading to important changes in the molecular excited-state width.

Figure 6.11   Averaged extinction cross-sections σ_{ext} (upper) and extinction coefficients σ_{ext}/(cross-sectional area) (lower) of the coupled pyridine–silver nanospheroid system as a function of excitation energy and plasmon resonance position. Explicit treatment of the repeated polarisation interaction between a single pyridine molecule and its image in a nearby metal spheroidal particle reveals that the pyridine polarisability and extinction (upper, grey) are sensitive to nanoparticle shape, LSPR position and width, and relative molecule–metal separation and orientation, here chosen so that molecule and image are collinear with each other and with the field-polarisation direction, and separated by $R = 1.6$ nm. For comparison, we also compute the pyridine extinction cross-section based upon an empirical and constant parametrisation of the molecule–metal interaction Γ_{exp} (upper, black) that is consistent with the earlier work.[24] Progression from a pyridine-coated oblate spheroid (right) with radii (50,1,50) nm, to a sphere (middle) with radii (50,50,50) nm, to a prolate spheroid (left) with radii (50,100,50) nm allows for the y-axis plasmon resonance to be tuned from nearly resonant with, to red-detuned from the lowest-lying molecular resonance of the interacting molecule. Extinction coefficients of the coupled system in the monolayer-coverage limit (lower, grey) and bare metal (lower, black) are shown with the excitation field polarised along the y-axis as well as polarised along the z-axis in the oblate case for the bare metal only (lower right, dotted black). Insets not drawn to scale. Reproduced with permission from Figure 2 of ref. 55.

Acknowledgement

NH, LKA and GCS acknowledge the DOE NERC EFRC (DE-SC0000989) for generous support. JMM was supported by AFOSR/DARPA Project BAA07-61 (Grant No. FA9550-08-1-0221), NSF MRSEC (Grant No. DMR-0520513) at the Materials Research Centre of Northwestern University and the Network for Computational Science. DJM gratefully acknowledges financial support from DOE (Grant No. DE-SC0001785).

References

1. C. F. Bohren and D. R. Huffman, *Absorption and Scattering of Light by Small Particles*, Wiley, Weinheim, 2004.
2. U. Kreibig and M. Vollmer, *Optical Properties of Metal Clusters*, Springer-Verlag, Berlin Heidelberg, 1995.
3. A. M. Schwartzberg and J. Z. Zhang, *J. Phys. Chem. C*, 2008, **112**, 10323.
4. M. J. Banholzer, N. Harris, J. E. Millstone, G. C. Schatz and C. A. Mirkin, *J. Phys. Chem. C*, 2010, **114**, 7521.
5. J. E. Millstone, S. J. Hurst, G. S. Metraux, J. I. Cutler and C. A. Mirkin, *Small*, 2009, **5**, 646.
6. C. Nova, A. M. Funston, I. Pastoriza-Santos, L. Liz-Marzan and P. Mulvaney, *J. Phys. Chem. C*, 2008, **112**, 3.
7. J. Boneberg, J. Konig-Birk, H. -J. Munzer, P. Leiderer, K. L. Shuford and G. C. Schatz, *Appl. Phys. A*, 2007, **89**, 299.
8. A. Rai, A. Singh, A. Ahmad and M. Sastry, *Langmuir*, 2006, **22**, 736.
9. S. S. Shankar, A. Rai, A. Ahmad and M. Sastry, *Chem. Mater.*, 2005, **17**, 566.
10. G. Mie, *Ann. Phys.*, 1908, **4**, 377.
11. H. C. van de Hulst, *Light Scattering by Small Particles*, Dover Publications, Inc. New York, 1981.
12. C.-T. Tai, *Dyadic Green's Functions in Electromagnetic Theory*, Intext Educational Publishers, Scranton San Francisco Toronto London, 1971.
13. C. L. Haynes, A. D. McFarland and R. P. Van Duyne, *Anal. Chem*, 2005, **77**, 338A.
14. G. C. Schatz, M. A. Young and R. P. Van Duyne, *Surface Enhanced Raman Scattering: Physics and Applications Topics App. Phys.*, Springer-Verlag, Berlin Heidelberg, 2006.
15. M. Fleischmann, P. J. Hendra and A. J. McQuillan, *Chem. Phys. Lett.*, 1975, **26**, 163.
16. D. L. Jeanmaire and R. P. Van Duyne, *Electroanal. Chem.*, 1977, **84**, 1.
17. M. G. Albrecht and J. A. Creighton, *J. Am. Chem. Soc.*, 1977, **99**, 5215.
18. M. Moskovits, *J. Raman Spect.*, 2005, **36**, 485.
19. L. K. Ausman and G. C. Schatz, *J. Chem. Phys.*, 2009, **131**, 084706.
20. M. Kerker, D.-S. Wang and H. Chew, *App. Opt.*, 1980, **19**, 4159.
21. P. B. Johnson and R. W. Christy, *Phys. Rev. B*, 1972, **6**, 4370–4379.
22. D. J. Masiello and G. C. Schatz, *Phys. Rev. A*, 2008, **78**, 042505.

23. A. Taflove and S. C. Hagness, *Computational Electrodynamics: The Finite-Difference Time-Domain Method*, Artech House, Boston, 2000.
24. J. M. Montgomery, T.-W. Lee and S. K. Gray, *J. Phys. Condens. Matter*, 2008, **20**, 323201.
25. J. Zhao, A. O. Pinchuk, J. M. McMahon, S. Li, L. K. Ausman, A. L. Atkinson and G. C. Schatz, *Acc. Chem. Res.*, 2008, **41**, 1710.
26. K. S. Yee, *IEEE Trans. Antennas Propag.*, 1966, **14**, 302.
27. P. D. Lax and B. Wendroff, *Comm. Pure Appl. Math*, 1960, **13**, 217.
28. J. M. McMahon, J. Henzie, T. W. Odom, G. C. Schatz and S. K. Gray, *Opt. Exp.*, 2007, **15**, 18119.
29. A. L. Atkinson, J. M. McMahon and G. C. Schatz, in *Self-Organization of Molecular Systems: From Molecules and Clusters to Nantubes and Proteins*, ed. N. Russo, Springer, Berlin, Edition edn., 2009, pp. 11–32.
30. K. A. Willets and R. P. Van Duyne, *Annu. Rev. Phys. Chem*, 2007, **58**, 267.
31. J. M. McMahon, S. K. Gray and G. C. Schatz, *Phys. Rev. Lett*, 2009, **103**, 097403.
32. G. S. Agarwal, D. N. Pattanayak and E. Wolf, *Phys. Rev. B*, 1974, **10**, 1447.
33. M. Anderegg, B. Feuerbacher and B. Fitton, *Phys. Rev. Lett*, 1971, **27**, 1565.
34. I. Lindau and P. O. Nilsson, *Phys. Lett.*, 1970, **31A**, 352.
35. W. E. Jones, K. L. Kliewer and R. Fuchs, *Phys. Rev.*, 1969, **178**, 1201.
36. S. Palomba, L. Novotny and R. E. Palmer, *Opt. Commun.*, 2008, **281**, 480.
37. B. B. Dasgupta and R. Fuchs, *Phys. Rev. B*, 1981, **24**, 554.
38. R. Chang and P. T. Leung, *Phys. Rev. B*, 2006, **73**, 125436.
39. R. Ruppin, *J. Opt. Soc. Am. B*, 1989, **6**, 1559.
40. J. M. McMahon, S. K. Gray and G. C. Schatz, *Phys. Rev. B*, 2010, **82**, 03547.
41. J. M. McMahon, S. K. Gray and G. C. Schatz, *J. Phys. Chem. C*, 2010, **114**, 15903.
42. J. M. McMahon, S. K. Gray and G. C. Schatz, *Nano Lett.*, 2010, **10**, 3473.
43. A. D. Boardman, *Electromagnetic Surface Modes*, Wiley, New York, 1982.
44. F. J. Garcia de Abajo, *J. Phys. Chem. C*, 2008, **112**, 17983.
45. E. M. Purcell and C. R. Pennypacker, *J. Astrophys.*, 1973, **186**, 705.
46. J. J. Goodman, B. T. Draine and P. J. Flatau, *Opt. Lett.*, 1991, **16**, 1196.
47. B. T. Draine and P. J. Flatau, *http://arxiv.org/abs/astro-ph/0409262*, 2004.
48. B. T. Draine and P. J. Flatau, *J. Opt. Soc. Am. A*, 1994, **11**, 1491.
49. B. T. Draine, *J. Astrophys.*, 1988, **333**, 846.
50. M. J. Collinge and B. T. Draine, *J. Opt. Soc. Am. A*, 2004, **21**, 2023.
51. N. Felidj, J. Aubard and G. Levi, *J. Chem. Phys*, 1999, **111**, 1195.
52. A. Brioude, X. C. Jiang and M. P. Pileni, *J. Phys. Chem. B*, 2005, **109**, 13136.
53. B. T. Draine and J. J. Goodman, *J. Astrophys.*, 1993, **405**, 685.
54. A. Rahmani, P. C. Chaumet and G. W. Bryant, *J. Astrophys.*, 2004, **607**, 873.
55. D. J. Masiello and G. C. Schatz, *J. Chem. Phys*, 2010, **132**, 064102.

56. A. B. Dahlin, S. Chen, M. P. Jonsson, L. Gunnarsson, M. Kall and F. Hook, *Anal. Chem*, 2009, **81**, 6572.
57. G. J. Nusz, A. C. Curry, S. M. Marinakos, A. Wax and A. Chilkoti, *ACS Nano*, 2009, **3**, 795.
58. A. J. Haes, S. Zou, G. C. Schatz and R. P. Van Duyne, *J. Am. Chem. Soc.*, 2006, **128**, 10905.
59. A. J. Haes and R. P. Van Duyne, *J. Am. Chem. Soc.*, 2002, **124**, 10596–10604.
60. A. J. Haes, L. Chang, W. L. Klein and R. P. Van Duyne, *J. Am. Chem. Soc.*, 2005, **127**, 2264–2271.
61. A. J. Haes, W. P. Hall, L. Chang, W. L. Klein and R. P. Van Duyne, *Nano Letters*, 2004, **4**, 1029–1034.
62. J. Zhao, L. Jensen, J. H. Sung, S. L. Zou, G. C. Schatz and R. P. Van Duyne, *J. Am. Chem. Soc.*, 2007, **129**, 7647–7656.
63. A. M. Kelley, *J. Chem. Phys.* 2008, **128**, 224702.
64. J. Gersten and A. Nitzan, *J. Chem. Phys*, 1980, **73**, 3023.
65. D. W. Lynch and W. R. Hunter, *Handbook of Optical Constants of Solids*, Academic Press, New York, 1985.
66. Y. Shao, L. Fusti-Molnar, Y. Jung, J. Kussman, C. Ochsenfeld, S. T. Brown, A. T. B. Gilbert, L. V. Slipchenko, S. V. Levchenko, D. P. O'Neill, R. A. DeStasio, R. C. Lochan, T. W. Wang, G. J. O. Beran, N. A. Besley, J. M. Herbert, C. Y. Lin, T. Van Voorhis, S. H. Chien, A. Sodt, R. P. Steele, V. A. Rassolov, P. E. Maslen, P. P. Korambath, R. D. Adamson, B. Austin, J. Baker, E. F. C. Byrd, H. Daschel, R. J. Doerksen, A. Dreuw, B. D. Dunietz, A. D. Dutoi, T. R. Furlani, S. R. Gwaltney, A. Heyden, S. Hirata, C.-P. Hsu, G. Kedziora, R. Z. Khaliullin, P. Klunzinger, A. M. Lee, M. S. Lee, W. Liang, I. Lotan, N. Nair, B. Peters, E. I. Proynov, P. A. Pieniazek, Y. M. Rhee, J. Ritchie, E. Rosta, C. D. Sherrell, A. C. Simmonett, J. E. Subotnik, H. L. Woodcock III, W. Zhang, A. T. Bell, A. K. Chakraborty, D. M. Chipman, F. J. Keil, A. Warshel, W. J. Hehre, H. F. Schaefer III, J. Kong, A. I. Krylov, P. M. W. Gill and M. Head-Gordon, *Phys. Chem. Chem. Phys.*, 2006, **8**, 3172.
67. L. Jensen, J. Autschbach and G. C. Schatz, *J. Chem. Phys.*, 2005, **122**, 224115.

CHAPTER 7

Electron Transport Theory for Large Systems

STEFANO SANVITO

School of Physics and Center for Research on Adaptive Nanostructures and Nanodevices (CRANN), Trinity College, Dublin 2, Ireland

7.1 Introduction

Electron transport is a ubiquitous phenomenon in modern technology. An electron current forms the signal of both a transistor and the read head of a hard-disk drive, in addition to being at the heart of the sensing mechanism of chemical and biological species.[1] At the same time, electron transfer plays a fundamental role in biology, being the tool enabling biological species to perform complex functionalities.[2] Finally, an electron current is the end product of photovoltaic energy conversion and it is the vehicle for delivering energy into every house on the planet.

When electron transfer occurs over a distance smaller than the mean free path, *i.e.* over a distance shorter than the one needed by the electrons to undergo an inelastic scattering process, the classical description of electricity based on the Ohm–Kirchhoff laws breaks down. In this limit, the electrical response of a device becomes materials- and, most importantly, device-specific, *i.e.* it is dominated by the details of the device, its electronic structure and its geometry at the atomic level. This is the limit in which quantum mechanics and electronic-structure theory should be applied to the electron-transport problem.

Importantly, such a limit is not simply an exotic academic curiosity, but it is very much on the near horizon of everyday devices. The microelectronic

RSC Theoretical and Computational Chemistry Series No. 4
Computational Nanoscience
Edited by Elena Bichoutskaia
© Royal Society of Chemistry 2011
Published by the Royal Society of Chemistry, www.rsc.org

industry for example in the next 5 to 10 years will move to transistors with a gate length approaching the 10-nm mark. At that length scale the device performances will be strongly hampered by quantum phenomena,[3] such as electron tunneling leading to current leakage. Furthermore, the devices are expected to become strongly sensitive to the precise atomic details of the dopant concentration profile[4] leading to complex manufacturing problems. However, quantum phenomena can also be exploited for improving device performances or for the design of novel device concepts. For instance, quantum-mechanical tunneling combined with band-structure engineering has led to the new generation of magnetic tunnel junctions (MTJs),[5–8] which now operate in any commercially available hard-disk drive. Moreover, proposals for powerless logic elements based on quantum-mechanical interference[9] are now looking for an experimental realisation.

From this landscape it becomes clear how important is a complete understanding of the electron-transfer process at the quantum level. This is difficult to achieve by experiments only, since the transport is determined by the device atomistic details, which are usually poorly known or at least difficult to infer. Therefore, *ab initio* parameter-free transport theory becomes a powerful investigative tool underpinning nano- and bioscience and quantum simulators are rapidly emerging as the natural replacement of conventional device simulators based on the drift-diffusion equation and parametrised electronic structure of materials. This chapter overviews the most recent progresses in constructing efficient and scalable quantum algorithms for electronic transport and presents some of the frontier devices/materials problems that these tools can tackle. In particular I will focus on the issue connected to the scalability of the computational methods to systems containing a large number of atoms, *i.e.* on understanding how quantum theory of electronic transport can be applied to real systems of interest.

And how big is a system of interest? A rough idea can be provided by estimating the number of active atoms present in a transistor with a gate length of 10 nm. Let us assume that the transistor active volume (including dopants, gate and connection to the electrodes) is approximately $10 \times 10 \times 10 \, nm^3$. Such a device contains approximately 45 000 Si atoms, so that 10^5 atoms represents the target for atomistic electronic transport theory: this is the *100k challenge*. Additional challenges may come from the necessity of describing supplementary degrees of freedom different from the electronic ones, like phonons, spinwaves, *etc*. Interestingly, however, this is a rare field where the race between theory and experiments for reaching new length scales is converging. In fact, theory is looking at describing increasingly larger systems, while the experimental activity is devoted to decreasing the typical device dimensions. Interestingly, the time for such a converge of scales is over the near horizon.

This chapter is organised as follows: after briefly overviewing the various computational transport schemes available I will focus my attention on the nonequilibrium Green's functions (NEGF) method combined with density-functional theory (DFT). NEGF + DFT is currently the workhorse of transport calculations and practically the only fully *ab initio* method that is currently capable of tackling systems with a large number of degrees of freedom. How this is achieved will be the subject of an entire section of this chapter, where I will also mention the pitfalls of such a method and the proposed remedies. Finally, I will provide a glance at possible applications where transport theory has been demonstrated to be useful and efficient.

7.2 Foundation of Electron Transport Theory

The typical setup for an electron transport experiment is schematically presented in Figure 7.1(a). Here, an "active region", for the moment defined as a region of space where most of the electrical resistance of the device originates from, is connected to a battery through two electrodes. A current, I, flows through the device across the active region and a resistance is measured. This is a closed system, whose dynamics at the quantum level is described by the quantum Liouville equation

$$i\hbar \frac{\partial \hat{\rho}}{\partial t} = [\hat{H}, \hat{\rho}], \tag{7.1}$$

where $\hat{\rho}$ and $\hat{H}$ are, respectively, the density matrix and the Hamiltonian of the entire device (active region, electrodes and battery). Note that all the degrees of freedom of the problem, both electronic and ionic, are included in eqn (7.1) so that the problem is, in principle, completely defined. Unfortunately, it is not solvable in practice for two main reasons. First, both the electrodes and the battery are macroscopic objects so that the number of degrees of freedom one has to deal with is prohibitively large. Secondly, Coulomb interaction makes the problem intrinsically correlated, so that calculating the dynamics of only a few electrons becomes already a formidable numerical task. The problem thus needs to be recast in an equivalent but more practical form.

An initial natural choice is that of removing both the battery and the largest portion of the electrodes from the problem, and concentrate only on the dynamics of the scattering region and a small part of the electrodes.[10] This essentially means mapping the full closed problem onto an effective one, which has the same dynamics of the original for the degrees of freedom of interest, but it is tractable. It is of course a nontrivial task and many schemes have been designed in the last several years. In general, there are two main strategies.

The first consists in solving an open problem, *i.e.* in replacing the battery and the electrodes with infinite electron reservoirs. These are kept at different chemical potentials, respectively, μ_1 and μ_2, and they are capable of exchanging electrons with the active region, producing the current flow. The extension to multi-electrodes is done in the same way and does not add any new conceptual step. In the linear response limit ($\mu_1 - \mu_2 \to 0$) for noninteracting electrons and under a number of very reasonable assumptions about the nature of the reservoirs, the transport problem can be formally mapped onto a quantum mechanical scattering problem[11,12] (schematically Figure 7.1(b)). One has then to calculate the quantum mechanical scattering matrix, *i.e.* the

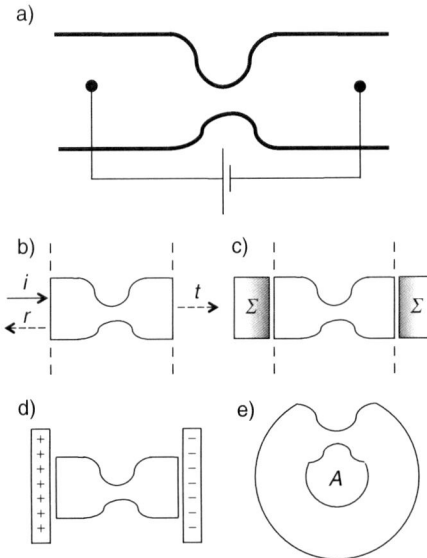

Figure 7.1   Schematic of the electron-transport problem. In (a) the typical transport
setup is shown, with a scattering region (the device) attached to two
electrodes, which in turn connect to the battery. The problem is to calculate the current across the device. This can be reduced, under different
approximations, to: (b) a quantum-mechanical scattering problem, (c) an
embedding potential problem, (d) a capacitor de-charging problem, (e) a
magnetic flux induced persistent current problem. In (c) Σ represents the
embedding potential (the self-energy) and in (e) A is the electromagnetic
vector potential.

quantum-mechanical wavefunction amplitudes for electrons to be reflected,
r, or transmitted, t, across the active region. This is the celebrated Landauer–
Büttiker formalism. Since a scattering problem is relatively simple to implement
numerically in combination with accurate electronic structure theories, it is
still today one of the most used quantum transport method available
for materials-specific calculations. Additionally, it has fewer interpretative
problems than other schemes, *i.e.* it is theoretically compatible with advanced
electronic structure methods. Of course this is limited to steady-state current
in the linear response, *i.e.* it provides only the zero-bias conductance, G, of a
given device.

   In fact the calculation of I–V characteristics and in general the inclusion of
an external bias voltage, V, in the theory presents substantial additional
complications, both at the fundamental and at the practical level. The nonequilibrium Green's functions scheme provides today the most practical solution
to such a problem. The foundations of the NEGF theory were laid down a
long time ago in the very general content of describing nonequilibrium
many-particle systems[13] and only later was the method applied to the transport

problem.[14,15] I will discuss the scheme in some detail in the next section, for the moment I just observe that in practice the NEGF scheme replaces the battery and the electrodes with appropriate embedding potentials, which carry the correct open boundary conditions (schematically Figure 7.1(c)). This remains in the spirit of the scattering theory and in fact in the linear response limit one can easily demonstrate the equivalence between the two.

The second general strategy for solving the electron-transport problem is that of mapping it onto an equivalent time-dependent problem for a closed system. Such a strategy has many variations depending on the general goal of the calculation and of the underlying electronic structure used. One possibility for instance is that of describing the decharging of a large capacitor that encloses the system of interest[16] (schematically Figure 7.1(d)). Alternatively, one can construct a ring and drive the electrical current by threading a magnetic flux across it[17] (schematically Figure 7.1(e)). Time-dependent approaches of course give access not only to the steady-state current but also to transients. The problem, however, is that the time evolution toward the steady state is often problematic and in fact under certain conditions a steady state cannot be reached at all.[18] This poses some questions of whether the time-dependent approach correctly reflects the original transport problem (the one with the battery and the electrode). In any case time-dependent transport can bring additional understanding on many aspects of the electron transfer such as the formation of negative differential conductance[19] or Coulomb blockade.[20] Finally, it is worth mentioning that there are additional schemes that uses time-dependent evolution in the context of the NEGF scheme.[10,21]

The additional element that one needs for a materials-specific transport theory is an accurate electronic-structure scheme[22] and indeed the choice of this sets the capabilities and the limits of the entire method. Furthermore, a mean-field like framework becomes imperative when the goal is that of treating large systems. Recently, density-function theory[23] has become the mainstream method of choice for electron transport. Thus, DFT-based transport schemes have rapidly developed in both the NEGF[24–28] and time-dependent[21,29,30] frameworks. However, the use of DFT in the context of quantum transport poses the fundamental question of whether the theory is rigorously funded.[31] This is usually the case for the time-dependent approach[21,30] but not for the NEGF one, which uses DFT effectively as a single-particle theory, *i.e.* by interpreting the Kohn–Sham eigenvalues[32] as the true electron-removal energies. Some of the consequences of such an interpretation and the possible remedies will be discussed later in this chapter.

In any case, even with such limitations, the NEGF + DFT approach to transport remains probably the most powerful tool available for materials-specific transport calculations. This is because it combines accuracy with undemanding computational overheads. In particular the method is well suitable for developing order-N algorithms, *i.e.* computational solvers that scale only linearly with the number of electrons, N, (or basis functions N_B).

7.3 The Nonequilibrium Green's Function Method

Here, the nonequilibrium Green's function method will be briefly presented and its implementation within the DFT framework for large systems will be discussed. I will not formally derive the theory but simply provide the necessary elements for understanding how it works. For a more formal discussion the reader should refer to the vast literature available.[13,15]

7.3.1 Theoretical Foundation and Simple Model

The main philosophy of the NEGF approach to transport can be understood from the simple energy scheme of Figure 7.2 for a device including a molecule sandwiched between two metallic electrodes. A given molecular level labelled α of energy ε_α is coupled to the electrodes with hopping integral γ_α^n ($n = 1, 2$), so that an electron can escape from the molecule at the rate $\gamma_\alpha^n/\hbar$. The state lifetime thus is $\hbar/\gamma_\alpha$. The two electrodes are kept (by the battery) at two different chemical potentials, μ_1 and μ_2, with $\mu_1 - \mu_2 = eV$, where e is the electron charge. The density of states (DOS) of such a level, $D_\alpha(E)$, is simply

$$D_\alpha(E) = \frac{1}{\pi} \frac{\gamma_\alpha}{(E - \varepsilon_\alpha)^2 + (\gamma_\alpha/2)^2}, \tag{7.2}$$

where $\gamma_\alpha = \gamma_\alpha^1 + \gamma_\alpha^2$ (note that this is for a spin-degenerate energy level). The level cannot be in equilibrium with both the electrodes unless one of the γ s vanishes, a situation describing a molecule in thermodynamic equilibrium with a metal. If both the γ s are not zero then a steady state will form by passing an electron current through the molecule.

The steady-state condition is found by equating the electron fluxes into and out from the molecular level originating from all the electrodes, *i.e.* by imposing that the molecular level electron occupation, n_α, does not change in time. Such a condition leads to

$$n_\alpha = \int\limits_{-\infty}^{+\infty} dE \, D_\alpha(E) \frac{\gamma_\alpha^1 f_1(E) + \gamma_\alpha^2 f_2(E)}{\gamma_\alpha}, \tag{7.3}$$

where $f_1(E)$ and $f_2(E)$ are the Fermi functions evaluated respectively at $E - \mu_1$ and $E - \mu_2$. The same condition provides an expression for the steady-state current, which reads

$$I = \frac{e}{\hbar} \int\limits_{-\infty}^{+\infty} dE \, D_\alpha(E) \frac{\gamma_\alpha^1 \gamma_\alpha^2}{\gamma_\alpha} [f_1(E) - f_2(E)] = \frac{2e}{h} \int\limits_{-\infty}^{+\infty} T_\alpha(E)[f_1(E) - f_2(E)]. \tag{7.4}$$

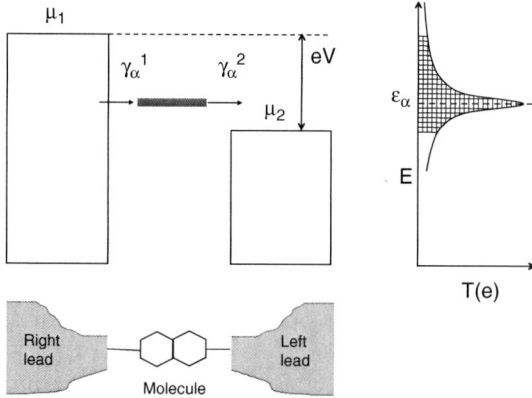

Figure 7.2 Schematic of the NEGF method setup for a device comprising a molecule sandwiched between two metallic leads. In the top left panel the energy-scheme diagram is shown: a molecular level of energy ε_α is coupled to the two electrodes with an escaping rate $\gamma_\alpha^n/\hbar$. The two electrodes are kept at the chemical potential $\mu_n(n=1,2)$. The resulting transmission coefficient as a function of energy is plotted in the right panel. The surface of the shadowed area is proportional to the current.

The second expression establishes a link with the Landaur–Büttiker formalism extended to finite bias, so that T_α (E) is interpreted as the transmission coefficient of the device

$$T_\alpha(E) = \pi \frac{\gamma_\alpha^1 \gamma_\alpha^2}{\gamma_\alpha} D_\alpha(E) = \frac{\gamma_\alpha^1 \gamma_\alpha^2}{(E-\varepsilon_\alpha)^2 + (\gamma_\alpha/2)^2}. \tag{7.5}$$

From eqn (7.5) it appears that the transmission coefficient is simply a Lorentzian function centred at the energy of the molecular level and with broadening given by the sum of the hopping integrals. This effectively describes a resonance so that the current is simply given by the integrals of all the resonances comprised into the bias window (see Figure 7.2).

The dynamics of this model is then rather simple. Consider the weak-coupling limit for simplicity where the γ s are smaller than any other energy of the problem. In this case D_α $(E) \sim 2\delta(E-\varepsilon_\alpha)$ and both the occupation and the current are solely determined by the position of the energy level with respect to the chemical potential of the leads. If ε_α is larger than both μ_1 and μ_2, then $n_\alpha \approx 0$ and no current flows (empty states do not carry current). In contrast, if the energy level is below the chemical potentials of both leads, then $n_\alpha \approx 2$, but the current is still zero (completely filled states do not carry current). Finally, if $\mu_1 < \varepsilon_\alpha < \mu_2$ or $\mu_2 < \varepsilon_\alpha < \mu_1$ the occupation is $0 < n_\alpha < 2$ and the current can flow. Note also that in the zero-bias ($|\mu_1 - \mu_2| \to 0$) and zero-temperature limit the current vanishes but the conductance, G, is well defined and reads

$$G = \frac{2e^2}{h} T_\alpha(E_F), \tag{7.6}$$

where E_F is the common equilibrium chemical potential of the leads (assumed to be identical), *i.e.* their Fermi energy.

This simple approach allows one to evaluate all the transport properties of a system when the energy of the molecular level is known. However, this is not the case in general. In first approximation, in fact, the energy position ε_α depends on the occupation of the molecule (in this case the single molecular level), *i.e.*

$$\varepsilon_\alpha = \varepsilon_\alpha(n_\alpha). \tag{7.7}$$

Equations (7.3) and (7.7) now form a closed system of equations, that can be solved, for instance in an iterative way, to give the stationary occupation of the molecular level and its energy ε_α. The entire scheme can be then generalised to a collection of N energy levels (the spectrum of a given molecule) where now $\varepsilon_\alpha = \varepsilon_\alpha(n_1, n_2, \ldots, n_N)$. This is the most general situation addressed by the NEGF method.

7.3.2 The NEGF Method: Basic Algorithm and Numerical Considerations

The setup for a NEGF calculation is identical to that of the simple model just described: two electrodes kept at different chemical potentials are electronically coupled to a central region, named the extended molecule (EM – or the "active region" as previously named in the introduction), with which they are capable of exchanging electrons (see Figure 7.3(a)). The spirit of such a tripartition is that of isolating the portion of the device, whose electronic structure and electron occupation are not affected by the nonequilibrium condition namely by the current flow, so that they can be treated at a statistical level. In a transport setup these are the macroscopical current/voltage electrodes. Whether such an approximation is always valid in general and whether corrections to the electronic structure of the leads need to be applied at the steady state is still a matter of debate.[33] When the electronic structure is described at a mean-field level an operative partition is that of defining as EM the region of space where the electrostatic potential drops. The external bias is then introduced simply as a rigid shift of the entire electrodes spectrum.

The central quantity of the NEGF formalism, which effectively replaces the DOS of the simple model [eqn (7.2)], is the nonequilibrium Green's function G for the extended molecule

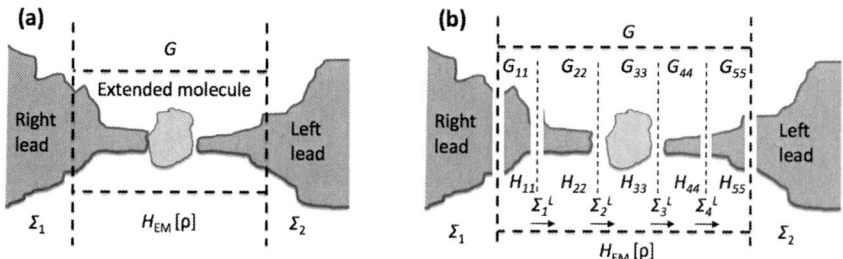

Figure 7.3 (a) Schematic two-terminal device: two leads are kept at different chemical potentials and they are able to exchange electrons with the central region (the extended molecule). In (b) the extended region is broken in small units and the associated Green's function, G, is constructed recursively (see text).

$$G(E) = \lim_{\eta \to 0^+} [(E + i\eta) - H_{EM} - \Sigma_1 - \Sigma_2]^{-1}. \tag{7.8}$$

Here, H_{EM} is the Hamiltonian for the EM in isolation, *i.e.* when it is decoupled from the electrodes, and Σ_n ($n = 1,2$) is the self-energy for the nth electrode. The self-energies contain information about the electronic structure of the electrodes, about their coupling to the extended molecule and about the external bias. These are energy-dependent non-Hermitian objects that should be computed separately from a ground-state electronic-structure calculation. There are several ways for calculating the Σs depending on the specific electronic-structure theory used and on the type of basis functions employed to expand the operators. In the case of a theory based on localised atomic orbitals the most efficient algorithms are obtained by semianalytic constructions[34,35] although recursive methods are still quite popular.[36,37]

A direct connection with the simple model presented before can be obtained by considering the spectral representation of G

$$G = \sum_\alpha \psi_\alpha \frac{1}{E - (\varepsilon_\alpha - i\gamma_\alpha/2) + i0^+} \tilde{\psi}_\alpha^\dagger, \tag{7.9}$$

where $\psi_\alpha, \tilde{\psi}_\alpha$ and $(\varepsilon_\alpha - i\gamma_\alpha/2)$ are, respectively, the right and left eigenvectors and the associated eigenvalues of H_{eff}, defined as

$$H_{eff} = H_{EM} + \Sigma_1 + \Sigma_2. \tag{7.10}$$

Importantly it is possible to show that

$$\gamma_\alpha^n = \tilde{\psi}_\alpha^\dagger [i(\Sigma_n - \Sigma_n^\dagger)] \tilde{\psi}_\alpha, \tag{7.11}$$

with $\gamma_\alpha = \gamma_\alpha^1 + \gamma_\alpha^2$. Note that the DOS of the EM can be calculated as $D(E) = -\frac{1}{\pi}\text{Im}G$, which returns eqn (7.2) for each single eigenvalue. Thus, the self-energies simply replace the γs of the simple model and describe the escape rate of the electrodes to and from the extended molecule.

The second element of the NEGF theory is an expression for the non-equilibrium occupation of the extended molecule. The convenient quantity in this case is the density matrix, ρ, and the expression that replaces eqn (7.3) reads

$$\rho = \frac{1}{2\pi} \int dE\, G[\Gamma_1 f_1 + \Gamma_2 f_2]G^+, \quad \text{where} \quad \Gamma_{1/2} = i\left[\Sigma_{1/2} - \Sigma_{1/2}^\dagger\right]. \quad (7.12)$$

In this case the system of eqns (7.8) and (7.12) can be closed, if H_{EM} has a functional dependence on the density matrix, *i.e.* $H_{EM} = H_{EM}[\rho]$. This is the case of the Kohn–Sham Hamiltonian from DFT, which makes DFT readily applicable to the NEGF method. I remark here again that this is an improper way of using DFT, since the DFT functional is defined only at the ground-state density and not at a steady-state one. Thus, effectively DFT is used as a single-particle theory with the Kohn–Sham states taking the place of the real quasi-particles. Before doing any calculation one should always be confident that for the particular problem under investigation the Kohn–Sham spectrum is a good approximation of the real one.

After the self-consistent density matrix and Hamiltonian for the EM are calculated then the current can be obtained by using a Landauer–Büttiker-like formula

$$I = \frac{2e}{h} \int dE\ \text{Tr}(\Gamma_1 G^\dagger \Gamma_2 G)[f_1 - f_2] = \frac{2e}{h} \int dE\ \text{T}(E;V)[f_1 - f_2]. \quad (7.13)$$

I now briefly review the numerical aspects of the NEGF + DFT algorithm. Let us assume, as in the vast majority of the quantum transport codes available, that the electronic structure (G, H_{EM}, Σ_n,...) is expanded over a spatially localised basis set. This might use atomic orbitals of some kind,[24–28] Wannier functions,[38] or a real-space numerical grid.[39] In any case let us assume that the extended molecule is described by N_B basis functions, so that G and the Σs are $N_B \times N_B$ matrices. Thus, in any self-consistent step leading to the steady-state density matrix one needs to invert a $N_B \times N_B$ matrix [eqn (7.8)] as many times as the energy points, N_E, needed to perform the integral in eqn (7.12). Matrix inversion is an operation that typically scales as N_B^3. This is the same scaling as standard matrix diagonalisation in mainstream ground-state DFT algorithms (better scaling can be obtained in DFT by particular choices of basis set, implementation, *etc.*). This means that, as it stands, the NEGF + DFT algorithm is N_E times more computational demanding than standard ground-state DFT. We will see in the following how it is possible to improve on this scaling and to reduce the general computational overheads. Interestingly parallelisation over energy is trivial and requires no communication, so that effectively a NEGF + DFT algorithm run over N_E processors has the same computational overheads as a ground-state DFT calculation run in serial mode.

7.3.3 Inelastic Effects

We have previously observed that the coupling of the molecule with the electrodes has the effect of broadening the molecule's DOS (see for instance eqn (7.2) and eqn (7.9)). This reflects the fact that when coupled to the electrodes the molecular levels acquire a finite lifetime since electrons can leak in and out. Apart from this there are no other interactions in the problem capable of producing a finite lifetime (*i.e.* to provide a purely imaginary contribution to the molecular level eigenvalue). This means that an electron populating a particular molecular orbital will occupy the orbital until it is absorbed by one of the two electrodes. In such a situation the transport is entirely elastic, meaning that the electron energy does not change when it passes across the device. Quantum interference is a direct consequence of the elastic electron transport process.

In reality, however, electrons can emit or absorb energy during scattering events by interacting with other degrees of freedom of the system. For instance, an electron of energy E_{in} can excite a lattice vibration of frequency ω and be absorbed by another electrode at an energy $E_{out} = E_{in} - \hbar\omega$. Such a process indeed introduces a second contribution to the lifetime, $\hbar/\gamma_{inelastic}$, which does not vanish if the interaction with the electrodes is switched off. The main question is how to relate $\gamma_{inelastic}$ to the microscopic description of the system. The NEGF method can handle this problem.

At a phenomenological level the inclusion of inelastic effects can be carried out by using an original idea from Büttiker[40] (the "Büttiker probe" method). The main observation is that, in a first approximation, the effects brought in by inelastic scattering processes are identical to those arising from the interaction of the molecule with the electrodes, *i.e.* the molecular levels acquire a finite lifetime. One can then map the system under investigation, which includes inelastic scattering, on a fictitious one where the transport is completely elastic. In the fictitious system, however, the molecule is coupled *via* a number of leads to several electron reservoirs (see Figure 7.4). The chemical potential of the fictitious reservoirs is determined by the condition that there is no net current to any reservoir. With this setup an electron that carries net current between the two real electrodes can be absorbed and re-emitted by the fictitious reservoirs. In doing so, information about its quantum mechanical phase is lost and an inelastic process is mimicked.

The "Büttiker probe" method is well suited for the NEGF scheme since the fictitious reservoirs enter into the formalism as additional lifetimes

$$G = \sum_\alpha \psi_\alpha \frac{1}{E - \left[\varepsilon_\alpha - \frac{i}{2}\left(\gamma_\alpha^1 + \gamma_\alpha^2 + \Sigma_n\gamma_{inelastic}^{fn}\right)\right] + i0^+} \tilde{\psi}_\alpha^\dagger, \qquad (7.14)$$

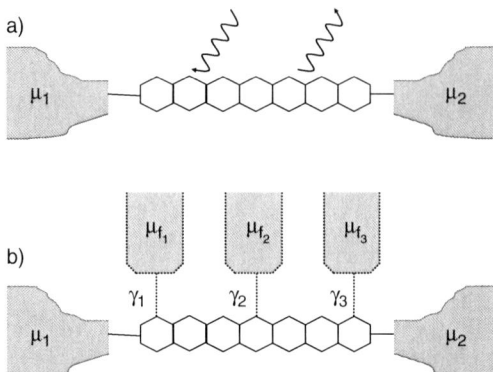

Figure 7.4 Schematic of the Büttiker probe approach to inelastic transport. A two-probe device where inelastic processes are active (a) is mapped onto a fictitious multiprobe device with elastic scattering only. The chemical potential of the fictitious reservoirs $\mu f_1, \mu f_2 \ldots$ is adjusted until no net current flows to the fictitious reservoirs.

where $\hbar/\gamma^{f_n}_{\text{inelastic}}$ represents the inelastic contribution to the lifetime of the αth level due to the nth fictitious reservoir. In this case ψ_α and ε_α form the solution of a Schrödinger-like equation with effective Hamiltonian given by

$$H_{\text{eff}} = H_{\text{EM}} + \Sigma_1 + \Sigma_2 + \sum_n \Sigma_{f_n}, \qquad (7.15)$$

where Σ_{f_n} is the self-energy associated to the nth fictitious reservoir. Therefore, the method essentially introduces additional self-energies to the definition of the nonequilibrium Green's function

$$G(E) = \lim_{\eta \to 0^+} \left[(E + i\eta) - H_{\text{EM}} - \Sigma_1 - \Sigma_2 - \sum_n \Sigma_{f_n} \right]^{-1}. \qquad (7.16)$$

Clearly, the Büttiker probe scheme lacks completely of any information concerning the nature of the scattering process leading to the loss of coherent transport. As such, beyond the fact that a large coupling between the fictitious electrodes and the molecule produces a large inelastic contribution to the current, the construction of the self-energies for the fictitious leads remains largely arbitrary. Despite this drawback the method, in combination with NEGF + DFT, has been useful for problems where a microscopic knowledge of the inelastic scattering processes was not necessary, although a detailed electronic structure was needed. This is for instance the case of magnetic tunnel junctions[35,41] or transport in molecular junctions.[42] The advantages of the method are clear, since the calculation presents only a minor additional complications with respect to the elastic case, namely the evaluation of the chemical potentials of the fictitious reservoirs.

A problem, however, arises when the microscopic details of the interaction leading to inelastic scattering are important. This is, for instance, the case in inelastic electron tunnel spectroscopy (IETS), where the knowledge of the I–V curve and its derivative allows one to extract information on the normal modes of oscillations of a molecule.[43] IETS has a magnetic counterpart where the excitations have a spin-nature and they are produced by an electron current over a system of spins.[44] There is no single way to tackle this problem. If one wants to remain within the NEGF + DFT framework, the convenient way is that of constructing an appropriate self-energy describing the inelastic process under consideration. Such a construction usually requires a number of approximations that need to be carefully monitored.[45]

Let us consider for example the case of electron–phonon interaction. This is a two-body problem, since the scattering of an electron generates the phonon. A common way to treat such an interaction is *via* a perturbative approach best known as the Born approximation, either in its first order or self-consistent form. The perturbation is about the electron–phonon coupling constant, *i.e.* the perturbative expansion is strictly valid for weak coupling between the electronic and the vibrational degrees of freedom. The self-energy is then constructed within the Hartree–Fock approximation in the electron–phonon coupling.[46] Most of the ingredients needed for constructing the electron-phonon contribution to the electronic self-energy can be calculated from equilibrium DFT, for instance using linear response theory (the quantities to calculate are the phonon spectrum and the electron–phonon coupling). However, some of them are crucially missing. In particular the evaluation of the nonequilibrium phonon population in the presence of an electron current is usually unknown, since it requires the precise knowledge of how the phonons excited in the molecule dissipate away into the electrodes. This essentially requires the simultaneous solution of some dynamical equations for the phonons.[47] Importantly, different conditions over the phonon dissipation may result in a different behaviour of the electron current.[47,48]

Still NEGF + DFT combined with the Born approximation to the electron–phonon interaction has led to important predictions about the IETS in molecular junctions[49] and metallic point contacts.[50] The entire perturbative approach, however, breaks down when the electron–phonon coupling becomes large,[45] and then it must be abandoned. This is a situation encountered in conducting polymers and in general in phenomena where the current density is high enough to induce electron migration. The NEGF + DFT method then becomes inappropriate and either nonperturbative approaches or time-dependent ones[51] are the only resource. Even in this framework the problem remains complex and the different levels of theory give access to different physical properties.[52] For instance time-dependent molecular dynamics in the presence of an electron current and treated at the Ehrenfest level can describe processes, where an ion relaxes energy into an electron flux. However, it cannot describe the reverse energy transfer leading to electron migration,[52] for which a proper time-dependent electron–ion correlated theory is needed. Interestingly, a recent comparison between static (NEGF and perturbative electron–phonon coupling) and dynamic inelastic transport calculations has revealed a substantial agreement over a wide range of conditions for quantities such as the local current-induced heating and inelastic corrections to the current.[53] This is an important result suggesting that the relatively computationally inexpensive NEGF + DFT scheme can be used for studying important thermoelectric properties of materials.

7.3.4   Large-Scale Simulations: Tackling the *100k Challenge*

I will now focus on the numerical algorithms needed to scale the NEGF + DFT scheme to large devices, *i.e.* on how the *100k challenge* can be tackled. Effectively, there are three main parts of the NEGF + DFT algorithm that can cause large computational overheads: (1) the matrix inversion leading to the nonequilibrium Green's function [eqn (7.8)], (2) the large energy sampling that one needs in order to obtain the nonequilibrium density matrix [eqn (7.12)], and (3) the Poisson solver needed for evaluating the electrostatic potential across the device. In what follows we will discuss these three aspects in some details, spending also some time in reviewing the level of parallelisation that the algorithm can achieve. For the entire discussion I will refer to a mean-field electronic-structure theory (say Kohn–Sham DFT) constructed over a localised basis set.

7.3.4.1   *Calculation of the Green's Function*

At present the most time-consuming operation in the NEGF scheme is the matrix inversion leading to G [see eqn (7.8)]. Direct matrix inversion becomes prohibitively expensive for $N_B > 10\,000$. In contrast, recursive algorithms perform effectively for extremely large matrices (N_B in the $1\,000\,000$ range), but they are usually slow. This means that the range $10\,000 < N_B < 1\,000\,000$ is a "no-algorithm land", *i.e.* there is no general computational strategy that universally performs well in that region. Physical considerations, however, help in simplifying the problem.

   The main concept behind all algorithms for order-N electronic-structure theory, including transport, is the concept of the "nearsightedness" of the electrons in many-atom systems.[54] This essentially establishes that for a fixed chemical potential, the local properties of an electron system, for instance the electron density at a point in space $\rho(\vec{r})$, depend significantly only on the effective external potential at close neighbouring points. This means that changes in the potential, even if very large, generated beyond a certain distance from the point under consideration have small effects on the local properties at that point. This is the concept underpinning very successful order-N algorithms such as the "divide and conquer" method.[55] The most immediate consequence of such a general property is that, in order to calculate the electronic structure of a large system, one can partition it into smaller units (called "cells") and calculate them independently from each other. Of course one needs to describe the possibility of electrons moving across cells, but still it is expected that only the neighbouring cells will affect the density matrix of any given cell. Suppose the system has been partitioned in M identical cells (this is not a stringent condition, in general, cells can have a different size) containing N_C basis function, so that $N_B = N_C \times M$, then an algorithm based on the concept of partitioning the space will scale as $N_C^3 \times M$ and not as $(N_C \times M)^3$.

There are a number of schemes for exporting such a concept to the NEGF method. Here, I will review the recursive embedding potential algorithm developed originally by Inglesfield[56] for surfaces and then extended to transport in the context of Huckel theory,[57,58] and finally only recently used within DFT.[59] In particular, I will only present the zero-bias limit, although an extension to finite bias is possible.[60]

Let us assume to have performed the aforementioned partitioning of the Hamiltonian of the extended molecule excluding the electrodes, as graphically illustrated in Figure 7.3. The strategy to partition the system is that the different cells should interact only with the neighbouring ones. As a matter on notation I will call H_{ii} the Hamiltonian matrix block defining the interaction within the ith cell and H_{ii+1} the Hamiltonian matrix block defining the coupling between the ith and the $(i+1)$th cells (note $H_{i+1i} = H_{ii+1}^+$). The Hamiltonian H_{EM} is then block-tridiagonal and writes as

$$H_{EM} = \begin{pmatrix} H_{11} & H_{12} & 0 & \cdots & 0 & 0 \\ H_{21} & H_{22} & H_{23} & \cdots & 0 & 0 \\ 0 & H_{32} & H_{33} & \cdots & 0 & 0 \\ \vdots & \vdots & \vdots & \ddots & \vdots & \vdots \\ 0 & 0 & 0 & \cdots & H_{M-1M-1} & H_{M-1M} \\ 0 & 0 & 0 & \cdots & H_{MM-1} & H_{MM} \end{pmatrix}. \tag{7.17}$$

Note that the overlap matrix S_{EM} has the same structure of H_{EM}. The Green's function for the EM [eqn (7.8)] can be written in the same block-form (however, in general G is not block-tridiagonal)

$$G = \begin{pmatrix} G_{11} & G_{12} & \cdots & G_{1M-1} & G_{1M} \\ G_{21} & G_{22} & \cdots & G_{2M-1} & G_{2M} \\ G_{31} & G_{32} & \cdots & G_{3M-1} & G_{3M} \\ \vdots & \vdots & \ddots & \vdots & \vdots \\ G_{M-11} & G_{M-12} & \cdots & G_{M-1M-1} & G_{M-1M} \\ G_{M1} & G_{M2} & \cdots & G_{MM-1} & G_{MM} \end{pmatrix}. \tag{7.18}$$

Note that at zero-bias $f_1 = f_2$ in eqn (7.12) so that the density matrix is defined only by the Green's function for the EM (see also the next section).

Importantly, when the Hamiltonian is defined over an atomic-like orbital basis set only a limited number of matrix elements of ρ are sufficient to determine $H_{EM}[\rho]$ in the self-consistent procedure. In fact, the real-space representation of the density matrix writes $\rho(\mathbf{r}) = \Sigma_{ij}\rho_{ij}\chi_i(\mathbf{r})\chi_j(\mathbf{r})$ with $\chi_i(\mathbf{r})$ the generic ith atomic-like basis function. This means that matrix elements corresponding to pairs of orbitals with zero or negligible overlap $S_{ij} = \int \chi_i(\mathbf{r})\chi_i(\mathbf{r}) \to 0$ do not contribute to $\rho(\mathbf{r})$. Practically, one can neglect all the matrix elements in the density matrix that correspond to vanishing matrix elements of the associated overlap matrix S. With our assumption over the form of S, one needs to calculate only the tridiagonal blocks of ρ and therefore of G. Furthermore, if the basis set is orthogonal, with no overlap between basis functions sitting at different atomic positions, then only the diagonal matrix elements are needed. The embedding scheme consists in calculating these diagonal block recursively.

The idea is that the diagonal blocks of the Green's function corresponding to the ith cell can be obtained simply by inverting the H_{ii} block plus an appropriate embedding potential. The starting point of the method is the calculation of the Green's function for the first cell in isolation, *i.e.* without considering its interaction with the rest of the system. This is simply done by inverting the Green's equation associated to the Hamiltonian block H_{11}, which also includes the self-energy of the left-hand side lead, *i.e.* by solving

$$(ES_{11} - H_{11})G_{11}^L = I, \tag{7.19}$$

where S_{11} is the corresponding block of the overlap matrix. One can then use Dyson's equation to construct the embedding potential of one cell into another. For instance, the effect of the first cell onto the second is described by the self-energy

$$\Sigma_2^L = (ES_{21} - H_{21})G_{11}^L(ES_{12} - H_{12}), \tag{7.20}$$

in such a way that the Green's function for the second cell can be now constructed as

$$G_{22}^L = (ES_{22} - H_{22} - \Sigma_2^L)^{-1}. \tag{7.21}$$

In eqn (7.21) the new Green's function for the cell 2, G_2^L, now includes the interaction with cell 1. This scheme can then be continued effectively providing a recursive algorithm for evaluating the Green's functions of all the partitions. The final recursive relation is

$$\Sigma_i^L = (ES_{ii-1} - H_{ii-1})G_{i-1i-1}^L(ES_{i-1i} - H_{i-1i}), \tag{7.22}$$

where the Green's function G_{ii}^L is the Green's function constructed by recursively embedding cells from the left-hand side

$$G_{ii}^L = (ES_{ii} - H_{ii} - \Sigma_i^L)^{-1}. \tag{7.23}$$

The same procedure can then be applied by starting from the last of the cells and recursively embedding from the right. The "right" self-energy for the ith cell is now

$$\Sigma_i^R = (ES_{ii+1} - H_{ii+1})G_{i+1i+1}^R(ES_{i+1i} - H_{i+1i}), \tag{7.24}$$

where the Green's function G_{i+1i+1}^R is that incorporating the right embedding potentials

$$G_{ii}^R = (ES_{ii} - H_{ii} - \Sigma_i^R)^{-1}. \tag{7.25}$$

Finally, once all the embedding potentials both from the left and from the right have been calculated, the Green's function of each individual cell can be evaluated

$$G_{ii} = (ES_{ii} - H_{ii} - \Sigma_i^L - \Sigma_i^R)^{-1}. \tag{7.26}$$

The final computational cost for obtaining all the diagonal blocks of the Green's function is proportional to $7 \times M \times N_C^3$. In fact, one needs to perform M matrix inversions and $2M$ matrix multiplications for constructing each of the embedding potentials and M final matrix inversions for solving eqn (7.26). Each of these operations scales as N_C^3, which gives the final scaling.

Since the starting point of this scheme is a matrix written in a block-tridiagonal form the method naturally applies to one-dimensional (or quasi-1D) objects. However, there is a second property that usually accompanies a DFT-derived Hamiltonian expanded over localised basis sets. This is the fact that usually those matrices are highly sparse, *i.e.* most of their matrix elements vanish. In this situation there are efficient algorithms to reduce H_{EM} into a block-tridiagonal form,[62,63] which allows one to tackle problems where the device has a complex geometry. The use of recursive algorithms with NEGF and electronic structures at various levels of approximation has mainly focused on graphene-based two-dimensional devices,[62–64] although examples of magneto-tunnel junctions[59] and three-terminal molecular devices[65] can also be found in the literature.

7.3.4.2 Density Matrix Energy Sampling

As mentioned before the second crucial point in the NEGF scheme is the integral leading to the charge density [eqn (7.12)], since the number of energy points N_E determines how many times one needs to calculate the Green's function $G(E)$ [eqn (7.8)] at every self-consistent step. It is thus crucial to keep N_E to a minimum. A standard way of manipulating the integral of eqn (7.12) is that of adding and subtracting the term $G\Gamma_2 G^+ f_1$ and to rewrite ρ as the sum of two contributions $\rho = \rho_{eq} + \rho_{neq}$, respectively

$$\rho_{eq} = -\frac{1}{\pi} \int \text{Im}[G] f_1 dE, \tag{7.27}$$

and

$$\rho_{neq} = -\frac{1}{2\pi} \int [G\Gamma_2 G^\dagger (f_2 - f_1)] dE. \tag{7.28}$$

We interpret the first contribution as the "equilibrium" density matrix, since it is the one resulting from the two electrodes having the same chemical potential (only one Fermi function enters in the definition of ρ_{eq}). The second one instead contains the nonequilibrium corrections. From a numerical point of view ρ_{neq} is bound by the two Fermi functions, *i.e.* it has to be calculated over a limited energy window roughly corresponding to the bias window. In contrast, ρ_{eq} is unbound, but the integral can be performed in the complex plane using a standard contour integral technique,[61] since $\text{Im}[G]$ is both analytical and smooth. Usually calculating ρ_{eq} is not numerically problematic and integrations over a few hundred points are enough to converge. In contrast, the integration leading to ρ_{neq} can be rather challenging if the DOS of the extended molecule is extremely sharp. This essentially means that the integration is problematic when the molecular orbitals of the extended molecule are weakly coupled to the electrodes, an almost unavoidable situation in large systems.

Let us consider for example the case of the two macromolecules that will be discussed in the next sections, namely the single-molecule magnet Mn_{12}[66] and polymeric poly(G)-poly(C) double-stranded DNA.[67] In the first case, both the highest-occupied molecular orbital (HOMO) and the lowest-unoccupied molecule orbitals (LUMO) manifolds originate from the Mn d-shell. This is intrinsically localised in space, so that one would expect a weak electronic coupling with the electrodes ($\gamma_\alpha \ll |\varepsilon_\alpha - E_F|$). Furthermore, the Mn ions are spatially quite far from the surface of the electrodes so that a further suppression of the electronic coupling is expected. A similar situation is encountered for DNA where electronic states associated to bases in the middle of the strand are only weakly coupled to the electrodes.

In these two situations we have encountered molecular states with a broadening in the range of 10^{-6} eV (the lifetime is of the order of 0.1 ns). When considering a bias window of 1 V, one then needs to evaluate ρ_{neq} over about 10^7 points (by assuming to expand the peaks over 10 energy grid points). This is clearly not possible. Naively, one would be tempted to simply neglect such narrow peaks, since their contribution to the current will be minimal [note that the transmission coefficient depends on the product of the γs of the two electrodes – see eqn (7.5)]. However, neglecting them will result in an erroneous electron counting, *i.e.* in a severely compromised electrostatic potential, which in turn will affect the transport across the broader states that conducts the most.

The crucial aspect of this problem is that although a very fine energy resolution is required for integrating the extremely localised energy levels corresponding to weakly coupled molecular orbitals, there are some substantial portions of the DOS that are smooth in energy and extended energy regions where there is no DOS at all. This suggests a possible solution to the problem, which was implemented in the last few years in the NEGF + DFT *Smeagol* code.[66,67] It consists in using an adaptive mesh algorithm in order to integrate accurately only those energy regions where there is nonvanishing DOS. The strategy is to introduce first an artificial broadening to the states of the extended molecule, by simply including an additional γ in the expression (7.9). This allows one to identify the peaks in the DOS with a coarse mesh. The mesh is then refined only around the peaks in the DOS and then the artificial broadening is reduced. The procedure is then repeated until the artificial broadening is smaller than the natural linewidth originating from the interaction with the electrodes. An example of how the adaptive mesh works is presented in Figure 7.5. Importantly, as mentioned before, the energy grid integration is an operation that can be easily parallelised and distributed over a large number of CPUs.

7.3.4.3 Poisson Solvers

The final main ingredient of a NEGF algorithm is a solver for extracting the electrostatic potential (the Hartree potential in DFT) of the extended molecule, once its nonequilibrium charge density has been calculated. One essentially needs to solve the Poisson equation with the appropriate boundary conditions imposed by the contact to the electrodes. Ground-state DFT codes solve the Poisson equation in reciprocal space by using the highly efficient fast Fourier transform (FFT) algorithm. This is sometime used in NEGF codes.[28] However, FFT essentially uses periodic boundary conditions, which can be adapted to the transport problem if one considers a two-terminal device. However, when the geometry of the device under investigation becomes more complex and the number of current/voltage probes increases FFT must be abandoned and one has to look for real-space methods.

This is the realm of the multigrid methods, which are a group of algorithms for solving differential equations using a hierarchy of discretisation steps. The main strategy of these approaches is to discretize the charge density and the electrostatic potential over a coarse real-space grid, which gets progressively refined. Essentially, they work on a principle similar to that of performing interpolation between coarser and finer grids. So far, the use of multigrid algorithms has been relatively limited, mostly because of a certain inertia from the electronic-structure community to abandon the very successful FFT

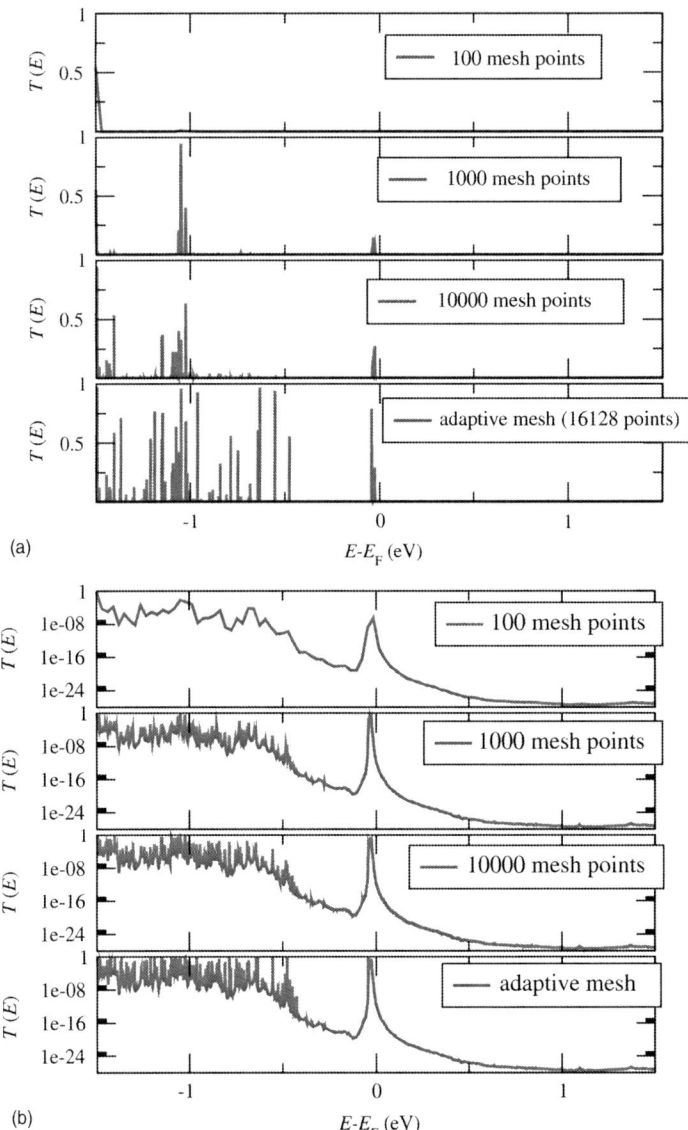

Figure 7.5 Demonstration of the resolution achievable with an adaptive mesh algorithm for integrating the charge density. The transmission coefficient is shown as a function of energy for a 12 base pair double-stranded poly(G)-poly(C) molecule attached to two Au electrodes *via* thiol groups. The first three panels from the top refer to an increasingly fine energy mesh, while the last one is for an adaptive mesh. In the picture to the left $T(E)$ is plotted on a linear scale, while in the one to the right the scale is logarithmic. Note that the adaptive mesh is capable of capturing a much larger number of peaks than those accessible by a uniform mesh with approximately the same number of points. The calculations presented here employ the LDA exchange and correlation potential.

algorithm. The situation is, however, rapidly changing and several mainstream electron-transport codes have moved to such real-space methods.[65,68] Interestingly multigrid approaches have been specifically developed for graphics processing units (GPUs), demonstrating a significant speedup as compared with standard CPUs[69] and paving the way for important hybrid computational architectures specifically designed for the quantum transport problem.

7.3.4.4 Parallelism

Typical NEGF + DFT algorithms can support four levels of parallelism: 1) bias-voltage (needed for a full *I–V* curve), 2) *k*-points (needed only if periodic boundary conditions are employed in the direction orthogonal to the transport), 3) energy grid (needed for integrating the density matrix), and 4) atomic orbitals. For this reason, NEGF + DFT algorithms are ideal for scaling to large computational infrastructures and demonstrations of efficient runs at petaflop facilities have been already produced for a NEGF code using empirical tight-binding electronic structure.[68] The first three levels of parallelism are somehow trivial since they all require very little communication between parallel running CPUs. Usually, the self-consistent cycle at finite bias converges faster if the starting density matrix for a given value of the bias voltage is the converged one calculated at a similar bias. This suggests that it is usually a good practice to run sequentially the cycle over the bias (alternatively, different segments of the *I–V* curve can be run in parallel). The *k*-point sampling is needed only if periodic boundary conditions in the transverse direction are applied. In some situations the sampling can be rather extended (see Section 7.4.1), enough to represent the largest portion of the calculation. Again, adaptive *k*-point mesh techniques can improve the quality of the sampling and keep down the computational overheads. The last two levels of parallelism (energy and orbitals) are present in any calculations, although for small systems parallelisation over orbitals does not produce speedup, since the large prefactor in the scaling of order-*N* algorithms makes them unfavourable against direct matrix inversion for small systems.

As an example let us estimate the typical CPU count for running a transport calculation for a relatively large molecule comprising about 1000 atoms. The electronic-structure calculation typically requires 15 000 basis orbitals, which can be partitioned in about 15 cells. If the lateral dimensions of the extended molecule are large only a few *k*-points are needed, typically around 4. For such molecules, experience suggests that one usually needs about 10 000 energy points (see Section 7.4.2). Finally, let us assume that the *I–V* curve is sampled over 20 bias points (for voltages in the range $[-1,1]$ V with a 0.1 V resolution). This gives us a run that can be parallelised over $15 \times 4 \times 10\,000 \times 20 = 12\,000\,000$ CPUs.

7.3.5 When Kohn–Sham DFT Goes Wrong

In this section I would like to briefly discuss the implications of the "improper" use of Kohn–Sham theory as a single-particle theory for electron transport.

In order to make the discussion simpler let us assume that for the device that we want to describe the transport is only through the HOMO of the molecule. In this case an exact result of DFT establishes that $\varepsilon_{\text{HOMO}}^{\text{KS}} = -I_N$, *i.e.* that the Kohn–Sham eigenvalue of the HOMO is the negative of the true ionisation potential on the N-electron molecule (the neutral molecule).[70–72] Unfortunately, for most of the local and semilocal approximations of the exchange and correlation functional, the local density and the generalised gradient approximations (LDA and GGA), this is not the case and $\varepsilon_{\text{HOMO}}^{\text{KS}}$ is predicted rather far from $-I_N$ (usually by several eV). This essentially means that $\varepsilon_{\text{HOMO}}^{\text{KS}}$ will incorrectly line up with the Fermi energy of the electrodes. In practice, the interaction between the molecule and the metal may improve the level alignment because of error cancellations, but still a substantial error remains.[73] The consequence of this fact is that the resonances in the transmission coefficient are in the wrong place.

As an example in Figure 7.6 the transport properties of a benzene-dithiol molecule sandwiched between two Au leads as calculated with the LDA are presented.[74] These are compared with the same calculated by using an approximated self-interaction correction (SIC) scheme, the atomic self-inter-action correction (ASIC) method,[75] which much better reproduces the condition $\varepsilon_{\text{HOMO}}^{\text{KS}} = -I_N$. The LDA transmission coefficient at zero-bias shows a broad resonance at about 1.5 eV below the Fermi level, corresponding to the position of the molecule HOMO. This results in a zero-bias conductance of about 0.23 G_0 ($G_0 = 2e^2/h$). The self-interaction corrections downshift $\varepsilon_{\text{HOMO}}^{\text{KS}}$ by about 1 eV, thus enlarging the HOMO–LUMO gap. Such a downshift removes the tail of the HOMO resonance in $T(E)$ from the Fermi level and now the conductance is 0.06 G_0, a value much closer to the experiments.[76] This is a rather common feature,[77] *i.e.* the LDA and GGA return highly conductive devices even when they should be insulating, a failure that reminds us of the inability of the LDA and GGA in describing Mott insulators. Note that in the case of electron transport such a problem is present both for strong and weak coupling between the molecule and the electrodes.

The second failure of LDA/GGA DFT applied to transport concerns how the Kohn–Sham eigenvalues change their energy as a function of the molecule occupation. This is a problem that appears mainly in the weak-coupling limit and it has to do with a second fundamental property of DFT, *i.e.* the presence of the derivative discontinuity of DFT potential.[70–72] In order to understand such a property let us consider a molecule containing N electrons in its neutral state weakly coupled to an electron bath and let us assume that some tiny fractional charge n is transferred from the bath to the molecule. Then, for $-1 < n \leq 0$, $\varepsilon_{\text{HOMO}}^{\text{KS}} = -I_N$, but for $0 < n \leq 1$, $\varepsilon_{\text{HOMO}}^{\text{KS}} = -I_{N+1}$, where I_{N+1} is the ionisation potential of the $(N+1)$-electron molecule. This means the Kohn–Sham eigenvalue of the HOMO has to jump discontinuously as its occupation crosses an integer value. To achieve this, the Kohn–Sham potential also has to change in a discontinuous step of $I_N - I_{N+1} = I_N - A_N$, where A_N is the electron affinity. The crucial point is that the LDA and GGA lacks the derivative discontinuity and typically the

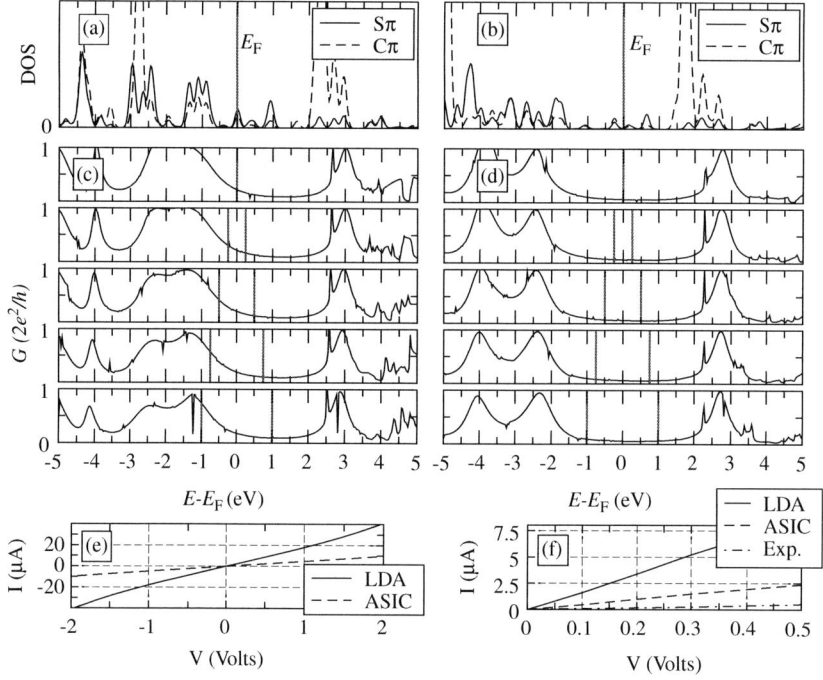

Figure 7.6 Transport properties of a BDT molecule attached to the gold (111) hollow site. The left plots correspond to LDA and the right ones to ASIC. The upper panels are the DOS of the S and C π orbitals ((a) and (b)), the middle are the transmission coefficients as a function of energy for various bias ((c) and (d)) and the lower are the I–V curves. Figure 7(f) is a zoom of (e) and compares our results with experiments from ref. 76. The vertical lines in (c) and (d) mark the bias window. C. Toher and S. Sanvito, *Phys. Rev. Lett.*, 2007, **99**, 056801. Copyright (2007) by the American Physical Society.

Kohn–Sham eigenvalues move almost linearly with the fractional occupation, at least close to neutrality.

What are then the consequences of the lack of such a derivative discontinuity on the transport? Clearly these emerge as the bias is applied and the electron occupation of a given molecular orbital changes. Figure 7.7 illustrates the different behaviours.[78] The figure reports results (I–V curve, level occupation and level position) for the simple model presented in Section 7.3.1, where $\varepsilon_\alpha (n_\alpha)$ depends on its own occupation n_α either: 1) linearly, $\varepsilon_\alpha(n_\alpha) = Un_\alpha$ mimicking the LDA, or 2) discontinuously (see Figure 7.7), mimicking the exact DFT. In the calculation it was assumed $\varepsilon(0)$ just below E_F ($|\varepsilon(0) - E_F| = 0.5\,\mathrm{eV}$) and both the weak and the strong-coupling limit were considered. For weak level-electrode coupling ($\gamma_\alpha = 0.02\,\mathrm{eV}$) at zero-bias ε_α pins at E_F and then follows the bias linearly for both the LDA and the discontinuous model. Importantly, as soon as ε_α shifts above the lower of the two chemical potentials (say μ_2), then $f_2 \approx 0$, and the current will be simply proportional to the level occupation

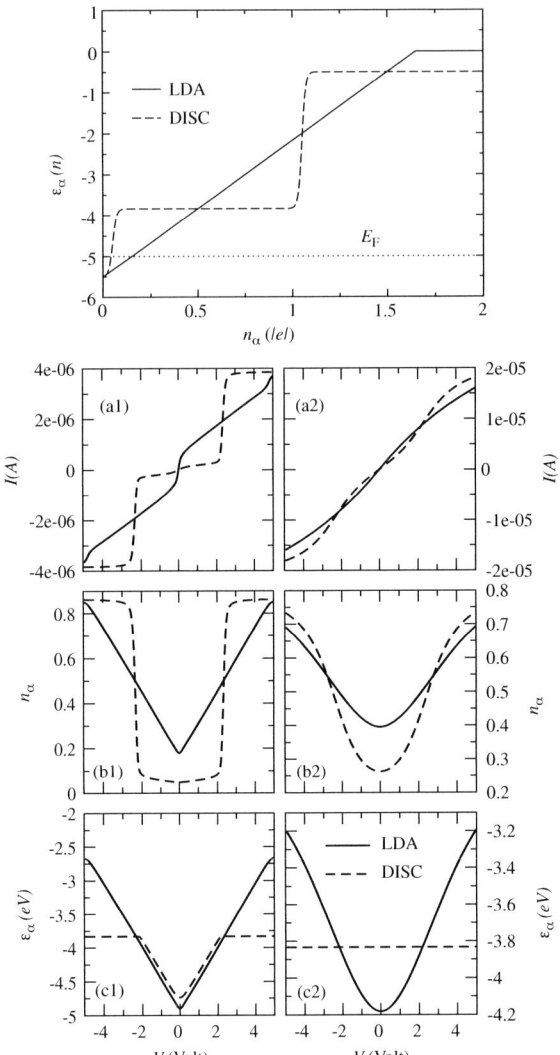

Figure 7.7   Transport properties for the simple level model illustrating the effects produced by the lack of derivative discontinuity in the energy-level eigenvalue. In the left-hand side panel the dependence of ε_α on the own occupation n_α is presented. The straight line corresponds to a typical LDA dependence and the step-like line to that of the discontinuous model (DISC). Notice that the LDA line becomes flat at $n_\alpha \sim 1.5$ [$\varepsilon_\alpha(n_\alpha) = 0$] since here the eigenstate is unbound. The dotted horizontal line denotes the position of the leads Fermi level E_F. In the right panels the transport properties are presented: (a) Current I, (b) occupation n_α and (c) position of the energy level ε_α as a function of the bias V. The parameters used here are $\varepsilon_\alpha(0) = -5.5$ eV, $U = 5$ eV, $E_F = -5.0$ eV and $T = 300$ K. The curves on the left-hand side are obtained in the weak-coupling limit ($\gamma_\alpha = 0.02$ eV) and those on the right-hand side in the strong-coupling limit ($\gamma_\alpha = 1.2$ eV). C. Toher, A. Filippetti, K. Burke and S. Sanvito, *Phys. Rev. Lett.*, 2005, **95**, 146402. Copyright (2005) by the American Physical Society.

$(I \approx \gamma_\alpha f_1)$. This is where the LDA and the discontinuous model behave differently.

The LDA-like potential in fact results in a linear dependence of the occupation on bias [see Figure 7.7(b1)], and consequently in a metallic conductance. In contrast, for the discontinuous model ε_α shifts upwards in energy without much change in the occupation, which is almost discontinuous when going from $n_\alpha = 0$ to $n_\alpha = 1$. As a consequence a gap opens in the I–V. This is as large as the one in the occupation, *i.e.* it is roughly A_N. Thus, the model illustrates that the lack of eigenvalue discontinuity causes a dramatic overestimation of the metallicity of a molecular junction. A similar analysis can be formulated for a three-terminal device,[31,79] concluding that both the position and the linewidth of the conductance resonance as a function of gate potential of given molecular level are incorrect in LDA. In particular, one can show that even for an infinitesimally small coupling to the electrodes the linewidth remains finite.

In contrast, the curves obtained for LDA and the discontinuous model in the strong-coupling limit [Figures 7.7(a2), (b2) and (c2)] look rather similar. This is because a considerable fraction of the level occupation and the current originates from the tail of the energy-level DOS. In the large-coupling limit, $\gamma_\alpha \sim A_N$, both n_α and the current are rather insensitive to the dependence of ε_α on n_α, and we find that standard continuous functionals give rather accurate I–V characteristics.

In concluding this section, I would like to remark that the drawbacks of NEGF + DFT discussed here all concern the use of DFT local and semilocal exchange and correlation functionals. The main message is that if one uses a functional describing both a correct level alignment and the derivative discontinuity, then the calculated transport properties will be in much improved agreement with the experiments. However, the problem of using DFT improperly still remains so that additional errors will be unavoidable. Some of them can be corrected[79] but some others will be simply intrinsic of the method. One should then be always careful when using the NEGF + DFT approach. I also wish to remark that some of the features discussed here transfer directly to other transport schemes. For instance, the derivative discontinuity appears to be a key element for describing Coulomb blockade in time-dependent approaches.[80]

7.4 Applications

We finally move on to discuss a number of applications of the NEGF + DFT method, focusing on physical systems presenting different computational complexity. This, of course, will not be an exhaustive discussion but will simply provide an overview on the capabilities of the method.

7.4.1 TMR Junctions: Large *k*-Point Sampling and Bound States

Magnetic tunnel junctions (MTJs) are today the key components in read heads of hard-disk drives. They are composed of two magnetic layers sandwiching an inorganic insulator. Their functioning principle is based on the fact that the tunneling current depends on the orientation of the magnetisation vectors of the two magnetic layers with respect to each other, with the parallel (P) configuration being usually less resistive than the antiparallel (AP). The figure of merit of such tunneling magnetoresistance (TMR) is the TMR ratio, defined as the difference between the resistances in the P and AP configuration divided by the smaller of the two. MTJ technology, however, did not really take off until a spectacularly large TMR ratio in Fe/MgO/Fe(100) MTJs was first predicted[5,6] and then demonstrated experimentally.[7,8] This is in fact a clear example of how predictive quantum transport theory can seriously impact a large technological area.

The main reason behind why the use of epitaxially grown Fe/MgO/Fe increases massively the TMR is rooted in the rôle played by the tunneling barrier. Before epitaxial junctions were made the tunnel barriers used were largely amorphous and the TMR signal was limited to the spin-polarisation of the DOS of the magnetic electrodes.[81] This is never too large for transition metals and the largest TMRs measured were of the order of 40–50%. In epitaxial junctions, however, the situation is different. The conservation of the transverse momentum (with respect to the device stack) makes the coherent tunneling amplitude sensitive to the wavefunction symmetry. In particular in MgO electrons with Δ_1 symmetry have a much larger probability to tunnel across. Crucially only in the Fe majority band are there electrons at the Fermi level with Δ_1 symmetry, meaning that the Fe/MgO interface effectively transmits only electrons of one spin. Thus, Fe behaves as a half-metal and the TMR becomes huge (now it is in excess of 600% at room temperature[82]).

Transport in epitaxial MTJs is an ideal problem for NEGF + DFT. Since the junctions are both epitaxial and almost defect free, there is very little scattering between electronic states with different k-vector component in the plane perpendicular to the device stack (the x-y plane), *i.e.* the transverse k-vector, (k_x, k_y), is conserved. As a consequence, the extended molecule is simply the primitive cell along the device stack (it comprises the tunnel barrier and several atomic layers of the metallic contacts) and periodic boundary conditions are applied in the x-y plane. All the relevant quantities become k-dependent so that the total transmission coefficient reads

$$\mathbf{T}^{\sigma}(E; V) = \frac{1}{N_k} \sum_{k_x, k_y} \mathbf{T}^{\sigma}(E, k_x, k_y; V), \qquad (7.29)$$

where the sum runs over the 2D Brillouin zone perpendicular to the transport direction and N_k is the number of k-points included in the sum. The crucial part of the calculation is that of performing an adequate sampling of the Brillouin zone, an operation that, however, is not trivial.

The complexity can be appreciated by looking at Figure 7.8, where we report the transmission coefficient calculated at the Fermi level for various Fe/MgO/Fe(100) MTJs of different thicknesses as a function of (k_x, k_y). Two main aspects are worth noticing. First, that $T(k_x, k_y)$ varies within the same graph by 10 orders of magnitude, and secondly that the average transmission coefficient

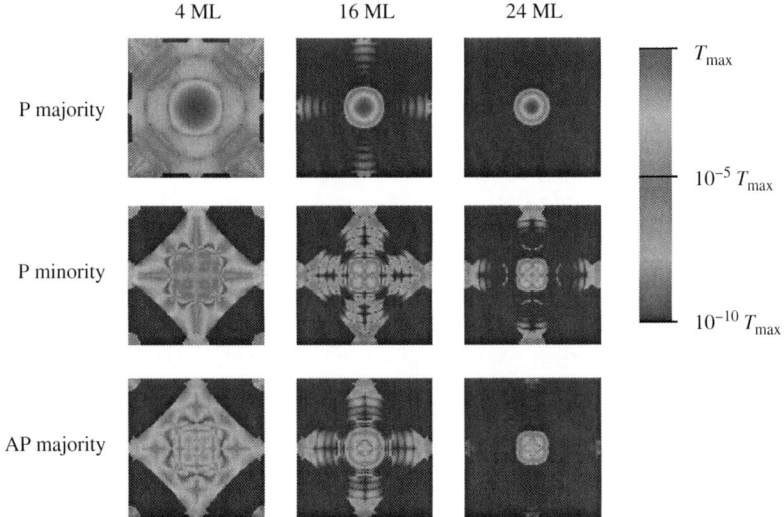

Figure 7.8 Transmission coefficient at the Fermi level as a function of the transverse *k*-vector (k_x, k_y) for three Fe/MgO/Fe with different length of the MgO barrier: left 4 monolayers (ML), middle 16 ML, right 24 ML. The two upper panels are, respectively, for the majority and minority electrons in the P configuration, while the lower one is for both spins in the AP configuration. Calculations are performed with the LDA functional. Note that for each panel $T(k_x, k_y)$ varies over 10 orders of magnitude, but the transmission maps for devices of different thicknesses are rescaled by the largest of the $T(k_x, k_y)$. On average, when going from 4 to 24 monolayers the transmission coefficient changes by 20 orders of magnitudes.

changes by about 20 orders of magnitude when going from 4 to 24 MgO monolayers. This essentially means that one needs an algorithm capable of calculating accurately a quantity (the transmission coefficient) that fluctuates over 30 orders of magnitude. Furthermore, this should be done over an extremely fine *k*-grid since the fluctuations of $T(k_x, k_y)$ are extremely rapid.

The first of these two requirements can be matched by constructing extremely accurate and fault-tolerant algorithms. A recently developed strategy is that of identifying the various singularities arising when computing the Green's function and then systematically removing them.[35] The second requirement concerns only the calculation of the current since the calculation of the density matrix typically requires a much smaller sampling to converge. Thus, one usually runs first a self-consistent calculation with a relatively small sampling (a few hundred *k*-points) and then calculates the current over a much larger *k*-grid (that can also be adaptive). This is a relatively inexpensive operation since the calculation is no longer self-consistent and can be carried out with massive parallelisation. However, there is also a second and more subtle problem that arises for MTJs when calculating the *I–V* curves. This concerns the calculation of the junction electrostatic potential.

The problem is rooted in the presence of bound states contributing to the nonequilibrium density matrix [see eqn (7.28)]. A bound state is an electronic state completely electronically decoupled from the electrodes, *i.e.* a state that, in the absence of any inelastic interaction, has an infinite lifetime. Mathematically, from the spectral representation for the Green's function (7.9) one can see that a bound state is characterised by having $\gamma_{\alpha} = 0$, *i.e.* it is a pole of $G(E)$. From a physical point of view a bound state is an electronic state, whose wavefunction decays exponentially in every direction [see Figure 7.9(b)]. This is indeed possible in a MTJ at the interface between the tunnel barrier and the electrode. In fact, although the electrodes are metals, *i.e.* they possess nonvanishing DOS over a large bandwidth around E_F, for a given (k_x, k_y) the electronic structure can be gapped. This essentially means that there exist electronic states with real (k_x, k_y) but imaginary k_z, *i.e.* states that cannot propagate in the z-direction in the metal (z is the direction of the MTJ stack). The wavefunction of such a state is localised at the metal/insulator interface and decays exponentially both in the metal and in the insulator.

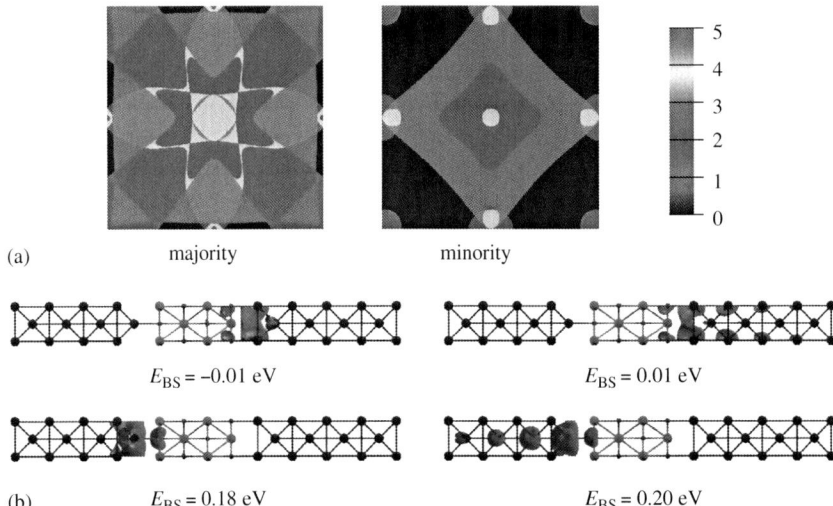

Figure 7.9 Bound-state analysis for *bcc* Fe along the (100) direction. In the top panel we show the k-dependent plot over the entire Brillouin zone of the number of real k_z wavevectors at the Fermi level [*i.e.* the number of times k_z (k_x, k_y, E_F) is real]. The grey scale code is in linear scale ranging from zero to 5. Note that for the minority bands there are large portions of the Brillouin zone with no real k_z values. Bond states can be found in these regions. In the bottom panel we present the charge-density distribution of the four minority spin-bound states close to E_F for an Fe/MgO/Fe MTJ, where the MgO has a thickness of 4 monolayers. The states are calculated at $k_x = k_y = 0.58 \, \pi/a$ (a is the Fe lattice parameter) and at a bias voltage of 0.2 V. All the energies of the bound states, E_{BS} are taken with respect to E_F. Note the bound states decay both in the insulator and in the metal. Data are from ref. 83, Copyright Trinity College Dublin (2008).

Clearly, since $\gamma_\alpha = 0$, bound states do not contribute to the current. So how can they affect the transport properties? The crucial point is that they do contribute to the charge density, but their occupation is unknown at finite bias. This can be clearly understood already at the simple model level. A look at eqn (7.3) immediately reveals that for $\gamma_\alpha^n \to 0 \, (n = 1, 2)$ the occupation is not defined, a situation that simply reflects the fact that an electronic state completely decoupled from an electron reservoir is not in equilibrium with such a reservoir. Numerically, however, the limit $\eta \to 0^+$ in the definition of the Green's function is achieved by setting η to a small positive value, which means that every electronic level is artificially broadened by η. This is true also for the bond states, which become artificially coupled to the electrodes in an arbitrary way. When this happens there is a contribution to ρ_{neq} that is simply arbitrary and essentially driven by the numerics of the calculation. Since the charge density defines the electrostatic potential a spurious contribution to the charge density translates into a nonphysical contribution to the electrostatic potential.

In order to illustrate this feature in Figure 7.10 the potential drop across a Fe/MgO/Fe MTJ is presented.[83] In particular, data are shown for three different occupations of the bound states: *correct occupation*: bound states located

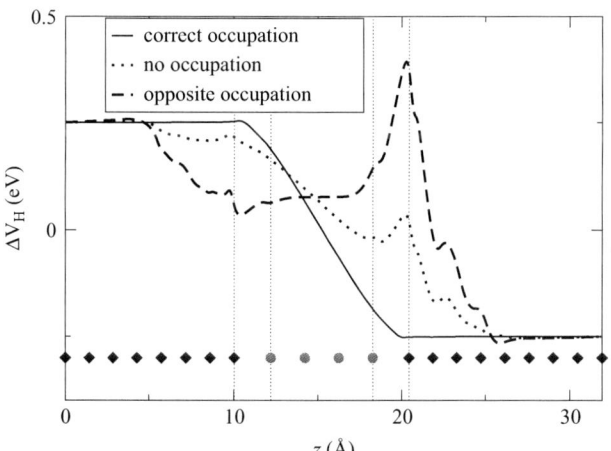

Figure 7.10 Planar average along the *x-y* plane of the difference between the electrostatic potential calculated at 0.5 V and that calculated at zero bias as function of position *z* (this is effectively the potential drop along the junction stack). Different strategies for occupying the bound states are presented: *correct occupation*: bound states located to left-hand (right-hand) side of the junction are in equilibrium with the left-hand (right-hand) electrode; *opposite occupation*: bound states located to left-hand (right-hand) side of the junction are in equilibrium with the right-hand (left-hand) electrode; *no occupation*: all bound states are empty. The diamonds indicate the position of the Fe layers and the circles indicate the position of the MgO layers. Data are from ref. 83, Copyright Trinity College Dublin (2008).

to left-hand (right-hand) side of the junction are in equilibrium with the left-hand (right-hand) electrode; *opposite occupation*: bound states located to left-hand (right-hand) side of the junction are in equilibrium with the right-hand (left-hand) electrode; *no occupation*: all bound states are empty The figure clearly reveals that for some types of occupation the potential profile does not look at all like the linear drop expected from the elementary theory of a par-allel-plate capacitor.

A number of aspects need to be remarked. First, it is important to remember that formally the occupation of a bound state is truly undefined. In practice, however, one expects that nonelastic effects (say lattice vibrations) eventually couple the bound states to the electron reservoirs. A proper treatment of the problem will require the explicit inclusion of all the relevant inelastic con-tributions, *i.e.* it will add substantial complexity. In the case of a MTJ, where bound states are localised at the metal/insulator interface, common sense suggests that regardless of the nature of the inelastic excitation a bound state will eventually be in equilibrium with the closest electrode. Figure 7.10 reveals that this is a successful and practical strategy. In general, however, when a bound state is not spatially localised near any of the electrodes no simple strategy exists and unless one uses a higher-level theory the problem is not solvable.

The second observation is that in many situations the bound-state problem is not really significant since the total number of bound states is much smaller than that of all the available states. In MTJs this unfortunately might not be true because of the electronic structure of the electrodes. In fact, one can observe in Figure 7.9 (top panel) that for the minority spin of Fe at E_F, the states with a purely imaginary k_z are distributed over approximately half of the Brillouin zone. This means that half of the Brillouin zone can potentially contribute bound states to the charge density at E_F. This situation is usually not encountered for simple metals (Au, Al,...) but it is common in transition metals where there is a substantial contribution of the d electrons to the DOS at the Fermi level.

It is also important to remark that at zero bias, *i.e.* in the linear response limit, the problem of bound states is immaterial. In this case, in fact, there is only one chemical potential common to both the electrodes and all the energy levels below such a potential are occupied. The same is valid for bound states and their correct occupation is guaranteed by performing the integral (7.27) over the complex contour (note that in this case one does not even need to identify the bound states). This puts on safe grounds the multitude of linear-response calculations for MTJs available in the literature. In reality, often for MTJs linear response calculations provide enough useful information. Furthermore, since all the prominent features of the *I–V* are given by the move-ment with bias of the electrodes band-edge for the bands with Δ_1 symmetry, a good approximation to the real *I–V* characteristics can be obtained by approximating the true potential drop with a linear slope across the tunnel barrier.[84] This approximation, however, does not hold true when the barrier has more complex polarisation effects than those in simple insulators. This is for instance the case of organic barriers,[85] double junctions[86] or ferroelectric materials,[87] where the self-consistent calculation of the potential drop is unavoidable.

In conclusion, I wish to remark that the tunneling problem and in particular in MTJs is indeed an aspect where the NEGF + DFT method has made and still makes substantial impact. Despite the complications described in this section the calculations in general are well grounded on accurate band-structures: usually the bands of transition metals are well reproduced by LDA/GGA and the true gap of the insulator is not crucial.

7.4.2 Large Macromolecules: Mn_{12} and Negative Differential Resistance

In this section, I will review some aspects of transport in large-molecular junctions and in particular I will discuss how the *I–V* curve is affected by orbital rehybridisation, an aspect that the NEGF + DFT method captures rather well. As an example, I take the case of transport across a Mn_{12} single magnetic molecule.[66] This system is characterised by HOMO levels consisting of rather rich multiplets with weak coupling to the electrodes. In the case of Mn_{12} this is formed by the occupied e_g *d*-electrons of the eight Mn^{3+} ions. The main characteristic of such situation is that the electronic coupling with the electrodes may drastically change with bias, *i.e.* $\gamma_\alpha (E, V)$. Let us investigate this aspect by looking at Figure 7.11 where the *I–V* curve of a Mn_{12} molecule attached to two Au electrodes *via* thiol-terminated C_6H_4 ligands is presented.

Results are presented for two different configurations of the molecule, namely the ground state (GS) and a spin-flip state (SF) obtained from the GS by reversing the magnetic moment of a Mn^{3+} and of a Mn^{4+} ion. Let us focus on the GS. The most striking feature that immediately appears is the presence of a number of negative differential resistances (NDRs), *i.e.* portions of the *I–V* curve where the current decreases as the voltage is increased. NDRs for molecules have been predicted in the past as the result of particular features in the electrodes DOS,[88,89] of strong electron–phonon coupling,[90] or because of configurational changes.[91] In the case of Figure 7.11 these originate from the dynamical coupling to the electrodes of the various components of the HOMO multiplet.

Let us consider again the simple model of Section 7.3.1 and the expression for the occupation of an energy level [eqn (7.3)]. Clearly, if the level is more strongly coupled to one of the two electrodes (say electrode 1, *i.e.* $\gamma_1 \gg \gamma_2$) then its occupation will be determined by the chemical potential of that particular electrode. Since the single particle level of a given molecular orbital will move upwards in energy as its occupation increases and will move downwards in energy as the occupation is reduced, one can conclude that in a situation of very asymmetric electric coupling the energy level will follow the position of the chemical potential it is more strongly coupled to. This is the situation in a scanning tunnel microscopy (STM) experiment. In any case, if the coupling of the level with the electrodes does not depend on bias, the width of the

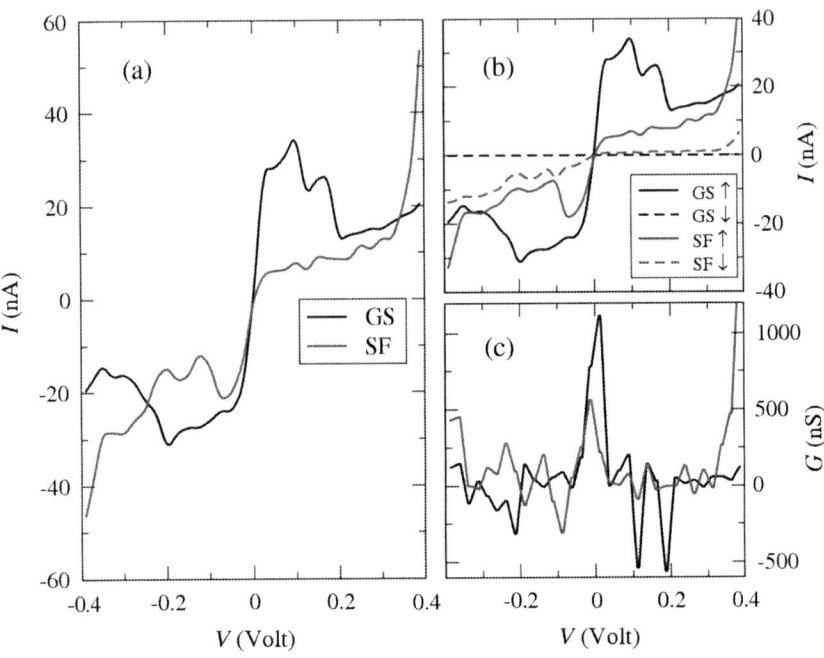

Figure 7.11 Transport properties of a Mn$_{12}$ molecule attached to two Au electrodes *via* thiol-terminated C$_6$H$_4$ ligands. The *I–V* curves for both the ground state (GS) and a spin-flip (SF) state are presented in (a), while (b) and (c) show the *I–V* spin decomposition and the differential conductance, respectively. C.D. Pemmaraju, I. Rungger and S. Sanvito, *Phys. Rev. B*, 2009, **80**, 104422. Copyright (2009) by the American Physical Society.

transmission peak will not change with *V*. As a consequence the current will be solely determined by the *position* of the peaks in *T(E)*. Since upon increasing the bias charging alone cannot expel from the bias window a molecular level that is already inside at a lower bias, no NDRs are expected.

The situation is, however, different for Mn$_{12}$, as it can be seem in Figure 7.12 where *T(E; V)* calculated at different voltages is presented for the GS. At zero-bias the first four molecule HOMOs are less than 50 meV from EF and approximately 300 meV from the LUMO, hence they determine the low bias current. As the bias is increased (*V* > 0) the current is initially determined by the first HOMO, which is pinned at E_F. This HOMO moves into the bias window at about 100 mV and then roughly maintains its energy position at any larger positive *V*. The fact that $T(\varepsilon_{HOMO}) \sim 1$ for *V* < 100 mV suggests that $\gamma^1_{HOMO} \sim \gamma^2_{HOMO}$ [see eqn (7.5)]. However, as soon as the level enters completely the bias window the amplitude of its transmission peak drastically reduces. This means that the coupling to the electrodes has changed. We then conclude that the HOMO coupling to the electrodes has become asymmetric with bias.

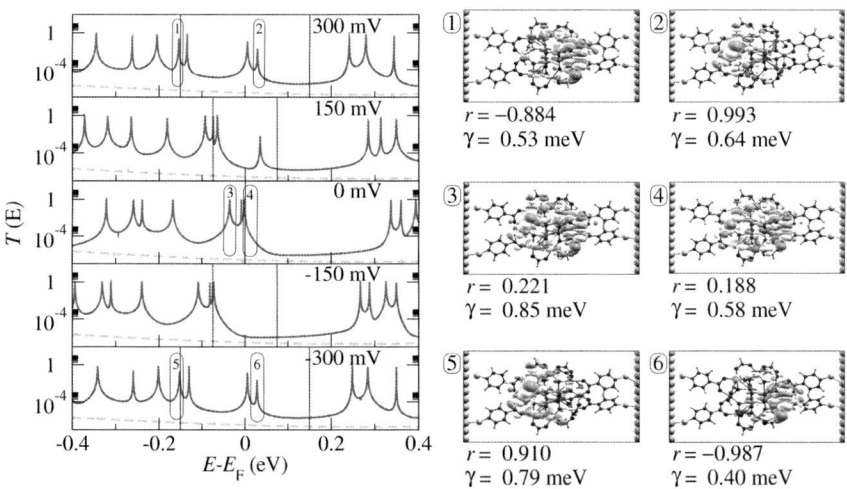

Figure 7.12 Spin-resolved transmission coefficient, $T(E)$ as a function of bias for the GS configuration of Mn_{12}. Solid (dashed) curves show the transmission of the majority (minority) spins, while the two vertical lines mark the bias window. The isosurfaces display the wavefunctions of a number of selected molecular levels at different V. For these we also report the value of both r_α and γ. Note the dramatic rehybridisation with bias leading to highly asymmetric wavefunctions. C.D. Pemmaraju, I. Rungger and S. Sanvito, *Phys. Rev. B*, 2009, **80**, 104422. Copyright (2009) by the American Physical Society.

Such asymmetry can be quantified by defining the coupling asymmetry parameter

$$r_\alpha = \frac{\gamma_\alpha^1 - \gamma_\alpha^2}{\gamma_\alpha^1 + \gamma_\alpha^2}. \tag{7.30}$$

This is $+1$ (-1) for molecular orbitals more strongly coupled to electrode one (electrode two) and 0 for equal coupling to both the electrodes. Note that in the weak-coupling limit the current is essentially given by the individual contributions of those peaks in the bias window, *i.e.* it is obtained by integrating $T(E)$ of eqn (7.5). Thus, the individual peak contribution to the current is simply

$$I_\alpha = 2\pi \frac{e}{h} \frac{\gamma_\alpha(1 - r_\alpha^2)}{4}, \tag{7.31}$$

so that an increase in the coupling asymmetry reduces the current. Importantly, the electronic coupling γ_α^n can be calculated from the NEGF *via* the eqn (7.11), so that a quantitative analysis can be carried out.

This is carried out in Figure 7.13, where the position of the various ε_α (for the GS) is traced as a function of V. In the figure the grey scale encodes the values of r_α and γ_α, while the two straight black lines mark the bias window. As noted previously at small bias the first HOMO is pinned to the lower bound of the bias window and it is roughly equally coupled to the electrodes, $r_\alpha \sim 0$. However, for $V \sim 100$ mV it suddenly jumps into the bias window. In doing so its coupling to the leads becomes extremely asymmetric, $r_\alpha \sim 1$, while γ_α remains approximately constant. Considering that its contribution to the current is I_α [eqn (7.31)], one can conclude that as soon as the peak enters the bias window the total current actually decreases. This creates the NDR at 100 mV shown in Figure 7.11(c). Similarly, the second HOMO enters the bias window at 200 mV, again with a drastic change of r_α producing the second NDR at 200 mV.

The question is now, why the electronic coupling changes so drastically as soon as a molecular level enters the bias window. In order to answer to this question let us go back to Figure 7.12 and take a look at the isosurfaces displaying the electronic wavefunctions of given molecular levels at finite bias. In particular, let us follow the evolution of the HOMO level with bias by looking at its wavefunction at 300 mV, 0 and –300 mV (states labeled with 2, 4 and 6). As the bias is applied there is a significant polar distortion of the wavefunction.

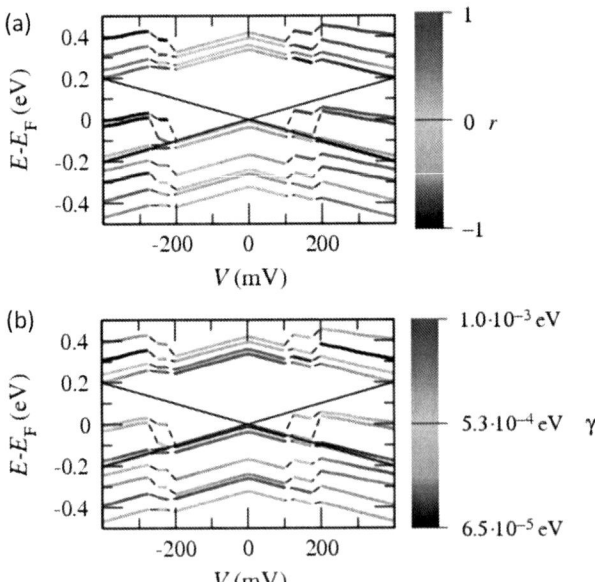

Figure 7.13   Trace of (a) r_α and (b) γ_α as a function of energy and bias. The grey lines denote the position of the peaks in the transmission coefficient and the grey scale encodes the magnitude of r_α and γ_α. The two solid straight black lines mark the boundaries of the bias window. C.D. Pemmaraju, I. Rungger and S. Sanvito, *Phys. Rev. B*, 2009, **80**, 104422. Copyright (2009) by the American Physical Society.

The electron clouds changes from a uniform distribution at $V=0$ to one that is considerably localised either to the left- or to the right-hand side of the junction. Such a polar response under bias originates from molecular orbital rehybridisation and always involves a number of molecular states. For instance, when the HOMO and HOMO-1 enter the bias window at $V>0$ the rehybridisation involves the first several HOMOs, as one can deduce from the fact that for most of them there is a change in r_α [see Figure 7.13(a)]. The net result of such a rehybridisation is the induced asymmetric coupling to the electrodes, which allows molecular levels to enter the bias window and conduct without charging the molecule. Thus, this NDR has nothing to do with Coulomb blockade, which typically occurs at a higher-energy scale, but simply with the polar response of the molecule to an external electric field.

In concluding, I wish to make a few remarks. First, it is important to note that the mechanism for NDR described here is not specific to magnetism but it will occur in situations where a molecule, weakly coupled to the electrodes, presents a closely spaced HOMO (or LUMO) multiplet. In this case the calculation becomes technically challenging since the electrostatic problem must be solved in an extremely precise way. For instance, the calculations presented here would not have been possible without an efficient and accurate mesh-refinement algorithm for integrating the charge density. The second consideration concerns the DFT aspects of calculations in the weak-coupling limit. These are situations where electron charging becomes important so that the choice of exchange correlation functionals is crucial. Ideally, one would like to use functionals including the derivative discontinuity, which, however, usually require extra computational overheads, so they are currently not compatible with large-scale calculations. Thus, the use of approximated corrective schemes like the ASIC method may be a reasonable compromise.

7.4.3 Environmental Effects

In this conclusive section, I will briefly discuss a new emerging line of calculations where one attempts to describe electron transport in molecules kept in solution. This is of course of paramount importance for molecular devices in biological conditions. For instance, the interaction with the solvent is believed to play a fundamental rôle in transport along DNA molecules.[92] This is indeed another area where the NEGF + DFT method can make significant contributions. However, efficient and fast algorithms are still needed, so that the field is in its infancy and the published calculations are to date scarce.[93–96]

The main concept behind electron transport across molecules in solution is that the solution provides dynamical charge screening. This means that the electrostatic potential experienced by the molecule may be rather different from that in a dry situation. Just consider water. H_2O is a polar molecule so that a specific arrangement of water molecules around the one to measure will produce a dipolar screening field. Unfortunately, the relevant field is not simply that of a static configuration of the solvent but an average over its thermal motion. One then assumes that in a steady-state condition the *I–V* curve can be obtained simply as the average curve over those calculated for a statistically representative number of frozen solvent configurations. The calculation then comprises two parts. First, one performs molecular dynamics simulations either at the force field[93] or at the *ab initio* level.[94] These provide the trajectories of the solvent around the molecule to measure over a period of time, *i.e.* they provide a number of instantaneous snapshots of the device geometry.

Then, the second part of the calculation consists in computing the transmission coefficient and the *I–V* curves of such geometries. Since a typical molecular dynamics simulation runs with time steps of around 1 fs and usually simulations run for several nanoseconds, one has around 10^6 configurations available. Clearly, it would be impossible to perform so many transport calculations. Then, the strategy is that of considering only an appropriate number of time-uncorrelated configurations and averaging over this restricted number. One has also to check that the addition of further configurations does not change significantly the averaged quantities.

As an example of this procedure in Figure 7.14, together with the simulation cell, I present results for a benzene-dithiol molecule sandwiched between Au electrodes in water. In the figure both the transmission coefficient and the DOS projected on the H_2O molecules only are shown. In this case the calculation needs to be carried out at the ASIC level in order to position correctly the water HOMO at the negative of the H_2O ionisation potential. Note that also for this problem the choice of DFT functionals is rather important. In particular, now one needs a good description of the alignment of the molecule levels with the electrodes E_F, but also with the energy level of the molecules composing the solvent. An erroneous description of such levels alignment may result in artificial conductance through the solvent. This is not the case of Figure 7.14, where the HOMO of H_2O appears at a considerably lower energy than that of the benzene-dithiol.

The most important feature of the figure is the significant difference between the average $T(E)$ calculated over the solvent configurations (200 in this case) and that calculated in dry conditions. In particular, one can immediately notice a shift of about 0.5 eV towards lower energies of the transmission peak corresponding to the benzene-dithiol HOMO. A rather similar energy drift is also found for the LUMO, so that one has to conclude that the effect of the solvent is that of effectively gating the molecule. This is explained by looking at the average distribution of H_2O molecules around the benzene, which reveals a preferential orientation around the S atoms of the linking thiol groups.[93] Thus, water on average acts as an effective gate for the junction. Importantly, the shift of the transmission spectra toward lower energies repositions the electrodes' E_F right in the middle of the Kohn–Sham HOMO–LUMO gap. As such, the zero-bias conductance gets reduced, in better agreement with experiments.[76] Interestingly, the same calculations repeated for a C-nanotube, a hydrophobic material, reveals no gating effect, so that the transport in solution and in the dry is rather similar.

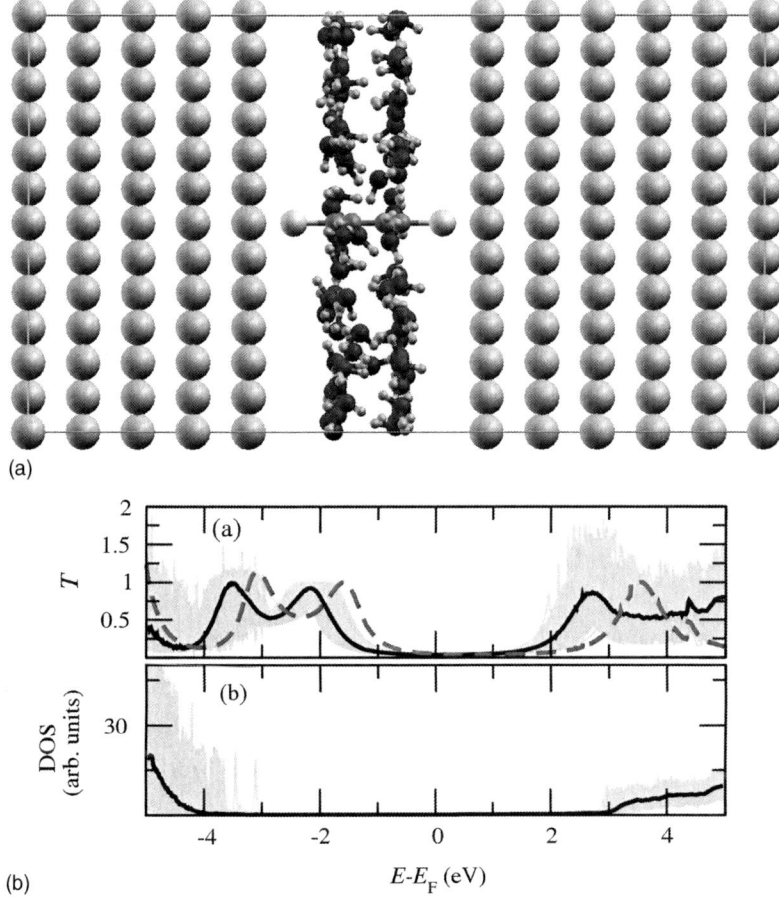

Figure 7.14 Unit cell used for the BDT molecule in water attached to Au electrodes. Time-averaged (a) transmission coefficient and (b) DOS projected onto the H₂O molecules for BDT attached to Au. The calculations are obtained with ASIC; the solid black curves are the time averages over 201 time steps while the grey ones forming a shadow are for each of the 201 time steps. The dashed curve is for the dry situation. I. Rungger, X. Chen, U. Schwingenschlögl and S. Sanvito, *Phys. Rev. B*, 2010, **81**, 235407. Copyright (2010) by the American Physical Society.

The case of the benzene-dithiol molecule just discussed is somehow simple, since the molecule is rather compact and the solvent acts as an approximately uniform gate. The situation for macromolecules is rather different. This is illustrated in Figure 7.15 where results are presented for a 6-base poly(G)-poly(C) polymeric A-DNA duplex connected to Au electrodes *via* thiol groups attached at the C3′G3′ positions. In the lower panel the average electrostatic potential at zero bias is plotted along the transport direction for all the molecular dynamics snapshots considered. One can immediately note the large

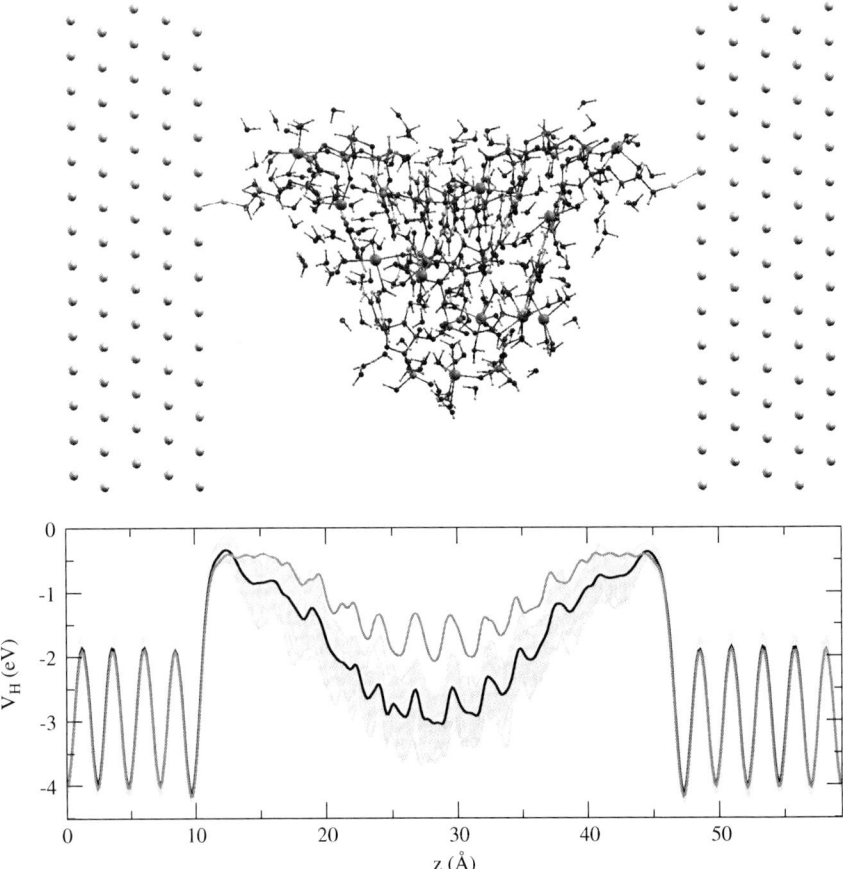

Figure 7.15 A 6-base-pair pGpC molecule in wet conditions. In the top-left panel show the EM setup is presented. This consists of a 6 base-pair pGpC A-DNA strand surrounded by a few layers of solvent including water molecules and Na$^+$ counterions. In the bottom panel the planar average of the electrostatic potential across the system at zero-bias is presented. The grey line shows the electrostatic potential in the absence of the solvent, while the black line shows the potential averaged over 100 time snapshots of the system taken from the classical molecular dynamics simulations. The grey lines show the electrostatic potential for the individual snapshots. In the panel on p. 217 is presented a histogram showing the number of times, N, a certain molecular energy level is located within the energy interval given by the position and bin width of the corresponding histogram bar. The light grey lines in panel (a) represent the HOMO and LUMO levels of the DNA strand in the absence of the solvent. The dashed lines represent the averages taken over the histograms. Panels (b)–(d) show the distribution of the highest (lowest) occupied (unoccupied) molecular levels localised on individual Guanine (Cytosine) bases. The Guanine bases labelled G1, G2, *etc.* are arranged from left to right with Guanine G1 at the 5′ end and G6 at the 3′ end. C.D. Pemmaraju, I. Rungger, X. Chen, A.R. Rocha and S. Sanvito, arXiv:1007.0035. Copyright (2010) by the American Physical Society.

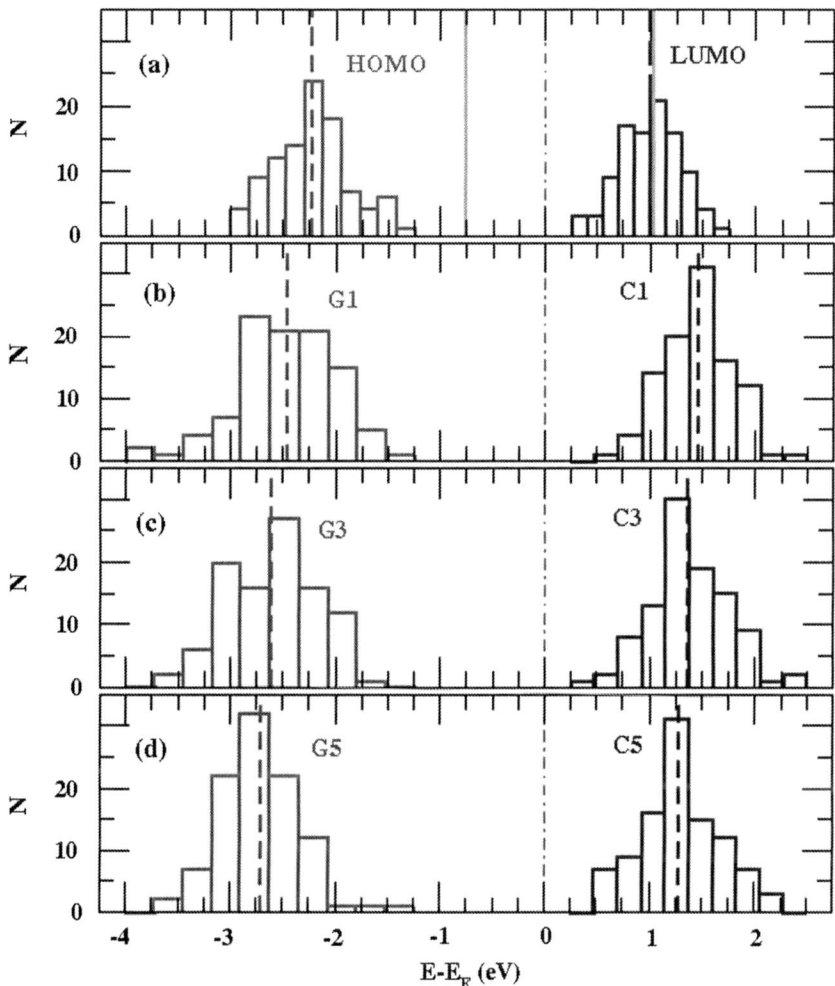

Figure 7.15 Continued.

fluctuations about the average potential caused by the dynamics of the polar solvent molecules. This time, the gate acts locally on the electronic energy levels of the bases along the DNA strand. As a result, the energy eigenvalues corresponding to molecular orbitals localised on the bases spread across a range of energies.

In the panels of Figure 7.15 on p. 217 this spread is quantified by presenting a histogram of the frequency at which a certain molecular level is located in a given energy range. First, one can note that the average HOMO in the presence of the solvent is shifted down in energy by approximately 1.5 eV relative to its position in dry conditions, while no changes can be found for the LUMO. Furthermore, the HOMO and LUMO energies fluctuate about the mean value with a standard deviation of 0.36 eV and 0.28 eV, respectively. However, let us also consider the fluctuation of other energy levels, namely the HOMO and the LUMO for Guanine and Cytosine along the DNA strand. The large amplitude of the fluctuations relative to the energy spacing between the molecular levels in the dry means that unlike in the static case, states localised on different bases can cross each other. In fact, while in the dry the HOMO is always localised on the Guanine base at the 5′ end of the DNA strand, the dynamic energy disorder induced by the solvent rearranges the order of the states in the HOMO manifold. This means that at a given moment in time (*i.e.* for a given water configuration) any of the energy levels associated to the six Guanine bases can be the HOMO. There is even the possibility that the instantaneous HOMO may be delocalised across several Guanine bases whose energy levels are close to resonant. Full transport calculations have not been carried out yet for a statistically representative ensemble of configurations due to the large size of the molecule. However, the current knowledge helps in constructing effective models for DNA transport.

These two examples show how transport calculations in environmental conditions can help in understanding transport measurements in solution and to improve the agreement with the experiments. In addition to the complications that large systems usually have (localised states, dynamical electrostatic coupling, large number of basis functions needed, accurate electrostatic needed), such a research line requires large ensemble averages, so that the same calculations must be repeated several times. Of course, this is an intrinsic parallel operation, but still it requires substantial machine time and then computational resources. For this reason efficient order-N codes, with substantial speedup for large systems, are highly desirable.

7.5 Conclusion and Outlook

In this chapter, I have provided a brief overview of the rapidly growing field of first-principles material-specific electron-transport theory, in particular focusing on the NEGF + DFT method applied to large electronic systems. I have reviewed the foundation of the theory, provided a simple model useful as an interpretative tool, explained the technical aspects of going to large systems and discussed a few selected examples. The emerging picture is that of a theoretical framework, which can really tackle important problems, both from the fundamental and the applied points of view. Moreover, the example of the MTJs clearly demonstrates how quantitative theory can impact directly onto mainstream technology. Still several limitations remain with the method. Let us review some of them.

First, there is the problem of performing the calculations with the Kohn–Sham spectrum, whose eigenvalues do not necessarily correspond to the real removal energies. A possible solution is that of employing corrective quasiparticle theories based on many-body methods such as the GW scheme.[73,97] An alternative is that of using nonlocal functionals, such as DFT exact exchange.[98] These are mostly self-interaction free so that the Kohn–Sham energy levels are usually in a better position with respect to those calculated with LDA/GGA. Other advantages are that the derivative discontinuity is usually well reproduced and that the polar response of molecules is improved.[99-101] These methods, however, are rather computationally intensive and most of the time they do require the calculation of the individual Kohn–Sham wavefunctions, which are not readily available in the NEGF + DFT method. As such a challenge is that of developing "effective" functionals specific for the quantum transport problem, that act as a reliable single-particle theory and also are computationally efficient.

One valid alternative for tackling problems including strong electron-electron interaction, where the NEGF + DFT method either fails or is completely inadequate, is that of using NEGF + DFT simply for extracting parameters to be used in a higher-level many-body theory. For instance the master equation approach to transport[102] is rapidly growing as the method of choice for problems like Coulomb blockade and the Kondo effect, and most generally for problems where the electron transport changes the internal electronic configuration of the molecule. Here, there is the need of constructing projection schemes where the NEGF + DFT results are used for extracting parameters for an effective Hamiltonian to be solved with the Master equation method. One then will benefit from both worlds, NEGF + DFT can provide accurate electrode–molecule coupling constants (not accessible by the Master equation method), and the Master equation solver can tackle the many-electron transport calculation.

A number of limitations also remain when considering the interaction between the transport electrons and other quasiparticles. The inclusion of electron–phonon coupling within NEGF + DFT is well established and in general performs well for small coupling where perturbation theory works. Away from this limit the theory breaks down[45] and one has to move to many-body-based schemes. These also are numerically heavy and so far they have not been implemented in the NEGF + DFT context. Additionally, there remains the problem of calculating the nonequilibrium phonon distribution, a problem still largely unsolved. Finally, even when perturbation theory works, efficient algorithms are still needed to tackle large electron systems. In this regard, it would be useful to construct a hierarchic theory where accurate electron–phonon calculations are mapped over effective lifetimes to be used in simple schemes such as the Büttiker probes method.

In addition, it would be desirable to construct efficient theories for interaction with other degrees of freedom. In particular, the interaction between current-carrying electrons and spin excitations is of paramount importance in spin-electronics and its investigation may help in the design of new devices. It is also important to remark on the rapidly growing field of spin-flip inelastic spectroscopy,[44] which promises to open access to a detailed understanding of the elementary electron-spin excitations and does play an important rôle in helping the development of the theory. Some advances in that direction have been already made,[103,104] but the theory is still rather young. In particular the adiabatic approximation, which allows one to well separate the electronic and the phononic degrees of freedom is less obvious for spins and the theory still needs to be developed. This is a problem somehow similar to that of spin-transfer torque.[105]

Finally, the theory of electron transport in environmental conditions (in solution) needs to be further developed. In particular, the effects of both an electrostatic potential and of the current over the molecular dynamics have never been explored. In the first case one may still use molecular dynamics at the force field level, but the second will require a full *ab initio* theory. This appears as a fascinating new challenge since one needs to perform the molecular dynamics using current-induced forces. Initial calculations based on simple tight-binding Hamiltonian have already revealed intriguing phenomena such as ionic motion driven by an electrical current.[106] This has implications also on the development of a quantitative theory for electron migration.

In summary, electron-transport theory is rapidly evolving and it has already become a multifaceted research area. The availability of increasingly powerful computational resources together with the development of accurate and efficient new algorithms will soon allow us to simulate entire nanodevices in their working environment. This will drive technological advances and will establish theory as a core activity for many technology sectors.

Acknowledgment

I wish to thank Chaitanya Das Pemmaraju, Ivan Rungger and Clotilde Cucinotta for the critical reading of the manuscript and for having provided some of the graphics. This work is supported by the Science Foundation of Ireland.

References

1. F. Patolsky, G. Zheng and C. M. Lieber, *Anal. Chem.*, 2006, **78**, 4260.
2. E. M. Boon, A. L. Livingston, N. H. Chmiel, S. S. David and J. K. Barton, *Proc. Natl. Acc. Sci.*, 2003, **100**, 12543.
3. *International Technology Roadmap for Semiconductors,* www.itrs.net.
4. G. P. Lansbergen, R. Rahman, C. J. Wellard, I. Woo, J. Caro, N. Collaert, S. Biesemans, G. Klimeck, L. C. L. Hollenberg and S. Rogge, *Nature Phys.*, 2008, **4**, 656.
5. W. H. Butler, X.-G. Zhang, T. C. Schulthess and J. M. MacLaren, *Phys. Rev. B*, 2001, **63**, 054416.
6. J. Mathon and A. Umerski, *Phys. Rev. B*, 2001, **63**, 220403(R).
7. S. Yuasa, T. Nagahama, A. Fukushima, Y. Suzuki and K. Ando, *Nature Mater.*, 2004, **3**, 868.
8. S. S. P. Parkin, C. Kaiser, A. Panchula, P. M. Rice, B. Hughes, M. Samant and S.-H. Yang, *Nature Mater.*, 2004, **3**, 862.
9. Z. Qian, R. Li, X. Zhao, S. Hou and S. Sanvito, *Phys. Rev. B*, 2008, **78**, 074710.
10. E. J. McEniry, D. R. Bowler, D. Dundas, A. P. Horsfield, C. G. Sánchez and T.N. Todorov, *J. Phys.: Condens. Matter*, 2007, **19**, 196201.
11. R. Landauer, *Philos. Mag.*, 1970, **21**, 863.

12. M. Büttiker, Y. Imry, R. Landauer and S. Pinhas, *Phys. Rev. B*, 1985, **31**, 6207.
13. H. Haug and A. P. Jauho, *Quantum Kinetics in Transport and Optics of Semiconductors*, Springer, Berlin, 1996.
14. C. Caroli, R. Combescot, P. Nozieres and D. Saint-James, *J. Phys. C*, 1972, **5**, 21.
15. S. Datta, *Electronic Transport in Mesoscopic Systems*, Cambridge University Press, Cambridge UK, 1995.
16. D. R. Bowler, A. P. Horseld, C. G. Sánchez and T. N. Todorov, *J. Phys.: Condens. Matter*, 2005, **17**, 3985.
17. A. Kamenev and W. Kohn, *Phys. Rev. B*, 2001, **63**, 155304.
18. G. Stefanucci, *Phys. Rev. B*, 2007, **75**, 195115.
19. C. G. Sánchez, M. Stamenova, S. Sanvito, D. R. Bowler, A. P. Horsfield and T. N. Todorov, *J. Chem. Phys.*, 2006, **124**, 214708.
20. S. Kurth, G. Stefanucci, E. Khosravi, C. Verdozzi and E.K.U. Gross, *Phys. Rev. Lett.*, 2010, **104**, 236801.
21. S. Kurth, G. Stefanucci, C.-O. Almbladh, A. Rubio and E. K. U. Gross, *Phys. Rev. B*, 2005, **72**, 035308.
22. R. M. Martin, *Electronic Structure*, Cambridge University Press, Cambridge UK, 2004.
23. H. Hohenberg and W. Kohn, *Phys. Rev.*, 1964, **136**, B864.
24. J. Taylor, H. Guo and J. Wang, *Phys. Rev. B*, 2001, **63**, 245407.
25. Y. Xue, S. Datta and M. A. Ratner, *Chem. Phys.*, 2002, **281**, 151.
26. M. Bradbyge, J. Taylor, K. Stokbro, J.-L. Mozos and P. Ordejón, *Phys. Rev. B*, 2002, **65**, 165401.
27. J. J. Palacios, A. J. Pérez-Jiménez, E. Louis, E. SanFabián and J. A. Vergés, *Phys. Rev. B*, 2002, **66**, 035322.
28. A. R. Rocha, V. M. García Suárez, S. W. Bailey, C. J. Lambert, J. Ferrer and S. Sanvito, *Phys. Rev. B*, 2005, **73**, 085414.
29. R. Gebauer and R. Car, *Phys. Rev. B*, 2004, **70**, 125324.
30. K. Burke, R. Car and R. Gebauer, *Phys. Rev. Lett.*, 2005, **94**, 146803.
31. M. Koentopp, C. Chang, K. Burke and R. Car, *J. Phys.: Condens. Matter*, 2008, **20**, 083203.
32. W. Kohn and L. J. Sham, *Phys. Rev.*, 1965, **140**, A1133.
33. H. Mera, P. Bokes and R. W. Godby, *Phys. Rev. Lett.*, 2005, **72**, 085311.
34. S. Sanvito, C. J. Lambert, J. H. Jefferson and A. M. Bratkovsky, *Phys. Rev. B*, 1999, **59**, 11936.
35. I. Rungger and S. Sanvito, *Phys. Rev. B*, 2008, **79**, 035407.
36. B. Wenzien, J. Kudrnovský, V. Drchal and M. Šob, *J. Phys.: Condens. Matter*, 1989, **1**, 9893.
37. M. B. Nardelli, *Phys. Rev. B*, 1999, **60**, 7828.
38. A. Calzolari, N. Marzari, I. Souza and M. B. Nardelli, *Phys. Rev. B*, 2004, **69**, 035108.
39. K. Hirose, T. Ono, Y. Fujimoto and S. Tsukamoto, *First-Principles Calculations in Real-Space Formalism*, Imperial College Press, London, 2005.

40. M. Büttiker, *Phys. Rev. B*, 1986, **33**, 3020.
41. I. Rungger, O. Mryasov and S. Sanvito, *Phys. Rev. B*, 2009, **79**, 094414.
42. J. Maassen, F. Zahid and H. Guo, *Phys. Rev. B*, 2009, **80**, 125423.
43. R. C. Jaklevic and J. Lambe, *Phys. Rev. Lett.*, 1966, **17**, 1139.
44. C. F. Hirjibehedin, C.-Y. Lin, A. F. Otte, M. Ternes, C. P. Lutz, B. A. Jones and A. J. Heinrich, *Science*, 2007, **317**, 1199.
45. W. Lee, N. Jean and S. Sanvito, *Phys. Rev. B.*, 2009, **79**, 085120.
46. A. Fetter and J. D. Walecka, *Quantum Theory of Many-Particle Systems*, Dover, New York, 2003.
47. M. Galperin, M. A. Ratner and A. Nitzan, *J. Chem. Phys.*, 2004, **121**, 11965.
48. H. Nakamura, K. Yamashita, A. R. Rocha and S. Sanvito, *Phys. Rev. B*, 2009, **78**, 235420.
49. T. Frederiksen, M. Paulsson, M. Brandbyge and A.-P. Jauho, *Phys. Rev. B*, 2007, **75**, 205413.
50. T. Frederiksen, N. Lorente, M. Paulsson and M. Brandbyge, *Phys. Rev. B*, 2007, **75**, 235441.
51. A. P. Horsfield, D. R. Bowler, A. J. Fisher, T. N. Todorov and C. G. Sanchez, *J. Phys.: Condens. Matter*, 2004, **16**, 8251.
52. A. P. Horsfield, D. R. Bowler, H. Ness, C. G. Sanchez, T.N. Todorov and A. J. Fisher, *Rep. Prog. Phys.*, 2006, **69**, 1195.
53. E. J. McEniry, T. Frederiksen, T.N. Todorov, D. Dundas and A. P. Horsfield, *Phys. Rev. B*, 2008, **78**, 035446.
54. W. Kohn, *Phys. Rev. Lett.*, 1996, **76**, 3168.
55. W. Yang and T.-S. Lee, *J. Chem. Phys.*, 1995, **103**, 5674.
56. J. E. Ingleseld, *J. Phys. C: Solid State Phys.*, 1981, **14**, 3795.
57. O. R. Davies and J. E. Inglesfield, *Prog. Surf. Sci.*, 2003, **74**, 161.
58. O. R. Davies and J. E. Inglesfield, *Phys. Rev. B*, 2004, **69**, 195110.
59. D. Waldron, L. Liu and H. Guo, *Nanotechnology*, 2007, **18**, 424026.
60. I. Rungger and S. Sanvito, in preparation.
61. A. R. Williams, P. J. Feibelman and N. D. Lang, *Phys. Rev. B*, 1982, **26**, 5433.
62. M. Wimmer and K. Richter, *J. Comput. Phys.*, 2009, **228**, 8548.
63. K. Kazymyrenko and X. Waintal, *Phys. Rev. B*, 2008, **77**, 115119.
64. D. A. Areshkin and B. K. Nikoliæ, *Phys. Rev. B*, 2010, **81**, 155450.
65. A. Pecchia, G. Penazzi, L. Salvucci and A. Di Carlo, *New. J. Phys.*, 2008, **10**, 065022.
66. C. D. Pemmaraju, I. Rungger and S. Sanvito, *Phys. Rev. B*, 2009, **80**, 104422.
67. C. D. Pemmaraju, I. Rungger, X. Chen, A. R. Rocha and S. Sanvito, arXiv: 1007.0035.
68. M. Luisier and G. Klimeck, *Parallel Computing*, 2010, **36**, 117.
69. Z. Feng and P. Li, *Proc. IEEE/ACM DAC*, 2008, **647**.
70. J. F. Janak, *Phys. Rev. B*, 1978, **18**, 7165.

71. J. P. Perdew, R. G. Parr, M. Levy and J. L. Balduz Jr, *Phys. Rev. Lett.*, 1982, **49**, 1691.
72. J. P. Perdew and M. Levy, *Phys. Rev. Lett.*, 1983, **51**, 1884.
73. J. B. Neaton, M. S. Hybertsen and S. G. Louie, *Phys. Rev. Lett.*, 2006, **97**, 216405.
74. C. Toher and S. Sanvito, *Phys. Rev. Lett.*, 2007, **99**, 056801.
75. C. D. Pemmaraju, T. Archer, D. Sánchez-Portal and S. Sanvito, *Phys. Rev. B*, 2007, **75**, 045101.
76. X. Xiao, B. Xu and N. J. Tao, *Nano Lett.*, 2004, **4**, 267.
77. C. Toher and S. Sanvito, *Phys. Rev. B*, 2008, **77**, 155402.
78. C. Toher, A. Filippetti, K. Burke and S. Sanvito, *Phys. Rev. Lett.*, 2005, **95**, 146402.
79. M. Koentopp, K. Burke and F. Evers, *Phys. Rev. B*, 2006, **73**, 121403(R).
80. S. Kurth, G. Stefanucci, E. Khosravi, C. Verdozzi and E. K. U. Gross, *Phys. Rev. Lett.*, 2010, **104**, 236801.
81. M. Julliere, *Phys. Lett. A*, 1975, **54**, 225.
82. S. Ikeda, J. Hayakawa, Y. Ashizawa, Y. M. Lee, K. Miura, H. Hasegawa, M. Tsunoda, F. Matsukura and H. Ohno, *Appl. Phys. Lett.*, 2008, **93**, 082508.
83. I. Rungger, *Computational methods for electron transport and their application in nanodevices*, (PhD Thesis, University of Dublin, Trinity College, 2008).
84. C. Heiliger, P. Zahn, B. Y. Yavorsky and I. Mertig, *Phys. Rev. B*, 2006, **73**, 214441.
85. A. R. Rocha, V. M. García Suárez, S. W. Bailey, C. J. Lambert, J. Ferrer and S. Sanvito, *Nature Mater.*, 2005, **4**, 335.
86. J. Peralta-Ramos, A.M. Llois, I. Rungger and S. Sanvito, *Phys. Rev. B*, 2008, **78**, 024430.
87. J. P. Velev, C. -G. Duan, J. D. Burton, A. Smogunov, M. K. Niranjan, E. Tosatti, S. S. Jaswal and E.Y. Tsymbal, *Nano Lett.*, 2009, **9**, 427.
88. Y. Xue, S. Datta, S. Hong, R. Reifenberger, J. I. Henderson and C. P. Kubiak, *Phys. Rev. B*, 1999, **59**, R7852.
89. Ti. Rakshit, G.-C. Liang, A. W. Ghosh and S. Datta, *Nano Lett.*, 2004, **4**, 1803.
90. M. Galperin, M. A. Ratner and A. Nitzan, *Nano Lett.*, 2005, **5**, 125.
91. E. G. Emberly and G. Kirczenow, *Phys. Rev. B*, 2001, **64**, 125318.
92. D. M. Basko and E. M. Conwell, *Phys. Rev. Lett.*, 2002, **88**, 098102.
93. I. Rungger, X. Chen, U. Schwingenschlögl and S. Sanvito, *Phys. Rev. B*, 2010, **81**, 235407.
94. A. Tawara, T. Tada and S. Watanabe, *Phys. Rev. B*, 2009, **80**, 073409.
95. E. Leary, H. Hobenreich, S. J. Higgins, H. van Zalinge, W. Haiss, R. J. Nichols, C. M. Finch, I. Grace, C. J. Lambert, R. McGrath and J. Smerdon, *Phys. Rev. Lett.*, 2009, **102**, 086801.
96. H. Cao, J. Jiang and Y. Luo, *J. Am. Chem. Soc.*, 2008, **130**, 6674.
97. K. S. Thygesen and A. Rubio, *Phys. Rev. B*, 2008, **77**, 115333.

98. S.-H. Ke, H. U. Baranger and Weitao. Yang, *J. Chem. Phys.*, 2007, **126**, 201102.

99. S. Kümmel, L. Kronik and J. P. Perdew, *Phys. Rev. Lett.*, 2004, **93**, 213002.

100. C. D. Pemmaraju, S. Sanvito and K. Burke, *Phys. Rev. B*, 2008, **77**, 121204(R).

101. T. Körzdörfer, M. Mundt and S. Kümmel, *Phys. Rev. Lett.*, 2008, **100**, 133004.

102. F. Haake, *Statistical treatment of open systems by generalized master equations*, Springer, Berlin, Germany, 1973.

103. A. A. Yanik, G. Klimeck and S. Datta, *Phys. Rev. B*, 2007, **76**, 045213.

104. L. Michalak, C. M. Canali, M. R. Pederson, M. Paulsson and V.G. Benza, *Phys. Rev. Lett.*, 2010, **104**, 017202.

105. D. C. Ralph and M. D. Stiles, *J. Magn. Magn. Mater.*, 2008, **320**, 1190.

106. D. Dundas, E. J. McEniry and T. N. Todorov, *Nature Nanotechnol.*, 2009, **4**, 99.

Theoretical Strategies for Functionalisation and Encapsulation of Nanotubes

GOTTHARD SEIFERT,[1] MATTEO BALDONI,[1,2] FRANCESCO MERCURI[3] AND ANDREY ENYASHIN[1,4]

[1] Physikalische Chemie, Technische Universität Dresden, 01062 Dresden, Germany; [2] Department of Chemistry, University of Perugia, 06123 Perugia, Italy; [3] ISTM-CNR c/o Department of Chemistry, University of Perugia, 06123 Perugia, Italy; [4] Institute of Solid State Chemistry, 620990 Ekaterinburg, Russia

8.1 Introduction

New technologies that allow manipulation of matter at atomic resolution hold the potential to meet many of the greatest global challenges, bringing revolutions in science, medicine, energy, and industry. To face such challenges the creation of new kind of "smart" materials is required. This need has led to an increasing interest of the scientific community in the field of materials science. However, to obtain functional materials with novel properties, the attention of material scientists had to move in the range of the nanoscale. Building complex structures with atom-by-atom control is a hard task. The suggestive nanoworld in which the concepts of function and shape are indissoluble could become extremely chaotic if combined with the elusive character of "quantistic objects" that are ruled by the Indetermination Principle. Nano-objects are

RSC Theoretical and Computational Chemistry Series No. 4
Computational Nanoscience
Edited by Elena Bichoutskaia
© Royal Society of Chemistry 2011
Published by the Royal Society of Chemistry, www.rsc.org

extremely sensitive systems in which subtle variations in structural features could induce dramatic alterations of electronic structure and therefore in the overall properties. These revolutionary concepts should be customary for a chemist, who is used to facing the task of an extremely precise definition of nuclear and electronic structures. Hence, chemical knowledge is a prerequisite in the study of nanotechnological materials. Nanotubes (NTs) based on carbon (CNTs) but also on many other compounds of other elements (inorganic nanotubes – INNTs) are one of the most promising materials for nano-technological applications. Nevertheless, the development of large-scale technological applications based on NTs often had to face intrinsic difficulties related to the accurate and unambiguous experimental determination of their properties. Hence, computational modeling constitutes a favourable alternative to experimental techniques and, in many cases, a preferential investigation route. Chemical functionalisation of CNTs can be regarded as a possibility of modifying and tailoring their reactive, electrical and mechanical behaviour, leading to new nanostructured materials with interesting properties.

Theoretical methods based on the description of the electronic structures of molecular systems constitute particularly useful tools in the modelling of nanostructured objects, as for example in the study of functionalisation processes. In particular, methods based on density-functional theory (DFT) can be applied even in the case of relatively extended systems. However, the precise definition of the model system is one of the critical steps of theoretical modeling. As will be discussed in detail the Clar aromatic sextet theory, developed to study of polycyclic aromatic hydrocarbons (PAHs), can be applied with success in the case of CNTs to identify building blocks representative of real systems. Another possible field of application of NTs concerns the development of novel materials by making use of NTs as nanosized templates. Indeed, template synthesis is one of the most promising routes for the fabrication of nanomaterials. This general methodology offers the possibility to create spatially confined reaction environments in which reactions take place in a controlled manner and lead to specific reactions products. Especially CNTs but also INNTs have shown the ability to encapsulate different kinds of materials (ionic salts, metals and molecular solids) inside their hollows, inducing the stabilisation of quasi-one-dimensional (1D) phases not usually observed under ordinary conditions. Due to intrinsic difficulties in the experimental characterisation of nanostructured materials, many microscopic details of the filling process and of the resulting structures remain, to a large extent, unclear. Computer simulation methods, in particular molecular dynamics (MD) simulations that can take into account the time-dependent evolution of systems constituted of thousands of atoms, represent an ideal tool for the investigation of the NTs encapsulation process.

The present review is divided into two main parts. In the first part, the theoretical background is presented. In the first section of this part the main properties and structural features of CNTs and INNTs are described. The second section concerns in particular DFT, since it proved to be a very useful tool for the study of the properties of extended systems.

Nevertheless, more qualitative concepts are not outdated and can be very useful to describe properties of complex systems in a clear way. Such a concept is the Clar sextet theory, originally developed for the study of the PAHs. This formalism and its application to CNTs are discussed as well.

Molecular dynamics (MD) simulations that can take into account the time-dependent evolution of systems constituted of thousands of atoms, represent an ideal tool for the investigation of the NTs encapsulation process. For such simulations the application of *first-principles* methods on the basis of DFT and even approximate methods of DFT are still computationally too demanding. Therefore, the application of empirical force fields for the description of the interatomic forces have to be applied. The main aspects of the MD and the force fields, used in the studies of NT encapsulation are also briefly described.

In the second part the functionalisation of CNTs and the encapsulation in CNTs and INNTs are outlined. In particular, the results from MD simulations concerning confined growth of quasi-one-dimensional structures of AgI, CsI, PbI_2, KI inside the CNTs and INNTs are discussed.

8.2   Methodology

8.2.1   Structural Concepts of Nanotubes

A SWCNT can be described as a graphene sheet rolled into a cylindrical shape so that the structure is one-dimensional with axial symmetry, and in general exhibiting a spiral conformation, which gives rise to peculiar symmetry properties generally referred to as *chirality*. The chirality, as defined in this section, is defined in terms of a single vector called the chiral vector.[1]

An interesting and essential fact about the structure of a CNT is the orientation of the six-membered carbon rings (hereafter called hexagons) in the honeycomb lattice respect to the axis of the CNT. Three examples of SWCNTs are shown in Figure 8.1. From the figure, it can be seen that many different orientations of the hexagons in the honeycomb lattice with respect to the CNT axis are compatible with the cylindrical structure, without any distortion of the hexagons except for the distortion due to the curvature of the CNT. This fact provides many possible structures for CNTs, even though the basic shape of the CNT sidewall is a cylinder. In their idealised representation, the CNTs are infinitely long objects in which the symmetry of the sidewall is maintained along the CNT axis. In more realistic situations, like in experiments, the edges of CNT fragments exhibit a variety of terminations. They can be constituted of caps (or end-caps) that consist of a "hemisphere" of a fullerene. Each cap contains an appropriate number and placement of pentagons and hexagons that are selected to fit the cylindrical section.[1]

The primary symmetry classification of a CNT is as either being achiral (symmorphic) or chiral (nonsymmorphic). An achiral CNT is defined by a CNT whose mirror image has an identical structure to the original one. There are only two cases of achiral CNT: armchair and zigzag CNTs, as shown in Figures 8.1(a) and (b), respectively. The names of armchair and zigzag arise

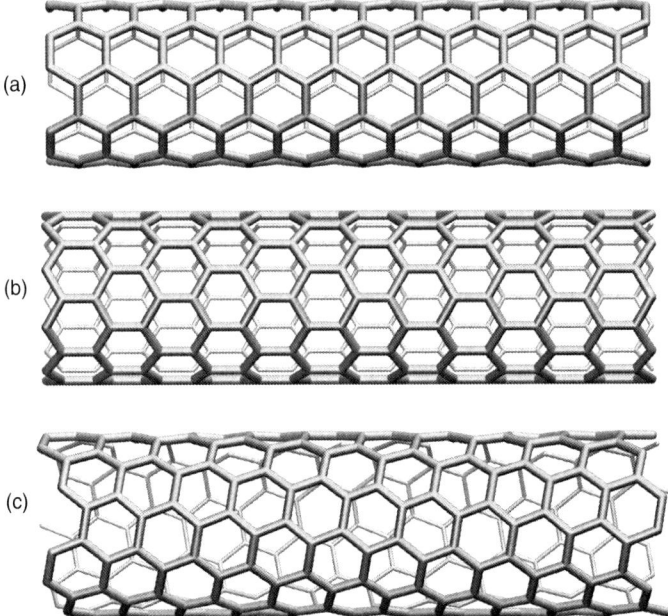

Figure 8.1   Symmetry classification of CNTs: (a) armchair, (b) zigzag, (c) chiral
nanotube. From the figure it can be seen that the orientation of the
hexagons in the honeycomb lattice relative to the CNT axis can be taken
almost arbitrarily.

from the shape of the cross-sectional ring, as shown in Figures 8.1(a) and (b),
respectively. Inversely, chiral CNTs exhibit a spiral symmetry whose mirror
image cannot be superimposed onto the original one. This tube is called a chiral
CNT, since such structures are called axially chiral in the chemical nomenclature. We have thus a variety of geometries in CNTs that can change
diameter, chirality and cap structures.

The structure of a SWCNT is conveniently defined in terms of a parent
lattice, which is constituted by a planar sheet of graphene. In Figure 8.2, the
unrolled honeycomb lattice of the CNT is shown. The structure of a SWCNT is
specified by the vector $\mathbf{C_h}$ (**OA** in Figure 8.2) which corresponds to a section of
the CNT perpendicular to the direction of the CNT axis **OB** (we shall hereafter
define this section as the equator of the nanotube).[1] By considering the crys-
tallographically equivalent sites O, A, B and B'', and by rolling the honeycomb
sheet so that points O and A coincide (and points B and B'' coincide), a 3D
model of a CNT can be constructed. The vectors **OA** and **OB** define the chiral
vector $\mathbf{C_h}$ and the translational vector of a CNT, respectively, as further
explained below. The chiral vector $\mathbf{C_h}$ can be expressed by the real space unit
vectors $\mathbf{i}$ and $\mathbf{j}$ (see Figure 8.2) of the hexagonal lattice as:

$$\mathbf{C_h} = n\mathbf{i} + m\mathbf{j} \equiv (n, m), \quad (n, m \text{ are integers}, 0 \leq |m| \leq n), \qquad (8.1)$$

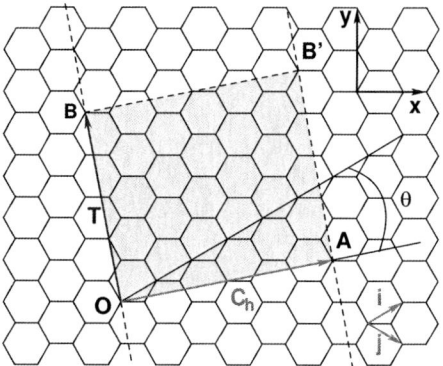

Figure 8.2 The unrolled honeycomb lattice of a CNT. When we connect the sites O with A and B with B'', a CNT can be constructed. **OA** and **OB** define the chiral vector $\mathbf{C_h}$ and the translational vector $\mathbf{T}$ of the CNT, respectively. The rectangle $OAB'B$ defines the unit cell for the nanotube (highlighted in grey). The vector $\mathbf{R}$ denotes a symmetry vector. The example depicted corresponds to $\mathbf{C_h} = (4,2)$ and $\mathbf{T} = (4\text{-}5)$. Adapted from ref. 1.

with $\mathbf{i} = \left(\frac{\sqrt{3}}{2}i, \frac{i}{2}\right)$, $\mathbf{j} = \left(\frac{\sqrt{3}}{2}j, -\frac{j}{2}\right)$ where $i = j = |\mathbf{i}| = |\mathbf{j}| = 1.42 \times \sqrt{3} = 2.46$ Å is the lattice constant of the two-dimensional graphene layer.

For example, the specific chiral vectors $\mathbf{C_h}$ shown if Figure 8.1 are, respectively, a (5,5), (b) (9,0) and (c) (10,5), and the chiral vector shown in Figure 8.2 is (4,2). An armchair CNT corresponds to the case of $n = m$, that is $\mathbf{C_h} = (n,m)$ and a zigzag nanotube correspond to the case of $m = 0$, or $\mathbf{C_h} = (n,0)$. All other (n,m) chiral vectors correspond to chiral CNTs. Because of the hexagonal symmetry of the honeycomb lattice, we need to consider only $0 \leq |m| \leq n$ in $\mathbf{C_h} = (n,m)$ for chiral CNTs.

The CNT diameter, d_t, is given by L/π, in which L is the circumferential diameter of the CNT:[1]

$$d_t = L/\pi \quad L = |\mathbf{C_h}| = \sqrt{\mathbf{C_h} \cdot \mathbf{C_h}} = i\sqrt{n^2 + m^2 + nm}. \quad (8.2)$$

It must be noted here that $\mathbf{i}$ and $\mathbf{j}$ are not orthogonal to each other and that the inner product between $\mathbf{i}$ and $\mathbf{j}$ yields:

$$\mathbf{i} \cdot \mathbf{i} = \mathbf{j} \cdot \mathbf{j} = i^2 = j^2, \quad \mathbf{i} \cdot \mathbf{j} = \frac{i^2}{2} = \frac{j^2}{2}, \quad (8.3)$$

where the lattice constant $i = j = 1.44 \times \sqrt{3}$ Å $= 2.49$ Å. The chiral angle θ (see Figure 8.2) is defined as the angle between the vectors $\mathbf{C_h}$ and $\mathbf{i}$, with values of θ in the range of $0 \leq |\theta| \leq 30°$, because of the hexagonal symmetry of the honeycomb lattice.[1] The chiral angle θ determines the tilt angle of the hexagons with respect to the direction of the nanotube axis, and the angle θ specifies the spiral symmetry. According to this definition, the chiral angle θ is given by the inner product of $\mathbf{C_h}$ and $\mathbf{i}$:

$$\cos(\theta) = \frac{C_h \cdot i}{|C_h||i|} = \frac{2n + m}{2\sqrt{n^2 + m^2 + nm}}. \tag{8.4}$$

Thus, relating θ to the integers (n,m) defined in eqn (8.1).

The translation vector $\mathbf{T}$ is defined as the unit vector of a 1D CNT. The vector $\mathbf{T}$ is parallel to the CNT axis and is normal to the chiral vector $\mathbf{C_h}$ in the unrolled honeycomb lattice in Figure 8.2.[1] The lattice vector $\mathbf{T}$ shown as $\mathbf{OB}$ in Figure 8.2 can be expressed in terms of the basis vectors $\mathbf{i}$ and $\mathbf{j}$ as:

$$\mathbf{T} = t_i\mathbf{i} + t_j\mathbf{j} \equiv (t_i, t_j), \quad (\text{where } t_i, t_j \text{ are integers}). \tag{8.5}$$

The translation vector $\mathbf{T}$ corresponds to the first lattice point of the 2D graphene sheet through which the vector $\mathbf{OB}$ (normal to the vector $\mathbf{C_h}$) passes. From this fact, it is clear that t_i and t_j do not have a common divisor except for unity.

Using $\mathbf{C_h} \cdot \mathbf{T} = 0$ and eqns (8.1), (8.3) and (8.5) we obtain the expressions for t_i and t_j:

$$t_i = \frac{2m + n}{d_R}, \quad t_j = -\frac{2n + m}{d_R}, \tag{8.6}$$

where d_R is the greatest common divisor (gcd) of $(2m+n)$ and $(2n+m)$.

As carbon nanotubes, most of the inorganic nanotubes known at present were synthesised from layered – quasi-two-dimensional – bulk compounds. The characteristic peculiarity of these compounds is the clearly expressed aniso-tropy of the strong (covalent) and weak (Van-der-Waals) bonding within and between the layers, respectively. Experimentally obtained nanotubes usually have complex structures: they can be composed of various numbers of coaxial cylindrical layers and the bonds within the layers exhibit different orientations relative to the tube axis.[2,3] Nanotubes can be open or closed at the ends. They can have not only cylindrical, but also a scroll-like morphology. Never-theless, all the tubular forms of matter may be structurally characterised by using the classification developed for the single-walled nanotubes, which are produced by rolling up of monolayers. The structure of the crystalline monolayer can be characterised using the terms of a 2D lattice spanned by the translation vectors ($\mathbf{i}$ and $\mathbf{j}$) and the angle between them (φ). Five types of 2D Bravais lattices exist: oblique ($\mathbf{i} \neq \mathbf{j}$, $\varphi \neq 90°$), square ($\mathbf{i} = \mathbf{j}$, $\varphi = 90°$), hex-agonal ($\mathbf{i} = \mathbf{j}$, $\varphi = 120°$), primitive rectangular ($\mathbf{i} \neq \mathbf{j}$, $\varphi = 90°$), and centred rectangular ($\mathbf{i} \neq \mathbf{j}$, $\varphi = 90°$). Cylindrical surfaces may be constructed by rolling up these lattices as in the case of CNTs. A lot of layered inorganic compounds, which were discovered to form nanotubes, also exhibit layers with a hexagonal atom arrangement (see Figure 8.3).

Therefore, the classification of the carbon nanotubes explained above may often be used for the classification of the inorganic nanotube structure. Every

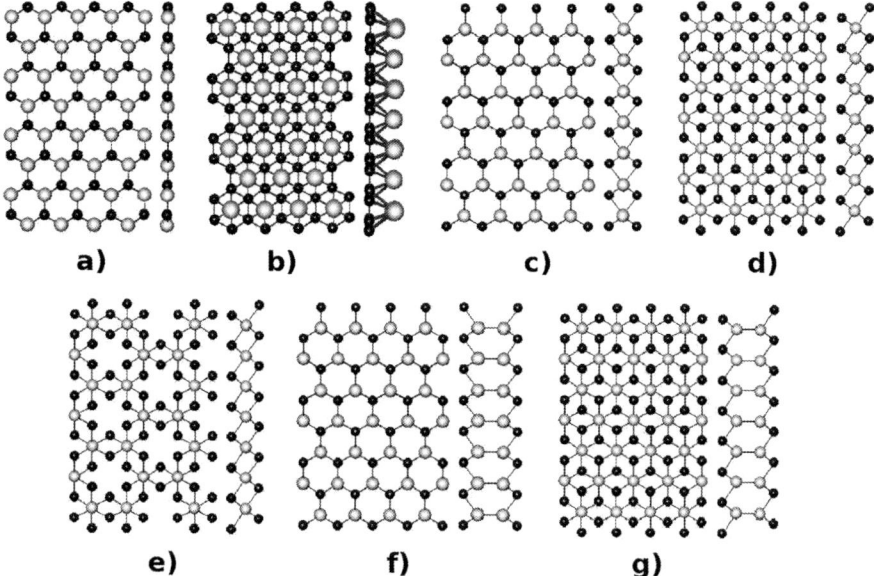

Figure 8.3 Structures of the hexagonal monolayers of some (quasi)layered compounds: (a) BN; (b) AlB_2, MgB_2; (c) MoS_2, WS_2, $NbSe_2$; (d) TiS_2, ZrS_2, anatase TiO_2, $NiCl_2$; (e) $FeCl_3$, MoO_3; (f) GaS, GaSe; (g) metastable InS. Adapted from ref. 2.

cylindrical nanotube is characterised by the chiral vector C_h by which it is possible to describe the basic geometry parameters of a tubulene, which is produced by rolling of a ribbon sliced from a monolayer (see Figure 8.2).

As for CNTs the diameter d_t and the chiral angle θ are associated uniquely with **i** and **j** using the integer indexes n and m, these indices may be used for the structural classification of the inorganic nanotubes in the same way as for carbon nanotubes – (n,m). Then, depending on the values of n and m all such nanotubes can be subdivided into two groups: chiral tubes with $0° < \theta < 30°$ and the nonchiral ones, the so-called "zigzag" and "armchair" tubes with $\theta = 0°$ and $30°$, respectively.

Single-walled tubular structures based on nonhexagonal layers may analogously be described by the primitive vectors of the respective lattice (see Figure 8.4).[2,3]

In those cases, R and θ will be associated with n and m by other correlations.

It is important to note that single-walled carbon nanotubes are based on a planar, almost ideally two dimensional graphitic layer, which is just one carbon atom thick. In contrast, in most of the layered inorganic compounds a single layer has a more complicated atom arrangement and is several atoms thick. Therefore, inorganic single-walled tubes are composed of several concentric and interconnected cylinders of atoms (Figures 8.2 and 8.4). This fact gives *a priori* a basis to suppose, that inorganic tubulenes are less stable than carbon ones with the same radii, since a rolling of a thicker layer to a cylinder is energetically less preferable.

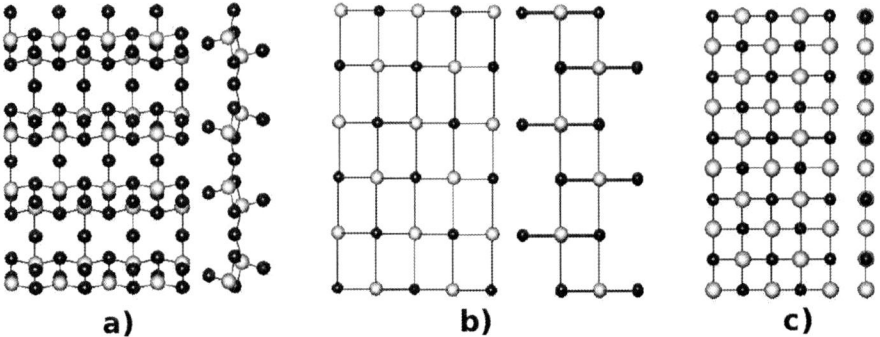

Figure 8.4 Structures of the nonhexagonal monolayers of some (quasi)layered compounds: (a) rectangular V_2O_5; (b) centred rectangular lepidocrocite TiO_2; (c) square MgO. Adapted from ref. 2.

The model monolayers of the inorganic compounds described so far represent only a small part of the possible structural variety. For a more detailed discussion see refs. 2 and 3.

8.2.2 Density Functional Theory Based Approach

The electronic and structural properties of molecules as well as their stabilities can be calculated, in principle, by solving the (stationary) Schrödinger equation in connection with the Born–Oppenheimer approximation. The consideration of translational symmetry, using the Bloch theorem, also allows calculations for extended (bulk) systems. However, in practice, such "*ab initio*" or "first principles" treatment is fundamentally hampered by the many-particle (electron) problem. Several approximate schemes were developed and successfully applied to small and medium-sized molecules.[4–9] These methods suffer several computational difficulties whenever accurate results are needed, thus using correlated methods, especially for extended systems with a large number of basis functions. Post-Hartree–Fock methods, like MP, CI, and so on, are characterised by a computational scaling of order N^5, N^6 or N^7, giving rise to their impracticability in simulations of relatively large systems.[4–9]

More details about theoretical chemistry methods can be found in the specialised literature.[4–9]

As an alternative, especially for extended systems, Hohenberg, Kohn and Sham formulated the density-functional theory (DFT).[10,11]

The DFT can give an accuracy to the same level as the second-order Møller–Plesset perturbation theory (MP2) method.[12] Efficient algorithms of the Kohn–Sham equations have been implemented scaling to N^3 with respect to the size of the basis sets and, hence, being much more efficient than the N^5 of the MP2 methods. DFT is the method chosen for a huge range of applications, including nanotube systems.

Density-functional theory has been extensively reviewed. In this section a very brief review of DFT is given.

The Hohenberg–Kohn (HK) theorems[10] have rigorously made the electronic density acceptable as the basic variable to electronic-structure calculations. However, development of practical DFT methods only became relevant after W. Kohn and L.J. Sham published their famous set of equations: the so-called Kohn–Sham (KS) equations.[11]

The use of the electronic density within the KS scheme allowed for a significant reduction of the computational demand involved in quantum calculations. Furthermore, the KS method paved the way for studying systems that could not be investigated by conventional *ab initio* methods (which use the wavefunction as the basic variable).

Within the KS scheme the electron density is written as a sum of orbital densities $\rho(\vec{r}) = \sum_{i}^{M} |\psi_i(\vec{r})|^2$. Therefore, the HK energy functional

$$E[\rho] = \int v_{ext}(\vec{r}) d\vec{r} + F_{HK}[\rho] \tag{8.7}$$

can be written as

$$F_{HK}[\rho] = T_s[\rho] + J[\rho] + E_{xc}[\rho], \tag{8.8}$$

where T_s represents the kinetic-energy functional for the reference system of M noninteracting electrons, given by

$$T_s[\rho] = \sum_{i}^{M} \left\langle \psi_i \left| -\frac{1}{2}\nabla_i^2 \right| \psi_i \right\rangle; \tag{8.9}$$

J represents the classic Coulomb interaction functional

$$J = \frac{1}{2} \iint \frac{\rho(\vec{r})\rho(\vec{r'})}{|\vec{r} - \vec{r'}|} d\vec{r} d\vec{r'}; \tag{8.10}$$

and the remaining interactions are summarised in E_{xc}, the exchange-correlation functional, which contains the difference between the exact kinetic energy T and T_s, besides the nonclassic part of V_{ee}, *i.e.* $E_{xc}[\rho] \equiv T[\rho] - T_s[\rho] + V_{ee}[\rho] - J[\rho]$. The application of a variation principle leads to effective single particle equations, the KS equations:

$$\left[-\frac{1}{2}\nabla_i^2 + v_{KS}(\vec{r}) \right] \psi_i = \varepsilon_i \psi_i, \tag{8.11}$$

with the KS effective potential

$$v_{KS}(\vec{r}) = v_{ext}(\vec{r}) + \int \frac{\rho(\vec{r})}{|\vec{r} - \vec{r'}|} d\vec{r'} + v_{xc}(\vec{r}). \tag{8.12}$$

The exchange-correlation potential v_{xc} is defined as

$$\nu_{xc}(\vec{r}) = \frac{\delta E_{xc}[\rho]}{\delta \rho(\vec{r})}. \tag{8.13}$$

Since the effective potential v_{KS} depends on ρ through v_{xc}, the KS equations must be solved iteratively using a self-consistent procedure. The most difficult part of the KS scheme is to calculate v_{xc} in eqn (8.12). Although the existence of an exact density functional is assured by the first HK theorem, the exact form of the E_{xc} functional remains unknown. However, many approximations of this functional have been described in the literature over the last 30 years. For an overview see, *e.g.*, ref. 12.

8.2.3 Clar's Concept

Apart from merely structural characteristics, many interesting features of CNTs, including technological applications, rely on their electronic properties. Due to the peculiar structure of the sidewall constituted by a rolled hexagonal network of carbon atoms, aromaticity is a central aspect in the study of the electronic properties of CNTs.

Even though the concept of aromaticity was introduced in 1865,[13] it still continues to be of central importance in physical organic chemistry for the rationalisation of structure, stability and reactivity of a large group of molecules.[14–19] As we will see later, such studies have been extended to new classes of molecules like fullerenes and CNTs. In particular, CNTs exhibit striking structural similarities with polycyclic aromatic hydrocarbons (PAHs) that are constituted of condensed rings of carbon atoms, saturated at the edges with hydrogen atoms. Therefore, one can think to extend the concepts usually applied in the study of the aromaticity properties of PAHs to the case of CNTs. This hypothesis will be one of the central aspects of this discussion. Hence, we first need to introduce suitable theoretical approaches to the study of PAHs and their aromaticity.

Among the numerous attempts made to extend the Hückel $4n + 2$ rule[20–23] to polycyclic systems, one of the most successful was the Clar model of the extra stability of $6n$ π-electron benzenoid species.[24–26] The Clar VB model[24,25] employs both conventional two-electron π-bonds, schematically represented by lines, and aromatic sextets (six-electron π-cycles), represented by circles. According to Clar's rule, the most important Kekulè resonance structure for the characterisation of the properties of PAHs, is that with the largest number of disjoint aromatic-sextets, *i.e.* benzene-like moieties. Aromatic-sextets are defined as six electrons localised in a single benzene-like ring separated from adjacent rings by formal C–C single bonds. For instance, application of this

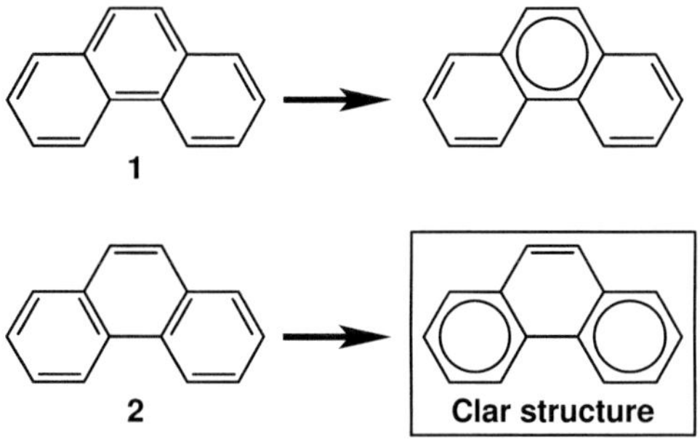

Figure 8.5   Two of the five Kekulè resonance structures of phenanthrene and their corresponding Clar aromatic π-sextets, indicated with a circle. The structure with the largest number of aromatic π-sextets is the so-called Clar structure.

rule to phenanthrene indicates that the resonance structure **2** (see Figure 8.5) is more important than the resonance structure **1**. Therefore, the outer rings in phenanthrene are expected to have a larger local aromaticity than the central ring. This result has been confirmed by several measures of local aromaticity.[27,29] Moreover, recent experimental results on the distribution of π-electrons in large PAHs,[30] valence bond calculations[31] and nucleus-independent chemical shift (NICS) studies on pericondensed benzenoid PAHs[32] have provided extensive support for Clar's rule. The Clar structure (see Figure 8.5) of a given PAH is the resonance structure having the maximum number of isolated and localised aromatic π-sextets, with a minimum number of localised double bonds. For example the Clar structure of phenantrene corresponds to the structure depicted in bottom right of Figure 8.5. In general, a PAH with a given number of aromatic π-sextets is kinetically more stable than its isomers with fewer aromatic π-sextets.[24–26,32] Moreover, aromatic π-sextet rings are considered to be the most aromatic centres in the PAH. The other rings are less aromatic and show different chemical reactivity. The number of Clar structures depends on the particular PAH considered.[26]

As previously suggested by Ormsby and King, the most convenient way to apply the Clar theory to CNTs consist in defining the two-dimensional carbon atom network resulting from the unrolling of the sidewall in terms of two translation basis vectors, which reflect the translation symmetry of a fully benzenoid graphene sheet.[33] This basis is connected to the conventional (*n,m*) roll-up vector basis[1] (see Section 8.2.1) by simple algebraic relationships (see below) and allows the Clar representation of the sidewall. Moreover, the Clar vector basis leads to a new definition of a primitive cell for the CNTs, which is usually smaller than the conventional (*n,m*) cell. Hence, if we make use of the CNTs basis **(i,j)**, defined in eqn (8.1) (see Figure 8.2) and define a new basis **(p,q)** as in Figure 8.6, we can express a generic roll-up vector as:[33]

$$n\mathbf{i} + m\mathbf{j}, \tag{8.14}$$

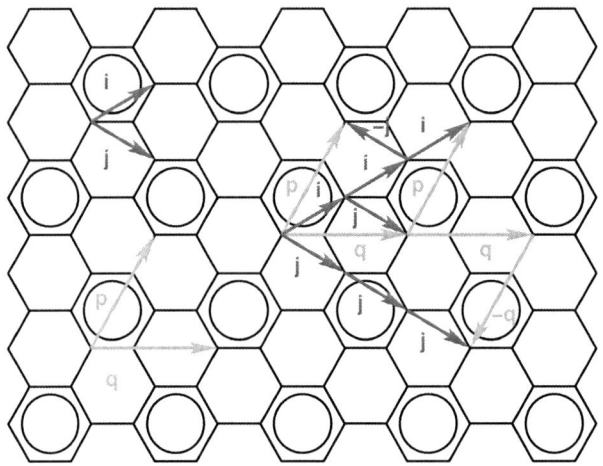

Figure 8.6 Conventional roll-up vectors (highlighted in dark grey) and Clar basis
vectors (highlighted in light grey). Adapted from ref. 33.

and define a generic Clar vector as:

$$r\mathbf{p} + s\mathbf{q}. \tag{8.15}$$

As shown in Figure 8.6, we have:

$$\mathbf{p} = 2\mathbf{i} - \mathbf{j} \qquad \mathbf{q} = \mathbf{i} + \mathbf{j}, \tag{8.16}$$

and

$$\mathbf{i} = \frac{\mathbf{p} + \mathbf{q}}{3} \qquad \mathbf{j} = \frac{2\mathbf{q} - \mathbf{p}}{3}. \tag{8.17}$$

Substituting eqn (8.16) into eqn (8.15) we obtain:

$$r\mathbf{p} + s\mathbf{q} = r(2\mathbf{i} - \mathbf{j}) + s(\mathbf{i} + \mathbf{j}) \tag{8.18}$$

By putting the second member of eqn (8.18) equal to eqn (8.14) and
rearranging we obtain:

$$n\mathbf{i} + m\mathbf{j} = (2r + s)\mathbf{i} + (s - r)\mathbf{j}. \tag{8.19}$$

The relation becomes an equality only for:

$$n = (2r + s) \quad m = (s - r). \tag{8.20}$$

Symmetrically substituting eqn (8.17) into eqn (8.14) we obtain:

$$n\mathbf{i} + m\mathbf{j} = n\frac{\mathbf{p} + \mathbf{q}}{3} + m\frac{2\mathbf{q} - \mathbf{p}}{3}. \tag{8.21}$$

By putting the second member of eqn (8.21) equal to eqn (8.15) and rearranging we obtain:

$$r\mathbf{p} + s\mathbf{q} = \frac{n - m}{3}\mathbf{p} + \frac{2m + n}{3}\mathbf{q}. \tag{8.22}$$

Again the relation becomes an equality only for:

$$r = \frac{n - m}{3} \quad s = \frac{n + 2m}{3}. \tag{8.23}$$

Hence, the relationship (8.23) allows us to obtain the values of r and s, which specify the topological properties of the sidewall in the Clar (benzenoid) vector basis, in terms of the more general chiral indexes n and m. In the relationship (8.23), r and s assume integer values only for particular choices of n and m.[33] Therefore, a fully benzenoid network can be only constructed when mod(n–m,3) = 0. From this consideration, a new parameter for the classification of the CNTs can be conveniently defined as:[33]

$$R(n, m) = \mathrm{mod}(n - m, 3). \tag{8.24}$$

Most notably, the parameter R is in strict relationship with the electronic properties of CNTs. In particular, both theory and experiments show that the value of the parameter R largely determines the conductivity of defect-free CNTs:[1,34–37] if $R(n,m) = 0$, the CNT is metallic; otherwise, the CNT is semiconducting. These observations agree with the Clar VB model for CNTs. Indeed, the value for the parameter R gives rise to three different situations for the electronic structure of the sidewall[33] R (see Figure 8.7).

If $R(n,m) = 0$, then the CNT is fully benzenoid and potentially metallic (see Figure 8.8).[33]

This behaviour contrasts with that of planar PAHs, in which the fully benzenoid structures are known to have large HOMO–LUMO.[38] If $R(n,m) = 1$, then a row of double bonds wraps about an otherwise fully benzenoid CNT at the chiral angle 60°, parallel to the unit vector $\mathbf{p}$ (see Figure 8.9).

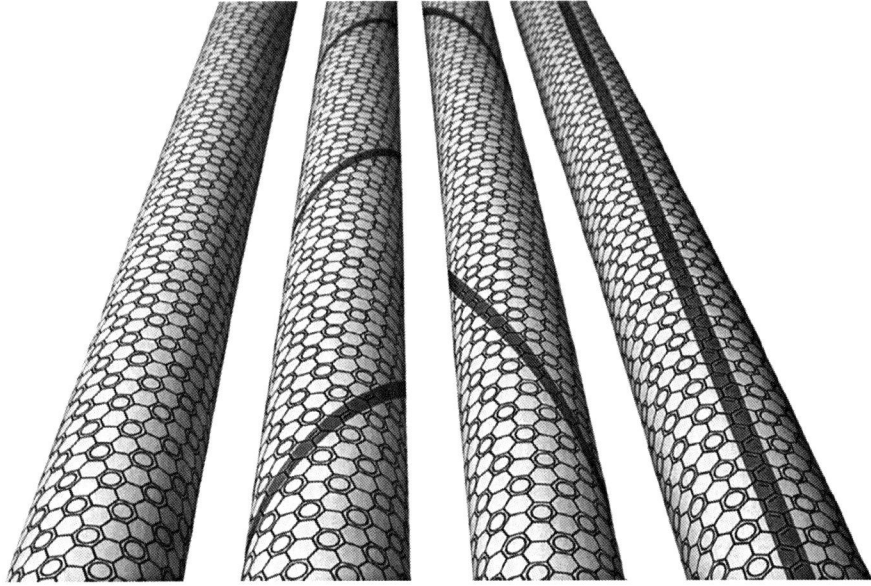

Figure 8.7   Clar VB representation of (12,9), (12,8), (12,7) and (19,0) CNTs, with $R(n,m) = 0, 1, 2$ and 1, left to right. Adapted from ref. 33.

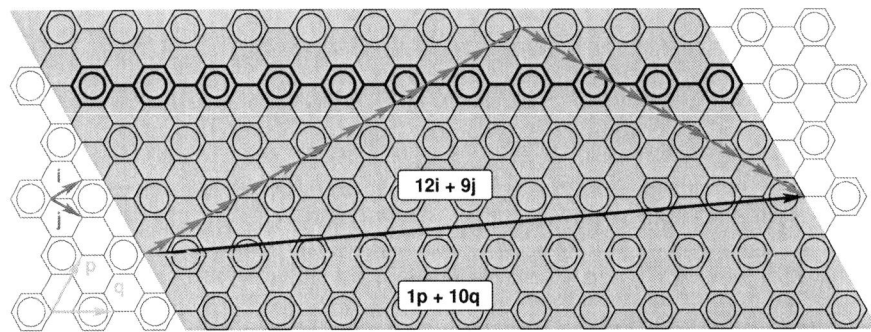

Figure 8.8   Roll-up vector (black arrow) for (12,9) CNT $R = 0$ in the conventional basis **i**, **j** (dark grey) and the Clar basis **p**, **q** (light grey). Atoms in the unit cell are bold; the parent graphene strip is grey. Adapted from ref. 33.

If $R(n,m) = 2$, then a seam of double bonds wraps around the otherwise fully benzenoid CNT at the chiral angle, parallel to the vector **p**–**q** (see Figure 8.10).[33]

Let us first consider fully benzenoid CNTs, which are necessarily derived from fully benzenoid graphite (see Figure 8.8). The introduced Clar basis (p, q), unlike the conventional basis (i, j), reflects the translational symmetry of fully benzenoid graphite.[33] Any roll-up vector $rp + sq$, where r and s are integers, will generate a fully benzenoid CNT. Changing from the Clar basis to

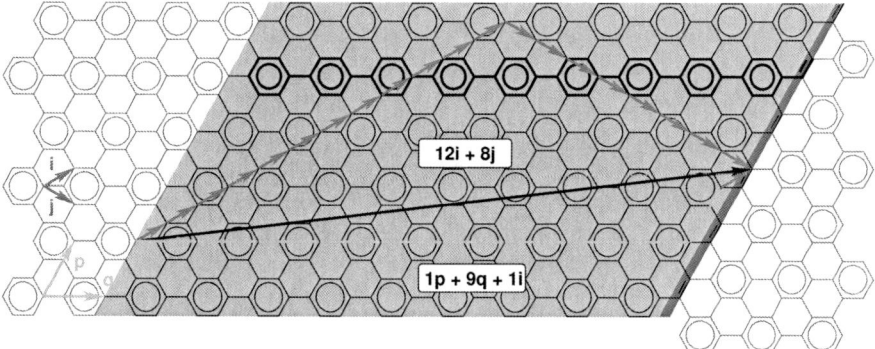

Figure 8.9 Roll-up vector (black arrow) for (12,8) CNT $R = 1$ in the conventional basis **i**, **j** (dark grey) and the Clar basis **p**, **q** (light grey). Atoms in the unit cell are bold; the parent graphene strip is grey. Adapted from ref. 33.

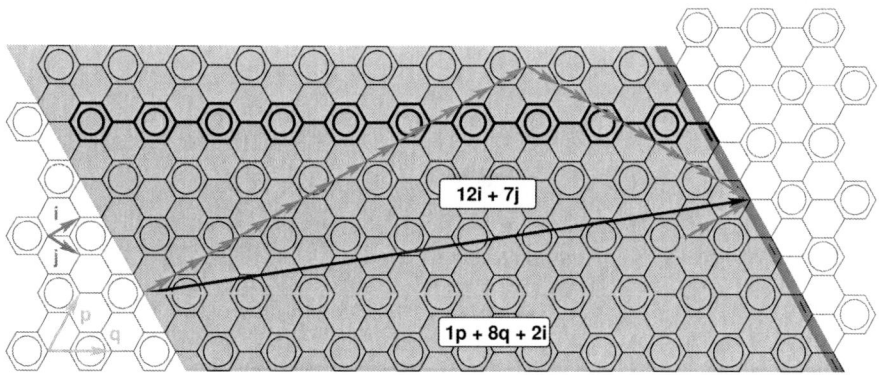

Figure 8.10 Roll-up vector (black arrow) for (12,7) CNT $R = 2$ in the conventional basis **i**, **j** (dark grey) and the Clar basis **p**, **q** (light grey). Atoms in the unit cell are bold; the parent graphene strip is grey. Adapted from ref. 33.

the conventional basis determines which CNTs are fully benzenoid (see eqn (8.23)). Moreover, since r is an integer, the expression $r = (n-m)/3$ is identical to the criterion for conductivity in CNTs. Similarly, only those CNTs where $(n-m)/3$ is an integer can be expressed in the Clar basis.[33] The condition for a fully benzenoid structure is fulfilled if and only if $(n-m)/3$ is integer, a necessary criterion for metallic CNTs. Having defined the properties of fully benzenoid CNTs, one may wonder whether particular units of the whole periodic system exist and which are the symmetry relationship among them. In particular, we look for the minimal constituting units of the rolled hexagonal network that (i) allow the generation of the periodic structure of the sidewall, by application of simple symmetry operations and (ii) and are related to the Clar representation of the sidewall.[33] This latter point constitutes a crucial requisite for connecting structural and electronic properties of the sidewall, as will be discussed in detail. A unit cell that fully contains the Clar VB structure can be chosen parallel to the unit vector $\mathbf{q}$, and generates the CNT by action of a screw axis operator. This Clar unit cell is preferable to the conventional translational unit cell1 because it is smaller and fully represents the Clar VB structure. A similar cell has previously been described using reciprocal space.[39] Since each of the unit vectors $\mathbf{p}$ and $\mathbf{q}$ spans exactly one aromatic sextet, the unit cell of a fully benzenoid CNT contains exactly $r + s = (2n - m)/3$ aromatic sextets. These ideas provide a physical picture for the value of $r = (n - m)/3$: it is the degree of helicity of a fully benzenoid CNT constructed from p-polyphenylene strips wrapped around a hypothetical cylinder. If $r = 0$, the CNT is not helical; if $r = 1$, the CNT consists of a single right-handed helix of p-polyphenylene; if $r = 2$, the CNT consists of a double right-handed helix, *etc.* The (12,9) CNT (see Figure 8.8) is a single helix. Achiral CNTs with $R(m,n) = 0$ and $r \neq 0$ can be described as either a left- or right-handed helix.[33]

If $R(n,m) = 1$, the CNT is not fully benzenoid, but a best Clar VB structure, which has the smallest ratio of double bonds to aromatic sextets, can still be constructed (see Figure 8.9).[33] Since the termini of the roll-up vector must be identical, a seam containing double bonds must exist in the otherwise fully benzenoid graphene sheet, and the roll-up vector must cross at least one double bond. Therefore, a roll-up vector that crosses exactly one double bond provides the best Clar VB structure. The existence of a seam of double bonds parallel to the unit vector $\mathbf{p}$ satisfies this condition when $R(n,m) = 1$. The unit cell contains $(2m + n - 1)/3$ aromatic sextets and one double bond. Likewise, if $R(n,m) = 2$, again the CNT is not fully benzenoid, but a best Clar VB structure can be constructed (see Figure 8.10). In this case, the seam of single bonds is parallel to the vector $\mathbf{p} - \mathbf{q}$. The unit cell contains $(2n + m - 1)/3$ aromatic sextets and one double bond.[33]

Of course, a single VB representation from a degenerate set alone does not provide a chemically realistic structure: a closer approximation is the resonance combination of all degenerate structures.[33] Hence, it is important to describe the degree of degeneracy of the Clar VB representation for a given CNT. Finite-length fully benzenoid CNTs having an integral number of Clar unit cells possess exactly one best Clar VB representation. Infinite fully benzenoid CNTs possess exactly three degenerate representations. For infinite $R(n,m) = 1$ or 2 semiconducting CNTs, the number of resonance structures equals the ratio of hexagons to double bonds in the structure.[33] This is shown in Figure 8.11.

In the figure, the double bond marked with the triangle is shifted into each consecutive hexagon along the armchair vector parallel to the unit cell until it reaches a hexagon (marked with the square) adjacent to that one which is topologically exactly equivalent to the starting hexagon. The sum of the hexagons occupied by the grey, shifting double bond is the number of the best Clar resonance structures. For $R(n,m) = 1$ and 2 this number is given by $n + 2m$

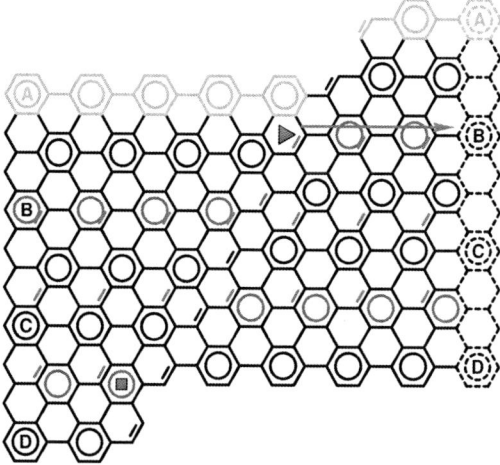

Figure 8.11 Resonance structures for a (9,5) $R=1$ CNT. Hexagons in light grey represent one unit cell, corresponding hexagons are dashed and the translating double bond is grey. The translation path follows the direction of the grey arrow and it is included between the triangle and the square. The sextets along the path of the translating double bond are grey. Adapted from ref. 33.

and $2n+m$, respectively[33] (the (9,5) $R=1$ CNT in Figure 8.11 possesses $9+2\times5=19$ resonance structures). The rest of the π-valence resonance structure would also be shifted, but only the double bond is shown for the sake of clarity. Moving the double bond into a hexagon identical to that from which it started is equivalent to a symmetry operation constituted of a translation and a rotation. Like their degenerate Clar VB contributors, the resonance hybrids have distinct structures depending on the value of $R(n,m)=1$. If $R(n,m)=0$, all C–C bonds are equivalent in that their π-bonding is represented exclusively by aromatic sextets. If $R(n,m)=1$, then the C–C bonds parallel to the basis vector **p** have enhanced double-bond character. If $R(n,m)=2$, then the C–C bonds parallel to the vector **p–q** have enhanced double-bond character.[33] Recent measurement, based on STM images[36,37,39,40] show bond patterns consistent with this analysis.

8.2.4 Classical Force-Field Based Approach

The force field is the set of functions needed to define the interactions in a molecular system. These may have a wide variety of analytical forms, with some basis in chemical physics, which must be parametrised to give the correct energy and forces. Since a huge variety of forms is possible we will discuss hereafter only the implementations used in calculations. The total configuration energy of a molecular system may be written as:

$$U(\mathbf{r_1}, \mathbf{r_2}, \ldots, \mathbf{r_N}) = \sum_{i_{bond}=1}^{N_{bond}} U_{bond}(i_{bond}, \mathbf{r_a}, \mathbf{r_b})$$

$$+ \sum_{i_{angle}=1}^{N_{angle}} U_{angle}(i_{angle}, \mathbf{r_a}, \mathbf{r_b}, \mathbf{r_c})$$

$$+ \sum_{i_{dihed}=1}^{N_{dihed}} U_{dihed}(i_{dihed}, \mathbf{r_a}, \mathbf{r_b}, \mathbf{r_c}, \mathbf{r_d}),$$

$$+ \sum_{i=1}^{N-1} \sum_{j>1}^{N} U_{pair}(i, j, |\mathbf{r_i} - \mathbf{r_j}|)$$

$$\ldots$$

(8.25)

where U_{bond}, U_{angle}, U_{dihed}, U_{pair} are empirical interaction functions representing chemical bonds, valence angles, dihedral angles and pair-body. The first three are regarded as intramolecular interactions and the last as the intermolecular interaction. eqn (8.25) could hold many other terms such as inversion angles, three-body, Tersoff (many-body covalent), *etc.* Here, we will discuss only the terms needed for the calculations. The position vectors $\mathbf{r_a}$, $\mathbf{r_b}$, $\mathbf{r_c}$ and $\mathbf{r_d}$ refer to the positions of the atoms specifically involved in a given interaction (almost universally, it is the differences in position that determines the interaction). The numbers N_{bond}, N_{angle} and N_{dihed} refer to the total numbers of these respective interactions present in the simulated system, and the indices i_{bond}, i_{angle} and i_{dihed} uniquely specify an individual interaction of each type, all of which must be individually cited. The indices i, j appearing in the pair-body term indicate the atoms involved in the interaction. There is normally a very large number of these and they are therefore specified according to atom types rather than indices. In general, it is assumed that the pair-body terms arise from van der Waals and/or electrostatic (Coulombic) forces. The former are regarded as short ranged interactions and the latter as long ranged.

For the simulations of CNTs encapsulations the rigid-ion interaction potential of Parrinello *et al.*[41] was used for AgI:

$$V_{AgAg}(\mathbf{r}) = \frac{H_{AgAg}}{\mathbf{r}^{n(AgAg)}} + \frac{0.36}{\mathbf{r}},$$

$$V_{AgI}(\mathbf{r}) = \frac{H_{AgI}}{\mathbf{r}^{n(AgI)}} - \frac{0.36}{\mathbf{r}} - \frac{1.1736}{\mathbf{r}^4},$$

$$V_{II}(\mathbf{r}) = \frac{H_{II}}{\mathbf{r}^{n(II)}} + \frac{0.36}{\mathbf{r}} - \frac{2.3472}{\mathbf{r}^4} - \frac{6.9331}{\mathbf{r}^6},$$

(8.26)

where

$$H_{ij} = A(\sigma_i + \sigma_j)^{n(ij)} \quad \text{with Å and } e^2/(1 \text{ Å}^2) = 14.39 \text{ eV as units.} \quad (8.27)$$

The value of charge used is $Z = 0.6$ and σ_i, σ_i are the particles radii as expressed by means of the Pauling's concept of ionic radii.[42,43] All the parameters used in eqns (8.26) and (8.27) are summarised in Table 8.1.

This potential was developed in accordance with the (α–β) phase transition in AgI and has proven very reliable over a broad range of temperatures and pressure,[44] both in the bulk and for small clusters.[45] The potential was implemented in a customised version of the DL_POLY program package,[46] used throughout the simulations in this chapter.

The carbon nanotubes were described with the flexible potential of Walther et al.,[47] including terms for bonded (stretching, bending, torsion) and nonbonded (Lennard-Jones, LJ) interactions, thus providing a realistic picture of the structure of the sidewall and of its possible modifications upon filling. The functional forms of the terms are the following, as implemented in the DL_POLY program package.[46]

Morse potential for bond stretching:

$$U(r) = E_0 \left(\left(1 - e^{-k_r(r-r_0)}\right)^2 - 1 \right). \tag{8.28}$$

Harmonic cosine potential for angle bending:

$$U(\theta) = \frac{k_\theta}{2} \left(\cos(\theta) - \cos(\theta_0)\right)^2. \tag{8.29}$$

Cosine potential for dihedral angle torsion:

$$U(\phi) = A(1 + \cos(m\phi - \delta)). \tag{8.30}$$

Lennard-Jones potential for nonbonded interactions:

$$U(r) = 4\varepsilon \left[\left(\frac{\sigma}{r}\right)^{12} - \left(\frac{\sigma}{r}\right)^6 \right]. \tag{8.31}$$

Table 8.1 All parameters used in the AgI potential adapted from refs. 41, 42. Units for H_{ij} derive from eqn (8.27).

Parameter	Value	Parameter	Value
σ_{Ag} (Å)	0.63	$n(AgI)$	9
σ_I (Å)	2.20	A (14.39)	0.010248
$n(Ag)$	11	H_{AgAg}	0.014804
$n(I)$	7	H_{AgI}	114.48
$n(AgAg)$	11	H_{II}	446.64
$n(II)$	7	Z (e)	0.6

All the parameters used in calculations for the description of the CNTs are summarised in Tables 8.2 and 8.3 (adapted from refs. 47–49).

The ion–nanotube interactions were accounted for with a LJ potential with parameters derived from the atom/graphite values corresponding to the closed-shell neutral isoelectronic counterparts. For Ag^+/C, literature parameters for Pd/C^{50} were taken, whereas for I^-/C the values suggested by Wilson, based on Kr/C parameters, were employed.[51] All parameters used in the calculations are summarised in Table 8.3.

For the calculations we have employed an effective pair potential, which is based on the Born–Mayer model. This approach is widely used for the simulations of the stability and structure of many ionic compounds in bulk and nanostructures and considers the information about the formation and stability of KI, KBr, AgI nanowires and clusters in narrow CNTs.[28–30] The long-range Coulomb interaction between two atoms (i and j) with charges q_i and q_j at a distance r_{ij} is described by the pair potential

$$V_{ij} = \frac{e^2}{4\pi\varepsilon_0} \frac{q_i q_j}{r_{ij}} + \frac{A_{ij}}{r_{ij}^{12}} - \frac{B_{ij}}{r_{ij}^6}, \tag{8.32}$$

where A_{ij} and B_{ij} are the parameters specific to each type of pair interaction modeling short-range forces (repulsion of nuclei, repulsion of electron shells, van der Waals forces). In our calculations, we used the parameters that were fitted and successfully applied for bulk PbI_2, carbon and BN nanotubes.[31–33] Missing parameters for pair interactions between Pb and I ions with C, B, N,

Table 8.2 Parameters for bonded terms of the CNT potential used in calculations. Adapted from Refs 47,48.

Parameter	Value
E_0 (eV)	4.963320
r_0 (Å)	1.418000
k_r (eV Å^{-2})	2.186700
k_θ (eV)	5.826641
θ_0 (deg)	120.00
A (eV)	0.171318
δ (deg)	180.00
m	2

Table 8.3 Parameters for nonbonded (Lennard-Jones) terms of the potential used in calculations (adapted from refs. 49–51).

Parameter	C/C	Ag^+/C	I^-/C
ε (eV)	0.004556	0.033500	0.006842
σ (Å)	3.851	2.926	3.735

Mo and S atoms were chosen by applying the Lorentz–Berthelot mixing rules. To estimate the atomic charges q for PbI_2, BN and MoS_2 we applied the simple Pauling scheme using Pauling electronegativities.

8.2.5 Molecular Dynamics

In molecular dynamics (MD) simulations Newton's equations of motion (EOM) for a system of N interacting objects are solved:

$$m_i \frac{\partial^2 \mathbf{r}_i}{\partial t^2} = \mathbf{F}_i, \quad i = 1, \ldots, N, \tag{8.33}$$

where m_i and $\mathbf{F}_i$ are mass and position of the object i, respectively. The forces are the negative derivatives of a potential function $V(\mathbf{r}_1, \mathbf{r}_2, \ldots, \mathbf{r}_N)$:

$$\mathbf{F}_i = -\frac{\partial V}{\partial \mathbf{r}_i}. \tag{8.34}$$

In general, the complete definition of the dynamics for the mechanical system requires the simultaneous integration of eqns (8.33) and (8.38). In practice, eqns (8.33) and (8.34) can be solved by integration on small and finite time intervals. However, in practical MD simulations, the EOM of the systems under study are solved applying special methods and algorithms, derived from eqns (8.33) and (8.38).[52,53]

In MD simulations of molecular systems, the properties of the system under study are initially assigned accordingly to the thermodynamic ensemble (NVE, NVT or NPT) to simulate. These properties include atom velocities and the definition of the extended (thermostat and/or barostat) system. The initial conditions allow the integration of the EOM, as described above, on subsequent intervals of small and finite time steps. Repeating the procedure with the updated configuration allows following the dynamics of the system under study, thus obtaining trajectory, in its phase space, for an arbitrary interval of space. For nonreacting systems, after a short equilibration time, the conserved quantities of the mechanical system should average to the values of the reference thermodynamic ensemble. This observation derives from the ergodicity of the dynamical system,[54] which is at the basis of MD simulations. Moreover, as a further consequence of the ergodicity, the average of values obtained on single points of the trajectory of the mechanical system average to macroscopic properties.

8.3 Applications

8.3.1 Functionalisation of Carbon Nanotubes

The development of technological applications based on CNTs often had to face intrinsic difficulties related to the accurate and unambiguous experimental determination of their properties, mostly due to the nonuniform and heterogeneous nature of the material at hand.[55] Consequently, in the past few years, an intense and active field of research on CNTs and their properties has been represented by computational modeling, which constitutes a favourable alternative to experimental techniques and, in many cases, a preferential investigation route.[56,57] Indeed, electronic structure methods allow accurate investigations on structural, reactive, and magnetic properties of nano-structured materials at an atomic level of detail. In particular, the largest part of electronic structure studies in this field has been carried out by applying density-functional theory (DFT) (see Section 8.2.2), which couples the possibility of dealing with nanosized model systems with reasonable computing demands. Therefore, most previous DFT calculations on CNTs and related systems were based on finite-length clusters or translationally repeated models constituted of up to a few hundreds of carbon atoms. However, it has been already shown that the particular choice of either cluster or periodic models of CNTs, including those considered in hybrid quantum mechanics/molecular mechanics (QM/MM)[58,59] techniques, could strongly bias the results of electronic-structure calculations.[60,61] Indeed, both approaches could exhibit severe drawbacks. For example, the cluster and the periodic models could require a unit cell constituted of several hundred carbon atoms for the description of a chiral CNT.[62,63] Moreover, the use of finite-length cluster models, when applied through purely size-based criteria, provide contrasting results[64–66] and, as previously shown by Bettinger, slow convergence.[60] As we will see later, an alternative to this situation is given by the choice of model systems based on Clar sextet theory[24,25] (see Section 8.2.3), thus extending to CNTs many of the concepts widely applied to polycyclic aromatic hydrocarbons (PAHs).

In this section, we apply a general formulation of Clar sextet theory to the definition of finite-length models of zigzag, armchair and chiral CNTs.[67,68] As we will see later, the advantages of using a finite-length Clar cluster (FLCC) approach are manifold, in particular: (i) a consistent and uniform description of the electronic properties of different kinds of CNTs, often lacking in other finite-length models; (ii) fast and monotonic convergence of the electronic properties to the "bulk" (*i.e.* with an infinite number of carbon atoms) values; (iii) remarkable computational benefits, due to the reduction, up to one order of magnitude, in the number of atoms employed in calculations, with respect to standard translationally repeated models.

First, the definition of finite-length models of CNTs based on the FLCC approach is discussed. For the sake of clarity, we first recall the basic points concerning the correlation between the Clar theory of the aromatic sextet and the structural and electronic properties of the CNT sidewall. Subsequently, the results of the FLCC in different cases will be analysed and discussed. It is also noting the possibility of describing accurately the electronic structure of the sidewall, allowing the FLCC approach to be applied to several problems of interest in the field of CNTs and their applications, as for example in the study of their chemical reactivity and functionalisation processes.

As noted previously (see Section 8.2.3) the most convenient way to apply Clar theory to CNTs consists of defining the two-dimensional carbon-atom network resulting from the unrolling of the sidewall in terms of two translational basis vectors, which reflect the translational symmetry of a fully benzenoid graphene sheet.[33] This basis is connected to the conventional (n,m) CNT vector basis[1] by simple algebraic relationships and allows the Clar representation of the sidewall. Moreover, the Clar vector basis leads to a new definition of a primitive cell for the CNTs, which is usually smaller than the conventional (n,m) cell. From the primitive Clar cell, the infinite CNT is generated by application of a screw axis operation along the nanotube principal axis.[33] The definition of the sidewall topology in terms of the Clar cell basis leads to three different situations, depending on the value of the parameter R, defined as $R = \text{mod}(n - m, 3)$.[33] For $R = 0$, the Clar cell is fully benzenoid and the corresponding infinite nanotube is metallic. In contrast, for $R = 1$ or $R = 2$, the Clar cell contains six-membered rings and one double bond (or two double bonds for $(n,0)$ CNTs with $R = 1$, and the corresponding CNTs are semiconducting. Following these considerations, we applied the same classification scheme to finite-length CNTs, generated by a finite number of applications of the screw-axis operator to the Clar unit cell.

To demonstrate the consistency of the FLCC approach in general cases, several models of armchair, zigzag, and chiral CNTs, corresponding to different values of the parameter R, were taken into account.[67,68] The complete topology of Clar cells and FLCC models will be discussed later. It is, however, worth noting here that, in some cases, the Clar unit cells used throughout our work differ from their canonical representation, as for instance used in ref. 33. Indeed, it is possible to choose different Clar unit cells, each of which generates the same CNT by application of the screw-axis operator an infinite number of times.[67,68] From this point of view, the choice of the Clar unit cell is somewhat arbitrary. Therefore, in our calculations the choice of a particular Clar unit cell was driven by an adequate description of the geometry of finite-length models, even for a small number of repeated cells.[67,68]

As shown previously, the case of $R = 0$ corresponds to metallic CNTs. In Figure 8.12 a FLCC planar representation, constituted of three cells, for the (6,6) armchair CNT is shown. The primitive Clar cell (highlighted in light grey) is composed by a six-membered p-polyphenylene ring. In this case, the cell used in calculations coincides with the canonical Clar cell. A similar six-membered p-polyphenylene strip composes the canonical Clar primitive cell of a (9,0) metallic zigzag CNT[67,68] (see Figure 8.13).

The angle between the direction of the para-oriented bonds and CNT main axis is 60° wide. This results in a primitive cell with noncircular structure. Indeed, the FLCCs constituted by less than three canonical Clar cells show open structures, resulting in a huge distortion from the cylindrical ideal shape upon geometry relaxation.[67,68] To give a good description of the CNT cylindrical shape and, at the same time, preserving the proper Clar VB CNT picture, we considered a new structure of the Clar unit cell in which all the para-oriented bonds between the six-membered rings were changed in meta. As can be seen in Figure 8.13 both the canonical Clar cell and that used in calculations generate the hexagonal network corresponding to an infinite-length CNT

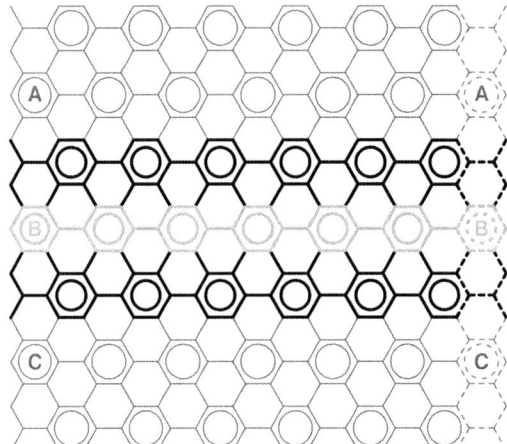

Figure 8.12 Schematic planar representation of the FLCC model for the (6,6) ($R=0$) armchair CNT. The model, constituted by three Clar cells (highlighted in black), is superimposed to the hexagonal network corresponding to an infinite-length CNT (in grey). The Clar unit cell is highlighted in bold lines. Equivalent hexagons are dashed. Adapted from refs. 67 and 68.

by means of the same screw-axis operator.[67,68] As we will see later this choice causes a slightly slower convergence of the electronic properties *vs.* the model size, but provides a good description of the CNTs structural features and makes the clusters suitable for the study of the CNTs chemical reactivity.

For the case of $R=1$, the examples of the chiral (6,5) and zigzag (7, 0) CNTs were considered. In Figure 8.14, a three Clar cell planar representation for a (6,5) CNT is shown. The primitive Clar cell (highlighted in light grey) contains five aromatic sextets and one double bond. As discussed previously (see Section 8.2.3), the case of (7,0) CNT is an exception. As can be noticed in Figure 8.15 the Clar unit cell for a (7,0) CNT contains four aromatic sextets and two double bonds.[67,68]

Once again the canonical Clar cell exhibits an open structure, thus resulting in strong geometrical strain from the cylindrical shape. Moreover, because of the intrinsic features of the canonical Clar cell, one C=C double bond (emphasised with an ellipse) is dangling at one edge of the FLCC. To avoid these problems, the unit cell has been rearranged as shown in the right side of Figure 8.15. The new unit cell contains one *naphtalene* unit (and three aromatic sextets) and has circular shape.[67,68]

The last case is that of $R=2$. In Figure 8.16 the planar representations of the (6,4) semiconducting chiral CNT, built up by canonical Clar cell and the cell used in calculations (left and right side, respectively) are reported. Similarly to previous cases, a new cell to be used in calculations has been defined to avoid the dangling C=C double bond (highlighted with an ellipse) at one edge of the canonical Clar cell. Consequently the orientation of one bond between a six-membered ring and a double bond needed to be changed in para. Both the

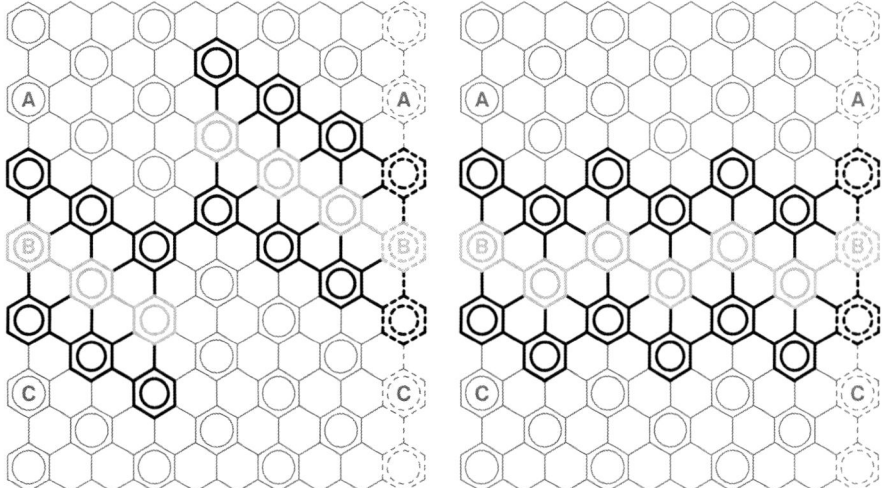

Figure 8.13 Schematic planar representations of the FLCC models for the (9,0)
($R = 0$) zigzag CNT: canonical Clar cell (left) and cell used in calculations
(right). Models constituted by three cells (highlighted in black) are
superimposed onto the hexagonal network corresponding to an infinite-
length CNT (in grey). The unit cell is highlighted in bold lines. Equivalent
hexagons are dashed. Adapted from refs. 67 and 68.

canonical Clar cell and that used in calculations contain five sextets and one
double bond. The case of (8,0) zigzag semiconducting CNT is very similar to
that of (7,0). As can be seen in Figure 8.17 (left side), the rolling direction of the
FLCC built from the canonical Clar cell, is very far from the normal to the
CNT axis, thus resulting in a dangling C–C bond. Also in this case the canonical Clar cell has been substituted with a new one in which the orientation of
some bonds was changed.[67,68]

To demonstrate the applicability of FLCC models, the energy of the highest-
occupied molecular orbital (HOMO), lowest-unoccupied molecular orbital
(LUMO), and corresponding HOMO–LUMO gap for a series of FLCC models
of CNTs were analysed.[67,68] Indeed, the HOMO and LUMO energies, as well
as the HOMO–LUMO gap, can be considered as good descriptors of many
electronic properties of molecular systems, including reactivity.[69]

In particular, we developed a set of models constituted of a finite number of
replica (from 2 to 6) of the Clar unit cell along a screw axis parallel to the
nanotube principal axis (see Figure 8.18). The dangling bonds at the edges of
the model systems were saturated with hydrogen atoms.[67,68] The structures of
all systems under study were optimised at the B3LYP/3-21G level[70–72] as
implemented in the Gaussian03 program package.[73] Orbital energies were
evaluated at the same level of theory.

For models of (6,6) and (9,0) CNTs ($R = 0$), see Figures 8.18(a) and (d)),
the HOMO and LUMO energies exhibit a regular increase and decrease, respectively, with the nanotube size (see parts a and d of Figure 8.19).

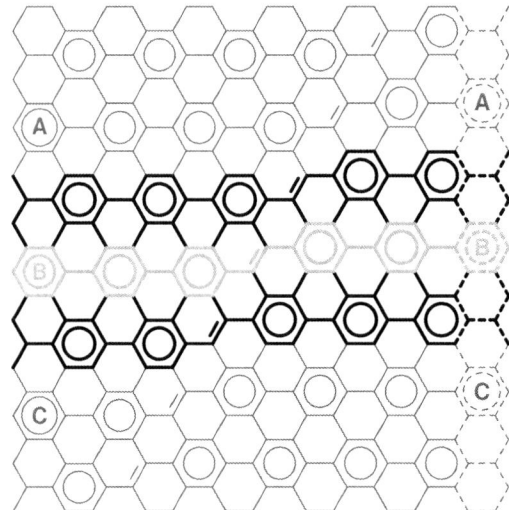

Figure 8.14   Schematic planar representation of FLCC model for the (6,5) ($R=1$) chiral CNT. Model constituted by three Clar unit cells (highlighted in black) is superimposed to the hexagonal network corresponding to an infinite-length CNT (in grey). The unit cell is highlighted in bold lines. Equivalent hexagons are dashed. Adapted from refs. 67 and 68.

In particular, the LUMO energy decreases more rapidly than the increase in the HOMO energy, leading to an overall monotonic decrease of the HOMO–LUMO gap with the length of the model. These trends are consistent with the properties expected for finite-length models of metallic CNTs, where the bandgap is zero at infinite nanotube length. Therefore, the use of Clar theory allows a consistent definition of the electronic properties of finite-length models for metallic CNTs, which can be particularly useful in the study of systems where a finite bandgap is introduced, as for example in the case of doping or chemical reactions at the sidewall.[67,68]

A different situation arises for semiconducting CNTs. As representatives of CNTs with $R=1$, finite length models of (6,5) and (7,0) nanotubes (see parts b and e of Figure 8.18) were taken into account. The FLCC models of the (6,5) CNT exhibit a remarkable convergence of HOMO, LUMO, and HOMO–LUMO gap energies with the number of Clar cells, as shown in Figure 8.19(b). The HOMO–LUMO gap converges to a value of 1.88 eV within a deviation of 0.06 eV already at a model length corresponding to 4 Clar unit cells (128 carbon atoms). Most notably, the use of a Clar unit cell drastically reduces the number of carbon atoms to take into account in calculations with respect to periodic or finite-length models based on the translational unit cell (364 C atoms for a (6,5) CNT). A similar trend can be observed for FLCC models of a (7,0) CNT (Figure 8.19(e)) where, on the other hand, a slightly slower convergence is noticed (0.12 eV within 5 Clar unit cells for the HOMO–LUMO gap). Analogous results were obtained for models of CNTs with $R=2$. In this case,

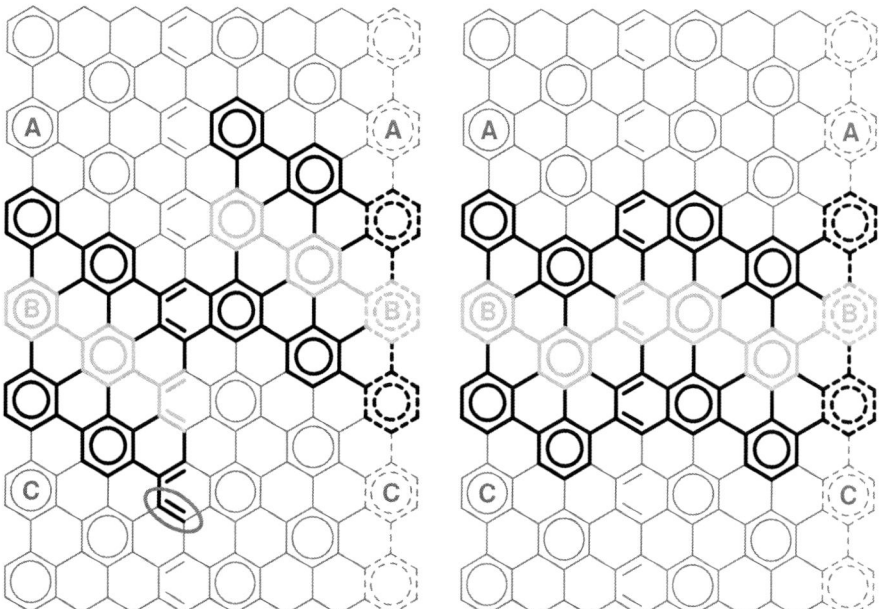

Figure 8.15 Schematic planar representations of FLCC models for the (7,0) $R=1$ zigzag CNT: canonical Clar cell (left) and cell used in calculations (right). Models constituted by three cells (highlighted in black) are superimposed to the hexagonal network corresponding to an infinite-length CNT (in grey). The unit cell is highlighted in bold lines. Equivalent hexagons are dashed. The C=C double bond dangling at one edge of the canonical Clar cell is emphasised with an ellipse. Adapted from refs. 67 and 68.

finite-length models of CNTs (6,4) and (8,0) nanotubes were considered (see Figures 8.18(c) and (f)). The convergence properties of HOMO, LUMO and gap for FLCC models of the (6,4) CNT (see Figure 8.19(c)) are remarkably similar to those obtained in the case of a (6,5) CNT. This is not surprising since the planar representations of Clar cells for (6,4) and (6,5) CNTs are mirror images of each other. Similar to the previous case, a slightly slower convergence is observed for FLCC models of the (8,0) CNT, where the HOMO–LUMO gap converges within a deviation of 0.10 eV for 5 Clar unit cells. As a further assessment of the applicability of FLCC models to the theoretical study CNTs, the dependence of HOMO and LUMO energies and HOMO–LUMO gap with the diameter of chiral nanotubes ($R=1$ and $R=2$) was investigated. As shown in Figure 8.20, both in the case of $R=1$ (Figure 8.20(a)) and $R=2$ (Figure 8.20(b)), a regular and monotonic relationship between the observed properties and the CNT diameter is observed.[67,68]

The orbital energies analysed here show the use of FLCC models to provide a unified description of the electronic properties of CNTs. In particular, we have shown that the electronic properties of FLCC models for metallic CNTs, such as HOMO and LUMO energies and the HOMO–LUMO gap, converge

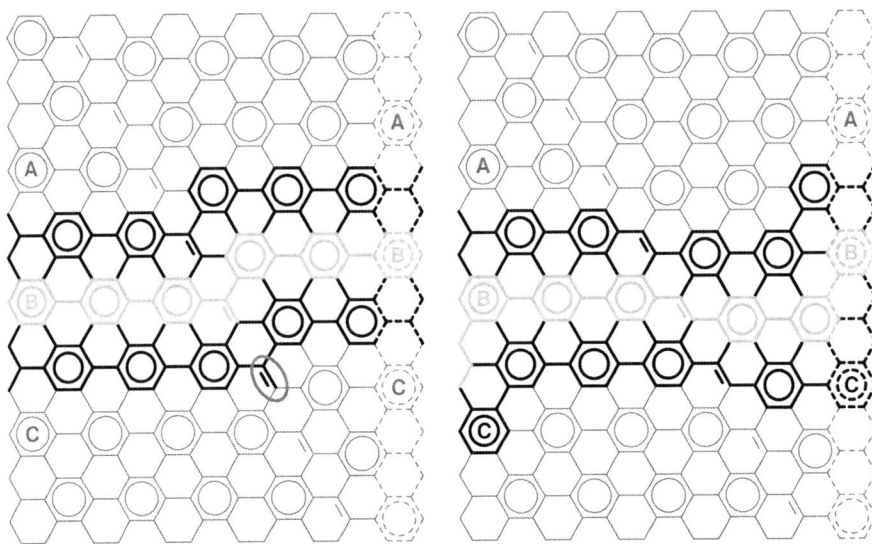

Figure 8.16   Schematic planar representations of FLCC models for the (6,4) ($R = 2$)
chiral CNT: canonical Clar cell (left) and cell used in calculations (right).
Models constituted by three cells (highlighted in black) are superimposed
onto the hexagonal network corresponding to an infinite-length CNT (in
grey). The unit cell is highlighted in bold lines. Equivalent hexagons are
dashed. The C=C double bond dangling at one edge of the canonical
Clar cell is emphasised with an ellipse. Adapted from refs. 67 and 68.

uniformly and monotonically to the values expected for quasiunidimensional
metallic systems. Moreover, FLCC models of semiconducting nanotubes
exhibit a prompt convergence of HOMO and LUMO energies and HOMO–
LUMO gap within a small number of carbon atoms. This results in remarkable
computational advantages in using FLCC models with respect to finite-length
or periodic models based on translationally repeated cells, especially for chiral
CNTs. As we will see later, an accurate description of such properties play a
crucial role in the consistent evaluation of the nanotubes reactivity.[67,68]

Then we explore, by means of DFT calculations and FLCC models, the
reactivity of (6,5) and (6,4) CNTs towards different functionalisation
agents.[68,74] Namely, reaction with a fluorine atom and carbene (CH_2) were
investigated. The case of the (6,5) and (6,4) chiralities analysed here can be
considered as representative of the more general case of semiconducting chiral
CNTs with different values of the parameter R, which accounts for the electronic structure of the CNTs. Our results confirm the possibility of using finite-
length CNT models based on FLCCs to describe correctly the electronic
properties of semiconducting chiral CNTs and, consequently, their reactivity.
FLCC models constituted of a finite number of replica (from 3 to 8) of the Clar
unit cell of (6,4) and (6,5) CNTs were created. For the evaluation of binding
energies, different adsorption sites (on top for fluorine and bridging for CH_2)
were taken into account, as shown in Figure 8.21. All calculations were

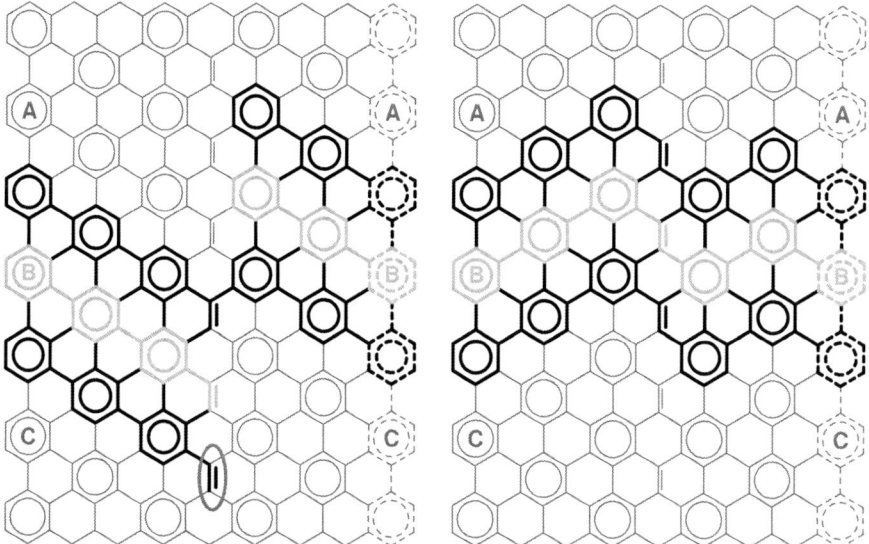

Figure 8.17 Schematic planar representations of FLCC models for the (8,0) ($R = 2$) zigzag CNT: canonical Clar cell (left) and cell used in calculations (right). Models constituted by three cells (highlighted in black) are superimposed to the hexagonal network corresponding to an infinite-length CNT (in grey). The unit cell is highlighted in bold lines. Equivalent hexagons are dashed. The C=C double bond dangling at one edge of the canonical Clar cell is emphasised with an ellipse. Adapted from refs. 67 and 68.

performed by applying the hybrid B3LYP functional[70–72] as implemented in the Gaussian03 program package.[73] Energies and optimised geometries of all systems were evaluated with the 3-21G basis set. The binding energy (BE) for the reaction:

$$CNT + X \rightarrow X@CNT \qquad (8.35)$$

was calculated following the equation:

$$BE = E_{X@CNT} - E_{CNT} - E_{X}, \qquad (8.36)$$

where $E_{X@CNT}$ is the energy of the adduct and E_{CNT} and E_{X} are the energies of the isolated species, respectively ($X = F$, CH_2). Calculations on fluorine atom and on fluorinated adducts, were performed in the doublet ground state. The lower-lying ground state of CH_2 is a spin triplet and the $CH_2@CNT$ adduct was optimised in the singlet ground state. In calculations, the basis set superposition error (BSSE) was taken into account using the counterpoise (CP) approach.[75,76]

   In an ideal (infinite) CNT, all carbon atoms are equivalent. Accordingly, the interaction energy for a single atom chemisorbed on top of a carbon atom of

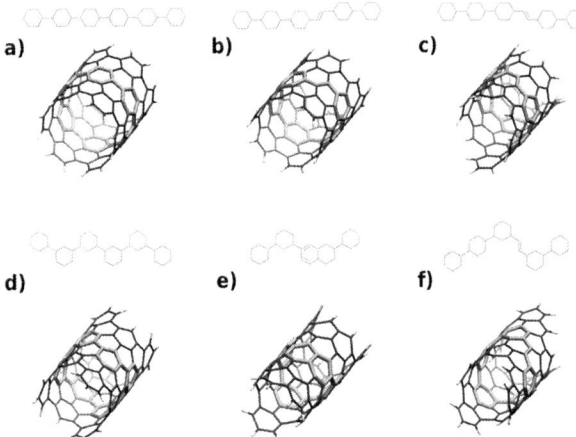

Figure 8.18 Optimised geometries of FLCC CNT models consisting of three unit cells used in calculations: (a) (6,6), (b) (6,5), (c) (6,4), (d) (9,0), (e) (7,0), and (f) (8,0) CNTs. The generating Clar cells and their corresponding planar representations are depicted in grey. Adapted from refs. 67 and 68.

finite-length models of CNTs should converge to an unique value with the model length, depending only on the theoretical level used in calculations. In Figure 8.23 the computed binding energy of fluorine to the sites of Figure 8.22 and its dependence from the FLCC model size is plotted.[68,74]

Indeed, very short-sized FLCC models exhibit a strong dependence of fluorination energy on the adsorption site considered. This evidences the different bonding situation for the coordinating site (belonging to benzenoid rings or double bonds) induced by edge effects and also indicates that the electronic structure of short FLCCs matches the valence-bond (VB) representation provided by Clar's theory. As expected, double-bond sites (dark grey sites in Figures 8.21(a) and (b) and 8.22) are more reactive towards fluorine atom (see Table 8.4), than the benzenoid ones (light grey and white sites in Figures 8.21(a) and (b) and 8.22). However, for longer FLCC models fluorine binding energies converge to the value of 25 kcal/mol for (6,5) CNTs and around 27 kcal/mol for (6,4) CNTs, respectively, within a deviation less than 1 kcal/mol already for models constituted of 6 Clar unit cells (see Figure 8.23). The (6,4) CNT is slightly more reactive than the (6,5) CNT, due to a more favourable orientation of C–C bonds.[60,77] Nonetheless, the convergence behaviour is similar for both the chiralities considered. Hence, our results indicate the fluorination of both the (6,5) and (6,4) CNTs as energetically favourable, even if slightly less exothermic than in the case of $(n,0)$ and (n,n) CNTs.[60,77]

In the case of reaction with the CH_2 biradical, ideal infinite (n,m) CNTs exhibit three intrinsically distinct "bridge" adsorption sites, depending on the orientation of the C–C bond involved in the chemisorption with respect to the CNT axis: equatorial sites (EQ), axial sites parallel to the seam of double bonds (AX2) and axial sites opposite oriented (see Figure 8.24). Figure 8.25 shows the

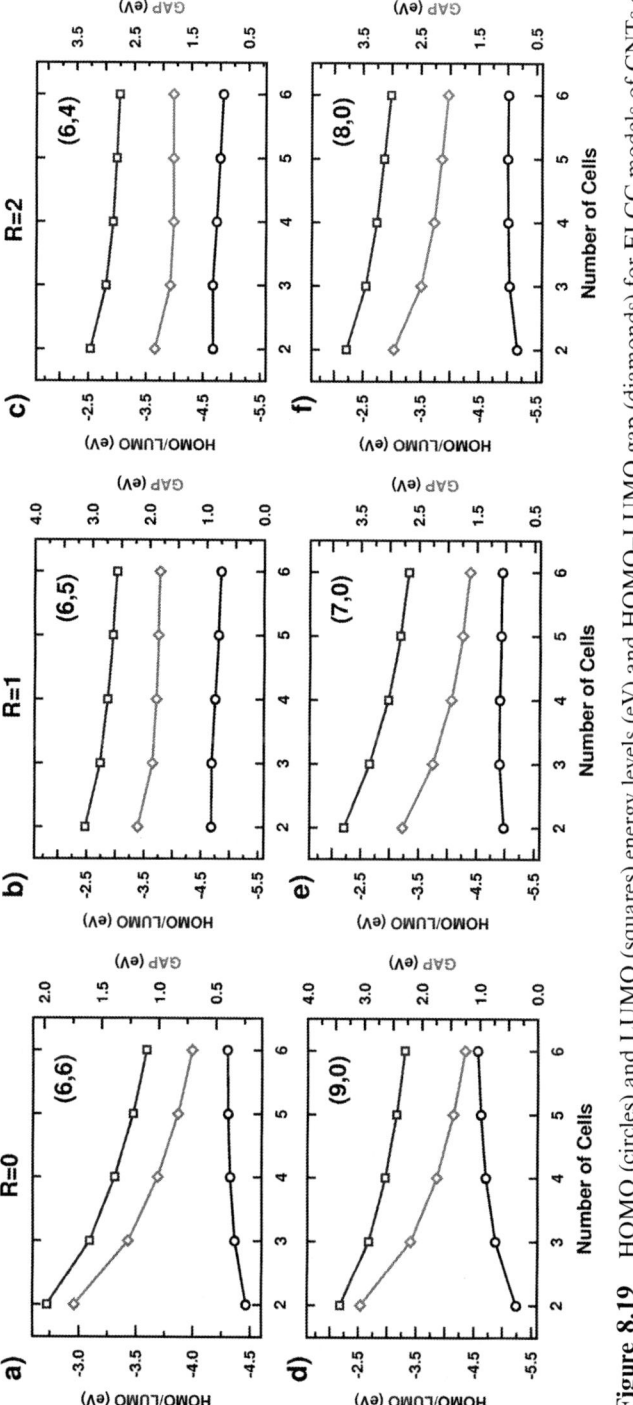

Figure 8.19 HOMO (circles) and LUMO (squares) energy levels (eV) and HOMO–LUMO gap (diamonds) for FLCC models of CNTs of different length as discussed as in the text: (a) (6,6), (b) (6,5), (c) (6,4), (d) (9,0), (e) (7,0), and (f) (8,0). Adapted from refs. 67 and 68.

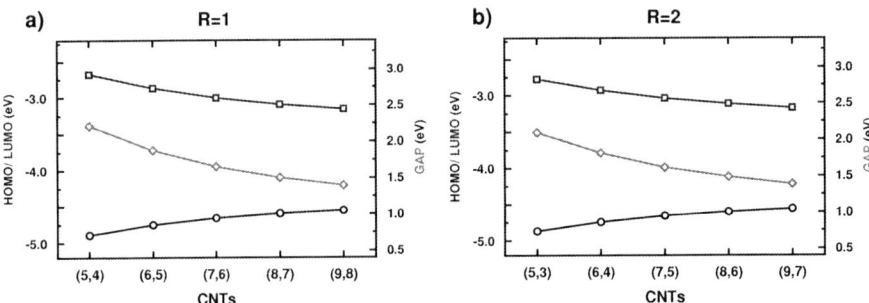

Figure 8.20   HOMO (circles), LUMO (squares) energy levels (eV) and HOMO–LUMO gap (diamonds) for FLCC models of CNTs of different diameters consisting of four unit cells: (a) $(n,n–1), R = 1$, and (b) $(n,n–2), R = 2$. Adapted from refs. 67 and 68.

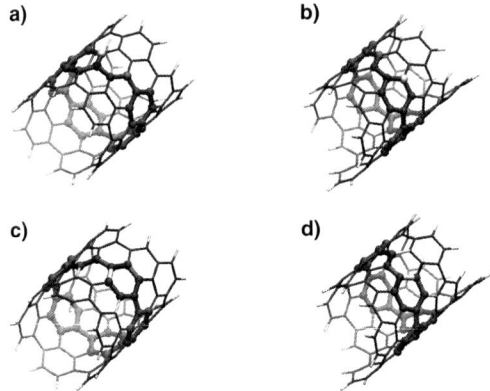

Figure 8.21   FLCC CNT models (three unit cells shown) and adsorption sites for the interaction with F and CH_2: (a) F@(6,5); (b) F@(6,4); (c) CH_2@(6,5); (d) CH_2@(6,4). Atoms in ball-and-sticks highlight the Clar unit cell. Adapted from refs. 68 and 74.

computed reaction energies of CH_2 to the sites of Figure 8.24 and their dependence from the FLCC model size. Analogously to the case of fluorination, the reaction energy for CH_2 addition for very short-sized models depends strongly on the adsorption site considered. This arises from both the intrinsic differences mentioned above, to be ascribed to the C–C orientation, and from the different electronic situation, as described by the Clar representation of the CNT model (double bond, bond belonging to benzenoid rings and bond belonging to nonbenzenoid rings).[68,74]

However, binding energies for CH_2 converge very quickly with respect to the model length (see Figure 8.25) for both (6,5) and (6,4) CNTs. In the case of the equatorial sites binding energies converge to 76 kcal/mol and 80 kcal/mol ((6,5)

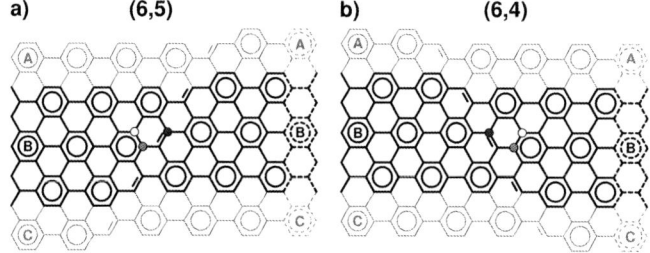

Figure 8.22 Planar representation of FLCC CNT models (three unit cells shown) and on-top adsorption sites for the interaction with F: (a) F@(6,5) (b) F@(6,4). Interaction sites are marked in dark grey for sites belonging to double bonds (D), light grey (B1) and white (B2) for different sites belonging to benzenoid rings. Tridimensional representations are shown in Figures 8.21(a) and (b). Adapted from refs. 68 and 74.

Table 8.4 Binding energies in kcal/mol for the chemisorption of fluorine atom on (6,5) ($R = 1$) and (6,4) ($R = 2$) semiconducting chiral CNTs. The number of Clar cells in the FLCC is reported in the first column. Chemisorption sites are labelled as in Figure 8.22.

	F@(6,5) chemisorption sites			F@(6,4) chemisorption sites		
n	D	B1	B2	D	B1	B2
3	− 31.5	− 29.1	− 27.5	− 34.4	− 30.6	− 31.2
4	− 27.3	− 26.2	− 26.8	− 30.8	− 28.1	− 29.6
5	− 26.5	− 25.8	− 25.6	− 28.5	− 27.6	28.3
6	− 25.4	25.1	− 25.5	− 27.5	− 26.9	− 27.6
7	− 25.1	− 25.0	− 25.0	− 27.0	− 26.7	− 26.9
8	− 25.0	− 25.0	− 24.8	− 26.7	− 26.5	− 26.8

and (6,4) CNT, respectively) already within 5 Clar cells in the FLCC. Namely, in both cases binding energies converge to three distinct values (39, 41 and 76 kcal/mol for (6,5) CNTs and 41, 44 and 80 kcal/mol for (6,4) CNTs, respectively), reflecting the different nature of the C–C adsorption sites. For reaction towards CH_2, convergence is achieved within less than 2 kcal/mol already for FLCC models constituted of 5 unit cells. In agreement with previous studies,[60,78] reaction of CH_2 on C–C bonds with orientation close to the nanotubes axis (AX1 and AX2) results in cycloaddition adducts, whereas reaction on C–C bonds with a larger mechanical strain (EQ) leads to the sidewall opening.[68,74]

The convergence of reaction energies with the model length reflects the fact that increasing the cluster size, as previously stated, different VB representations of the sidewall become energetically similar and tend to mix up together. Therefore, the electronic properties of sites that differ in the VB representation of the sidewall, but topologically equivalent in the infinite nanotube, tend to

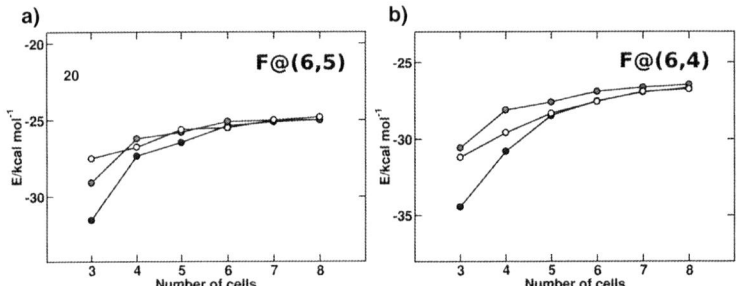

Figure 8.23   Binding energies for the addition of F to FLCC models of (6,4) and (6,5) CNTs: (a) F@(6,5); (b) F@(6,4). Dark grey, light grey and white code different reaction sites according to the scheme in Figure 8.22. Adapted from refs. 68 and 74.

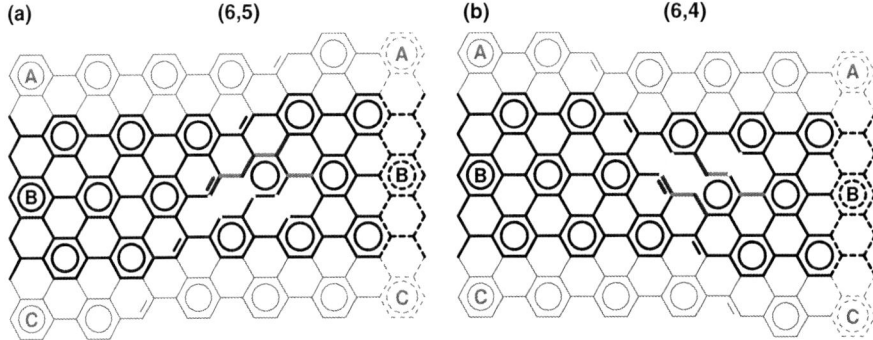

Figure 8.24   Planar representation of FLCC CNT models (three unit cells shown) and bridge adsorption sites for the interaction with CH_2: (a) $CH_2@(6,5)$ (b) $CH_2@(6,4)$. Tridimensional representations are shown in Figures 8.21(c) and (d). Adapted from refs. 68, 74.

converge with the model length. Hence, even in the case of chemical reactions, the analogy between FLCC models and the VB representation of the sidewall guarantees the consistency of the electronic properties of CNT clusters with those of ideal (infinite) systems and ensures fast convergence to the asymptotic value. It is worth noting that binding energies for addition of CH_2 converge much better with the model length with respect to the case of fluorine atoms. This can be ascribed to the drastic change in the electronic structure induced by the adsorption of a fluorine atom, which leads to a spin-doublet state for the chemisorbed adduct. In terms of VB representations, the bonding pattern of the pristine FLCC model is strongly affected by fluorine adsorption that, consequently, change the overall aromatic properties, as already pointed out by Bettinger.[60] Inversely, reaction of CH_2 on a C–C bond at the sidewall induces practically only local modifications, due to the formation of new carbon–carbon bonds that do not markedly affect the bonding situation at the sidewall.

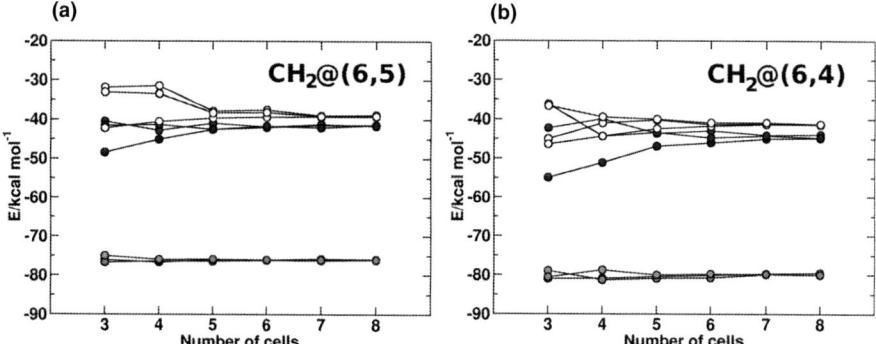

Figure 8.25 Binding energies for the addition of CH_2 to FLCC models of (6,4) and (6,5) CNTs: (a) $CH_2@(6,5)$; (b) $CH_2@(6,4)$. Dark grey, light grey and white code different reaction sites according to the scheme in Figure 8.24. Adapted from refs. 68, 74.

Hence, reaction energies for CH_2 are expected to depend more on local effects and, consequently, are less affected by the size of finite-length models.[68,74]

In conclusion, FLCC models allow us to describe quantitatively the reactivity of semiconducting chiral CNTs, as evidenced in the cases of reaction of fluorine and CH_2 with (6,5) and (6,4) CNTs. However, the chemical considerations behind the definition of the computational models allow us to conclude that the FLCC approach can in principle be applied to other reactions at the sidewall and to different chiralities. It is worth mentioning that calculations on FLCC models of chiral CNTs are computationally much less demanding with respect to the use of periodic or finite-length models based on the periodically translated unit cell.[67,68,74]

8.3.2 Encapsulation of Carbon Nanotubes

The ability to encapsulate different materials has constituted one of the first applications of CNTs. The pioneering work of Ajayan and Iijima[79] has been followed by several studies on this topic, leading to one of the most active fields of research concerning nanostructured materials.[55,80–83] Besides the possibility of creating intrinsically one-dimensional materials through nanotube-driven template syntheses, one of the most interesting aspects of encapsulation concerns the formation of ordered structures in the hollow of CNTs, in the form of nanocrystallites, as often observed in experiments. The one-dimensional (1D) confined growth of different compounds has been viewed as a suitable route to the development of new low-dimensional materials, like nanowires. Moreover, the structure of the nanotube usually plays a crucial role in the growth process by selectively determining the stable atomic configuration of the filler.[80] As a result of the filling process, the properties of both the filler and the nanotube are generally different from their bulk and pristine forms, respectively.

Table 8.5 Binding energies in kcal/mol for the chemisorption of CH_2 on (6,5) ($R = 1$) and (6,4) ($R = 0$) semiconducting chiral CNTs. The number of Clar cells in the FLCC is reported in the first column. Chemisorption sites are labelled as in Figure 8.24.

	$CH_2@(6,5)$ chemisorption sites								
$n°$	*AX1*	*AX1*	*AX1*	*AX2*	*AX2*	*AX2*	*EQ*	*EQ*	*EQ*
3	− 31.9	− 42.2	− 33.1	− 48.4	− 40.4	− 41.5	− 76.0	− 76.7	− 75.0
4	− 31.5	− 40.5	− 33.5	− 45.1	− 42.8	− 41.3	− 76.6	− 76.3	− 75.9
5	− 37.9	− 39.7	− 38.3	− 42.6	− 40.9	− 42.5	− 76.0	− 76.4	− 75.9
6	− 37.6	− 39.3	− 38.2	− 42.0	− 41.9	− 41.7	− 76.0	− 76.1	− 76.0
7	− 39.0	− 39.3	− 39.2	− 41.5	− 41.2	− 42.1	− 75.8	− 76.2	− 76.0
8	− 39.0	− 39.3	− 39.1	− 41.4	− 41.7	− 41.4	− 75.9	− 76.1	76.1
	$CH_2@(6,4)$ chemisorption sites								
$n°$	*AX1*	*AX1*	*AX1*	*AX2*	*AX2*	*AX2*	*EQ*	*EQ*	*EQ*
3	− 45.0	− 46.4	− 36.6	− 54.9	− 42.3	− 36.2	− 81.0	− 78.9	− 80.6
4	− 41.0	− 44.4	− 39.5	− 51.1	− 39.9	− 44.4	− 80.9	− 81.3	− 78.7
5	− 40.2	− 42.4	− 40.0	− 46.9	− 43.6	− 43.4	− 80.4	− 80.9	− 80.0
6	− 41.3	− 41.7	− 40.9	− 46.0	− 43.0	− 44.8	− 80.1	− 80.7	− 79.7
7	− 41.1	− 41.4	− 40.9	− 45.0	− 44.1	− 44.2	− 79.7	− 79.9	− 79.9
8	− 41.5	− 41.4	− 41.3	− 44.9	− 44.0	− 44.8	− 79.5	− 80.0	− 80.0

In recent experimental studies, metal halides were inserted into the nanotube hollow with excellent yield[84–89] (see Figures 8.26 and 8.27). In particular, Bendall *et al.* demonstrated the encapsulation of silver iodide (AgI) into CNTs with diameters as small as 1 nm.[84] The filling process for small-diameter nanotubes induces a structural deformation of the sidewall, which, in turn, modifies the nanotube properties. Notably, the same authors related the observed changes to barely structural factors, owing to the nonbonding interactions between d^{10} guest metal ions and the carbon atoms on the sidewall.[84] The most significant structural deformation is found in small diameter AgI-filled nanotubes in the form of an ovalisation of the nanotube section. The deformation of the CNTs induced by filling and the resulting valence orbital rehybridisation can be expected to induce remarkable changes in their reactivity.

However, due to intrinsic difficulties in the experimental characterisation of nanostructured materials, many microscopic details of the filling process and of the resulting structures remain, to a large extent, still unclear. In contrast to experimental observations, computer simulation methods, in particular molecular dynamics (MD) simulations (see Section 8.2.5), represent an ideal tool for the investigation of reactions at the atomic level. In this respect, the computational studies carried out by Wilson *et al.* during the past few years provided a fundamental approach to the field.[51,90–95] In particular, Wilson *et al.* focused on the structural and dynamical aspects concerning the behaviour of model systems[90,91] or KI crystallites[51,92–94] confined by rigid CNTs.

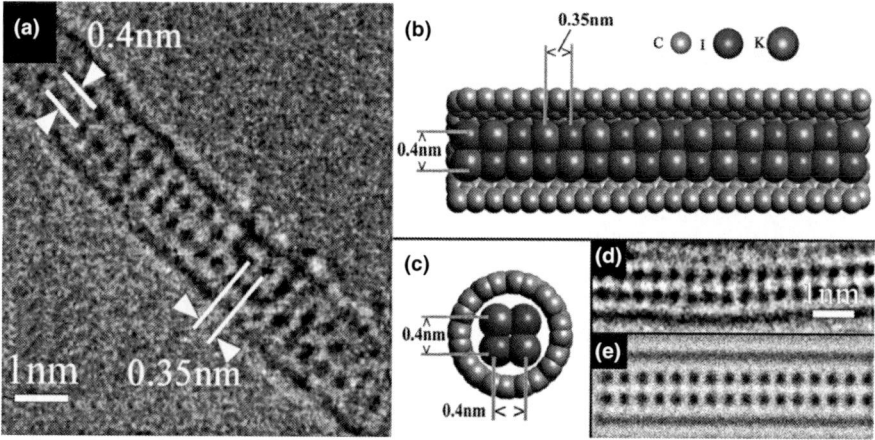

Figure 8.26 (a) Conventional HRTEM image of a KI crystal formed within a 1.4-nm diameter SWNT. (b) and (c) side-on and end-on structural representations of the KI crystal within a (10,10) SWNT showing measured lattice distortions. Adapted from ref. 102.

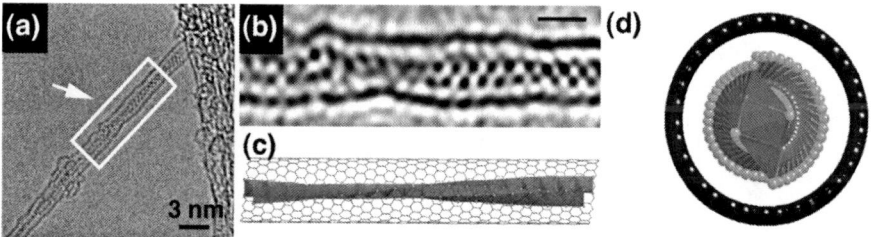

Figure 8.27 (a) and (b) HRTEM micrograph and detail (scale bar = 1 nm) of encapsulated CdC_2; (c) idealised representation of a twisted 1D polyhedral chain of $CdCl_2$ within a 1.1-nm chiral SWNT; (d) end-on view of (c). Adapted from ref. 102.

In this section, we discuss the results of MD simulations on the formation, stability and structure of AgI nanocrystallites encapsulated into single-walled (SW) (n,n) CNTs of different diameters.[68,96] Motivated by the experimental evidences cited above[84] and by theoretical predictions[90] we adopt flexible CNT models able to account for the deformation of the sidewall upon filling, thus proceeding a step further with respect to previous simulations. As will be discussed in detail later, we used two strategies for the confined crystallisation of AgI into CNTs: direct insertion of AgI clusters (ranging from 64 to 124 atoms) into (n,n) CNTs of finite length ($n = 8$–10) and diffusion of ions into CNTs from a bath of molten AgI. The diameter range of the chosen CNT system is of the order of that observed in encapsulation experiments for AgI.[84]

In MD simulations a suitable force field, for the definition of the system under study, must be first defined (see Section 8.2.4). In the MD simulations of finite systems the equations of motion were integrated in the canonical ensemble using a time step of 2 fs. For periodic systems an isothermal–isobaric ensemble was used.[97] Long-range electrostatic interactions were treated with Ewald summation. (See also Section 8.2.5.)

In a first set of simulations, sections of suitable diameter were recut from equilibrated bulk structures and subsequently inserted into the hollow of CNTs, previously equilibrated at room temperature.[68,96] Sectioning was performed along crystallographic directions (*e.g.*, [100] in NaCl, or [001] in wurtzite-type structures) in order to produce charge-neutral clusters of composition $(AgI)_n$. As bulk structures we considered the hexagonal β phase of wurtzite-type structure,[98] which is stable under ambient conditions, and the high-pressure polymorph of rocksalt type structure (NaCl type).[99,100] In fact, the latter may become relevant in confined situations in which additional pressure is applied on the guest aggregates by the CNT walls. The size of the CNT models was chosen with an approximate length ratio of 1:1 between the filled part of the nanotube and the unfilled edges. This resulted in approximate CNT lengths ranging from 15 to 20 nm. The AgI-filled systems were equilibrated at 300 K and propagated for 400 ps in the canonical ensemble. In all simulations, the AgI clusters underwent sensible structural rearrangements irrespective of the initial (wurtzite or rocksalt) structure, converging to similar configurations in both cases. Hereafter, we will only discuss structures originating from clusters recut from the rocksalt structure. All equilibrated structures of AgI clusters encapsulated into (n,n) CNT models of different radii ($n = 8$–10) exhibit structural order, as shown in Figures 8.28 and 8.29, and will be hereafter discussed in detail.

The insertion of an $(AgI)_{32}$ cluster, initially cut as a 2×2 (Ag_2I_2) square-section parallelepiped (cut parallel to [100] directions in NaCl) into the cavity of a (8,8) CNT and the subsequent thermal equilibration led to the formation of an unprecedented lowest dimensional crystal structure. In contrast with the initial configuration, the relaxed structure exhibits a one-dimensional (2×1) section along the nanotube axis, giving rise to a ribbon-like lattice with long and short connections (see Figure 8.28(a)), comparable to the atomic arrangement in (110) planes of the rocksalt structure. By extending the (n,m) index nomenclature usually applied to CNTs,[1] this structure corresponds formally to a (1,1) nanotube fragment with alternating silver and iodine atoms on the CNT hexagonal net. The narrow diameter of (8,8) nanotubes does not allow accommodation of crystal structures with a 2D transversal section. In contrast, the formation of the AgI nanoribbon induces a remarkable deformation of the sidewall, as shown in Figure 8.28(a) (right panel). Moreover, the 2×1 AgI wire is twisted along the CNT axis (Figure 8.28(a), bottom panel), giving rise to a helical propagation of the distorted nanotube cross-section, similar to CoI_2-filled CNTs.[89] Although there is no strict one-to-one correspondence between the atomic positions of the guest aggregate and the sidewall carbon atoms sites, the CNT deformation propagates commensurably with the twist of

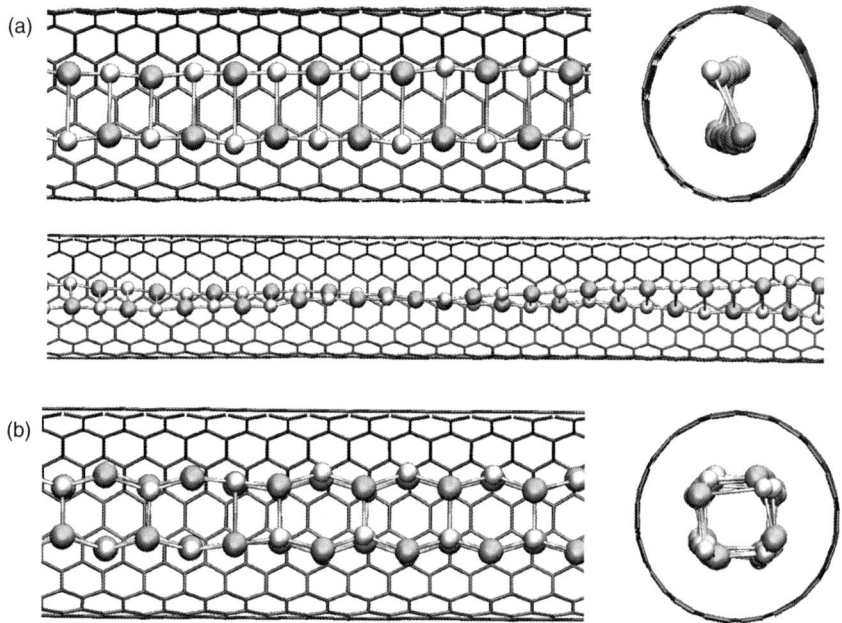

Figure 8.28 Snapshots from the MD simulation at 300 K of an AgI cluster (a) encapsulated in a (8,8) CNT; (b) encapsulated in a (9,9) CNT. Silver and iodine ions are shown as white and grey spheres, respectively. Adapted from refs. 68 and 96.

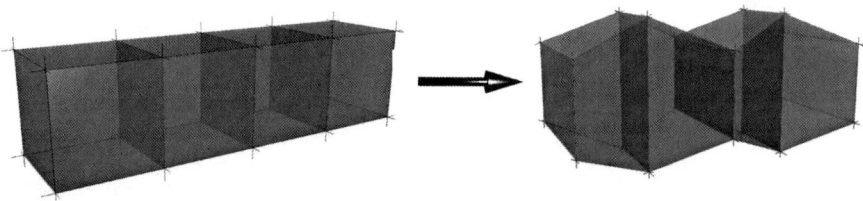

Figure 8.29 Schematic image of the geometrical transformation leading from a cubic (rocksalt-like) structure to a nanotube-like arrangement. In the final state, the occurrence of hexagonal patterns is observed. Adapted from refs. 68 and 96.

the internal ribbon. For the encapsulated structure, the low coordination number of Ag and I is only partially amended by the helicity of the ribbon along the CNT axis. Hence, an additional means to achieve a denser local packing is the flattening of the CNT sidewalls towards the ribbon surface. The chemical interactions are thus retuned to stabilize the whole system. This finding confirms the experimental evidence[84] of the deformation of the CNT cross-section upon filling with AgI for nanotube diameters of around 1 nm, and is thus consistent with the size of the (8,8) model. This result also underlines the importance of using flexible CNT models to simulate the filling process, pinpointing the active role of the container to stabilize lower-dimensional structures that are not limited to the bare confining function.

In the following, a larger $(AgI)_{66}$ cluster was placed into the hollow of a (9,9) CNT.[68,96] Relaxation of this system led to a new configuration with a 2D cross-section, in contrast with the 1D case of the (8,8) nanotube. The resulting crystal structure (Figure 8.28(b)) can be described as a folded (CNT-like) hexagonal sheet of alternating Ag and I ions and can be described as an armchair (2,2) AgI inorganic nanotube. The larger (9,9) CNT diameter allows for more space for accommodating an ordered structure and the cross-section of the crystallite is 2D. Owing to the reduced anisotropy of the crystal cross-section (Figure 8.28(b)) the filled (9,9) CNT remains cylindrical. The equilibrium configurations result from the initial cubic motifs of NaCl structure type by distorting the square cross-sections along directions perpendicular to the nanotube axis, as depicted in Figure 8.29. This results in an arrangement of alternating rectangular Ag_2I_2 units rotated relative to each other by 90°. The alternance of hexagonal and square motifs is common in binary systems based on tetrahedral structures. Apart from AgI, chalcogenides of the copper group like CdSe or main-group pentelides crystallize in the wurtzite or sphalerite structure types, and also transform into the rocksalt type under pressure.[101] Intermediate structures often display a coordination in between, exhibiting compressed six-ring or skewed cubes. In confined systems the sequence is reversed with small tubes favouring the formation of compact structures consisting of square-cubic motifs, with the appearance of six rings on enlarging the tube diameter. In this respect, the structure crystallised inside the (9,9) CNT represents an intermediate packing situation. The distorted hexagonal motifs are sometimes spaced by defects containing AgI four- and eight-rings (Figure 8.28(b), left panel).

For encapsulation into a (10,10) CNT, the internal cavity was large enough for different starting AgI cluster sizes to be considered.[68,96] This offered the opportunity to investigate the dependence of the final crystallite morphology on the initial conditions. The total number of inserted atoms was kept constant, while the way of occupying the internal volume was modified. To this end, AgI nanorods carved from rocksalt with 2×2 or 3×3 cross-sections were inserted into the nanotube. Both initial clusters exhibit some form of mismatch within the nanotube. The 2×2 nanorod is elongated and consequently locally less dense. The 3×3 rod is shorter and locally very close to the internal surface of the CNT. Both initial clusters evolve, at 300 K, towards intermediate aggregates of similar density, the former by shortening, the latter on expanding along the CNT axis. However, they result in distinct equilibrium structures.

The system evolved from the 2×2 nanorod crystallizes into a (4,0) AgI nanotube (Figure 8.30(a)).

In contrast with the case of encapsulation into a (9,9) CNT, the AgI nanotube is larger and rolled according to chiral vectors corresponding to a zigzag sidewall. The pseudo-octagonal section of the confined AgI lattice does not

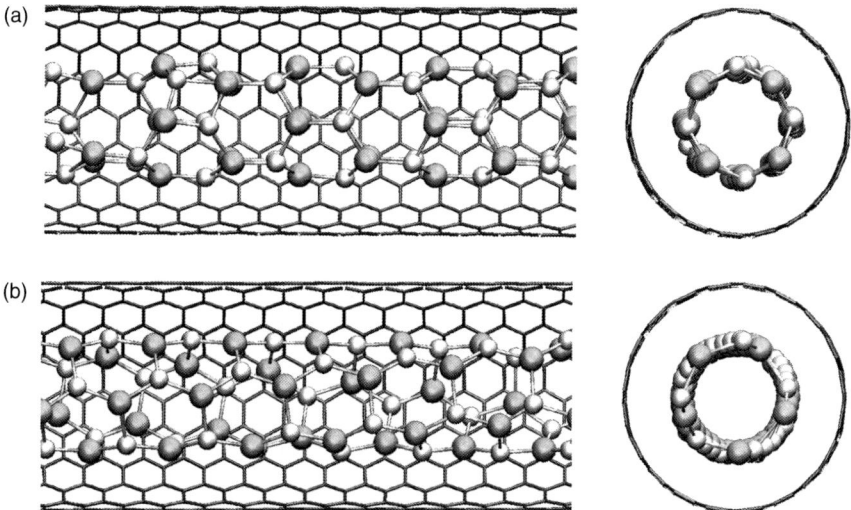

Figure 8.30 Snapshots from the MD simulation at 300 K of AgI encapsulated in a (10,10) CNT starting from (a) a 2×2 cross-section AgI cluster and (b) a 3×3 cross-section AgI cluster. Silver and iodine ions are shown as white and grey spheres, respectively. Adapted from refs. 68 and 96.

cause ovalisation of the nanotube shape, as shown in Figure 8.30(a) (right panel). Also, in this case, lattice defects with four- and eight-ring motifs do form during crystallisation, as shown in the snapshot of Figure 8.30(a) (left panel). In contrast, the equilibration started from a 3×3 cross-sectional nanorod converged to a AgI configuration formally equivalent to a chiral (4,2) nanotube (see Figure 8.30(b)). The AgI nanotube exhibits a quasicircular, decagonal cross-section that leaves the CNT cross-section practically undeformed. Notably, the same structure is obtained from a starting cluster of wurtzite structure (cut along the [001] direction). The emerging polymorphism suggests on the one hand different ways to novel forms of aggregation in the nanoregime. On the other hand, it hints at concurring factors in the formation of crystallites within CNTs. Encapsulated AgI nanotubes with a larger diameter can grow if the additional energy costs caused by a larger surface area are compensated by the lowering of the chemical potential, associated with a denser packing. The (4,0) AgI nanotube is thinner and less dense, whereas the (4,2) nanotube has a larger cross-section but is more compact. The interplay of these two factors should allow for several free energy local minima, corresponding to possible metastable phases that can be quenched by encapsulation.

All geometries were optimised by simulated annealing (heating above 500 K, cooling at a rate of –0.1 K/ps) and the final structures are shown in Figure 8.31.[68,96]

The optimised structures of the four AgI-CNT systems discussed above exhibit essentially the same features as their counterparts equilibrated at room temperature, shown in Figures 8.28 and 8.30. The confined AgI clusters are arranged in regular CNT-like (n,m) geometries with a characteristic silver iodide bond length of between 2.74 and 2.81 Å (see Figures 8.31(b)–(d)), with the exception of the (8,8) CNT, in which the Ag–I distance shrinks to 2.60 Å (Figure 8.31(a)). The structure of the optimised AgI-filled (8,8) CNT is also twisted, but to a lesser extent with respect to the system equilibrated at 300 K. Moreover, the annealing procedure confirmed the presence of structural defects (left side of Figure 8.31(c)) and competing structures (right side of Figure 8.31(b)) based on the alternation of hexagonal and squared patterns.

The dynamics of the formation of CNT-confined AgI structures was also investigated by simulating the filling process from the immersion of a nanotube in a bath of molten silver iodide.[68,96] Only (8,8) and (9,9) CNTs were considered, since they account for both two- and three-dimensional confined crystal structures. A simulation setup similar to that implemented by Wilson *et al.*[51,90–95] was used. A 35×35×57 Å box of molten AgI was initially equilibrated for 800 ps at 1000 K. A sufficiently large cylindrical region of AgI was removed from the melt and substituted by initially capped (8,8) or (9,9) CNTs. As the filling process depends less critically on nanotube edge effects, the calculations were carried out with shorter CNTs, consisting of 288 and 324 carbon atoms for the (8,8) and (9,9) systems, respectively. After an additional equilibration time of around 600 ps at 1000 K the CNT caps were removed. A filling time ranging from 20 to 40 ps was observed, after which the system was equilibrated for at least 100 ps.

The dynamical evolution of the filling process for (8,8) and (9,9) CNTs was followed by analysing the MD trajectories.[68,96] The initial steps of the filling of (8,8) and (9,9) nanotubes share mechanistic similarities. Silver and iodine ions

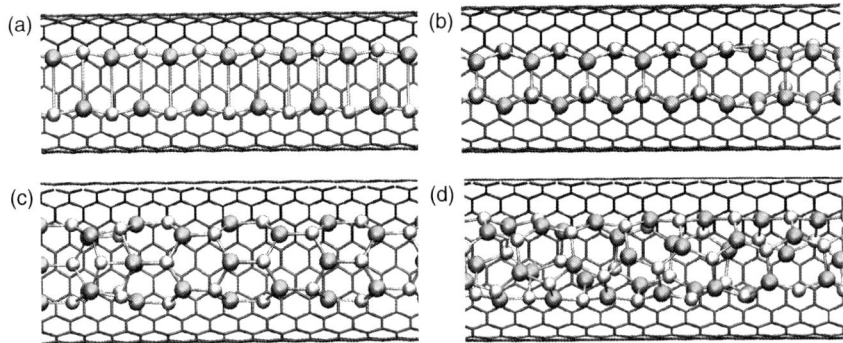

Figure 8.31   Optimised geometry of an AgI cluster encapsulated in (a) an (8,8) CNT, starting from a 2×2 section of an equilibrated rocksalt structure; (b) a (9,9) CNT, starting from a 2×2 section of an equilibrated rocksalt structure; (c) a (10,10) CNT, starting from a 2×2 section of an equilibrated rocksalt structure, and (d) a (10,10) CNT, starting from a 3×3 section of an equilibrated rocksalt structure. Silver and iodine ions are shown as white and grey spheres, respectively. Adapted from refs. 68 and 96.

enter the CNT from both ends in the form of Ag–I linear chains, as also observed in ref. 90. The relaxation towards the final configurations begins when the two chains come closer to each other. Representative snapshots of the equilibrated CNT-filled systems are shown in Figure 8.32.

In the case of the (8,8) CNT, typical configurations (see Figure 8.32(a)) correspond to the arrangement obtained from relaxing nanorods recut from bulk structures inside the CNTs. In particular, the formation of a 12×1 Ag–I twisted nanoribbon is confirmed. The latter forms from AgI filaments approaching each other from opposite tube ends. No alternative structures appear. The ovalisation of the CNT takes place while the nanoribbon is formed, underlying the active role of the CNT in the stabilisation of the forming guest aggregate. The filling of the (9,9) CNT results in a confined crystal structure with 2×2 cross-section (Figure 8.32(b)) reproducing the (2,2) nanotube found from relaxing clusters derived from bulk structural motifs, as described above. The nanotube shape is practically undeformed during and upon filling (see Figure 8.32(b), right). However, the dynamics of the filled (9,9) CNT exhibit an intense interplay between AgI structures based on square and hexagonal patterns. Indeed, the high temperature facilitates the processes, as sketched in Figure 8.29, leading to the coexistence of different structural (stable and metastable) AgI phases in nanometer-sized regions of space.

In summary, MD simulations provided clear evidence for novel silver iodide phases upon confinement in small-diameter SWCNTs. Such structures exhibit remarkable order and stability at room temperature and are formed spontaneously by immersion of open-end CNTs into molten AgI. All crystal structures are formally (n,m) AgI nanotubes with chiral vectors n and m depending on the CNT diameter and on the local environment. Confined growth in very small diameter CNTs, such as the (8,8) nanotube, gives rise to ribbon-like AgI arrangements. This configuration – a lower limiting case for formation of AgI ordered structures into the hollow of CNTs – exhibits a significant deformation of the nanotube shape. Slightly larger diameter CNTs, such as the (9,9) and (10,10) CNTs, allow the confined growth of AgI structures with 2D cross-sections. This finding agrees with experimental observations that indicate pseudocircular sections of encapsulated AgI crystallites for CNT diameters larger than or equal to 1.15 nm. Moreover, the filling of large-diameter CNTs can eventually lead to the concurrent formation of different AgI aggregates, depending on the nanotube diameter and on environmental factors such as local instantaneous pressure. Interfaces between different confined AgI nanocrystallites are likely to be observed, as well as a variety of defective structures. Indeed, the interplay between different stable and metastable phases is observed in the course of MD simulations on the encapsulation of AgI from the melt into a (9,9) CNT. Work is ongoing to assess the detailed dynamics of the filling process and the properties of the resulting encapsulated crystal nanowires in different environmental and growth conditions.

Experimental evidences have confirmed the existence of junctions between CNTs of different diameters.[1] Therefore, the study of CNTs junctions as

(a)

(b)

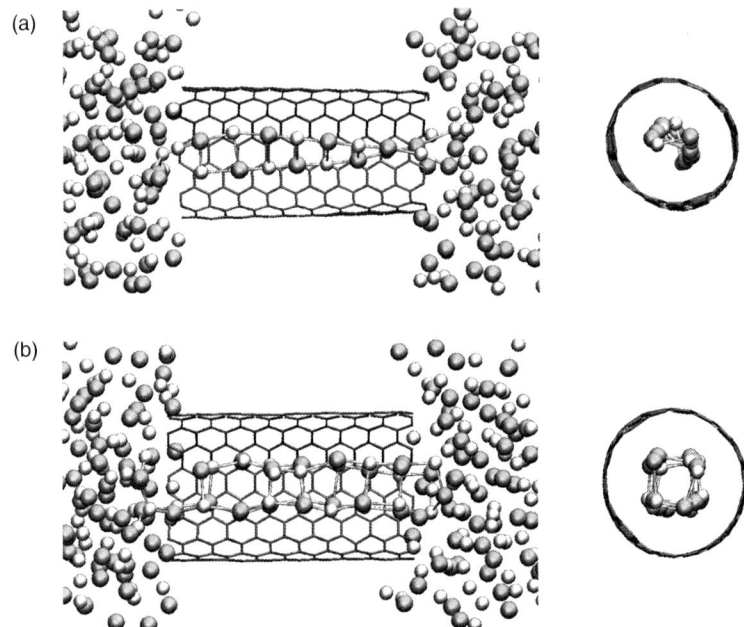

Figure 8.32 Snapshot from the MD simulation of the filling of (a) an (8,8) CNT
inserted in a bath of molten AgI at 1000 K, and (b) a (9,9) CNT inserted
in a bath of molten AgI at 1000 K. Silver and iodine ions are shown as
white and grey spheres, respectively. Only the relevant part of the AgI
bath exterior to the nanotube is shown, for clarity. Adapted from refs. 68
and 96.

templates for the confined growth of nanocrystallites would constitute a natural
extension of the modeling approach discussed in this chapter. Calculations were
performed to simulate the growth of 1D nanocrystallites of AgI inside CNTs
junctions. Both the procedures of direct insertion of AgI clusters inside the
CNTs hollow and the direct filling from a bath of molten salt were applied in
the case of different CNTs junctions. Namely, the cases of (8,8)–(9,9), (9,9)–
(10,10), and (8,8)–(10,10) junctions were taken into account. The resulting
structures, as obtained by preliminary calculations, are depicted in Figure 8.33
that shows optimised geometries of nanocrystallites obtained after insertion of
2×2 cross-sectional AgI clusters inside the hollow of different CNTs junctions.
As can be seen from the figure, in all cases the 1D AgI nanocrystallites exhibit
polymorphic structures. Structural features very similar to those observed in
the case of simple CNTs can be recognised relatively far from the junction
between the two CNTs of different diameter. More interestingly, in the region
of the CNTs junctions, new interphase structures that allow the transaction
between different AgI phases are obtained.

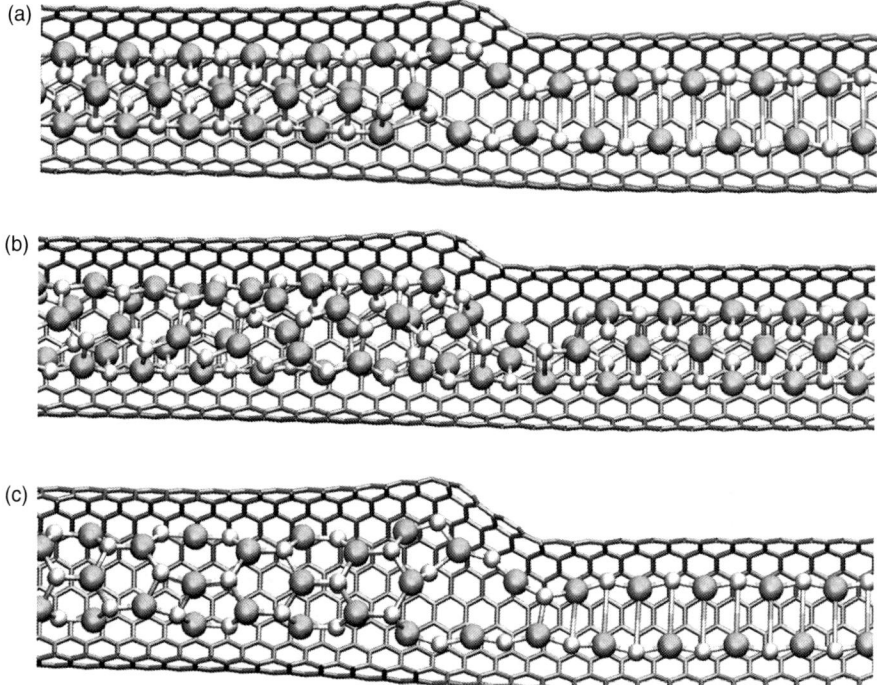

Figure 8.33   Snapshots from the MD simulation at 300 K of an AgI cluster (a) encapsulated in a (8,8)–(9,9) CNTs junction; (b) encapsulated in a (9,9)–(10,10) CNTs junction; (c) encapsulated in a (8,8)–(10,10) CNTs junction. Silver and iodine ions are shown as white and grey spheres, respectively. Adapted from refs. 68 and 96.

8.3.3   Encapsulation of Inorganic Nanotubes

In contrast to the wettability and capillary characteristics of carbon nanotubes, these properties are hardly studied for inorganic nanotubes. BN nanotubes (BNNTs) of hexagonal BN, which is isoelectronic and isostructural to graphite, have been filled by potassium halides.[103] The filling techniques for carbon nanotubes do not work efficiently for BN nanotubes, whereas wetting properties like contact angles with various liquids and surface tension of BNNTs are comparable with or only slightly different from those for CNTs.[104] This different behaviour may be explained in many cases by the high chemical endurance of BN, which hampers the opening of the BNNT caps.

The capillary properties of inorganic nanotubes were outlined recently by *Kreizman et al.* studying the filling of tungsten disulfide WS_2 nanotubes by molten lead iodide.[105,106]

The process of capillary insertion into this kind of nanotube shows some peculiarities in comparison with CNTs. Chalcogenide nanotubes have larger strain energies than CNNTs or BNNTs.[107] Therefore, the diameters of such inorganic nanotubes are typically an order of magnitude larger than the diameters of CNTs. The high strain energy of chalcogenide hollow nanostructures also leads to the formation of mainly open nanotubes, because corresponding fullerenic structures, which could form the caps of chalcogenide nanotubes, are less stable than carbon fullerenes.[108]

Thus, chalcogenide nanotubes are genuine nanocapillars, and the imbibition of a liquid would be a one-step process enabled without any preliminary oxidation.[106]

The final product after capillary filling and cooling of molten PbI_2 within WS_2 nanotubes may differ from the corresponding bulk structure. Even the formation of multiwall nanotubes of layered PbI_2 coaxial to the WS_2 nanotubes could be observed.[105]

Evidently, the diameters of chalcogenide nanotubes are large enough to provide a nanocoating of the inner surface by a halide and not only the formation of halide nanocrystal or nanowire in the cavity of the nanotube. The structure formation inside of nanotubes by capillar filling of BN and WS_2 nanotubes by the imbibition of PbI_2 and KI was studied by MD simulations.[106,109] The insertion of PbI_2 was also experimentally studied on carbon[87] and inorganic nanotubes.[105]

Already, a first view at the insertion of molten PbI_2 into carbon, BN and MoS_2 nanotubes shows a clear difference of this process for CNT, BNNT and MoS_2 NTs (see Figure 8.34). In contrast to CNTs and BNNTs, the penetration of PbI_2 melt is accompanied by the partial adsorption of Pb and I ions on the outer surface of the tube. It starts with a rapid adsorption of Pb and I ions at the edge of the inner nanotube wall, and a concave meniscus edge occurs almost immediately. However, there is no acceleration of the imbibition at a certain stage as is observed for CNTs and BNNTs.[106] Therefore, the penetration of the PbI_2 drop into the MoS_2 nanotubes takes ~ 1 ps, considerably longer than in the case of the CNT and the BNNT.

The phase state of PbI_2 is nearly kept after its insertion into the nanotubes. The radial distribution functions $g_{ij}(r)$ for Pb–I, Pb–Pb and I–I of PbI_2 in a nanotube and in the free drop shows that the PbI_2 melt still remains as a liquid and is not affected by the character of the nanotube wall (Figure 8.35). There is also no marked difference in the shape and intensity profiles between $g_{ij}(r)$ for the free PbI_2 drop and inside the nanotubes.

The cross-sections of the final atomic structure for PbI_2 melts along the tube axes indicate a shell-like structure of the Pb and I ions in the nanotubes (Figure 8.36). This is well established across the whole nanotube for the BNNT, whereas for the CNT and especially for the MoS_2 NT such an effect is seen only close to the tube walls.

The MD simulations model the filling of nanotubes by a drop of finite size. Enyashin *et al.*[106] also investigated the validity of macroscopic phenomenological equations, known for a steady-state capillary flow in the case of nanotube filling. They applied the Lucas and Washburn theory for a capillary flow,[110] which was derived for an incompressible Newtonian fluid in a circular

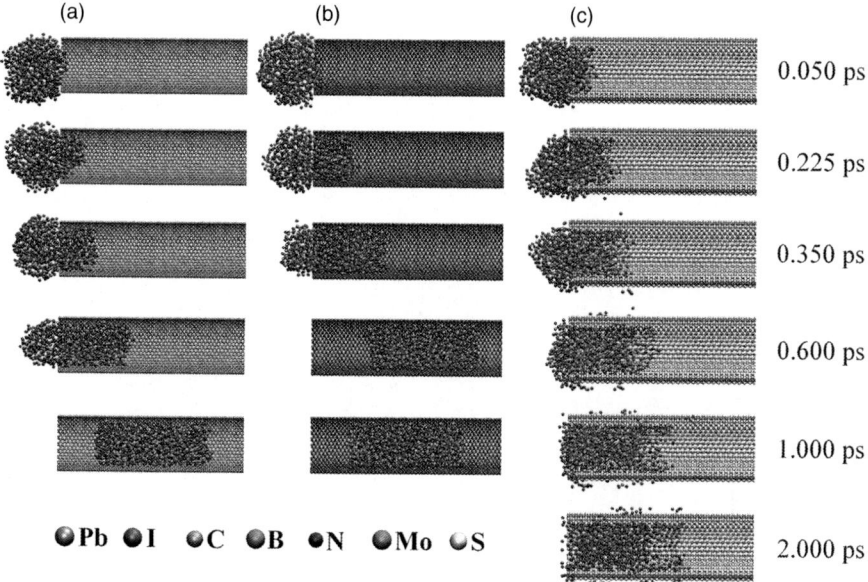

Figure 8.34 Snapshots for imbibition of molten PbI_2 drop by (25,25) carbon (a), (25,25) BN (b) and (21,21) MoS_2 nanotubes (c) at different times. For clarity, the front walls of nanotubes are removed. Adapted from ref. 106.

tube. It gives the dependence of the height h of the adsorbed liquid column (penetration length) at time t as $h \sim \sqrt{t}$. The MD simulations show that the filling of nanotubes by a drop of liquid PbI_2 are characterised by the formation of separate Pb and I shells within the drop. In the case of carbon and BN nanotubes this formation does not follow the Lucas–Washburn equation ($\langle N \rangle \sim \sqrt{t}$), but can also not be described by a $\langle N \rangle \sim t^2$ dependence for an initial flow behaviour. However, as a good description for the dependence of $<N>$ on t a conciliating function can be used:

$$\langle N \rangle = \frac{at^2}{b + t^{3/2}} \tag{8.37}$$

which at small t behaves like t^2 and for large t it approaches the $\sqrt{t}$ dependence. This expression reflects the initially slow penetration of the ionic melt into a CNT or a BNNT. Formally, eqn (8.37) can be used also for the case of the MoS_2 nanotube, but with the parameter b equal to 0, which means also that the penetration of molten PbI_2 drop into a MoS_2 nanotube is well described by the classical Lucas–Washburn theory.

A similar study of the melt penetration into the sulfide MoS_2 nanotube was performed also for potassium iodide KI.[109] Already, a first view at the molten KI inside the MoS_2 nanotube indicates the weaker interaction of the KI with

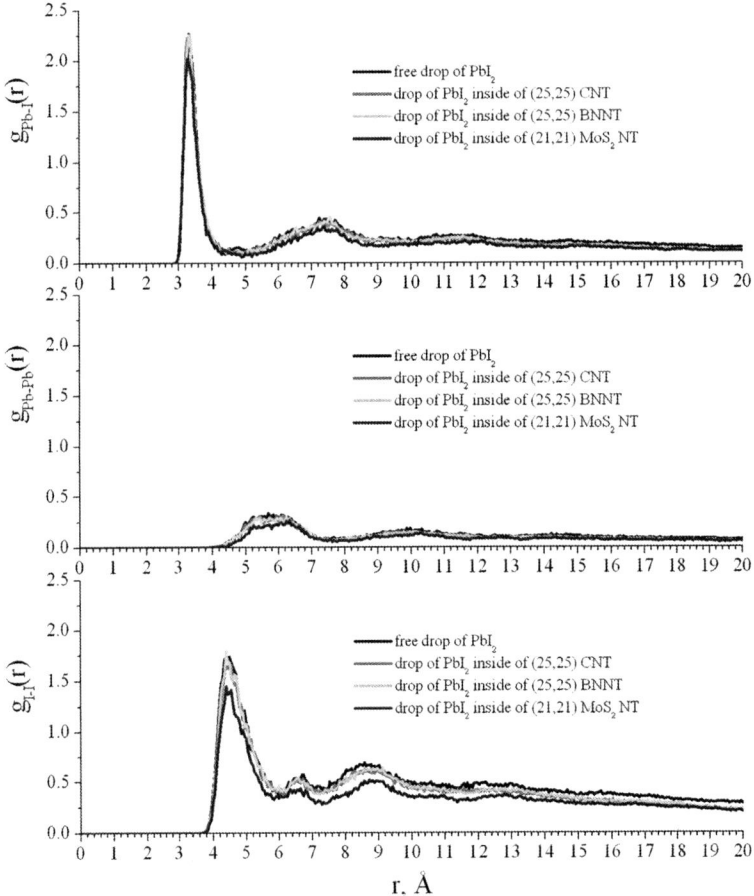

Figure 8.35 Distribution functions $g_{ij}(r)$ for interatomic distances in molten PbI$_2$ as a free drop and the drops within different nanotubes after 1 ps. Adapted from ref. 106.

the sulfide wall. In contrast to PbI$_2$, the liquid KI does not show a concave meniscus, and no wetting layer of the salt is seen at the inner side of the MoS$_2$ tube wall.

The radial distribution functions for K–I, K–K and I–I distances inside of a nanotube and in the free molten KI drop show evidence that the liquid state of the KI melt within the nanotube is unchanged. However, in contrast to the case of PbI$_2$, there are no indications of a shell-like structuring of the melt in the case of KI.

The difference in the inner structure of the KI and PbI$_2$ melts embedded within the cavity of the MoS$_2$ nanotube may be also an indirect evidence for their different behaviour regarding the preference in formation of either a nanowire or a nanotube of the halide. There exists a correlation between the simulated structure of the melt and the phase observed experimentally after the

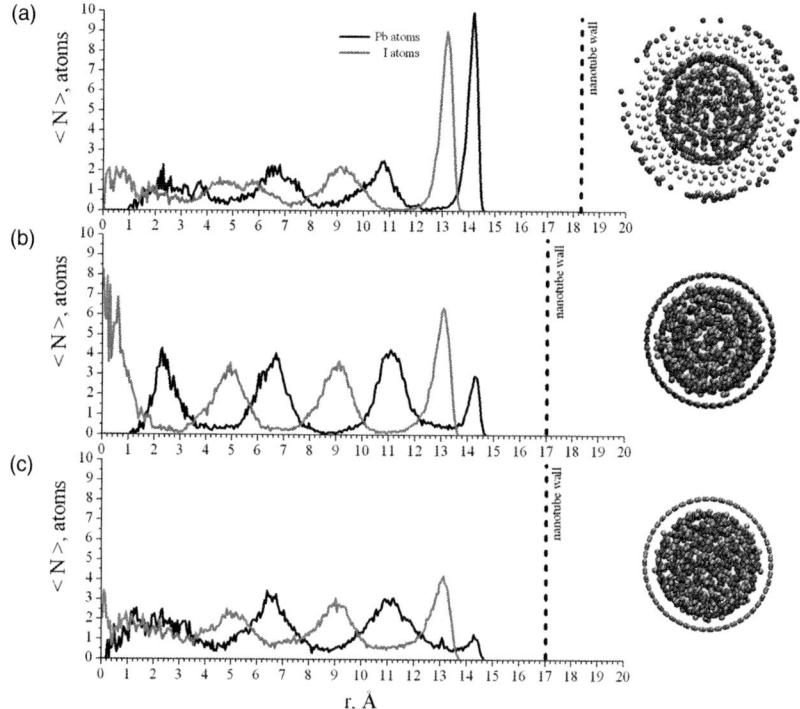

Figure 8.36 Average number of Pb and I atoms $<N>$ found at a certain distance from the tube axis in the drop of molten PbI_2 penetrated into (21,21) MoS_2 (a), (25,25) BN (b) and (25,25) carbon nanotubes (c) after $t = 1$ ps. The views on cross-sections along the tubes' axes demonstrate the shell-like character of the liquid PbI_2 within these nanotubes. Adapted from ref. 106.

crystallisation of the molten salt within the nanotube. The simulated PbI_2 liquid was shown to exhibit a shell-like structure. This is consistent with the observation of PbI_2 nanotubes after the cooling of a mixture of the molten PbI_2 in WS_2 nanotubes.[105]

Thus, the structure of coaxial shells of an ionic liquid could predetermine the formation of tubular structures inside the nanotube. In contrast, the simulated KI liquid has shown a random distribution of the K and I atoms along the radial direction of a sulfide nanotube. Evidently therefore, a bulk nanowire could be expected to occur upon cooling of the related system, which was observed for CsI within WS_2 nanotubes.[111]

The results of the computer simulations show that the main difference in the capillary filling process of different types of nanotubes with the same inner radius is caused by the structure of walls. Graphene-like surfaces of carbon and BN nanotubes, having weaker interaction with the melt than the sulfur layer of MoS_2 nanotubes. Therefore, the smooth surfaces of carbon and BN nanotubes provide a fast and friction-free salt-drift into these nanotubes, while the sulfur surface of MoS_2 does not support the imbibition of molten PbI_2 into the nanotube. All nanotubes have complex filling dynamics, which may lead, for example, to a shell-like structure of the embedded liquid. The filling of MoS_2 NT obeys a Lucas–Washburn equation within every shell, whereas the filling of CNT and BNNT needs a modified expression, as discussed in detail by Enyashin *et al.*[106]

8.4   Conclusions

The strength of interaction with the surface of a nanotube, plays an important role for the final morphology and the location of a "sucked" drop. In the case of a CNT, one may expect that a drop of an ionic melt can move in a tube far from the entrance point. In the case of a BNNT a drop of a salt will be located inside of the tube, but close to the edge. Due to the weak interaction with the tube wall, convex menisci are formed. After cooling, in both cases a structure closely related to the bulk structure (nanowire or nanocrystal) will be formed. A different situation may be obtained in the case of a MoS_2 NT due to the stronger interaction with the tube walls concave menisci occur. Good wettability can lead to the total spread of melt over the inner surface of the nanotube, which can lead to cylindrical nanostructures. That is, MoS_2 nanotubes can be used as templates for the formation of other tubular structures.[112]

Acknowledgment

The authors are grateful to the support from European Commission (ERC grant INTIF 226639)

References

1. M. Dresselhaus, G. Dresselhaus and P. Eklund, in *Science of Fullerenes and Carbon Nanotubes*, Academic Press, San Diego, 1996.
2. A. N. Enyashin, S. Gemming, G. Seifert in *Materials for Tomorrow: Theory, Experiments and Modelling*, ed. S. Gemming, M. Schreiber, J. Suck, Springer Series in Materials Science, **93**, Springer Berlin, Heidelberg, 2006, p. 33.
3. R. Tenne, M. Remskar, A. N. Enyashin, G. Seifert in *Carbon Nanotubes: Advanced Topics in the Synthesis, Structure, Properties and Applications*, ed. A. Jorio, G. Dresselhaus, M. S. Dresselhaus, Topics in Applied Physics, **111**, Springer Berlin, Heidelberg, 2008, p. 631.
4. R. McWeeny and B. T. Sutcliffe, in *Methods of Molecular Quantum Mechanics*, Academic Press, London and New York, 1976.
5. I. Shavitt in *Methods of Electronic Structure Theory*, ed. H.F. Schaefer III, Plenum, New York, 1977, pp. 189–275 and references therein.

6. A. C. Wahl, G. Das in *Methods of Electronic Structure Theory*, ed. H. F. Schaefer III, Plenum, New York, 1977, p. 51–78 and references therein.

7. W. J. Hehre, L. Radom and P. Schleyer and J. A. Pople, in *Ab-initio Molecular Orbital Theory*, Wiley-Interscience, London, 1986.

8. D. M. Hirst, in *A Computational Approach to Chemistry*, Blackwell Scientific Publications, Melbourne, 1990.

9. A. Szabo and N. L. Ostlund, in *Modern Quantum Chemistry*, McGraw-Hill, New York, 1989.

10. P. Hohenberg and W. Kohn, *Phys. Rev. B*, 1964, **136**(3B), B864.

11. W. Kohn and L. J. Sham, *Phys. Rev.*, 1965, **140**(4A), 1133.

12. S. F. Sousa, P. A. Fernandes and M. J. Ramos, *J. Phys. Chem. A*, 2007, **111**(42), 10439–10452.

13. A. Kekulè, *Bull. Soc. Chim. Fr.*, 1865, **3**, 98–110.

14. P. v. R. Schleyer, *Chem. Rev.*, 2001, **101**, 1115–1118.

15. P. J. Garrat, in *Aromaticity*, Wiley, New York, 1986.

16. V. I. Minkin, M. N. Glukhovtsev and B. Y. Simkin, in *Aromaticity and Antiaromaticity: Electronic and Structural Aspects*, Wiley, New York, 1998.

17. P. v. R. Schleyer and H. Jiao, *Pure Appl. Chem.*, 1996, **68**, 209–218.

18. T. Krygowski, M. K. Cyrañski, Z. Czarnocki, G. Häfelinger and A. R. Katritzky, *Tetrahedron*, 2000, **56**, 1783–1796.

19. L. T. Scott and J. S. Siegel, *Tetrahedron*, 2001, **57**, ix.

20. E. Z. Hückel, *Z. Phys.*, 1931, **70**, 104–186.

21. E. Z. Hückel, *Z. Phys.*, 1931, **72**, 310–337.

22. E. Z. Hückel, *Z. Phys.*, 1932, **76**, 628–648.

23. E. Z. Hückel, *Z. Elektrochem. Angew. Phys. Chem.*, 1937, **42**, 752–788.

24. E. Clar, in *The Aromatic Sextet; Wiley*, London, 1972.

25. E. Clar, in *Polycyclic Hydrocarbons*, Academic Press, New York, 1968.

26. M. Randic, *Chem. Rev.*, 2003, **103**, 3449–3606.

27. M. K. Cyranski, B. T. Stepien and T. M. Krygowski, *Tetrahedron*, 2000, **56**, 9663–9667.

28. J. Schulman and R. Disch, *J. Phys. Chem. A*, 1999, **103**, 6669–6672.

29. G. Portella, J. Poater, J. Bofill, P. Alemany and M. Sola, *J. Org. Chem.*, 2005, **70**, 2509–2521.

30. I. Gutman, Z. Tomovic, K. Mullen and J. P. Rabe, *Chem. Phys. Lett.*, 2004, **397**, 412–416.

31. R. Havenith, J. van Lenthe, F. Dijkstra and L. Jenneskens, *J. Phys. Chem. A*, 2001, **105**, 3838–3845.

32. Y. Ruiz-Morales, *J. Phys. Chem. A*, 2004, **108**, 10873–10896.

33. J. Ormsby and B. King, *J. Org. Chem.*, 2004, **69**, 4287–4291.

34. C. L. Kane and E. J. Mele, *Phys. Rev. Lett.*, 1997, **78**, 1932–1935.

35. T. W. Odom, J. L. Huang, P. Kim and C. M. Lieber, *J. Phys. Chem. A*, 2000, **104**, 2794–2809.

36. T. W. Odom, J. Huang, P. Kim and C. M. Lieber, *Nature*, 1998, **391**, 62–68.

37. J. W. G. Wildöer, L. C. Venema, A. G. Rinzler, R. E. Smalley and C. Dekker, *Nature*, 1998, **391**, 59–62.
38. D. Moran, F. Stahl, H. Bettinger, H. Schaefer and P. Schleyer, *J. Am. Chem. Soc.*, 2003, **125**, 6746–6752.
39. W. Clauss, D. J. Bergeron, M. Freitag, C. L. Kane, E. J. Mele and A. T. Johnson, *Europhys. Lett.*, 1999, **47**, 601–607.
40. W. Clauss, D. J. Bergeron and A. T. Johnson, *Phys. Rev. B*, 1998, **58**, R4266–R4269.
41. M. Parrinello, A. Rahman and P. Vashishta, *Phys. Rev. Lett.*, 1983, **50**, 1073–1076.
42. P. Vashishta and A. Rahman, *Phys. Rev. Lett.*, 1978, **40**, 1337–1340.
43. L. Pauling, in *The Nature of the Chemical Bond*, Cornell University Press, Ithaca, New York, 1960.
44. J. Tallon, *Phys. Rev. Lett.*, 1986, **57**, 2427–2430.
45. A. Wootton and P. Harrowell, *J. Phys. Chem. B*, 2004, **108**, 8412–8418.
46. W. Smith, T. R. Forester, *DL_POLY package of molecular simulation routines*, copyright The Council for the Central Laboratory of the Research Councils, Daresbury Laboratory at Daresbury, Nr. Warrington (1996).
47. J. Walther, R. Jaffe, T. Halicioglu and P. Koumoutsakos, *J. Phys. Chem. B*, 2001, **105**, 9980–9987.
48. Y. Quo, N. Karasawa and W. A. Goddard, *Nature*, 1991, **351**, 464–467.
49. A. K. Rappe, C. J. Casewit, K. S. Colwell, W. A. Goddard and W. M. Skiff, *J. Am. Chem. Soc.*, 1992, **114**, 10024–10035.
50. S. K. R. S. Sankaranarayanan, V. R. Bhethanabotla and B. Joseph, *Phys. Rev. B*, 2005, **72**, 195405.
51. M. Wilson and P. Madden, *J. Am. Chem. Soc.*, 2001, **123**, 2101–2102.
52. D. Chandler, in *Introduction to Modern Statistical Mechanics*, Oxford University Press, New York, 1960.
53. D. Frenkel and B. Smit, in *Understanding Molecular Simulation: from Algorithms to Applications*, Academic Press, London, 1996.
54. M. P. Allen and D. J. Tildesley, in *Computer Simulation of Liquids*, Clarendon Press, Oxford, 1960.
55. Y. Xia, P. Yang, Y. Sun, Y. Wu, B. Mayers, B. Gates, Y. Yin, F. Kim and H. Yan, *Adv. Mater.*, 2003, **15**, 353–389.
56. F. Mercuri and A. Sgamellotti, in *Chemistry of Carbon Nanotubes*, American Scientific Publishers, Stevenson Ranch, CA, 2007.
57. F. Mercuri and A. Sgamellotti, *Inorganica Chimica Acta*, 2006, **360**, 785.
58. S. Dapprich, I. Komaromi, K. S. Byun, K. Morokuma and M. J. Frisch, *J. Mol. Struct.*, **461**, 1–21.
59. M. Svensson, S. Humbel, R. Froese, T. Matsubara, S. K. Sieber and Morokuma, *J. Phys. Chem.*, 1996, **100**, 19357–19363.
60. H. Bettinger, *Org. Lett.*, 2004, **6**, 731–738.
61. Z. Chen, S. Nagase, A. Hirsch, R. C. Haddon, W. Thiel and P. v. R. Schleyer, *Angew. Chem. Int. Ed. Engl.*, 2004, **43**, 1552–1558.

62. G. Jia, J. Li and Y. Zhang, *Chem. Phys. Lett.*, 2006, **418**, 40–45.
63. V. Barone, J. Heyd and G. E. Scuseria, *Chem. Phys. Lett.*, 2004, **389**, 289–292.
64. A. Rochefort, D. Salahub and P. Avouris, *J. Phys. Chem. B*, 1999, **103**, 641–646.
65. Z. Zhou, M. Steigerwald, M. Hybertsen, L. Brus and R. Friesner, *J. Am. Chem. Soc.*, 2004, **126**, 3597–3607.
66. A. Galano, *Chem. Phys.*, 2006, **327**, 159–170.
67. M. Baldoni, A. Sgamellotti and F. Mercuri, *Org. Lett.*, 2007, **9**, 4267–4270.
68. M. Baldoni in *Inside and outside the sidewall of carbon nanotubes: theoretical strategies for functionalization and encapsulation*, Ph. D. thesis, University of Perugia, Perugia, Italy, 2008.
69. K. Fukui, *Science*, 1982, **218**, 747–758.
70. A. D. Becke, *J. Chem. Phys.*, 1993, **98**, 1372–1377.
71. P. J. Stephens, F. J. Devlin, C. F. Chabalowski and M. J. Frisch, *J. Phys. Chem.*, 1994, **98**, 11623–11627.
72. C. Lee, W. Yang and R. G. Parr, *Phys. Rev. B*, 1988, **37**, 785–789.
73. M. J. Frisch *et al.*, *Gaussian 03, Revision C.02*, Gaussian Inc., Wallingford, CT, 2008.
74. M. Baldoni, D. Selli, A. Sgamellotti and F. Mercuri, *J. Phys. Chem. C*, 2009, **113**, 862.
75. S. Simon, M. Duran and J. J. Dannenberg, *J. Chem. Phys.*, 1996, **105**, 11024–11031.
76. F. Boys and S. F.. Bernardi, *Mol. Phys.*, 1970, **19**, 553–566.
77. H. F. Bettinger, *ChemPhysChem*, 2003, **4**, 1283–1289.
78. H. F. Bettinger, *Chem. Eur. J.*, 2006, **12**, 4372–4379.
79. P. M. Ajayan and S. Iijima, *Nature*, 1993, **361**, 333–338.
80. R. R. Meyer, J. Sloan, R. E. Dunin-Borkowski, A. I. Kirkland, M. C. Novotny, S. R. Bailey, J. L. Hutchison and M. L. H. Green, *Science*, 2000, **289**, 1324–1326.
81. W. Han, S. Fan, Q. Li and Y. Hu, *Science*, 1997, **277**, 1287–1289.
82. G. Hummer, J. C. Rasaiah and J. P. Noworyta, *Nature*, 2001, **414**, 188–190.
83. K. Koga, G. T. Gao, H. Tanaka and X. C. Zeng, *Nature*, 2001, **412**, 802–805.
84. J. Bendall, A. Ilie, M. Welland, J. Sloan and M. L. H. Green, *J. Phys. Chem. B*, 2006, **110**, 6569–6573.
85. J. Sloan, D. M. Wright, S. Bailey, G. Brown, A. P. E. York, K. S. Coleman, M. L. H. Green, D. M. Wright, J. L. Hutchison and H. Woo, *Chem. Commun.*, 1999, **8**, 699–700.
86. J. Sloan, M. Terrones, S. Nufer, S. Friedrichs, S. Bailey, H. G. Woo, M. Ruhle, J. L. Hutchison and M. L. H. Green, *J. Am. Chem. Soc.*, 2002, **124**, 2116–2117.
87. E. Flahaut, J. Sloan, S. Friedrichs, A. Kirkland, K. Coleman, V. Williams, N. Hanson, J. Hutchison and M. L. H. Green, *Chem. Mater.*, 2006, **18**, 2059–2069.

88. A. Ilie, J. Bendall, D. Roy, E. Philp and M. L. H. Green, *J. Phys. Chem. B*, 2006, **110**, 13848–13857.

89. E. Philp, J. Sloan, A. I. Kirkland, R. R. Meyer, S. Friedrichs, J. L. Hutchison and M. L. H. Green, *Nature Mater.*, 2003, **2**, 788–791.

90. M. Wilson, *J. Chem. Phys.*, 2006, **124**, 124706.

91. M. Wilson, *Nano Lett.*, 2004, **4**, 299–302.

92. M. Wilson, *J. Chem. Phys.*, 2002, **116**, 3027–3041.

93. M. Wilson, *Chem. Phys. Lett.*, 2002, **366**, 504–509.

94. M. Wilson, *Chem. Phys. Lett.*, 2004, **397**, 340–343.

95. M. Wilson and S. Friedrichs, *Acta Crystallogr., Sect. A: Found. Crystallogr.*, 2006, **62**, 287–295.

96. M. Baldoni, S. Leoni, A. Sgamellotti, G. Seifert and F. Mercuri, *Small*, 2007, **3**, 1730–1738.

97. S. Melchionna, G. Ciccotti and B. L. Holian, *Mol. Phys.*, 1993, **78**, 533–548.

98. S. Hoshino, *J. Phys. Soc. Jpn.*, 1954, **9**, 295.

99. S. Hull and D. A. Keen, *Phys. Rev. B*, 1999, **59**, 750–761.

100. R. C. Hanson, T. A. Fjeldly and H. D. Hochheimer, *Phys. Status Solidi, B*, 1975, **70**, 567–576.

101. A. Mujica, A. Rubio, A. Muñoz and R. Needs, *J. Rev. Mod. Phys.*, 2003, **75**, 863–912.

102. J. Sloan, A. I. Kirkland, J. L. Hutchison and M. L. H. Green, *Chem. Commun.*, 2002, **13**, 1319–1332.

103. W. Q. Han, C. W. Chang and A. Zettl, *Nano Lett.*, 2004, **4**, 1355–1357.

104. D. Golberg, Y. Bando, C. Tang and C. Zhi, *Adv. Mater.*, 2007, **19**, 2413–2432.

105. R. Kreizman, S. Y. Hong, J. Sloan, R. Popovitz-Biro, A. Albu-Yaron, G. Tobias, B. Ballesteros, B. G. Davis, M. L. H. Green and R. Tenne, *Angew. Chem. Int. Ed.*, 2009, **48**, 1230–1233.

106. A. N. Enyashin, R. Kreizman and G. Seifert, *J. Phys. Chem. C*, 2009, **113**, 13664–13669.

107. G. Seifert, H. Terrones, M. Terrones, G. Jungnickel and T. Frauenheim, *Phys. Rev. Lett.*, 2000, **85**, 146–149.

108. A. N. Enyashin, S. Gemming, M. Bar-Sadan, R. Popovitz-Biro, S. Y. Hong, Y. Prior, R. Tenne and G. Seifert, *Angew. Chem. Int. Ed.*, 2007, **46**, 623–627.

109. S. Y. Hong, R. Kreizman, A. Zak, J. Sloan, A. N. Enyashin, B. G. Davis, G. Seifert, M. L. H. Green, R. Tenne, *Eur. J. Inorg. Chem.*, 2010, **27**, 4233–4243.

110. E. W. Washburn, *Phys. Rev.*, 1921, **17**, 273–283.

111. S. Y. Hong, R. Popovitz-Biro, G. Tobias, B. Ballesteros, B. G. Davis, M. L. H. Green and R. Tenne, *Nano Res.*, 2010, **3**, 170.

112. R. Kreizman, A. N. Enyashin, F. L. Deepak, A. Albu-Yaron, R. Popovitz-Biro, G. Seifert, R. Tenne, *Adv. Funct. Mater.*, 2010, **20**, 2459–2468.

CHAPTER 9

Density Functional Calculations of NMR Chemical Shifts in Carbon Nanotubes

EVA ZUREK AND JOCHEN AUTSCHBACH

Department of Chemistry, University at Buffalo, State University of New York, Buffalo, NY 14260-3000, USA

9.1 Introduction

Since the discovery of single-walled carbon nanotubes (SWNTs)[1,2] and multi-walled carbon nanotubes[3,4] the properties, separation, and potential applications of these fascinating systems have been intensely studied. As illustrated in Figure 9.1, a SWNT can be thought of as a rolled up strip taken out of a graphene sheet. The tube may be uniquely classified by a pair of numbers (n, m) which denote the two carbons that meet upon rolling. There are three distinct classes of SWNT structures: armchair, zigzag and helical (chiral) characterised by (n, n), $(n, 0)$ and (n, m) with $n \neq m$, respectively. One may also classify SWNTs by their electronic structure. For instance, in the early 1990s tight–binding calculations on a graphene sheet model predicted that SWNTs would be metallic if $n - m = 3q$ where q is an integer.[5,6] However, when $sp^2 - sp^3$ rehybridisation was taken into consideration, it was found that the (9,0), (12,0) and (15,0) systems are in fact narrow gap semiconductors.[5-7] The larger $(3n,0)$ tubes were predicted to be metallic since their increased diameter resulted in a reduction of the rehybridisation. The small diameter zigzag tubes with $n = 3-6$ are once again found to be metallic due to σ^* and π^* hybridisation.[8-10] All

RSC Theoretical and Computational Chemistry Series No. 4
Computational Nanoscience
Edited by Elena Bichoutskaia
© Royal Society of Chemistry 2011
Published by the Royal Society of Chemistry, www.rsc.org

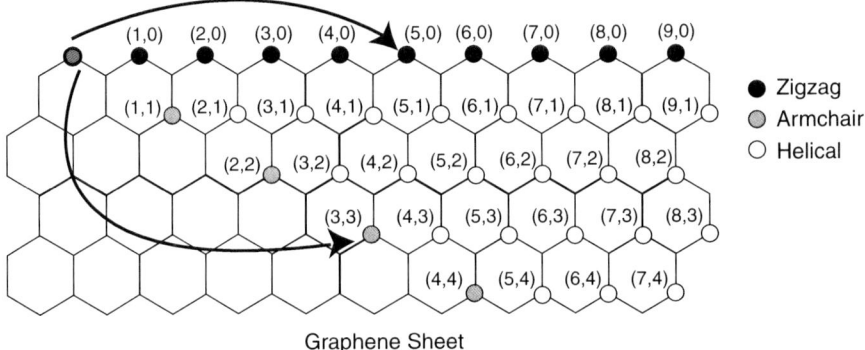

Graphene Sheet

Figure 9.1 Constructing a (5,0) and (3,3) single-walled carbon nanotube (SWNT) from a graphene sheet. Rolling the sheet so that the origin meets with a black/grey/white circle gives rise to a zigzag/armchair/helical nanotube.

isolated armchair tubes are metallic because of symmetry.[5–7,11] However, the symmetry may be broken by packing the tubes in bundles, thereby allowing pseudogaps to form.[12] The electronic structure of these nanosystems is also known to be dependent upon the presence of functional groups and defects.[13,14]

The sensitivity of the physical and chemical properties of SWNTs to geometry, structural defects, interactions with other tubes, and chemical modifications has resulted in a wide range of proposed applications in fields as diverse as molecular electronics, sensing, and in biotechnology.[15–18] However, this diversity has also resulted in a barrier towards rational materials design. For example, an outstanding problem in carbon-nanotube research is the full characterisation of heterogeneous bulk samples.[15,19] Atomic force microscopy (AFM), scanning tunnelling microscopy (STM) or transmission electron microscopy (TEM) may be employed to interrogate individual species. Scanning electron microscopy (SEM), X-ray diffraction, optical absorption and Raman scattering may be used to obtain information about the bulk. Unfortunately, even a combination of these techniques does not fully characterize a given sample.[15,19]

Within the last few years a significant number of experimental studies have appeared in the scientific literature that utilised NMR towards the characterisation of carbon nanostructures. In some instances NMR was the main characterisation technique, whereas in others it was one of various methods employed. Part of the reason for the increased usage of NMR is the advent of a number of experimental advances that have enabled the measurement of well-resolved spectra, and identification of NMR signals from the nanotube sidewall and functional group separately. As will be outlined below, some of these experimental studies have relied upon the results of first-principles computations for the assignment. It is conceivable that given a sufficient amount of accurate theoretical data it may be possible to predict the shapes and widths of NMR signals for different sample compositions computationally.

Within this chapter we illustrate that computations of NMR spectroscopic parameters of carbon nanosystems has the potential to aid characterisation. A related account of progress made in this area has also been given in ref. 20, where the reader can find additional details. In Section 9.2 we first sketch the methodology used to calculate the magnetic response properties of both solids and molecules from first principles. Investigations of the chemical shifts of finite nanotubes terminated with hydrogen atoms and fullerene hemispheres are reviewed in Section 9.3. Studies on pristine and bundled infinite SWNTs (Section 9.4), and double-walled carbon nanotubes (Section 9.5) are described subsequently. Results from recent investigations that have considered the aromaticity of both finite and infinite pristine systems are discussed in Section 9.6. The chemical shifts of tubes functionalised with amine groups and fluorine atoms are discussed in Sections 9.7 and 9.8, respectively. In Section 9.9 we focus on how Stone–Wales defects affect the ^{13}C NMR chemical shifts. Finally, tensor properties of the nuclear magnetic shieldings in representative nanotube systems are described in Section 9.10.

9.2   Methodology

Conceptually, one may distinguish between theoretical methods developed for isolated molecules and those developed for extended periodic systems. Isolated carbon nanotubes and many other carbon nanostructures can be considered as periodic extended systems in one or two dimensions, while being of a "molecular" extension and not periodic in the remaining dimension(s). First-principles methods to compute NMR shielding tensors have been developed for infinite periodic structures and for finite molecular species. In this section, we give a brief overview of different theoretical approaches. In subsequent sections it will be illustrated how both of these have been applied to carbon nanostructures.

The spin Hamiltonian for the nuclear spin–external field interaction is

$$H = -\mathbf{m}_A[1 - \bar{\sigma}_A]\mathbf{B}^{\text{ext}}, \tag{9.1}$$

where the dimensionless *nuclear magnetic shielding tensor* $\bar{\sigma}_A$ for a nucleus A describes the shielding of the external fields by the electrons near the nucleus. In eqn (9.1), $\mathbf{m}_A$ is the spin magnetic moment of nucleus A in a molecule or solid, and $\mathbf{B}^{\text{ext}}$ the amplitude of the static external magnetic field induction. The isotropic shielding σ_A is the rotational average of the shielding tensor. In the absence of electron paramagnetism the elements of the shielding tensor are typically small, on the order of a few ppm (10^{-6}) for hydrogen to several thousand ppm for heavy metal nuclei. The absolute shielding tensor $\bar{\sigma}_A$, or the shielding constant cannot be measured directly. However, differences in shielding tensors between the probe nucleus of interest and a reference nucleus can be measured accurately. The *chemical shift* tensor of nucleus A with respect to the reference can be defined for small values of the isotropic shielding as

$$\bar{\delta}_A = \bar{\sigma}^{\text{ref}} - \bar{\sigma}_A, \tag{9.2}$$

with a corresponding equation for the isotropic shift δ_A in terms of a difference of the isotropic shielding constants for the reference and the probe nucleus. (For large shielding, the correct definition of the chemical shift is $\delta_A = (\sigma^{\text{ref}} - \sigma_A)/(1 - \sigma^{\text{ref}})$.) The isotropic chemical shift is one of the main observables in NMR. For solids or molecules that are aligned by fields or in liquid crystals, one can measure some of the tensor properties of the chemical shift tensor by NMR experiments.[21] Since the shielding tensors are usually small, the approximation $\delta_A = \sigma^{\text{ref}} - \sigma_A$ is often used. A chemical shift tensor can be defined accordingly. For an isolated molecule, shielding tensors are usually calculated from

$$\bar{\sigma}_A = \frac{\text{d}^2 E}{\text{d}\mathbf{m}_A \text{d}\mathbf{B}^{\text{ext}}}, \tag{9.3}$$

where E is the electronic energy of the molecule in the presence of the external field and the magnetic field created by the nuclear spin magnetic moment but does not include the nuclear Zeeman term $-\mathbf{m}_A \cdot \mathbf{B}^{\text{ext}}$. The derivative is usually taken analytically at $\mathbf{m}_A = 0$, $\mathbf{B}^{\text{ext}} = 0$, for example by employing variational or nonvariational second-order perturbation theory.[22,23] As can be seen, the shielding tensor is a property of the electronic system. Alternative to eqn (9.3), the nuclear magnetic shielding tensor is sometimes formulated in terms of the linear response of the electronic current density with respect to the external magnetic field, $\mathbf{j}^{(\mathbf{B}^{\text{ext}})}$, as[24]

$$\bar{\sigma}_A = \frac{\mu_0}{4\pi} \int \text{d}^3 r \cdot \mathbf{j}^{(\mathbf{B}^{\text{ext}})}(\mathbf{r}) \times \frac{\mathbf{r} - \mathbf{R}_A}{|\mathbf{r} - \mathbf{R}_A|^3}. \tag{9.4}$$

For closed-shell systems, eqns (9.3) and (9.4) are equivalent and can be recast in the same form.[24] For molecules, eqns (9.3) or (9.4) are usually applied directly, based on the approximations to describe the electronic structure in the calculation. In an infinite periodic system, the shielding tensor can be calculated via

$$\bar{\sigma}(r) = \sum_G \tilde{\bar{\sigma}}(\mathbf{G}) e^{i\mathbf{G} \times \mathbf{r}}, \tag{9.5}$$

from components $\tilde{\bar{\sigma}}(\mathbf{G})$ that are determined in reciprocal space $\mathbf{G}$. The $\tilde{\bar{\sigma}}(\mathbf{G})$ are the Fourier components of the real-space shielding tensor $\bar{\sigma}(\mathbf{r})$.[25] The contribution from $\mathbf{G} = 0$ requires some consideration. It depends on the macroscopic shape of the sample and represents a shielding contribution due to an induced current density on the surface of the sample.[25,26] It is therefore necessary to specify the shape of the sample in NMR computations for periodic extended structures.

Software to calculate shielding tensors for molecules is widely available in commercial as well as in noncommercial program packages. The range of correlated electronic structure methods available for molecular NMR calculations includes density functional theory (DFT), perturbative approaches (Møller–Plesset), configuration interaction (CI), and coupled-cluster (CC) wavefunction based approaches.[22,27] Distributed gauge-origin methods such as "gauge-including atomic orbitals" (GIAOs)[28,29] or related methods are available in conjunction with some classes of approximate wavefunctions, and in DFT, in order to ensure that the calculated shielding tensors do not exhibit an unphysical dependence on the origin of the coordinate system used for a computation. In applications to nanoscale systems, NMR methods designed for molecules are usually used in conjunction with finite cluster models of the system.

For *ab initio* calculations on extended periodic systems DFT is ubiquitous, and typically a plane-wave (PW) basis set is used. To keep the energy cutoff for such a basis set at a manageable level, pseudopotentials are commonly applied. All-electron DFT response methods for calculating NMR shielding tensors using a projector-augmented PW basis have been developed.[25,26,30] As with a finite atomic orbital basis, such a formalism requires us to address the gauge-origin problem in order to avoid an origin dependence of the calculated shielding tensors. GIAO-like phase factors can be attached to the operators used to construct the projectors, leading to "gauge-including projector-augmented plane-wave" (GIPAW) methods.[26] Implementations for ultrasoft pseudopotentials are available.[31,32] Recently, an all electron GIAO approach for NMR shielding tensor calculations with AO basis sets, instead of plane waves, in computations with periodic boundary conditions has also been implemented.[33]

Traditionally, both in molecular and extended periodic systems a set of coupled perturbed equations for the linear response of the orbitals is solved for the magnetic-field perturbation, which subsequently allows the efficient calculation of all chemical shifts in the system. A converse approach in which the order of the perturbations is interchanged has also been proposed and applied to solids as well as molecules.[34,35] For ^{13}C chemical shifts of carbon nanotubes and other light-atomic systems, relativistic effects are generally small and will not be considered in this chapter, but we note in passing that methodology for relativistic NMR calculations on heavy atomic systems is quite well developed, in particular for molecules.[36]

Metallic systems can afford significant Knight shifts, a manifestation of an interaction of electron with nuclear spins.[37] For most metallic nanotubes the density of states at the Fermi level, $n(\varepsilon_F)$, is small. Spin-unrestricted DFT calculations with periodic boundary conditions in one dimension have indicated that as a consequence of the small values of $n(\varepsilon_F)$ the Knight shift of all but the narrowest carbon nanotubes is probably only on the order of -1 to -2 ppm.[38] In ultranarrow nanotubes, a rehybridisation of the conduction band states causes a higher density of states at the Fermi level and significant Knight shifts (up to a few hundred ppm) were found instead. It was pointed out in ref. 38 that an increase of $n(\varepsilon_F)$ and rehybridisation of the carbon orbitals upon chemical functionalisation might also give rise to a Knight shift in chemically modified (functionalised) SWNTs.

9.3 ^{13}C NMR Chemical Shifts of Finite Single-Walled Carbon Nanotubes

Early experimental ^{13}C NMR studies of SWNTs have been plagued by the large magnetic inhomogeneity caused by residual catalyst from the growth process. Before purification and annealing, very broad lines with about 1500 ppm anisotropy have been observed.[39] High-resolution MAS NMR on these samples resulted in isotropic shifts of 126 ppm with sidebands ranging over 300 ppm and a linewidth at half-maximum height of about 50 ppm. Other early studies have reported chemical shifts of 124 ppm[40] and 116 ppm.[41] Advances in experimental techniques have led to the production of samples with a low metal content and few defects exhibiting resonances of 125 ppm,[42] 124 ppm,[43] and 121 ppm[44] with linewidths of only 9–10 ppm. Since the composition of these samples was not determined, it is unclear if the difference in the shifts is a result of a different diameter distribution, presence of defects, or ratio of semiconducting to metallic tubes. Remarkably, very recent studies have been able to acquire the NMR spectra of tubes with diameters of ~1.3 nm and a well-defined ratio of metallic and semiconducting species.[45] In single-component samples, the NMR signals assigned to metallic and semi-conducting nanotubes were centred at 123 and 122 ppm, respectively. Mixtures of the two showed that the shifts of the semiconducting entities changed little, whereas the metallic resonances were clearly affected by the presence of other tubes, shifting by up to 5 ppm.

Before describing the results of first-principles computations, it is appropriate to mention the findings of semiempirical model calculations on infinite length metallic and semiconducting SWNTs that were carried out nearly a decade ago.[46] In these computations any curvature-induced $sp^2 - sp^3$ rehybridisation and the Knight shift for metallic species were neglected. It was concluded that NMR may be able to discriminate between SWNTs with different electronic structures based on a predicted 11–12 ppm splitting between metallic and semiconducting tubes. However, the model predicted that further structural properties (diameter, defects) could not be resolved by NMR. Recently, both of these findings have been called into question. As illustrated below, first principles computations have shown that the chemical shifts of semiconducting species show a distinct diameter dependence. Moreover, the experimental work of ref. 45 described above is in stark contrast to the predicted 11 ppm splitting.

Due to the balance between accuracy and speed, most first principles calculations of the NMR chemical shifts of carbon nanosystems have employed

DFT. The first of these studies used finite models, and focused on (9,0) zigzag SWNTs in which the dangling bonds were saturated either with hydrogen atoms or with half of a C_{60} hemisphere.[47] Capping can lead to SWNTs with either D_{3h}/D_{3d} or D_3 symmetry, depending upon the relative orientation of the fullerene caps. Focus was placed on the systems with higher symmetry, since earlier work showed them to be more stable.[48] Computations were performed on finite capped SWNTs and the convergence of the properties with increasing tube length was monitored. Due to the computational cost involved, the hydrogen- and C_{30}-terminated systems contained up to 162 and 222 carbon atoms, respectively. In subsequent work,[49] computational speed-ups allowed for the study of an even larger 312-atom system, and for the symmetry nonequivalent atoms in the two longest tubes from ref. 47.

It was found that finite size and capping have an effect on the electronic structure, and NMR chemical shifts of these SWNTs. The hydrogen-terminated systems containing more than ninety carbon atoms were already found to be metallic due to a vanishing HOMO–LUMO gap (HOMO = highest occupied, LUMO = lowest unoccupied molecular orbital). The HOMO–LUMO gaps of the C_{30} capped tubes, on the other hand, seemed to converge to a finite value in agreement with band-structure calculations[5–8] and experiments[12] indicating that the (9,0) tube is in fact a small-gap semiconductor. This led to the conclusion that the tubes with carbon hemispheres were better models for the infinite systems, and below we will focus on their chemical shifts. For both sets of tubes the frontier orbitals had carbon π–character. However, whereas for the C_{30}-saturated SWNTs they were delocalised over the whole fragment, for the hydrogen-capped SWNTs they were localised at the ends of the tubes. Similar "edge states" have also been computed for one dimensional graphene sheets with zigzag edges,[50–52] zigzag triangular nanographenes[53] and finite zigzag SWNTs[54] where the dangling bonds have been saturated with hydrogen.

The ^{13}C NMR chemical shifts of the central symmetry nonequivalent carbons of the finite fullerene-hemisphere-capped tubes studied in refs. 47 and 49 are illustrated in Figure 9.2. The shifts are given using benzene as an internal computational reference. They are ~3 ppm lower than those provided in ref. 47, since in the original study an internal reference was not used, and this was the amount by which the computational methodology overestimated the experimental shift of benzene and C_{60}. Even though the average central chemical shift changes little when comparing data for the 132 and the 222 atom fragment, it decreases by 2 ppm when going from the 222- to the 312-atom nanotube. It appears that the computed shifts are not converged with respect to tube length and capping plays a significant role even for comparatively long fragments. Currently, it is not clear if the shifts of the central carbons in a finite tube converge smoothly to those of the infinite species for any finite tube of reasonable length. For example, for the 312-atom finite species the average central shift of 124.5 ppm is still about 4 ppm higher than for an isolated (9,0) SWNT of infinite length (see Section 9.4). Nonetheless, the shift of the central carbons is in good agreement with previous experimental estimates of 126 and

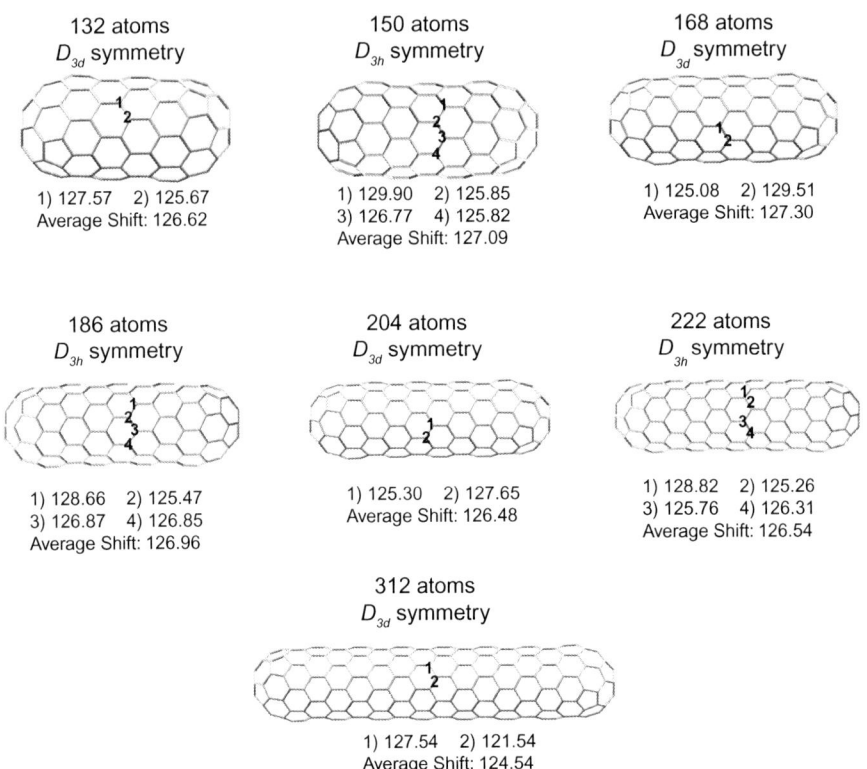

1) 127.57   2) 125.67
Average Shift: 126.62

1) 129.90   2) 125.85
3) 126.77   4) 125.82
Average Shift: 127.09

1) 125.08   2) 129.51
Average Shift: 127.30

1) 128.66   2) 125.47
3) 126.87   4) 126.85
Average Shift: 126.96

1) 125.30   2) 127.65
Average Shift: 126.48

1) 128.82   2) 125.26
3) 125.76   4) 126.31
Average Shift: 126.54

1) 127.54   2) 121.54
Average Shift: 124.54

Figure 9.2 ^{13}C NMR chemical shifts for the central symmetry inequivalent carbons in C_{30}-capped D_{3d}/D_{3h} (9,0) SWNT fragments calculated with the revPBE functional and a TZP basis.[49] The shifts are given with respect to TMS (tetramethylsilane), using benzene as an internal computational reference.

124 ppm obtained for heterogeneous samples.[40,43] The shifts in the tips of the caps were found to be about 10 ppm larger than those of C_{60}.

Besley *et al.* have computed the isotropic ^{13}C NMR chemical shifts of finite (6,0), (9,0), (10,0), (4,4), (5,5) and (6,6) SWNTs terminated with fullerene hemispheres.[55] Benchmark calculations on a number of "fullerene–like" molecules using Hartree–Fock theory and different basis sets were performed. The minimal basis STO-3G was chosen for the nanotube calculations because of the low computational cost. It was argued that the calculations benefit from a fortuitous but consistent cancelation of errors. However, due to the highly approximate character of these uncorrelated minimal-basis calculations the predictive power of the study is limited.

These initial studies on finite systems provided the first computational predictions of the chemical shifts of semiconducting SWNTs with respect to a standard experimental reference. The shifts of metallic species were estimated based upon the 11–12 ppm predicted difference by early model calculations, but not independently confirmed.[46] It was shown that the electronic structure and magnetic properties are dependent upon finite size, and capping, which was found to be important even for very long finite tubes. Unfortunately, the studies were limited to computations on one, or only a few species, due to the fact that testing for the convergence of the chemical shifts with respect to the tube length is computationally rather demanding. Thus, the next important step was to undertake first-principles computations of NMR chemical shifts using codes designed for infinite, periodic systems, as described below. However, it should be noted that a recent synthesis and characterisation of C_{90} has shown that the most abundant isomer is a short armchair nanotube capped with a fullerene hemisphere.[56] Perhaps the calculations for finite systems might become useful if other ultrashort capped SWNTs can be synthesised in sufficient quantity and purity.

9.4 ^{13}C NMR Chemical Shifts of Infinite Single-Walled Carbon Nanotubes

GIPAW DFT calculations of the ^{13}C NMR chemical shifts of infinite, pristine SWNTs have been undertaken by several research groups.[49,57–59] Most of these investigations focused on non–metallic zigzag species because they have small unit cells. In refs. 58 and 59 selected helical systems were also considered. Armchair tubes could not be examined, since they are metallic and only recently has a methodology been put forward that may be employed to calculate orbital and Knight shift contributions to the shieldings of metallic species.[60] The studies on semiconducting SWNTs varied in a number of respects such as the choice of the pseudopotential, the density functional, the intertube distance and shape of the lattice, as well as technical aspects of the implementation. Even though the chemical shifts obtained in each study are somewhat different, the observed trends and conclusions reached generally agree with each other.

It has been shown that using a chemically similar compound such as benzene or C_{60} as the internal reference leads to systematic error cancellation, yielding nanotube shifts that are closer to those measured experimentally.[49] This may be accomplished by reporting the shifts with respect to tetramethylsilane (TMS), $\delta^{TMS}_{C_6H_6}$, as

$$\delta^{TMS}_{C_6H_6}(\text{tube}) = \{\sigma(C_6H_6) - \sigma(\text{tube})\} + \delta(C_6H_6), \qquad (9.6)$$

where $\delta(C_6H_6)$ is the experimental chemical shift of benzene with respect to TMS, and σ is the computed shielding constant. The shifts given below were computed in this way.

Zigzag SWNTs can be subdivided into three families which are characterised by $\lambda = \text{mod}(n, 3) = 0, 1,$ or 2. In ref. 49 it was assumed that the small-bandgap (9,0), (12,0) and (15,0) species from the $\lambda = 0$ family do not afford a Knight shift, whereas in ref. 58 the Knight shift for the $(n,0)$; $n = 9, 12, 15, 18, 21$ tubes was taken to be -1.5 ppm based on previous estimates.[38] The chemical shifts $\delta \equiv \delta^{TMS}_{C_6H_6}$ of isolated $(n, 0)$ tubes can be fitted by the function

$$\delta/\text{ppm} = A(\lambda)/D^\alpha + B, \tag{9.7}$$

where D is the tube diameter, B is the chemical shift limit for infinite diameter and $A(\lambda)$ is a constant that depends on the nanotube family. In refs. 57 and 49, α was taken to be unity, whereas in ref. 58 α was used as an additional fitting parameter.

Figure 9.3 illustrates the calculated shifts and fits to eqn (9.7). Marques et al.[57] did not consider the small bandgap $\lambda = 0$ families. For the others they found that $(\alpha = 1)$:

$$A(1) = 107.0 \text{ Å} \; ; \; B(1) = 116.0 \tag{9.8a}$$

$$A(2) = 98.3 \text{ Å} \; ; \; B(2) = 116.0. \tag{9.8b}$$

Zurek et al.[49] considered all three families and obtained the following parameters $(\alpha = 1)$:

$$A(0) = 109(4) \text{ Å} \; ; \; B(0) = 105.4(5) \tag{9.9a}$$

$$A(1) = 169(8) \text{ Å} \; ; \; B(1) = 106(1) \tag{9.9b}$$

$$A(2) = 151(5) \text{ Å} \; ; \; B(2) = 106.5(6). \tag{9.9c}$$

Even though Lai et al.[58] computed shifts for all three families, a fit was only reported for the $\lambda = 1$, 2 sets since the $\lambda = 0$ species are known to become metallic with increasing diameter. The fits were:

$$A(1) = 306 \text{ Å}^{\alpha(1)} \; ; \; B(1) = 116.2 \quad ; \; \alpha(1) = 1.5 \tag{9.10a}$$

$$A(2) = 548 \text{ Å}^{\alpha(2)} \; ; \; B(2) = 117.7 \quad ; \; \alpha(2) = 1.9. \tag{9.10b}$$

These results confirmed the expectation that significant differences in the chemical shifts will only be seen for small- to medium-diameter systems, because the vanishing curvature of the larger-diameter tubes must ultimately lead to a convergence of their properties to those of a single graphene sheet. The chemical-shift range between small- and large-diameter tubes predicted in ref. 57 is smaller than that found in the other two investigations. The fitted B values obtained in refs. 57 and 58 of 116–117 ppm were in reasonable agreement with estimates of 128 ppm and 121.5 ppm for a perfect graphene layer. The B values in ref. 49 (106 ppm) were a little lower but we have found that they would increase to 114 and 112 ppm for $\lambda = 1$, 2 families, respectively, if α were fitted too.

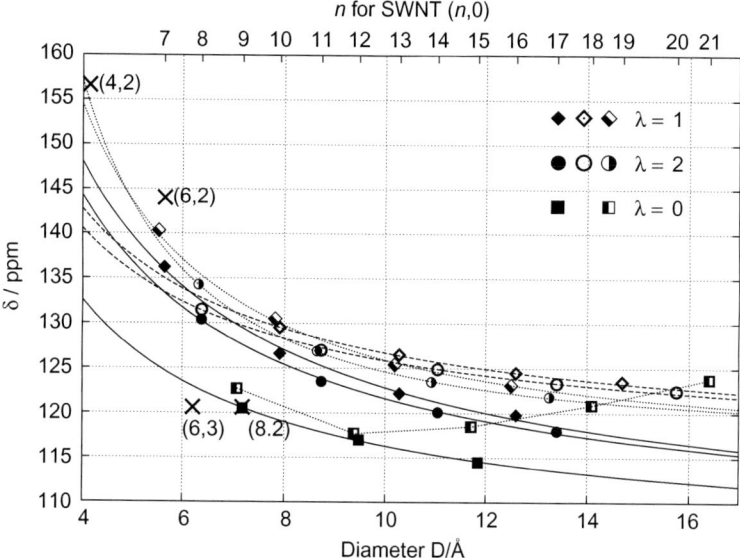

Figure 9.3 Calculated chemical shifts δ of isolated SWNTs as a function of the tube diameter. The data for the filled, open, and half-filled symbols was taken from refs. 49, 57, and 58, respectively. The crosses correspond to data for helical tubes from ref. 58. The lines are a fit to the calculated data according to eqn (9.7), using the parameters given in eqns (9.8), (9.9) and (9.10). Note that for the $\lambda = 0$ family a fit was not given in ref. 58, and the line is drawn to guide the eye.

The origins of the discrepancies remain unclear since the calculations varied in a number of respects.

Nonetheless, in all cases it was found that the ^{13}C NMR chemical shifts of pristine, isolated, infinite, zigzag SWNTs display a pronounced family behaviour, and that they decrease with increasing tube diameter. This conclusion is in stark contrast to the results of semiempirical model calculations.[46] Various first-principles calculations predicted a chemical shift range of between ~ 15–40 ppm between the smallest SWNTs studied and those of infinite diameter. For larger tubes, the difference between $(n, 0)$ and $(n + 1, 0)$ systems was predicted to be small, however, due to a decrease of the $sp^2 - sp^3$ rehybridisation. It was suggested that ^{13}C NMR may be used to identify small-, medium- and large-diameter SWNTs in a heterogeneous sample. Interestingly, recent experiments have shown that metallic and semiconducting SWNTs with diameters of ~ 1.3 nm have chemical shifts of about 123 and 122 ppm, respectively.[45] Examination of Figure 9.3 shows this to be inline with the shifts predicted by first-principles computations.

The two studies that also looked at the small-gap $\lambda = 0$ family showed that the (9,0), (12,0) and (15,0) tubes yielded shifts that were significantly lower than members of the other two families (with similar diameters).[49,58] For the larger (18,0) and (21,0) systems on the other hand, ref. 58 reported shifts that were equal to and larger than (respectively) similarly wide members of the other two families. These tubes are known to be metallic, and it is therefore not clear how meaningful the results are. Comparison of the shifts of the helical (4,2), (6,2), (8,2) and (6,3) SWNTs with the zigzag systems indicated that the chemical shift is virtually independent of the helicity of the tube, and mainly depends on its diameter and family.[58] Moreover, it was found that the average isotropic shifts of bundled tubes are not very different from those of the isolated species.[49,57]

Sebastiani and Kudin have also calculated the NMR chemical shifts of helical and nonhelical systems with diameters ranging between 5 and 14 Å.[59] Chemical shifts fell in the range of 110 to 130 ppm. However, since the geometries were not optimised, and bundled tubes with small intertube separations were considered, these results cannot be compared directly to those discussed above. For example, the chemical shifts of the large-diameter zigzag SWNTs reported in ref. 59 were noticeably lower than those shown in Figure 9.3.

We also note in passing that whereas previous work indicated that the magnetic susceptibility, χ, was independent of the SWNT family,[61,62] the first-principles calculations of ref. 57 could clearly distinguish between the $\lambda = 1$, and $\lambda = 2$ families. Moreover, a combined experimental and theoretical investigation of the helicity dependence of the magnetic susceptibility anisotropy, $\Delta\chi$, illustrated that it shows asymmetry between helical nanotubes of opposite chiral indices $v = \pm 1$.[63]

9.5 ^{13}C NMR Chemical Shifts of Infinite Double-Walled Carbon Nanotubes

Currently, only a few authors have computed the NMR parameters of double-walled carbon nanotubes (DWNTs).[57,59] The work of Marques *et al.*[57] was inspired by the experiments of ref. 64 that measured an isotropic shift of 111 ppm with a residual linewidth of ~35 ppm for the inner tube in a DWNT. This shift is significantly smaller than those obtained for typical SWNT samples. It was suggested that this may be explained either by the stronger curvature of the inner tube, or diamagnetic shielding caused by the outer one.

In excellent agreement with these experiments, calculations on an infinite, isolated (8,0)@(16,0) DWNT yielded isotropic shifts of 111, and 126 ppm for the inner and outer tubes, respectively.[57] A shift of 132 ppm was obtained for the pristine (8,0) species, implying that the presence of the outer tube leads to a diamagnetic shift of 21 ppm. The outer tube, on the other hand, was found to be only slightly perturbed (2 ppm) by the inner one. It was shown that simple classical models are able to reproduce these effects. DWNTs were also considered in ref. 59, as described in the next section.

9.6 Aromaticity, and NICS of SWNTs and DWNTs

Another aspect of carbon nanotubes that has fascinated chemists and physicists is their aromaticity. It is well known that aromatic and antiaromatic ring currents can lead to remarkable NMR properties.[65] A robust theoretical measure to quantify the aromatic or antiaromatic character of a molecule or an extended system computationally is by evaluating the shielding constant in the centre of a cyclic system.[66] This is known as the nucleus-independent chemical shift, NICS.

Ormsby and King[67] applied the Clar aromatic sextet valence-bond (VB) model[68] to carbon nanotubes. For a short, finite, hydrogen-capped (12,12) SWNT, they found calculated DFT NICS values to be consistent with the qualitative predictions of the Clar VB model. Recently, Clar's theories have been applied successfully to other conceptually related systems, such as graphene nanoribbons terminated with hydrogen groups.[69,70]

In order to study the aromatic stabilisation of different finite SWNTs, Zhao and Balbuena carried out DFT NICS computations on finite tubes capped with hydrogen, and also those containing hydrogen on one end, and fullerene hemispheres on the other end.[71,72] A correlation was found between the aromatic stabilisation indicated by the NICS values and the ring-formation energy in the first ring segment at the open end of a SWNT.[71] This was thought to provide evidence for a proposed SWNT growth mechanism. In a follow-up study, it was concluded that the detailed structure of the cap does not have a significant effect on the NICS patterns at the open end of the tube.[72]

Maps of the NICS and magnetic-field-induced current densities in a number of infinite SWNTs and DWNTs were recently generated.[59] In order to predict possible effects on the NMR parameters of molecules trapped inside a tube, the current densities in the tube sidewall, and the resulting shielding in the tube centres was investigated. The NICS values were found to be between -60 and -30 ppm in the centre of the tubes. For a (10,0)@(19,0) DWNT the NICS was found to be approximately equal to the sum of the values for the individual SWNTs.

Besley and Noble have carried out NMR computations on molecules trapped inside finite H-capped SWNT fragments.[73] Ring currents in the nanotube were found to induce a change in the chemical shifts of the nuclei of the encapsulated molecules (He, and the C, O, and H nuclei of small molecules were considered). The change in the chemical shift was found to depend upon the nanotube, but was relatively independent of the molecule itself.

9.7 ^{13}C and ^{1}H NMR of Single-Walled Nanotubes Functionalised by Amine Groups

The NMR chemical shifts of SWNTs covalently functionalised by organic groups have been measured.[42–44,74] Evidence for significant interaction between shortened SWNTs and phenol groups was found using 2D ^{1}H–^{13}C heteronuclear correlation spectroscopy.[42] The shift of the SWNT resonances to slightly lower values of $\sim$118 ppm was attributed to a change in hybridisation, from sp^2 to sp^3, upon functionalisation. In another study, the solution phase ^{13}C NMR spectrum of SWNTs functionalised with diamine-terminated oligomeric poly(ethylene-glycol) (PEG$_{1500\,\mathrm{N}}$) has been recorded.[74] The signal from the carbons in the tube was deconvoluted into a large and small peak centred at around 128 ppm and 144 ppm, respectively. A solid-state MAS spectrum yielded a double peak at 128 and 136 ppm, instead. Based upon the theoretical estimates available at the time[46,47] the former was tentatively assigned to semiconducting, and the latter to metallic SWNTs. On the other hand, recent results have shown that resonances of metallic and semi-conducting pristine tubes vary by no more than a few ppm.[45] Currently, it is not clear if the discrepancy between the two sets of experimental studies is a result of the chemical environment, the sample purity, composition, or pre-paration. Even though it is also possible to measure the ^{1}H NMR chemical shifts of covalently functionalised SWNTs, such experiments provide limited information since the signals are typically broad and weak, especially for protons in close proximity to the tube sidewall, resulting in uncertainties of several ppm.[75]

GIPAW DFT calculations on zigzag, infinite SWNTs functionalised with various N-R groups have been carried out.[76,77] Products typical of $[2+1]$ cycloaddition and 1,3-dipolar cycloaddition reactions of an amine group to a SWNT sidewall were considered, so that the results could be compared with those from solution NMR studies where the tubes were derivatised with $PEG_{1500 N}$.[74] Initial calculations on NH, NCH_3, NCH_2OH and CH_2NHCH_2 demonstrated that the shifts of the SWNT carbons are essentially independent of R.[76] Therefore, a detailed study of how the diameter of the tube, and the orientation of the functionalised bond influence the ^{13}C and ^{1}H NMR chemical shifts was carried out using NH – the simplest model.[77]

There are two distinct C–C bonds in a zigzag SWNT, one parallel (P-site) and one diagonal (D-site) to the tube axis. As shown in Figures 9.4(a)–(d) the distance between the functionalised carbons C_1 and C_6 (d_{16}) was found to depend strongly on the orientation of the bond that was functionlised, and for the D-site, the diameter of the tube. For the P-site, d_{16} was about the same (~ 1.5 Å) for all of the tubes considered, and the functionalised carbons were formally sp^3 hybridised. For the D-site, on the other hand, the C_1–C_6 bond broke ($d_{16} > 2.1$ Å) for narrower tubes with $n \leq 14$, and the functionalised carbons remained formally sp^2 hybridised. The difference between P- and D-site functionalisation was explained by noting that the strain in the diagonal C–C bonds is dependent upon the radius of the tube. At a critical diameter, the curvature-induced strain is small enough so that reaction with NH does not break, but only elongates, the C_1–C_6 bond, as illustrated in Figure 9.4(c) for the widest system studied.

Since ^{13}C NMR chemical shifts depend strongly on the hybridisation of the carbon atoms, it is not surprising that the calculated resonances of the func-tionalised carbons depended upon the site and, for the diagonal bond, the radius of the tube that was functionalised. As shown in Figure 9.5(a) the shifts of the carbons at the P-site were rather constant at ~ 44 ppm, inline with the fact that d_{16} was about the same for all of the systems studied. The shifts of the D-site functionalised carbons, on the other hand, had a range of approximately 80 ppm. For the $\lambda = 1,2$ families they initially decreased roughly linearly with

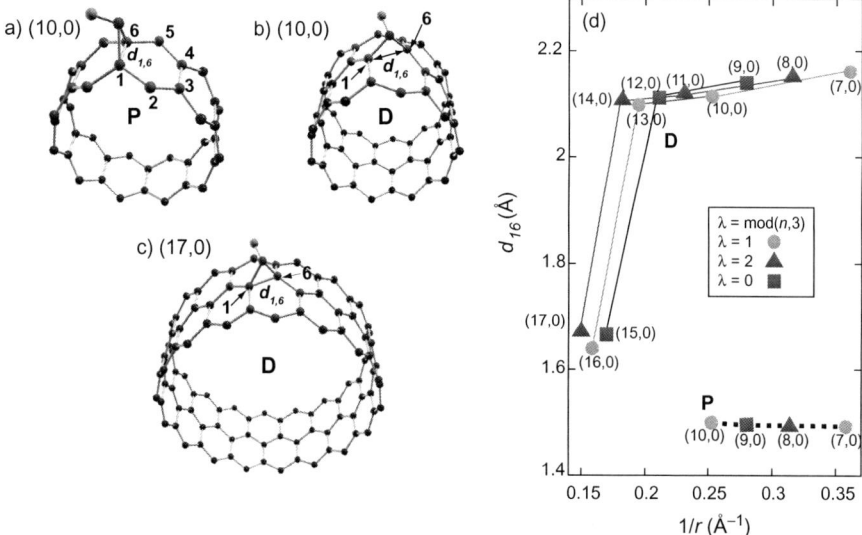

Figure 9.4 Optimised geometries of NH-functionalised zigzag SWNTs. The supercell consisted of two unit cells of the pristine SWNT and a single NH group. The functional groups were added to the C–C bonds (a) parallel (P-site) and (b,c) diagonal (D-site) to the tube axis. d_{16} denotes the distance between C_1 and C_6 of a [10]annulene framework. (d) The sidewall equilibrium bond distance (d_{16}), as a function of the inverse of the tube radius.[77] The lines are drawn to guide the eye.

decreasing $1/r$ up to $n = 14$. For wider systems, the C_1–C_6 bond was only somewhat elongated and the shifts approached those of the P-site. In the limit of infinite diameter, the resonances of the derivatised carbons of both sites were estimated to be around 45 ppm. Whereas the shifts of the functionalised carbons in the $\lambda = 0$ small-bandgap (15,0) tube were only slightly larger than those of the $n = 16,17$ systems, for the (9,0) tube they were significantly higher than any of the other species considered.

These results led to the suggestion that the shifts of the derivatised carbons could be used to determine the orientation of the bond that has been functionalised, and perhaps even be able to yield information about the diameter distribution of a sample of SWNTs. For example, the functionalised carbons ($\delta = 118$ ppm) whose resonances were seen by ^1H–^{13}C heteronuclear correlation spectroscopy in ref. 42 were predicted to be formally sp^2 hybridised, and their shift was thought to be indicative of D-site functionalisation of a (7,0) or (8,0) SWNT. Functionalisation significantly broadened the chemical shift range of the bulk carbons, but their average shifts did not differ much from those of the pristine systems. Thus, it was concluded that if low to medium degrees of PEG_{1500N} functionalisation were employed in ref. 74 the average shift ought to be similar to the one that would be obtained for a pristine sample.

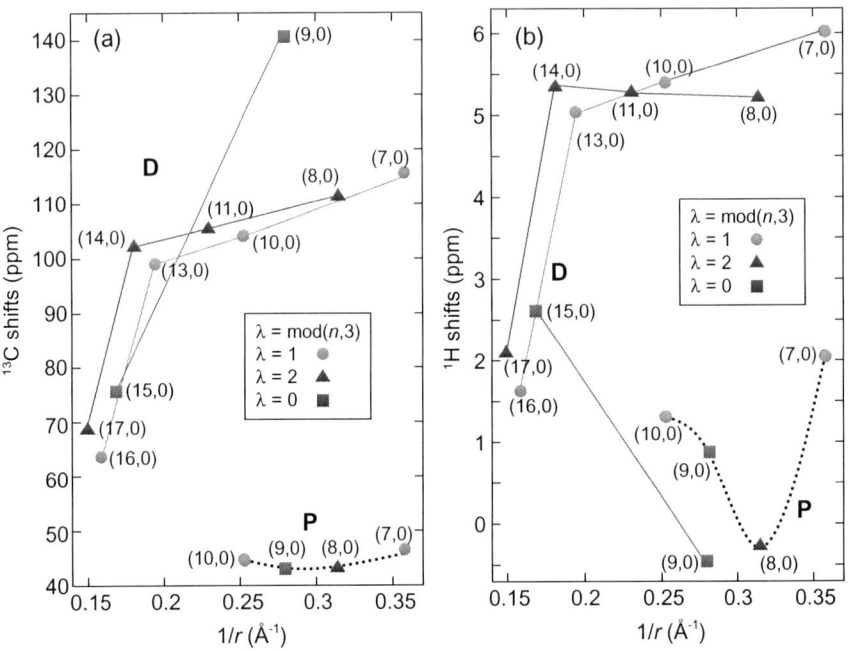

Figure 9.5   (a) The average ^{13}C NMR chemical shifts of the functionalised carbons, and (b) the ^{1}H NMR chemical shifts in zigzag ($n = 7$–17; $n \neq 12$) SWNTs functionalised with NH at a bond diagonal (D) and parallel (P) to the tube axis, as a function of the inverse of the tube radius.[77] The lines are drawn to guide the eye.

The trends of the chemical shifts of the amine functionalised carbons can be understood by considering a 1,6-X-[10]annulene, which represents a fragment of a functionalised SWNT[78,79] (as shown in Figure 9.4(a) the atoms labeled 1–6 comprise half of the annulene framework). Depending upon X, the annulene may exist as: (**1**) a localised cyclic polyene with alternating single and double bonds, (**2**) a delocalised aromatic annulene or (**3**) a bis-norcaradiene form.[80,81] For the first two cases, the C_1–C_6 bond is broken, whereas the third contains a strained three-membered ring. Computations were performed on 1,6-NH-[10] annulenes whose d_{16} was constrained to the optimised values of some of the D-site and P-site tubes studied. The chemical shifts compared well with those of the SWNTs, confirming that the shifts of the functionalised carbons in the SWNTs are primarily dependent upon d_{16} and the local environment of the functionalised carbons.

Despite the fact that it is difficult to measure the proton shifts in functionalised systems, it was noted that they may provide useful information about the aromaticity of the local environment. For the 1,6-NH-[10]annulene, calculations of the NICS have indicated that forms **3** and **2** are aromatic.[80] For functionalised SWNTs, there are two competing factors that might influence the aromaticity of the [10]annulene carbon framework. On the one hand, increasing tube diameter leads to smaller deviations from planarity. On the other hand, for the D-site, this also results in a shorter d_{16} for which the molecular computations yielded a less aromatic compound. The computed ^{1}H shifts are illustrated in Figure 9.5(b). They indicate that for the (8,0) and (9,0) tubes functionalised at the P- and D-sites, the local [10]annulene carbon framework may be aromatic.

9.8 ^{13}C NMR of Fluorinated Single-Walled Nanotubes

SWNTs may be functionalised with fluorine up to a saturation stoichiometry of C_2F without destroying the tubes.[82] The resulting fluoronanotubes (F-SWNTs) are versatile precursors for the preparation of covalently functionalised SWNTs *via* further derivatisation.[83,84] Barron and coworkers have employed ^{13}C NMR to study the fluorine–nanotube interaction.[85,86] The ^{13}C NMR shift for the functionalised sp^3 carbons in the nanotube sidewall was found to be at $\delta = 83.5$ ppm. Calculations on a (5,5) finite $C_{80}H_{20}F_2$ fragment led to the conclusion that this shift is indicative of 1,2-addition, rather than 1,4-addition of F_2. Importantly, it was found that for high degrees of fluorination NMR provides a better quantification of the degree of functionalisation than Raman scattering.[85]

In order to investigate if the results of ref. 85 can be extended to infinite F-SWNTs, GIPAW DFT calculations of the ^{13}C NMR chemical shifts of isolated zigzag (8,0) F-SWNTs were performed.[87] A number of fluorinated isomers, with varying degrees of fluorination were considered. Some of these are illustrated in Figure 9.6, along with the calculated ^{13}C NMR chemical-shift histograms. The influence of the nanotube curvature was studied by considering a few larger fluorinated (9,0) and (10,0) species.

In general, the shifts of the fluorinated carbons, C_α, ranged from 81.8– 83.9 ppm, in excellent agreement with the experimental results of ref. 85. In contrast to the findings on amine functionalised SWNTs presented in Section 9.7, the shifts of the fluorinated carbons were relatively insensitive to the pattern and degree of functionalisation because these factors did not affect the local environment around the C_α–F units significantly. In all cases, the C–C bonds were found to elongate by an amount less than 0.2 Å upon derivitisation, and the fluorinated carbons were formally sp^3 hybridised. Interestingly, the shifts of the C_α in a T_h-symmetric $C_{60}F_{24}$ functionalised fullerene were experimentally determined to be at 83.5 ppm,[88] almost identical to the SWNTs, suggesting that the shifts of the C_α do not depend upon the curvature of the carbon framework. Calculations on (9,0) and (10,0) F-SWNTs supported this conjecture.

For low degrees of fluorination the resonances of the unfunctionalised carbons were found to broaden significantly as indicated by the histogram in Figure 9.6(a). This was mainly a result of the geometrical distortion the tube undergoes. A few carbon atoms, denoted by the dots, gave rise to particularly large shifts at around 138 ppm. The average shifts of the other carbons were not

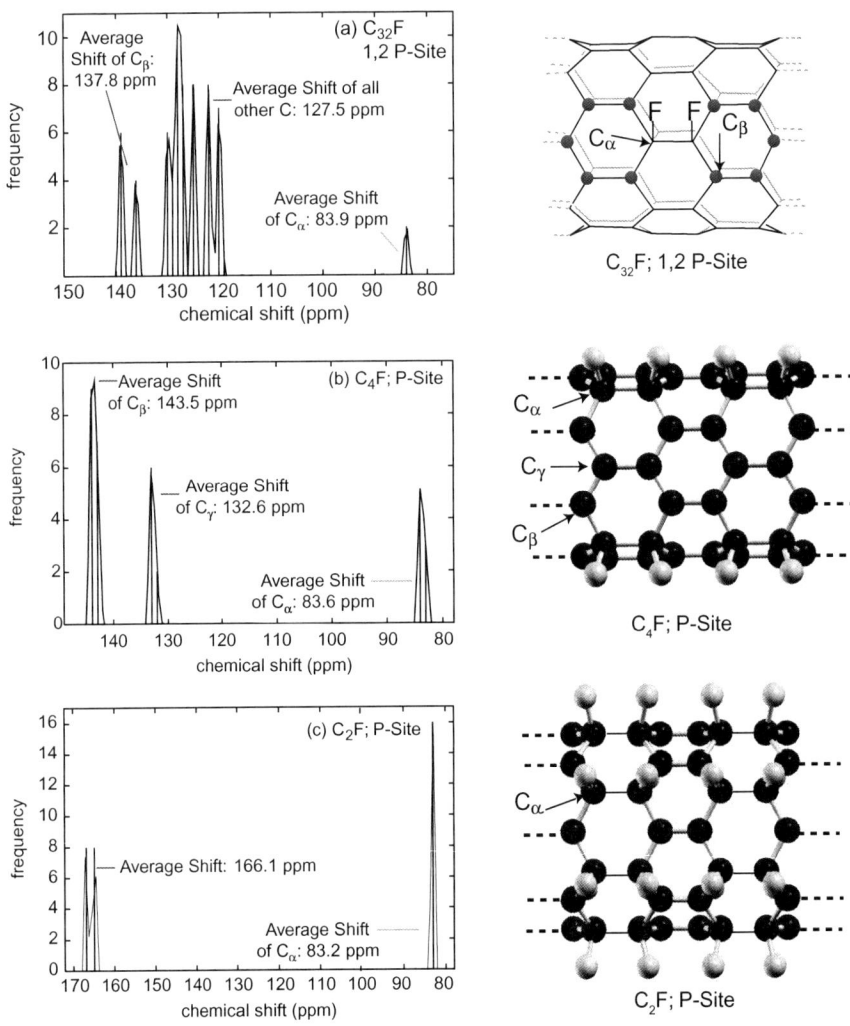

Figure 9.6 Calculated ^{13}C NMR chemical shifts histograms of some of the (8,0) F-SWNTs studied in ref. 87. (a) One of the four F-SWNTs with a high C:F ratio. The dots indicate C_β that had distinctly high chemical shifts. (b, c) The F-SWNTs with low C:F ratios. In all of the figures, the fluorines were added to the pair of carbons in a C–C bond parallel (P-site) to the tube axis, and the dashed lines indicate a periodic system. For comparison, the average shift of the carbons in a pristine (8,0) SWNT is 131.7 ppm.

too different from those of a pristine (8,0) SWNT. The systems with C_4F and C_2F stoichiometry contain 3 and 2 symmetry inequivalent carbons, whose shifts are pointed out on the histograms in Figures 9.6(b) and (c). For high degrees of functionalisation it was proposed that the shifts of carbons neighbouring C_α might be very useful experimental indicators of the functionalisation pattern.

9.9 ^{13}C NMR Chemical Shifts of Single-Walled Nanotubes with Stone–Wales Defects

Common defects in SWNTs include single or double vacancies, as well as pentagon–heptagon pairs.[89,90] Since these imperfections may affect the electronic structure,[91–93] chemical reactivity,[91–96] mechanical[97] and transport[98,99] properties of the tubes, it is important to be able to identify and quantify them in a given sample.

One type of rather low-energy defect, a Stone–Wales (SW) or 5/7/7/5,[100] may be formed by a 90° rotation of a C–C bond either parallel (P-SW) or diagonal (D-SW) to the axis of a zigzag tube, see Figure 9.7. In order to determine if NMR may potentially yield detailed information about the local electronic

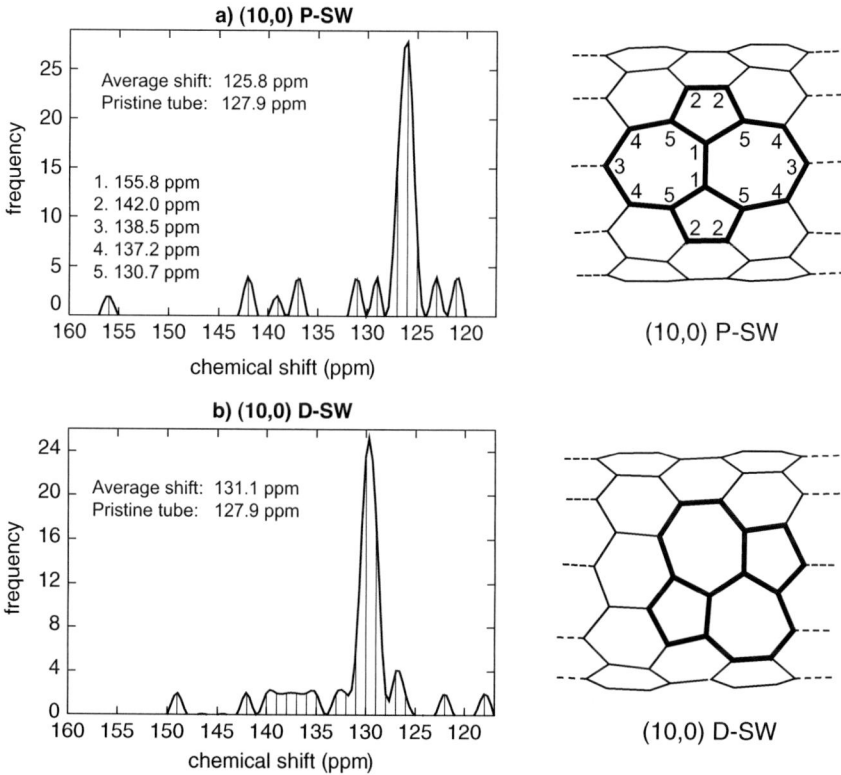

Figure 9.7   Stone–Wales, or 5/7/7/5, defects in zigzag SWNTs studied in ref. 101. A P-SW and D-SW defect can be obtained by rotating a C–C bond pair parallel or diagonal to the SWNT axis, respectively. Calculated histograms of the ^{13}C NMR chemical shifts of a (10,0) SWNT with (a) one P-SW defect, and (b) one D-SW defect per double unit cell. In (a) the shifts of the carbons within the defect site are number coded and the "average shift" does not include the shifts of the defect atoms. In (b) the average shift includes all atoms.

structure around these sites, GIPAW DFT computations were carried out on infinite $(n, 0)$; $n = 7$–10 tubes with both types of SW orientations.[101]

For all of the tubes studied, the defect gave rise to a substantial broadening of the ^{13}C NMR spectrum, as illustrated for the (10,0) system in Figure 9.7. Moreover, a signal well separated from the bulk arising from carbons within the defect site was found at around 150–160 ppm. For a P-SW defect the carbons with high shifts belonged to two heptagons and one pentagon, whereas for the D-SW they belonged to one pentagon and two hexagons. The different behaviour for the two defect orientations considered was attributed to the curvature-induced strain on the bonding pattern. Since the signals occurring at 150–160 ppm appeared to be independent of the defect concentration, it was suggested that they may be used to identify and perhaps even quantify the presence of SW defects. For the P-SW systems, the shifts of most of the carbons in the defect were well separated from the rest, and the average shift of the bulk carbons did not vary much from that of the pristine species. For tubes with D-SW defects, the separation between shifts belonging to the bulk and those of the defect was not as evident.

9.10   Shielding Tensors

The chemical shifts of nanotube carbons are obtained from the isotropic shieldings, which are an average of the shielding tensors' principal components. In the case of nanotubes the shielding anisotropy could be an important mechanism leading to line broadening, for instance in solution spectra of functionalised systems. Thus, the shielding tensors of a variety of molecules, finite and infinite SWNTs, as well as functionalised ones, and those with defects have been examined.[47,87,101] Graphical representations of the shielding tensors in the form of polar plots allow for the straightforward visualisation of the sign and relative magnitude of the principal components as well as the shielding-tensor orientation.

The tensors of a few representative systems are illustrated in Figure 9.8. The visualisations are based on the procedure described in ref. 20. An isotropic tensor would be represented by a sphere. The tensors in carbon nanotubes are typically strongly anisotropic. In a pristine SWNT, the three shielding tensor components can be labeled as axial (a, along the SWNT axis), radial (r, radially outward from the tube centre perpendicular to the sidewall), and orthoradial (o). The plot of the shielding tensor for the pristine (8,0) system shown in Figure 9.8 is representative for the finite-gap SWNTs investigated so far: there is a strongly dominating positive ("diamagnetic") radial component, along with rather negative axial and orthoradial components of comparable and small magnitude.[101] The average of the three components is therefore rather small compared to the magnitude of the radial component itself, leading to an overall deshielded carbon nucleus when comparing the isotropic shielding to that of the standard reference TMS. The shielding tensors for finite H-capped and C_{30}-capped SWNTs are very similar in the midsection of the tube and

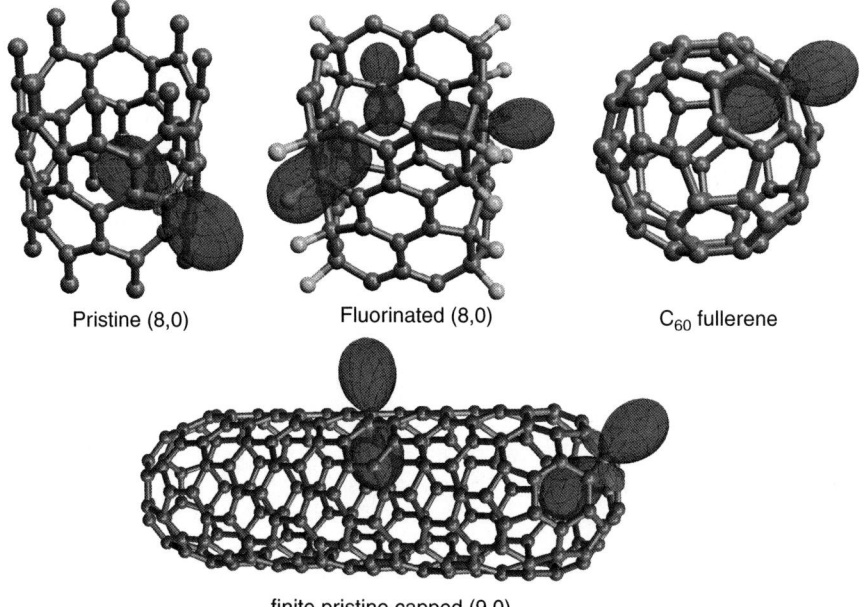

| Pristine (8,0) | Fluorinated (8,0) | C_{60} fullerene |

finite pristine capped (9,0)

Figure 9.8   Polar plots of selected computed nuclear magnetic shielding tensors in a pristine, and a fluorinated (8,0) SWNT.[87,101] For comparison, shielding tensors in C_{60} and a finite SWNT are also provided.[47]

compare well with the tensors computed for infinite pristine SWNTs. These tensors, as well as the shielding tensors at the tip of the caps in the carbon-capped system are also qualitatively similar to C_{60} in the sense that the large positive component is always pointing radially outward perpendicular to the carbon framework's surface.

Very different shielding tensors are obtained for carbon atoms in SWNTs in defect sites,[101] or for sidewall carbons that are covalently functionalised.[87] Shielding tensors in the two Stone–Wales defect sites indicated that for the carbons in the P–SW defect connecting the two heptagons the particularly high chemical shifts are caused by the occurrence of a large deshielding axial tensor component. As an example of a covalently functionalised SWNT, Figure 9.8 also displays shielding tensors for the three different carbon sites in a fluorinated (8,0) system with a C:F ratio of 4:1. The shielding tensor for C_γ (see Figure 9.6(b) for an illustration) is quite similar to that in the pristine SWNT both in shape/orientation as well as in magnitude. The tensor for C_β has a similar shape, too, but with more pronounced negative axial and orthoradial components and a reduced radial component, leading to a higher chemical shift. The shielding tensor for the fluorinated carbons (C_α) is very different, and as we have seen earlier these carbons consistently afford low isotropic chemical shifts. The tensor is much less anisotropic, with the three components of same sign reinforcing each other to produce a significantly more shielded C_α nucleus.

The largest component is still in the radial direction, however, which is similar to the shielding tensors of the other carbons in this fluorinated SWNT as well as the pristine nanotube.

9.11   Summary and Outlook

First-principles computations have shown that potentially a plethora of structural information may be extracted from the magnetic response of nanotubes.[20] These predictions supercede earlier model calculations that predicted a uniform chemical-shift magnitude for all semiconducting and metallic SWNTs, respectively. Moreover, in the last few years experimental advances in sample preparation, and separation have made it possible to obtain useful spectra, and correlate them with a particular type of tube, or modification of the nanotube sidewall. A few of these have employed the results of first-principles computations to aid their assignments.

The calculated ^{13}C NMR chemical shifts of pristine periodic zigzag SWNTs indicate that for small- to medium-diameter species the $sp^2 - sp^3$ rehybridisation is not negligible and NMR may perhaps be used to identify the diameter distribution within a homogeneous sample.[49,57,58] Finite size, and capping were found to substantially affect the electronic structure, and magnetic properties of even relatively long-tube fragments.[47,49,55] The effect of the ring current on the shifts of encapsulated tubes,[57,59] and molecules[73] has been examined. Aromaticity, and NICS have also been investigated.[59,67,71,72] With the exception of ultranarrow tubes, the Knight shift was found to be small (1–2 ppm).[38] Table 9.1 summarizes the ^{13}C chemical shifts and Knight shifts calculated so far for infinite, and finite pristine SWNTs. The full (orbital and Knight) shifts of metallic species have not yet been computed, but it is expected that this situation will soon be alleviated. Moreover, given the steady increase in computing resources, and ongoing development of faster methods, it is expected that studies on larger nonmetallic systems will be undertaken in the near future.

First-principles computations indicate that NMR may be of particular utility for studies of chemically modified nanotubes. For example, it was concluded that the ^{13}C NMR shifts of carbons functionalised by amine groups may be sensitive probes for the tube diameter, and they may provide information about the orientation of the bond that has been functionalised.[76,77] For fluorinated SWNTs, NMR may be useful in determining the degree, and pattern of fluorination.[87] Moreover, NMR may be employed to identify, and perhaps quantify the defects in a given sample.[101]

Thus, even though many challenges and open questions remain, a lot has been learned from first-principles calculations of the magnetic response of nanotubes. It is likely that computations of the NMR chemical shifts of other carbon-based nanostructures will also aid their characterisation. One example is graphite oxide (GO), an important precursor to chemically modified graphenes. Its structure is amorphous, nonstoichiometric, and may contain a variety of different entities such as sp^2 carbons, epoxide, carbonyl, lactol, and

Table 9.1   Calculated chemical shifts[*] and Knight shifts of pristine SWNTs (ppm).

(n,n)	Finite SWNTs	Infinite SWNTs	Knight shift
(3,0)			685[f]
(4,0)			214[f]
(5,0)			232, 310, 387[f]
(6,0)	137.9[b]		89[f]
(7,0)		136.2,[d] 136.7,[e] 140.3[g]	
(8,0)		131.5,[c] 130.4,[d] 131.7,[e] 134.3[g]	
(9,0)	130,[a] 130.1,[b] 124.5[d]	120.5,[d] 122.2,[e] 122.7[g]	−1.5[f]
(10,0)	130.3[b]	129.5,[c] 126.6,[d] 127.9,[e] 130.5,[g] 122[i]	
(11,0)		127,[c] 123.5,[d] 126.9,[g] 125.4,[h] 120[i]	
(12,0)		117.0,[d] 117.7[g]	−1.6[f]
(13,0)		126.5,[c] 122.2,[d] 125.4,[g] 123.9,[h] 116[i]	
(14,0)		124.8,[c] 120.1,[d] 123.5,[g] 122.3,[h] 115[i]	
(15,0)		114.5,[d] 118.5,[g] 118.1[h]	−1.5[f]
(16,0)		124.5,[c] 119.8,[d] 123.1,[g] 121.7,[h] 111[i]	
(17,0)		124,[c] 118.0,[d] 121.8,[g] 120.5,[h] 111[i]	
(18,0)		120.9[g]	
(19,0)		123.5,[c] 110[i]	
(20,0)		122.5[c]	
(21,0)		123.8[g]	
(3,3)			5.4[f]
(4,2)		156.6[g]	
(4,4)	132.2[b]		−0.1[f]
(5,5)	129.3[b]		−1.3[f]
(6,2)		144.0,[g] 130[i]	
(6,3)		120.6[g]	
(6,4)		127[i]	
(6,6)	127.4[b]		−1.6[f]
(8,2)		120.6[g]	
(8,4)		124[i]	
(8,8)			−1.5[f]
(10,5)		121[i]	
(10,10)			−1.4[f]
(12,4)		107[i]	
(12,12)			−1.2[f]

[*]All chemical shifts given in ppm with respect to TMS. The values from ref. 49 and refs. 76 and 77 differ slightly due to a different choice of the macroscopic shape of the sample. The values for the (9,0) SWNT fragment from refs. 47 and 49 differ since in the latter work a longer tube fragment could be studied, and the shifts were given using benzene as the internal reference.
[a]ref. 47, [b]ref. 55, [c]ref. 57, [d]ref. 49, [e]ref. 76, [f]ref. 38, [g]ref. 58, [h]ref. 77, [i]ref. 59.

carboxylic groups. Not surprisingly, a structural model for this system seemed elusive. However, recent advances in ^{13}C labelling[102] have led to a number of solid-state NMR studies of GO.[102–104] A few investigations have carried out first-principles calculations of the chemical shifts of proposed structural models,[104,105] thereby lending urgently needed theoretical support. For these, and other carbon-based nanostructures, it is expected that the synergy between experiment and theory will lead to a deeper understanding of the structure(s) of the nanosystems, and also to new, accurate characterisation schemes.

Acknowledgement

JA would like to acknowledge financial support from the NSF grant CHE-0952253.

References

1. S. Iijima and T. Ichihashi, *Nature*, 1993, **364**, 737–737.
2. D. S. Bethune, C. H. Kiang, M. S. Devries, G. Gorman, R. Savoy, J. Vazquez and R. Beyers, *Nature*, 1993, **363**, 605–607.
3. S. Iijima, *Nature*, 1991, **354**, 56–58.
4. M. Monthioux and V. L. Kuznetsov, *Carbon*, 2006, **44**, 1621.
5. C. T. White, D. H. Robertson and J. W. Mintmire, *Phys. Rev. B.*, 1993, **47**, 5485–5488.
6. R. Saito, M. Fujita, G. Dresselhaus and M. S. Dresselhaus, *Phys. Rev. B.*, 1992, **46**, 1804–1811.
7. N. Hamada, S. Sawada and A. Oshiyama, *Phys. Rev. Lett.*, 1992, **68**, 1579–1581.
8. X. Blase, L. X. Benedict, E. L. Shirley and S. G. Louie, *Phys. Rev. Lett.*, 1994, **72**, 1878–1881.
9. M. Machón, S. Reich, C. Thomsen, D. Sánchez-Portal and P. Ordejón, *Phys. Rev. B.*, 2002, **66**, 155410.
10. I. Cabria, J. W. Mintmire and C. T. White, *Phys. Rev. B.*, 2003, **67**, 121406.
11. J. W. Mintmire, B. I. Dunlap and C. T. White, *Phys. Rev. Lett.*, 1992, **68**, 631–634.
12. M. Ouyang, J.-L. Huang, C. L. Cheung and C. M. Lieber, *Science*, 2001, **292**, 702–705.
13. J. Zhao, Z. Chen, Z. Zhou, H. Park, P. von Raguè Schleyer and J. P. Lu, *ChemPhysChem*, 2005, **6**, 598–601.
14. J. Lu, D. Wang, S. Nagase, M. Ni, X. Zhang, Y. Maeda, T. Wakahara, T. Nakahodo, T. Tsuchiya, T. Akasaka, Z. Gao, D. Yu, H. Ye, Y. Zhou and W. N. Mei, *J. Phys. Chem. B.*, 2006, **110**, 5655–5658.
15. A. Minett, K. Atkinson, S. Roth in *Handbook of porous solids*, ed. F. Schiith, S. W. Sing, J. Weitkapmp, Wiley-VCH, Weinheim, 2002.
16. Z. Yao, H. W. C. Postma, L. Balents and C. Dekker, *Nature*, 1999, **402**, 273–276.
17. Y. B. Zhang, M. Kanungo, A. J. Ho, P. Freimuth, D. van der Lelie, M. Chen, S. M. Khamis, S. S. Datta, A. T. C. Johnson, J. A. Misewich and S. S. Wong, *Nano Lett*, 2007, **7**, 3086–3091.
18. D. Tasis, N. Tagmatarchis, A. Bianco and M. Prato, *Chem. Rev.*, 2006, **106**, 1105–1136.
19. S. Freiman, S. Hooker, K. Migler, S. Arepalli, *Measurement Issues in Single Wall Carbon Nanotubes*, Technical Report, National Institute of Standards and Technology (NIST), 2008, NIST recommended practice guides, http://www.nist.gov/public_affairs/practiceguides/practiceguides.htm.

20. E. Zurek and J. Autschbach, *Int. J. Quantum Chem.*, 2009, **109**, 3343–3367.
21. R. Wasylishen in *Calculation of NMR and EPR Parameters. Theory and Applications*, M. Kaupp, M. Bühl, V. G. Malkin (ed.), Wiley-VCH, Weinheim, 2004, pp. 433–447.
22. J. Autschbach and T. Ziegler, *Coord. Chem. Rev.*, 2003, **238/239**, 83–126.
23. T. Ziegler and J. Autschbach, *Chem. Rev.*, 2005, **105**, 2695–2722.
24. R. McWeeny, *Methods of molecular quantum mechanics 2nd edn*, Academic Press, London, 1992.
25. F. Mauri, B. Pfrommer and S. G. Louie, *Phys. Rev. Lett.*, 1996, **77**, 5300–5303.
26. C. J. Pickard and F. Mauri, *Phys. Rev. B.*, 2001, **63**, 245101–245113.
27. T. Helgaker, M. Jaszuński and K. Ruud, *Chem. Rev.*, 1999, **99**, 293–352.
28. F. London, *J. Phys. Radium.*, 1937, **8**, 397–409.
29. R. Ditchfield, *Mol. Phys.*, 1974, **27**, 789–807.
30. C. G. Van de Walle and P. E. Blöchl, *Phys. Rev. B.*, 1993, **47**, 4244–4255.
31. J. R. Yates, C. J. Pickard, M. C. Payne and F. Mauri, *J. Chem. Phys.*, 2003, **118**, 5746–5753.
32. J. Yates, Ph.D. thesis, University of Cambridge, London, UK, 2003.
33. D. Skachkov, M. Krykunov, E. Kadantsev and T. Ziegler, *J. Chem. Theory Comput.*, 2010, **6**, 1650–1659.
34. T. Thonhauser, D. Ceresoli, A. A. Mostofi, N. Marzari, R. Resta and D. Vanderbilt, *J. Chem. Phys.*, 2009, **131**, 101101.
35. T. Thonhauser, D. Ceresoli and N. Marzari, *Int. J. Quantum Chem.*, 2009, **109**, 3336–3342.
36. J. Autschbach and S. Zheng, *Annu. Rep. NMR Spectrosc.*, 2009, **67**, 1–95.
37. G. N. La Mar, W. D. Horrocks Jr., R. H. Holm (ed.) *NMR of paramagnetic molecules. Principles and Applications*, Academic Press, New York, 1973.
38. O. V. Yazyev and L. Helm, *Phys. Rev. B.*, 2005, **72**, 245416.
39. C. Goze Bac, S. Latil, L. Vaccarini, P. Bernier, P. Gaveau, S. Tahir, V. Micholet, R. Aznar, A. Rubio, K. Metenier and F. Beguin, *Phys. Rev. B.*, 2001, **63**, 100302–100304.
40. X.-P. Thang, A. Kleinhammes, H. Shimoda, L. Fleming, K. Y. Bennoune, S. Shinha, C. Bower, O. Zhou and Y. Wu, *Science*, 2000, **288**, 492–494.
41. S. Hayashi, F. Hoshi, T. Ishikura, M. Yumura and S. Ohshima, *Carbon*, 2003, **41**, 3047–3056.
42. L. S. Cahill, Z. Yao, A. Adronov, J. Penner, K. R. Moonoosawmy, P. Kruse and G. R. Goward, *J. Phys. Chem. B.*, 2004, **108**, 11412–11418.
43. H. Peng, L. B. Alemany, J. L. Margrave and V. N. Khabashesku, *J. Am. Chem. Soc.*, 2003, **125**, 15174–15182.
44. C. Engtrakul, M. F. Davis, T. Gennett, A. C. Dillon, K. M. Jones and M. J. Heben, *J. Am. Chem. Soc.*, 2005, **127**, 17548–17555.
45. C. Engtrakul, M. F. Davis, K. Mistry, B. A. Larsen, A. C. Dillon, M. J. Heben and J. L. Blackburn, *J. Am. Chem. Soc.*, 2010, **132**, 9956–9957.

46. S. Latil, L. Henrard, C. Goze-Bac, P. Bernier and A. Rubio, *Phys. Rev. Lett.*, 2001, **86**, 3160–3163.

47. E. Zurek and J. Autschbach, *J. Am. Chem. Soc.*, 2004, **126**, 13079–13088.

48. J. Cioslowski, N. Rao and D. Moncrieff, *J. Am. Chem. Soc.*, 2002, **124**, 8485–8489.

49. E. Zurek, C. J. Pickard, B. Walczak and J. Autschbach, *J. Phys. Chem. A.*, 2006, **110**, 11995–12004.

50. K. Nakada, M. Fujita, G. Dresselhaus and M. S. Dresselhaus, *Phys. Rev. B.*, 1996, **54**, 17954–17961.

51. Y. Miyamoto, K. Nakada and M. Fujita, *Phys. Rev. B.*, 1999, **59**, 9858–9861.

52. Y.-W. Son, M. L. Cohen and S. G. Louie, *Nature*, 2006, **444**, 347–349.

53. T. Yamamoto, T. Noguchi and K. Watanabe, *Phys. Rev. B.*, 2006, **74**, 121409.

54. G. Treboux, P. Lapstun and K. Silverbrook, *Chem. Phys. Lett.*, 1999, **302**, 60–64.

55. N. A. Besley, J. J. Titman and M. D. Wright, *J. Am. Chem. Soc.*, 2005, **127**, 17948–17953.

56. H. Yang, C. M. Beavers, Z. Wang, A. Jiang, Z. Liu, H. Jin, B. Q. Mercado, M. M. Olmstead and A. L. Balch, *Angew. Chem. Int. Ed.*, 2010, **49**, 886–890.

57. M. A. L. Marques, M. d'Avezac and F. Mauri, *Phys. Rev. B.*, 2006, **73**, 125433–125436.

58. L. Lai, J. Lu, W. Song, M. Ni, L. Wang, G. Luo, J. Zhou, W. N. Mei, Z. Gao and D. Yu, *J. Phys. Chem. C.*, 2008, **112**, 16417–16421.

59. D. Sebastiani and K. N. Kudin, *ACSNano*, 2008, **2**, 661–668.

60. M. d'Avezac, N. Marzari and F. Mauri, *Phys. Rev. B.*, 2007, **76**, 165122–165212.

61. H. Ajiki and T. Ando, *J. Phys. Soc. Jpn.*, 1995, **64**, 4382–4391.

62. J. P. Lu, *Phys. Rev. Lett.*, 1995, **74**, 1123–1126.

63. O. N. Torrens, D. E. Milkie, H. Y. Ban, M. Zheng, G. B. Onoa, T. D. Gierke and J. M. Kikkawa, *J. Am. Chem. Soc.*, 2007, **129**, 252–253.

64. F. Simon, C. Kramberger, R. Pfeiffer, H. Kuzmany, V. Zólyomi, J. Kürti, P. M. Singer and H. Alloul, *Phys. Rev. Lett.*, 2005, **95**, 017401.

65. H. Günther, *NMR Spectroscopy, 2nd edn*, John Wiley & Sons, Chichester, 1995.

66. P. v. R. Schleyer, C. Maerker, A. Dransfeld, H. Jiao and N. J. R. v. E. Hommes, *J. Am. Chem. Soc.*, 1996, **118**, 6317–6318.

67. J. L. Ormsby and B. T. King, *J. Org. Chem.*, 2004, **69**, 4287–4291.

68. E. Clar, *The Aromatic Sextet*, Wiley, London, 1972.

69. T. Wassmann, A. P. Seitsonen, A. M. Saitta, M. Lazzeri and F. Mauri, *Phys. Rev. Lett.*, 2008, **101**, 096402(1–4).

70. T. Wassmann, A. P. Seitsonen, A. M. Saitta, M. Lazzeri and F. Mauri, *J. Am. Chem. Soc.*, 2010, **132**, 3440–3451.

71. J. Zhao and P. B. Balbuena, *J. Phys. Chem. C.*, 2008, **112**, 3482–3488.

72. J. Zhao and P. B. Balbuena, *J. Phys. Chem. C.*, 2008, **112**, 13175–13180.
73. N. A. Besley and A. Noble, *J. Chem. Phys.*, 2008, **128**, 101102–101104.
74. A. Kitaygorodskiy, W. Wang, S.-Y. Xie, Y. Lin, K. A. S. Fernando, X. Wang, L. Qu, B. Chen and Y.-P. Sun, *J. Am. Chem. Soc.*, 2005, **127**, 7517–7520.
75. M. D'Este, M. De Nardi and E. Menna, *Eur. J. Org. Chem.*, 2006, 2517–2522.
76. E. Zurek, C. J. Pickard and J. Autschbach, *J. Am. Chem. Soc.*, 2007, **129**, 4430–4439.
77. E. Zurek, C. J. Pickard and J. Autschbach, *J. Phys. Chem. C*, 2008, **112**, 9267–9271.
78. Z. Chen, S. Nagase, A. Hirsch, R. C. Haddon, W. Thiel and P. von Raguè Schleyer, *Angew. Chem. Int. Ed.*, 2004, **43**, 1552–1554.
79. Y.-S. Lee and N. Marzari, *Phys. Rev. Lett.*, 2006, **97**, 116801.
80. H. Jiao, N. J. R. van Eikema Hommes and P. v. R. Schleyer, *Org. Lett.*, 2002, **4**, 2393–2396.
81. C. Gellini, P. R. Salvi and E. Vogel, *J. Phys. Chem. A.*, 2000, **104**, 3110–3116.
82. E. T. Mickelson, C. B. Huffman, A. G. Rinzler, R. E. Smalley, R. H. Hauge and J. L. Margrave, *Chem. Phys. Lett.*, 1998, **296**, 188–194.
83. V. N. Khabashesku, W. E. Billups and J. L. Margrave, *Acc. Chem. Res.*, 2002, **35**, 1087–1095.
84. E. P. Dillon, C. A. Crouse and A. R. Barron, *ACS Nano*, 2008, **2**, 156–164.
85. L. B. Alemany, L. Zhang, L. Zheng, C. L. Edwards and A. R. Barron, *Chem. Mater.*, 2007, **19**, 735–744.
86. L. Zeng, L. B. Alemany, C. L. Edwards and A. R. Barron, *Nano. Res.*, 2008, **1**, 72–88.
87. E. Zurek and C. J. Pickard, Autschbach, *J. Phys. Chem. A.*, 2009, **113**, 4117–4124.
88. N. I. Denisenko, S. I. Troyanov, A. A. Popov, I. V. Kuvychko, B. Zemva, E. Kemnitz, S. H. Strauss and O. V. Boltalina, *J. Am. Chem. Soc.*, 2004, **126**, 1618–1619.
89. Y. Fan, B. R. Goldsmith and P. G. Collins, *Nature Mater.*, 2005, **4**, 906–911.
90. J. C. Charlier, *Acc. Chem. Res.*, 2002, **35**, 1063–1069.
91. T. C. Dinadayalane and J. Leszczynski, *Chem. Phys. Lett.*, 2007, **434**, 86–91.
92. C. Wang, G. Zhou, H. Liu, J. Wu, Y. Qiu, B.-L. Gu and W. Duan, *J. Phys. Chem. B.*, 2006, **110**, 10266–10271.
93. S. H. Yang, W. H. Shin, J. W. Lee, S. Y. Kim, S. I. Woo and J. K. Kang, *J. Phys. Chem. B.*, 2006, **110**, 13941–13946.
94. H. F. Bettinger, *J. Phys. Chem. B.*, 2005, **109**, 6922–6924.
95. X. Lu, Z. Chen and P. v. R. Schleyer, *J. Am. Chem. Soc.*, 2005, **127**, 20–21.
96. J. Andzelm, N. Govind and A. Maiti, *Chem. Phys. Lett.*, 2006, **421**, 58–62.

97. S. L. Mielke, T. Belytschko and G. C. Schatz, *Annu. Rev. Phys. Chem.*, 2007, **58**, 185–209.

98. T. Zhou, J. Wu, W. Duan and B. L. Gu, *Phys. Rev. B.*, 2007, **75**, 205410.

99. H. J. Choi, J. Ihm, S. G. Louie and M. L. Cohen, *Phys. Rev. Lett.*, 2000, **84**, 2917–2920.

100. A. J. Stone and D. J. Wales, *Chem. Phys. Lett.*, 1986, **128**, 501–503.

101. E. Zurek, C. J. Pickard and J. Autschbach, *J. Phys. Chem. C.*, 2008, **112**, 11744–11750.

102. W. Cai, R. D. Piner, F. J. Stadermann, S. Park, M. A. Shaibat, Y. Ishii, D. Yang, A. Velamakanni, S. J. An, M. Stoller, J. An, D. Chen and R. S. Ruoff, *Science*, 2008, **321**, 1815–1817.

103. W. Gao, L. B. Alemany, L. J. Ci and P. M. Ajayan, *Nature Chem.*, 2009, **1**, 403–408.

104. L. B. Casabianca, M. A. Shaibat, W. Cai, S. Park, R. Piner, R. S. Ruoff and Y. Ishii, *J. Am. Chem. Soc.*, 2010, **132**, 5672–5676.

105. N. Lu, Y. Huang, H. B. Li, Z. Li and J. Yang, *J. Chem. Phys.*, 2010, **133**, 034502(1–7).

CHAPTER 10

Computational Study of the Formation of Inorganic Nanotubes

MARK WILSON

Department of Chemistry, Physical and Theoretical Chemistry Laboratory, University of Oxford, South Parks Road, Oxford OX1 3QZ, UK

10.1 Introduction

Since the discovery of carbon fullerenes,[1] followed by the closely related carbon nanotubes (C-NTs),[2–4] significant efforts have been made in finding technological applications. The richness of the extended carbon phase diagram, including the nanotubular and near-spherical structures, hints that analogous low-dimensional structures may be stable for other inorganic materials. Experimental investigations have demonstrated the existence of a range of inorganic nanotubular (INT) materials. Systems such as BN, GaN and GaSe form tubes similar to those adopted by elemental carbon (based on fullerene-like structures). More surprisingly perhaps, a range of MX_2 (M = metal cation, X = anion) systems have been shown to display a range of nanotube morphologies.[5] Sulphides such as MoS_2 and WS_2, which crystallize in the bulk in a MoS_2 structure, and chlorides such as $NiCl_2$, which crystallizes in a CdI_2 structure, form nanotubes. These INT structures have been most commonly observed by electron microscopy techniques (WS_2[6–8] MoS_2,[8,9] MSe_2 (M = W,Mo)[8]). Such images show how the layers that make up the bulk crystal structures (see below) may fold in order to form tubular structures

RSC Theoretical and Computational Chemistry Series No. 4
Computational Nanoscience
Edited by Elena Bichoutskaia
© Royal Society of Chemistry 2011
Published by the Royal Society of Chemistry, www.rsc.org

analogous to those formed by carbon, but also show that tubes with non-circular cross-sections are common. Raman spectroscopy has also been applied to probe the changes in the vibrational density of states (and hence atom dynamics) as a function of INT size for MoS_2.[10] These bulk crystal structures are examples of layered crystals consisting of repeating infinite approximately two-dimensional (charge-neutral) MX_2 "sandwiches". These layers (whose unit cells are closely related to that of a single graphene sheet) are effectively bound by relatively weak van der Waals forces (which has significant implications for their tribological applications). In hindsight, it is clear that older mass spectrometry experiments[11] indicate the presence of $(MgO)_n$ clusters constructed from percolating $(MgO)_3$ hexagons (as confirmed by potential model calculations).[12]

A critical part of any significant materials application is that the detail of any nanotube must be controllable at an atomistic level. Over the past two decades significant experimental effort has been focused on the study of the filling of carbon nanotubes (C-NTs). The C-NTs may act as nanoscale templates in the formation of unique (low dimensional) structures that may have useful practical physical and electronic properties. However, not all materials are available to directly fill the C-NTs as, if the liquid surface tension is too high, the C-NT may lose structural integrity.[13] Molten salts, therefore represent a useful potential class of filling materials as they often have a relatively low surface tension[14] and hence may fill the nanotubes rather than crushing the carbon shell. A wide range of different materials have been successfully inserted.[15-17] Molten salts inserted include NiO, Bi_2O_3, V_2O_5 and MoO_3,[18-22] UCl_4,[23] AgCl/AgBr,[24,25] KI,[26,27] BaI_2,[28] CoI_2,[29] Sb_2O_3,[30] metal trihalides,[31-33] PbI_2,[34] HgTe,[35] $ZrCl_4$[36] and PbO.[37] Once inserted the salts may be chemically altered in order to, for example, form metallic nanowires, thus circumventing the need to insert the metal directly. Alternatively, the carbon nanotube may be removed to isolate the low dimensional ionic crystallites (for example, for PbO[37]). Finally, templating may not be limited to the use of C-NTs as the confining environment. For example, INTs of WS_2 have recently been filled with CsI.[38] Furthermore, the C-NTs may be further functionalised to allow for the effective delivery of radioisotopes for medical application.[39] Significant additional experimental progress has been made to understand the thermal stability of the C-NT/inorganic system[40] and in probing the internal structure by alternative techniques to electron microscopy, such as Raman and optical absorption spectroscopy.[41] These experiments yield important information regarding the interatomic interactions between the confining C-NT and the inorganic adduct, for example, highlighting potential charge-transfer effects.[41]

Experimental methods, in particular high-resolution transmission electron microscopy (HRTEM), yield images that clearly show the presence of highly ordered (crystalline) ion environments within the C-NTs.[16] These experiments, whilst producing high-quality static structural information, yield only limited information in two significant areas; the filling mechanism and the underlying statistics of crystallite formation (*i.e.* which crystals are formed within a given C-NT, how, and how often?). These experimental limitations mean that theoretical (model) investigations are vital if the underlying factors controlling the crystallite formation are to be fully understood and hence fully exploited to facilitate genuine atomistic control over the internal structure. High level electronic structure calculations have been applied to a limited number of low dimensional structures, for example, for KI and CuI (both in the presence of the C-NTs)[42–44] and LiF, ZnSe, ZnO and MoS_2 (in isolation).[45–48] The computational demands associated with high-level methodologies precludes their effective use in the study of the direct C-NT filling. As a result, models in which the atomic interactions are expressed by relatively simple energy functions, may be significant in allowing for the simulation of both accessible length- and timescales. In addition, the ability to simulate multiple filling events may give insight into the statistics of nanotubes filling. Information of this sort is often lacking from experimental reports that tend to highlight specific examples of filling morphologies. To this end, such potential models (in which the C-NTs themselves are treated as fixed (atomistic) entities) have uncovered the potential existence of a range of crystallite structures extending beyond the published experimental structures.[49–59]

In the MX stoichiometry, for example, inorganic nanotube (INT) structures are observed whose morphologies may be thought of in terms of the folding of infinite sheets of either hexagonal- (three-coordinate ions) or square-nets (four-coordinate ions). The INTs may be characterised in terms of folding around a given chiral vector $C_h = (na_1 + ma_2)$, where a_1 and a_2 are the unit-cell vectors for the hexagonal or square net sheets allowing the use of an $(n,m)_X$ notation (X = sq or hex depending on the topology of the sheet) to describe the INT morphology, in a manner analogous to that employed to describe the C-NTs. The observed experimental structures are traditionally rationalised by reference to the *bulk* crystal structures formed by the filling material. However, it is not clear that such a *direct* relationship between the bulk structures and those formed within the pseudo-one-dimensional confining environments should exist. For example, it is well known that even simple hard spheres, packing into rigid cylinders, adopt complex (often chiral) structures, dependent upon the relative sphere and cylinder diameters.[60] For molten salts filling carbon nanotubes, computer modelling has shown how the structures formed by both KI[27] and AgCl/AgBr mixtures[24,25] rationalised with respect to the bulk rocksalt and wurtzite crystal structures, respectively, are simply special cases of general classes of structures formed by folding sheets of square nets and hexagonal nets, respectively,[56,58,61] but that coincide with sections of underlying bulk crystal. In addition, even if the structures formed internally appear trivially related to the underlying bulk crystal structure, they may show distortions specific to both the confined environment and to the low-dimensional nature of the crystallite. For example, Bichoutskaia and Pyper[62,63] have shown how these distortions may be understood in terms of the detail of the ionic pair potential.

The use of relatively simple models affords suitable simulation length- and timescales to allow the filling events to be observed *directly*.

To supplement this information ideal INT morphologies may be generated by folding the appropriate sheets, allowing for the construction of the underlying (internal energy) phase diagram. The advantage of running these two methods in parallel is that the direct filling studies may suggest structures to be investigated, rather than relying simply on a pre-existing (potentially incomplete) understanding of the underlying energy landscape. The potential model calculations indicate that the radii of the energetically preferred INTs depends critically on the C-NT radius for the case of the rigid encasing C-NT template.[54] As a result, the C-NT/INT energetics can be mapped onto a simple analytic model.[54] This dependence on the difference in radii between the C-NT and INT has further implications for the formed INT morphologies in that, for a given C-NT, there may be multiple INT structures whose radii satisfy the thermodynamic criterion. The implication is that filling events, nominally from liquids at the same $\{T,p\}$ state point, may show different INT structure. As a result, it is important to attempt to understand the basic statistics of the C-NT filling. The observed INT morphologies formed may follow a simple distribution (perhaps based on the difference in radii of the two tubes) or there may be additional complexity associated with the underlying entropy or the filling kinetics. Again, it is the use of the relatively simple models that allows these questions to be tackled.

It is important to emphasize that the expression of the total system energy incorporating an energy associated with folding a plane represents a thermodynamic model for understanding the relative energetics of the INTs. The use of this model does not imply that the formation *mechanism* involves such a folding. For example, recent work[58,59] shows that, for systems in which square-net-based INTs are favoured, the INTs appear to form by a "chain-by-chain" mechanism in which the INT is systematically built up by the ingress of charge-neutral anion–cation–anion–cation . . . chains.

This chapter is divided as follows. Section 10.2 describes the methods applied in the atomistic simulation of C-NT filling. Sections 10.3 and 10.4 describe the application of both direct-filling simulations and energy minimisation calculations to determine the underlying phase diagrams. In Section 10.5 the results of these calculations are brought together using the concept of the energy landscape.

10.2   Modelling Methodologies

In order to obtain *direct* information regarding the filling dynamics atomistic models (which utilise relatively simple potential energy functions to describe the interatomic interactions) are highly effective. Such models do not explicitly account for the electron density but their relative simplicity allows for the investigation of atom dynamics on relatively long length- and time-scales. Two basic modeling strategies may be applied to investigate the structure of the INTs. In the first, direct information regarding the filling processes is obtained by performing molecular dynamics simulations in which an initially empty cylindrical section of C-NT is filled from the bulk melt. In the second strategy ideal INT structures are explicitly constructed and their energetics, as a function of both INT and encasing C-NT morphology, are investigated. The direct filling approach is advantageous in the sense that the INTs may form inside the C-NTs with no bias. In addition, information that is difficult to obtain experimentally regarding filling kinetics and mechanisms can be obtained. Furthermore, more than one filling simulation can be performed at a given $\{p,T\}$ state point and, as a result, a statistical picture of the INT structures, the form of which can be constructed. Again, this information is difficult to obtain experimentally. The advantage of the energy-minimisation procedure is that an effective phase diagram (energy of INTs within a given C-NT) may be obtained rapidly. In practice, therefore, direct-filling simulations allow the classes of INTs of interest to be determined that then feed directly into the energy minimisations in order to obtain the underlying phase diagram.

10.2.1 Background

In the condensed phase the interaction between atoms may be expressed as a cluster expansion

$$U(\{R_i\}) = \sum_i \sum_j U_2(\{R_i\}, \{R_j\}) + \sum_i \sum_j \sum_k U_3(\{R_i\}, \{R_j\}, \{R_k\}) + \cdots,$$

where successive terms represents the two-, three-, four- . . . body interactions. In the ideal case, these terms represent the true interaction potential of an isolated atom dimer followed by the correction to represent the true three-body energy surface for the isolated trimer, *etc*. Potential models of this type are ultimately transferable in the sense that they would, by construction, be able to reproduce the system energetics over all state points. They would, however, become computationally intractable very rapidly with cluster size as the time to calculate the energy of a single n-body term effectively scales as N^n. A common computational procedure is to truncate the series at the three- or four-body level. Alternatively, higher-order terms are "folded" into the two-body term to give an *effective* pair potential (EPP). The advantage of this procedure is that the potentials are computationally tractable and scale well with system size (important if long length- and timescales are to be accessed). The disadvantage is that the transferability, both between state points and different (chemically related) systems, may be compromised. Furthermore, the potential parameters may lose their physical meaning as a result of accounting for significant higher-order terms making the development of new models more difficult.

In molecular dynamics (MD) simulation the atoms are moved along their lines of force (determined from the potential energy function) using finite-difference methods.[64,65] At each time step the energy and forces must be evaluated allowing the accelerations on each atom to be determined, which in turn allows the atom changes in both velocities and positions to be evaluated. The system temperature and pressure are easily determined from the atom velocities. Simulations can be performed in a number of ensembles of which constant temperature and volume (NVT ensemble) and constant temperature and pressure (NPT ensemble) are the most commonly applied. Constant temperature and pressure constraints can be imposed using thermostats and barostats in which additional variables are coupled to the system that act to modify the equations of motion.

Energy-minimisation calculations can, at the simplest level, be performed using the basic MD methodologies in a steepest descent algorithm in which the atom velocities are systematically quenched (for example when the kinetic energy reaches a maximum). Alternatively, more efficient and rapid methods, such as conjugate gradient minimisations, may be employed.[66]

10.2.2 Potential Models

In constructing potential models to study the range of INT structures that may form, two basic strategies are utilised. In the first, potential models are parameterised with respect to specific systems, allowing for the most direct comparison to experimental observation. In the second, a generic model may be systematically modified in order to observe the evolution of key system properties. An advantage of the potential model-based approach over more computationally demanding (but potentially more accurate) electronic-structure methods is that they may be systematically built up and the individual parameters altered in order to develop an understanding of the important interactions dominating specific processes.

The materials of interest are primarily ionic in nature. As a result, the potential model may be readily decomposed into ion–ion, ion–carbon and carbon–carbon interactions.

10.2.2.1 Ion–Ion Interactions

The most computationally tractable models are based on an effective pair potential as exemplified by the Fumi–Tosi potential (see ref. 67 and references therein) for which the interaction between a pair of ions labelled i and j is given by

$$U(r_{ij}) = B_{ij}e^{-a_{ij}r_{ij}} + \frac{Q_iQ_j}{r_{ij}} - \sum_{n=6,8,10\ldots} \frac{C_n^{ij}}{r_{ij}^n} f_n(r_{ij}),$$

where B_{ij} and a_{ij} are parameters representing the contribution of the ion radii to the repulsive wall and the rate of decay of the repulsion, respectively, C_n^{ij} are the dispersion parameters, $f_n(r_{ij})$ are the dispersion damping functions that mimic the loss of asymptotic behaviour in the dispersion energy at short ion–ion separations,[68] and $Q_{i(j)}$ is the (formal) charge on ion $i(j)$. The long-range electrostatic interactions are accounted for using an Ewald summation. The exponential dependence of the short-range energy with separation follows that suggested by Born and Mayer.[69] In the Fumi–Tosi potential the pre-exponent is further expanded in terms of the respective ion radii, $\sigma_{i(j)}$, such that $B_{ij} = A_{ij}\exp[a(\sigma_i + \sigma_j)]$. As a result, the unlike and like ion pre-exponents may be varied to change the coordination environment of the ground-state crystal structure (effectively radius-ratio rules). As a result, potential models may be readily adjusted to favour either hexagonal net- or square net-based INTs. These EPPs may reproduce the static crystal structures as the high site symmetries preclude the formation of (low-order) induced moments. Once the site symmetry is broken (*i.e.* at finite temperature) a description of ion-polarisation effects is required to reproduce key dynamic properties such as phonon mode frequencies. The inclusion of ion polarisation is also required to reproduce key liquid static structural and dynamic properties.[67] Polarisation effects may be incorporated using a polarisable-ion model (PIM),[70] the parameters for which can be obtained by reference to *ab initio* electronic-structure calculations.[71]

10.2.2.2 Ion–Carbon Interactions

The ion–carbon interactions are modelled using a (pairwise-additive) Lennard-Jones potential,

$$U(r_{iC}) = 4\varepsilon_{iC}\left[\left\{\frac{\sigma_{iC}}{r_{iC}}\right\}^{12} - \left\{\frac{\sigma_{iC}}{r_{iC}}\right\}^{6}\right].$$

The $\{\sigma_{iC}, \varepsilon_{iC}\}$ parameters used are obtained from Lorentz–Berthelot mixing rules. As a result, $\sigma_{iC} = (\sigma_{CC} + \sigma_{ii})/2$, where σ_{CC} and σ_{ii} are the respective carbon atom and ion i diameters, whilst $\varepsilon_{iC}^2 = \varepsilon_{ii}\varepsilon_{CC}$, where ε_{ii} and ε_{CC} are the potential-energy-well depths. These parameters can be obtained by reference to a carbon graphite surface interacting with isoelectronic noble gases, as described in ref. 72.

10.2.2.3 Carbon–Carbon Interactions

Within the principle of starting with a heavily simplified model (the application of Occam's razor) and then building up the complexity, the initial approach is to use a fixed atomistic description of the carbon atoms comprising the C-NT. The flexibility of the C-NT can be introduced using an atomistic potential model, for example using a three-body (Tersoff II) potential model.[73] Such models account for the relative stability of the bulk diamond and graphite crystal structures and account well for the basic mechanical and dynamic properties of the single-walled C-NTs.

10.2.2.4   Metallic Nanotubes

The use of either rigid or fixed C-NTs with Lennard-Jones pair potentials to describe their interactions with the ions means that the tube are effectively being treated as insulators despite real C-NTs being either semiconducting or metallic depending upon morphology.[74] The introduction of, in particular, metallicity into the model might be expected to have significant implications, particularly for the interactions between ions from within and outside of the C-NT. Again, the above models can be modified to take account of these interactions. For example, image charge-based models, in which charges are induced on the carbon atoms in order to constrain the tube to a constant potential,[75–80] are being developed for this purpose.[81] The development of such models has the added advantage that the use of C-NTs as potential electrodes (which may control the inner INT structure) can be investigated.

10.3   Direct Filling Simulations

10.3.1   Simulation Procedures

Direct filling simulations are initiated from configurations obtained from molecular dynamics simulations performed on the bulk liquid. A liquid configuration is extracted and a cylindrical section removed (retaining overall charge neutrality) in order to insert the C-NT. The C-NT ends are initially blocked using sections of graphene sheet and the system is allowed to re-equilibrate in the presence of these end groups. The blocking groups are then removed and the C-NT is left free to fill. Multiple filling events are observed by extracting liquid configurations performed at the required $\{T,p\}$ state point. The configurations can be considered as independent if the time taken in between successive extractions is significantly longer than the characteristic relaxation time (as monitored, for example, *via* an intermediate scattering function).

10.3.2   Results

10.3.2.1   The Filling of the Carbon Nanotubes

The progress in filling the C-NTs can be monitored by constructing *filling profiles* generated by counting the number of ions inside the cylindrical section defined by the C-NT as a function of time. For fixed C-NTs this procedure is effectively trivial as the appropriate cylindrical section is easy to identify. For flexible C-NTs, in which the nanotube may deform and both rotate and translate within the molten salt, some mathematical effort is required to generate the same information.[58] Figure 10.1 shows typical filling profiles obtained from MD simulations in which both fixed and flexible C-NTs are filled. For the fixed C-NT the filling profiles form a basically sigmoidal shape

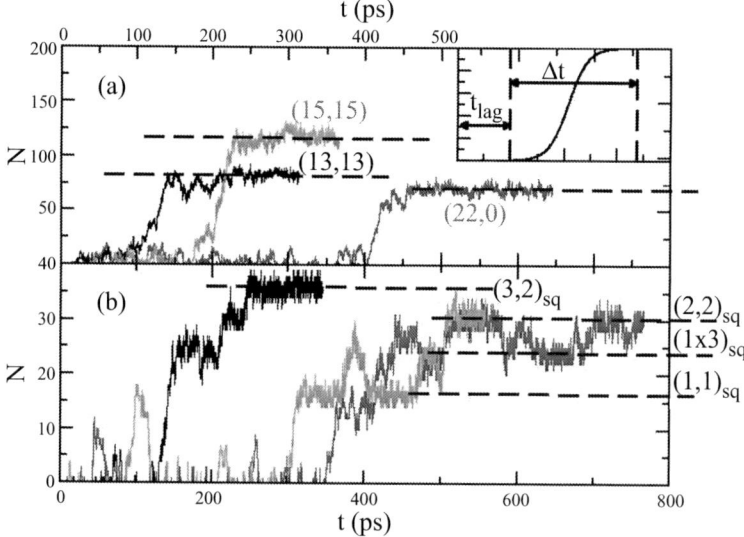

Figure 10.1 Filling profiles obtained for (a) a hexagonal sheet favouring potential model and fixed C-NTs and (b) a square net favouring potential model in flexible C-NTs. In (a) the three morphologies of the C-NTs being filled are indicated. In (b) the structures of the square net-based INTs being formed during the filling process are indicated. The plateau indicate the presence of well-defined stable intermediate structures. In both figures the dashed lines indicate the presence of stable structures. The inset to (a) shows the two time parameters, t_{lag} and Δt that define the sigmoidal filling profiles.

characterised by a time lag, t_{lag}, prior to the onset of significant tube filling and a further time for the tube to become fully filled (Δt). The three example C-NTs filled are the (22,0), (13,13) and (15,15) (diameters 17.2 Å, 17.6 Å and 20.3 Å, respectively) in which the final INTs formed contain around 72, 83 and 116 MX molecules indicating that the size of the forming INT is related to the C-NT pore diameter. Indeed, for low radius C-NTs the INT size is linear with respect to the pore diameter.[54] For the flexible tubes the filling profiles appear as successive near-sigmoidal curves. In all cases, the fully filled tubes can be readily identified by the plateau at long time. These plateau are indicative of the formation of relatively stable end structures, whilst those visible at earlier times are indicative of the formation of relatively stable intermediate structures.

10.3.2.2 Inorganic Nanotube Structures

Figure 10.2(a) show the final structures obtained using a hexagonal-net favouring potential model in a fixed (12,10) carbon nanotube of approximate diameter 15 Å. Figure 10.2(b) shows the analogous structures for a potential

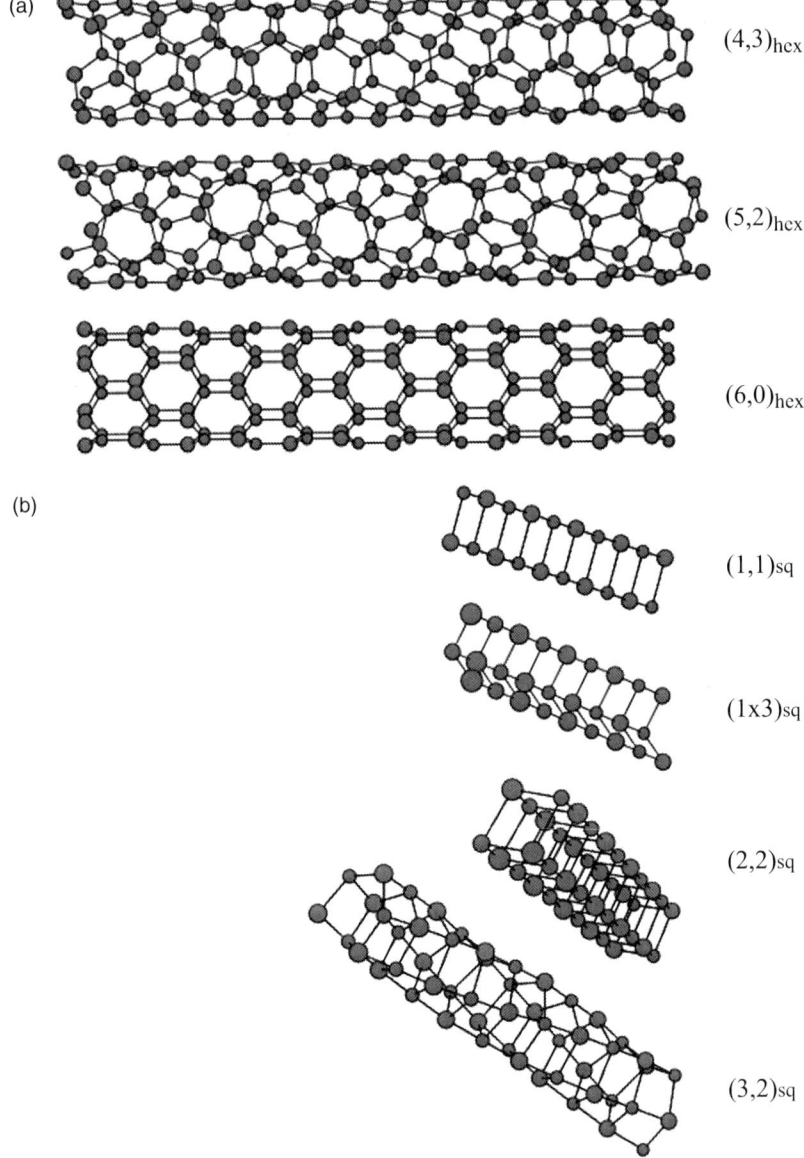

(a)

(4,3)$_{hex}$

(5,2)$_{hex}$

(6,0)$_{hex}$

(b)

(1,1)$_{sq}$

(1x3)$_{sq}$

(2,2)$_{sq}$

(3,2)$_{sq}$

Figure 10.2 (a) Molecular graphics "snapshots" of the three hexagonal INT structures formed within a (12,10) C NT. The large and small circles represent the anions and cations, respectively. (b) "Snapshots" of the two final square net-based structures [the (3,2)$_{sq}$ and (2,2)$_{sq}$] and the two stable intermediates, observed during the filling of a (10,10) C NT.

that favours square nets in the flexible (10,10) C NT of approximate diameter 13.5 Å. The INT structures formed are based on percolating hexagonal- and square-net motifs, respectively. A key observation is that filling events that begin with liquid configurations obtained from simulations performed at the same $\{p,T\}$ state point lead to a range of final INT structures. We will return to his observation in Section 10.5.

The hexagonal and square net based structures shown in Figure 10.2 may both be rationalised in terms of the folding of the respective infinite nets, as shown in Figure 10.3. In both cases the unit cells contain two atoms (corresponding to one MX stoichiometric unit). The folding of the hexagonal sheets are analogous to the folding of the graphene sheet to form the carbon nanotubes, with the exception that the two atoms in the unit cell have different chemical identities. In both cases, INTs are generated by selecting a chiral vector, $\mathbf{C}_h = (n\mathbf{a}_1 + m\mathbf{a}_2)$, where $\mathbf{a}_1$ and $\mathbf{a}_2$ are the respective unit-cell vectors. As a result, pairs of integers (n,m) can be used to define the final INT morphology. INTs formed from the hexagonal and square plane sheets, are distinguished by employing a nomenclature of the form $(n,m)_{hex}$ and $(n,m)_{sq}$, respectively. For the hexagonal net favouring potential three INTs of morphology $(4,3)_{hex}$, $(5,2)_{hex}$ and $(6,0)_{hex}$ are formed in the ratio 3:1:1 in this C-NT.

The fixed cylindrical C-NT favours the formation of highly symmetric cylindrical INTs. For the square net structures the final INT structure corresponds to either a $(2,2)_{sq}$ INT containing around 15 MX molecules, or a $(3,2)_{sq}$ INT containing around 18 MX molecules in the ratio of 1:1. Plateau are observed in the filling profiles (Figure 10.1) which correspond to the formation of well-defined intermediate (highly ordered) structures. The persistence times of these plateaux are significant, corresponding to many thousand molecular dynamics steps. The first intermediate structure formed is a $(1,1)_{sq}$ "ladder", which can be considered as formed from two parallel M-X-M-X . . . ion chains. The next intermediate is a $(1 \times 3)_{sq}$ "hinge", which can be considered as

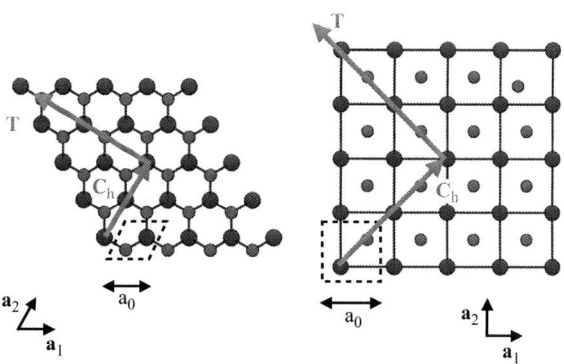

Figure 10.3 Two-dimensional hexagonal- (left) and square- (right) net structures shown for the MX stoichiometry. In both cases the two atom unit cells are highlighted by a dashed box, as are the respective unit-cell vectors, $\mathbf{a}_1$ and $\mathbf{a}_2$. Example chiral, $\mathbf{C}_h$, and translational, $\mathbf{T}$, vectors are also shown.

a "three-chain" extension of the $(1,1)_{sq}$ structure. The "hinge" "structure is unique in the present context as, although it can clearly be constructed from a section of square-net plane, it is not related *via* a simple folding transformation. For both the $(1,1)_{sq}$ and $(1 \times 3)_{sq}$ structures, with highly anisotropic cross-sections, the flexibility of the encasing C-NT, in the form of elliptical distortions, is critical in their stabilisation.[59] The dihedral angle of the $(1 \times 3)_{sq}$ structure of approximately $100°$ represents a balance between minimising the energy of the ionic structure (which would lead to a planar structure, minimising the like–like electrostatic repulsions between the outer chains) and the energy of interaction between the INT and the C-NT. The flexibility of the encasing C-NT is critical in stabilising these structures with highly anisotropic cross-sections.[59]

10.3.2.3   Why Inorganic Nanotube Structures Form in Confined Environments

A key question concerns why such highly ordered structures form within the C-NTs despite the elevated temperatures (which are above the ambient pressure crystal melting points). To understand their stability we consider a C-NT whose diameter is able to accommodate four ions in the cross-sectional plane perpendicular to the C-NT major axis (Figure 10.4). Once a single ion wets the internal surface of the C-NT the location of the three remaining ions in this plane are defined, forming a 2×2 square-net unit. The energetic penalty to adopting a structure that contains nearest-neighbour-like ion interactions is too great to be overcome even at the typically high melt temperatures.

This ion square will then percolate along the C-NT major axis with successive polyhedra rotated by $90°$ to maximise the attractive nearest-neighbour ion interactions to give a $(2,2)_{sq}$ INT. The formation of the more complex looking structures can be understood by a simple extension of this argument (Figure 10.4). If the radius of the confining C-NT is increased such that five ions can be accommodated within the plane then a pair of like ions must become nearest neighbours. To alleviate this, the ions are displaced along the major axis giving a helical chain structure equivalent to a folded plane of square-nets.

10.3.2.4   How Inorganic Nanotube Structures Form

The direct simulation of the filling events themselves yields important information regarding the *mechanisms* by which the C-NTs fill and the INTs form. Information of this type is difficult to obtain experimentally. For both classes of INTs discussed here the initial step is for a closed anion–cation . . . loop to enter the C-NT. For the hexagonal net-based INTs these loops would form a single row of percolating hexagons, whilst for the square net-based INTs the loop will form a single row of percolating squares [or a $(1,1)_{sq}$ "ladder" structure – Figure 10.2(b)]. For the square net-based INTs fully flexible C-NTs the development of the INT structure follows a "building up" pattern *via* the addition of further anion–cation . . . chains as evidenced by the "steps" in the filling profiles (Figure 10.1). In rigid tubes an alternative mechanism, in which

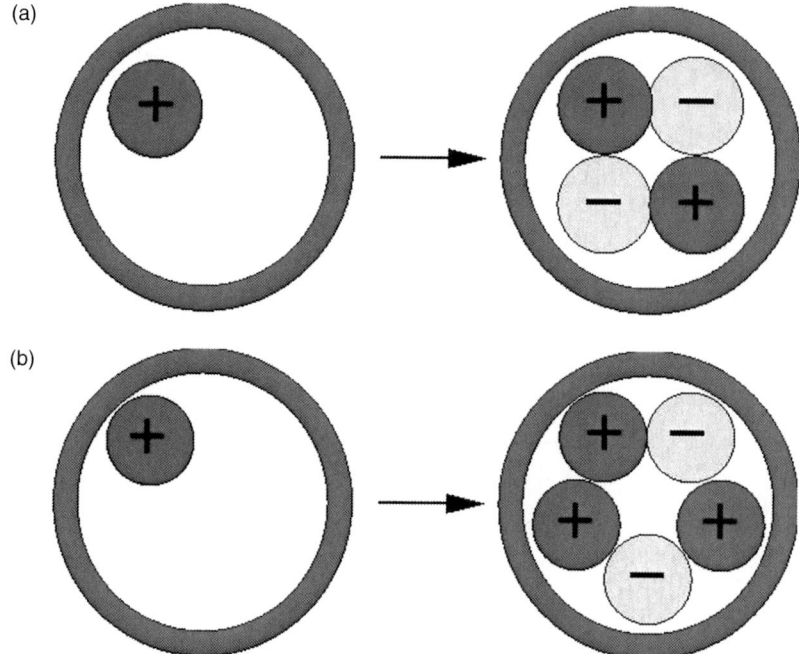

Figure 10.4 Shows why highly ordered (crystalline) geometries form within low radius C-NTs. In part (a) the encasing C-NT can accommodate four ions in the cross-section perpendicular to the major axis and so once a single ion of either charge wets an internal surface then the locations of the remaining three ions are defined. In part (b) the larger C-NT can accommodate five ions in the cross-section with the resulting repulsive nearest-neighbour electrostatic interactions should the five ions sit in a single plane. The solution is to pull one of these cations along the major axis leading to a helical anion–cation–anion . . . chain and hence the formation of the "twisted" INT structures.

MX molecules percolate along the C-NT, is promoted. In all cases the linking themes are that the ions move in a highly correlated manner, most significantly as charge neutral units (molecules, chains, loops . . .). The rationale for this behaviour is that the activation barriers associated with moving isolated charges would be expected to be high (as strong electrostatic forces would have to be overcome). However, typical activation barriers to moving collections of (charge neutral) ions may be significantly lower as the strong electrostatic interactions are nearly cancelled. The different mechanisms promoted in the flexible and rigid tubes highlight the potential role of the C-NTs in defining the filling mechanisms. The flexibility of the confining C-NT allows for the introduction of whole charge neutral chains as there is an associated deformation of the C-NT (which become eccentric in cross-section). In the rigid tube these deformations are effectively precluded, and hence these chain filling mechanisms are shifted to higher energy.

10.4 Energy Minimisations

The direct observation of the filling of the C-NTs by the molten salts of MX stoichiometry highlights the presence of two basic classes of INT whose structures may be rationalised in terms of the folding of hexagonal- and square-net sheets, respectively (see Section 10.3.2). Furthermore, repeated simulations at the same state point indicate the formation of a range of INT structures within a given C-NT. To rationalise this behaviour further it is necessary to fundamentally understand the factors that control the energetics of the INTs.

10.4.1 Hexagonal Net Sheets

Figure 10.5 shows the energies of a range of hexagonal and square net INTs plotted as a function of their ideal radii. In each case the ideal INT is constructed by folding the appropriate sheet (Figure 10.3) and the energy is then minimised using a steepest descent method. The hexagonal-net based tubes show a dependence on the final radius R of the formed INT but appear independent of the chiral vector. Furthermore, the energy is proportional to R^{-2},

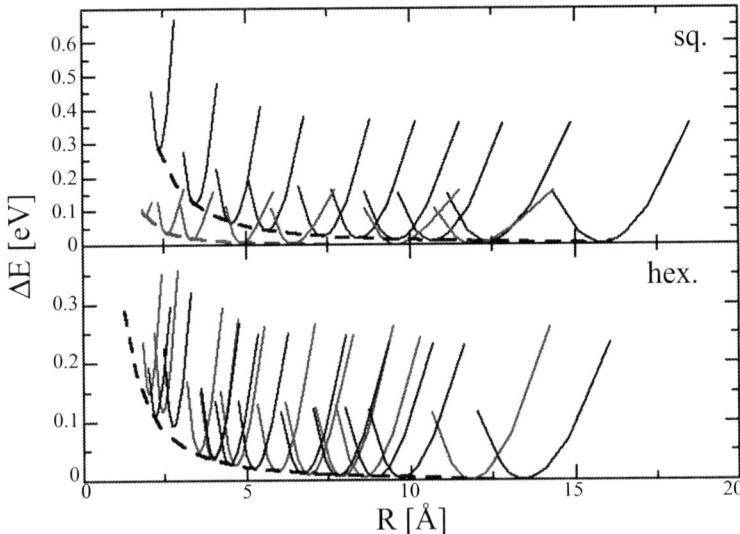

Figure 10.5 Energies of the square net- (upper panel) and hexagonal net- (lower panel) based INTs as a function of their ideal radii. Curves are shown for the high-symmetry $(n,0)_X$ and $(n,n)_X$ (X = sq, hex). Each curve represents the energy minimisation of a given INT as a function of the unit cell length along the major axis. The dashed curves link the respective energy minima. For the hexagonal net INTs the energy minima fall on the same curve independent of the final morphology. For the square net INTs the energies show a strong dependence on final morphology.

indicating that the hexagonal net behaves like a elastic continuum model.[60] If the two dimensional structures behave as simple macroscopic sheets then an elastic continuum model suggests a relationship between the INT energy and the radius of the form[60,82]

$$E_\sigma = \frac{\pi E T h^3}{12R},$$

where E is the elastic modulus and h is the thickness of the sheet being folded. T is the length of the final folded tube and R is the radius of the formed INT. T is related to the area of the folded tube $(A = 2\pi R T)$ leading to

$$E_\sigma = \frac{E h^3 A}{24R^2}.$$

For a INT generated from folded hexagons, the unit cell has area $A_{uc} = \sqrt{3}/2a^2$, giving the total area of the INT as $A = N A_{uc}$, where N is the number of unit cells making up the INT. As a result, for a hexagon-based INT

$$\frac{E_\sigma}{N} = \frac{\sqrt{3} E h^3 a^2}{48R^2} = \frac{F}{R^2}.$$

The subsuming of parameters into a single term F allows for a more direct comparison between systems as a somewhat arbitrary value of the sheet thickness (h) does not have to be selected. For these hexagonal nets the values of the parameter F (related to the elastic modulus of the sheet) are of the order of 0.60–0.73 eV Å^{-2} compared with values of 1.57–2.16 eV Å^{-2} for the analogous C-NTs.[58,83] As a result, the ionic hexagonal sheets are significantly easier to fold than the analogous graphene sheets.

10.4.2 Square Net Sheets

For the square net based INTs the dependence of the energies appears more complex (Figure 10.5) showing a clear dependence both upon the radius of the final INT but also on the chiral vector about which the sheet is folded.

The total energy of a given INT may be decomposed arbitrarily (by effectively changing the order of the summation over ion pairs). However, the "building" mechanism indicated by Figure 10.2(b) in Section 10.3.2 indicates that considering the energies of the INTs in terms of the interactions between charge neutral anion–cation–anion . . . chains may prove insightful.

In order to understand the relative energetics of the $(n,0)_{sq}$ and $(n,n)_{sq}$ INTs it is clear that we need to go beyond an elastic continuum model and involve the atomistic detail of their structures. Bichoutskaia and Pyper[62] consider the energetics of pseudo-one dimensional crystal structures that are directly related to an underlying bulk rocksalt structure. For an infinite square-net plane (Figure 10.3) the electrostatic (coulombic) energy may be expressed as

$$U^{coul}(a,b) = -\frac{2\ln 2}{b} + U^{ic}(a,b),$$

where a and b are the nearest-neighbour anion–cation separations parallel to and perpendicular to a chosen vector linking a nearest neighbour anion–cation pair chosen to correspond to the final INT major axis. The first term represents the sum of the ionic interactions in a single linear chain (for which the Madelung constant is 2ln2). The second term represents the interchain (ic) interaction of a given chain with all others. The interchain energy may be written as

$$U^{ic}(a,b) = 2\sum_{i=1}^{\infty}\left\{\frac{(-1)^i}{ia} + 2\sum_{j=1}^{\infty}\left[\frac{(-1)^{i+j}}{\sqrt{i^2a^2 + j^2b^2}}\right]\right\},$$

where i and j are indices labelling ions parallel and perpendicular to the major axis. Alternatively, taking $x = a/b$ (the ratio of the nearest-neighbour anion–cation separations parallel to and perpendicular to the final INT major axis),

$$U^{ic}(b,x) = \frac{2}{b}\sum_{i=1}^{\infty}\left\{\frac{(-1)^i}{ix} + 2\sum_{j=1}^{\infty}\left[\frac{(-1)^{i+j}}{\sqrt{i^2x^2 + j^2}}\right]\right\}.$$

The above equations for the interaction energy can be readily modified to deal with an $(n,n)_{sq}$ INT. In these cases, the interaction energy becomes

$$U^{ic}_{(n,n)}(b,x) = \frac{1}{b}\sum_{i=1}^{2n}\left\{\frac{(-1)^i}{K_i x} + 2\sum_{j=1}^{\infty}\left[\frac{(-1)^{i+j}}{\sqrt{K_i^2 x^2 + j^2}}\right]\right\},$$

where

$$K_i = \frac{\sin\left(\frac{i\pi}{2n_t}\right)}{\sin\left(\frac{\pi}{2n_t}\right)}$$

is a geometric factor that accounts for the folding of the square plane. The replacement of a sum to infinity with a sum to $2n$ reflects the change from considering an infinite sheet to a finite folded sheet. The $2n$ limit also reflects the presence of two anion–cation chains per unit cell (which run at an angle of $45°$ to both $\mathbf{a}_1$ and $\mathbf{a}_2$).

These equations can be used to define an effective Madelung constant,

$$M_b(x) = -2\ln 2 + \sum_{i=1}^{2n}\left\{\frac{(-1)^i}{K_i x} + 2\sum_{j=1}^{\infty}\left[\frac{(-1)^{i+j}}{\sqrt{K_i^2 x^2 + j^2}}\right]\right\}.$$

Figure 10.6 shows the value of the M_b as a function of the index n. The $(n,n)_{sq}$ Madelung constants rapidly tend to the two-dimensional square planar limit ($M_b = -1.618$) as n increases. The figure also shows the Madelung constants calculated for sections of the bulk rocksalt crystal structure[62] plotted against the number of molecules in a square cross-section perpendicular to the crystallite major axis. The bulk-like structures show Madelung constants that tend towards the bulk rocksalt limit ($M_b = -1.7476$). The $(3 \times 3 \times \infty)$ crystallite becomes energetically favourable with respect to a nanotube of comparable density (number of ions per unit length) reflecting the presence of fully (nearest-neighbour) coordinated ions running along the centre of the crystallite. However, in the confined (cylindrical) environments afforded by the C-NTs, the INT-like structures would be expected to be energetically favoured, at least for relatively small diameter structures.

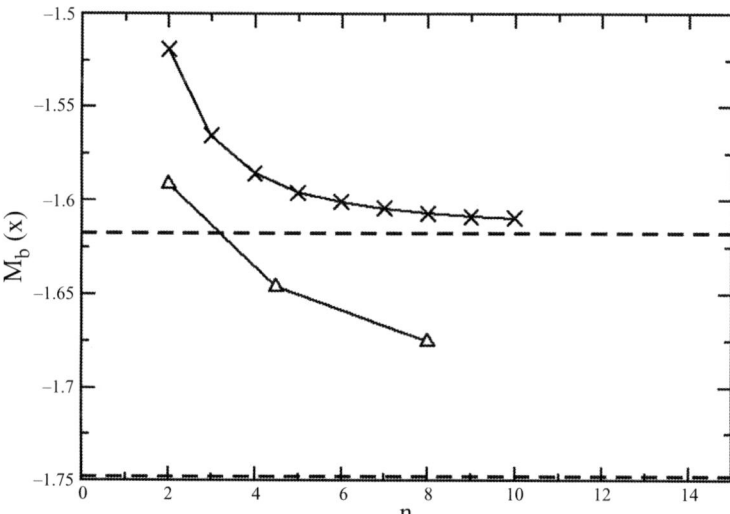

Figure 10.6 Effective Madelung constants, $M_b(x)$, for the $(n,n)_{sq}$ INTs (black crosses) compared with the values from the work of Bichoutskaia and Pyper[62] on rocksalt crystallites. The Δ points shows data from ref. 62 for the $(m \times m \times \infty)$ rocksalt crystallites. These data are plotted against the number of molecules in a plane perpendicular to the crystallite major axis. The dashed horizontal lines show the two dimensional plane (upper line) and bulk rocksalt (lower line) limits. In all cases the lines are drawn between points as a guide to the eye.

An analogous set of equations can be derived to account for the interactions in the $(n,0)_{sq}$ INTs. In this case the anion–cation chain energy is modified to reflect the twisting now present along the major axis,

$$U_{(n,0)}^{chain} = -\frac{2}{b} \sum_{m=1}^{m=\infty} \frac{(-1)^{m+1}}{\left[\left(D_{per}^m\right)^2 + \left(D_{par}^m\right)^2\right]^{1/2}},$$

where D_{par}^m and D_{per}^m are the components parallel and perpendicular to the final formed INT major axis. The INT energy is then calculated by constructing the tube from n chains.

Figure 10.7 shows the breakdown of the INT energies into inter- and intrachain interactions. Figure 10.7(a) shows the energies of an anion–cation chain making up part of a $(n,0)_{sq}$ INT, as a function of n. As n increases the energy of the chain approaches the linear chain limit of $-2\ln2/b$. The observation of a less-negative energy for the chains comprising the $(n,0)_{sq}$ INTs, compared with the linear chains (and hence the chains comprising the $(n,n)_{sq}$ INTs) reflects the effective "twisting" of the chain away from linear in the $(n,0)_{sq}$ INTS. The magnitude of this twist becomes more pronounced as n is reduced. Figure 10.7(b) shows the interchain energies, U^{ic}, for both the $(n,0)_{sq}$ and $(n,n)_{sq}$ INTs. As a result, both the interchain interaction energies and the chain energies (Figure 10.7) are relatively unfavourable for the $(n,0)_{sq}$ INTs.

As a final comment we note that the different dependencies of the INT energetics on their final diameter has implications for which class of INT may form within a given C-NT pore. For small radii C-NTs in particular the different dependencies of the INT energies on morphology may mean that square net-based INTs are thermodynamically stable over the hexagonal net-based analogues even if the hexagonal net sheet is thermodynamically favoured over the square net sheet.

10.5   The Energy Landscape Filter

The computational studies of the filling of the C-NTs indicate the process to be relatively facile. Furthermore, the strong electrostatic interactions dominate the structural properties of the filling material and, as a result, highly ordered structures are expected to be formed, at least in the lower pore diameter C-NTs. A clear implication, therefore, is that nanotubular structures may be ubiquitous. A key question, therefore, is to attempt to understand how such structures are accessed. A simple thermodynamic argument indicates that these nanotubular structures are simply crystal structures that are metastable with respect to the bulk crystal structures. A natural language exists in the form of the energy landscape (EL), At the simplest form the EL is often used to reflect simple reactions of the form $A \rightarrow B$ (see the inset to Figure 10.8). As a result,

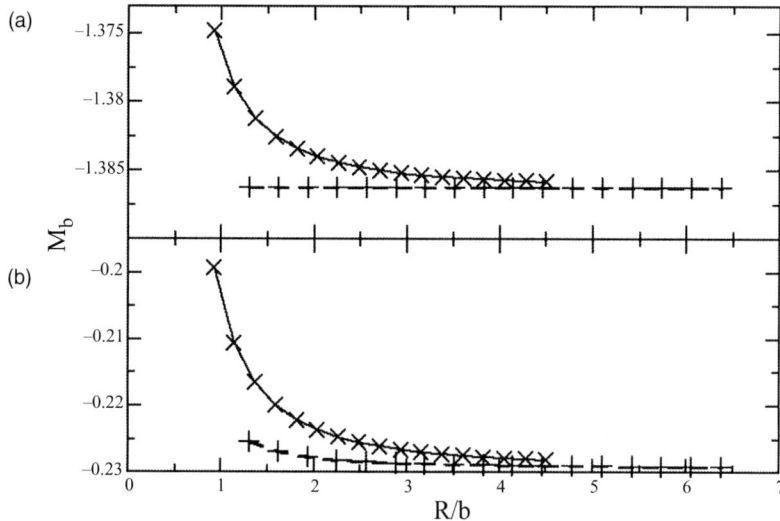

Figure 10.7 The breakdown of the system energies (displayed as effective Madelung constants) for the $(n,n)_{sq}$ (+ and dashed lines) and $(n,0)_{sq}$ (× and solid lines) INTs. The upper panel (a) shows the energy of a single anion–cation chain that is constant for the $(n,n)_{sq}$ INT. The lower panel (b) shows the interchain energy. The lines are drawn between points as a guide to the eye.

although relatively simple in nature, this representation contains information regarding both the free energy change for the reaction (thermodynamics) and the rate of the reaction *via* the height of the energy barrier (kinetics). More complex reactions can be understood in terms of more complex multi-dimensional energy landscapes.[84]

Figure 10.8 shows a section of an EL for elemental carbon. The bulk crystalline diamond and graphite structures appear at the most negative free energies. The C-NTs (and related fullerenes) appear at higher free energy (of the order of $6 \, \text{kJ mol}^{-1}$ above the graphite ground state energy minimum). The energy difference implies that these structures should be accessible at reasonable temperatures. However, in order to access this region of the EL extreme conditions (high temperature, electric arc . . .) must be applied to effectively vaporise the graphitic starting material. Once vaporised the material may condense to form a range of structures, both ordered and disordered, some of these may be nanotubular in nature. In terms of the EL there are a large number of competing pathways for condensations and, as a result, yields of the nanotubes may be low and their structures difficult to control.

Figure 10.9 shows the imagined EL for INT formation from molten salts. The bulk ground state crystal is melted and a resulting liquid may access a number of competing pathways with those returning to the crystalline state (crystallisation) or glassy state (vitrification) the most probable. The C-NTs act

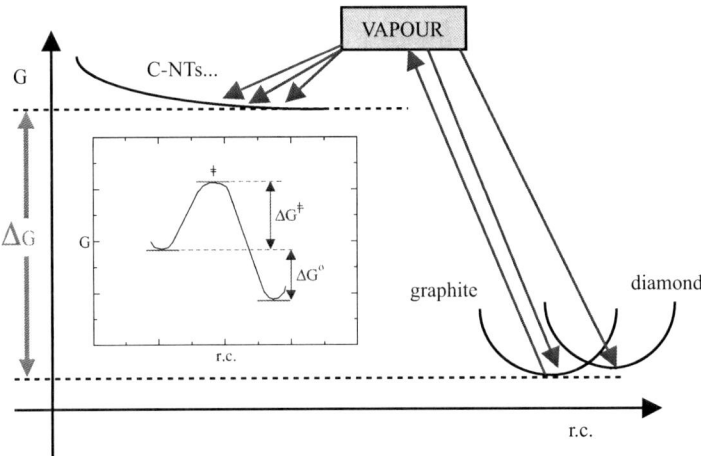

Figure 10.8   Schematic energy landscape highlighting the route of formation of C-NTs and related fullerenes. The thermodynamically stable graphite phase is vaporised under extreme conditions to produce a vapour. The pathways accessible to the vapour are also highlighted by arrows. The intense competition at extreme conditions hinders the controlled formation of C-NTs in high yield. The free energy gap, ΔG, between the graphite phase and the C-NTs is estimated to be of the order of $6\,kJ\,mol^{-1}$.

as effective energy landscape filters, in Figure 10.9(a) effectively blocking the pathways both back to the three-dimensional crystalline region of the EL and to the glassy states, as a result, promoting the pathways to the INT structures to becoming most probable. Note that the presence of the highly confining environment removes the usual disordered liquid state from the accessible region of the energy landscape (effectively a sphere packing effect[85]) whilst site disorder is precluded by the magnitude of the electrostatic interactions at typical molten salt temperatures.

Figure 10.9(b) shows a magnified section of the EL in order to highlight the detailed role of the C-NTs in selecting specific INTs. The INT energetics shown are for a model that favours square net INTs at low radius and hexagonal-nets at high radius. A low radius C-NT acts to select a small range of INTs whose radii lie in a critical range such that the attractive interactions within the encasing C-NT are maximised. A larger-radius C-NT will select a range of INTs of larger radius. Note that, in the example highlighted, the low radius C-NT will tend to select square net INTS despite the hexagonal sheet being thermodynamically favoured over the square net sheet. Furthermore, the larger the C-NT the more INTs are "available" for selection. The concept of the energy landscape filter, which acts to select a range of INT structures, ties in with the distribution of structures obtained in Section 10.3.2 (Figure 10.2). For the fixed (12,10) C-NT (diameter approximately $15\,\text{Å}$) three hexagonal INTs are observed with morphologies $(4,3)_{hex}$, $(5,2)_{hex}$ and $(6,0)_{hex}$, respectively.

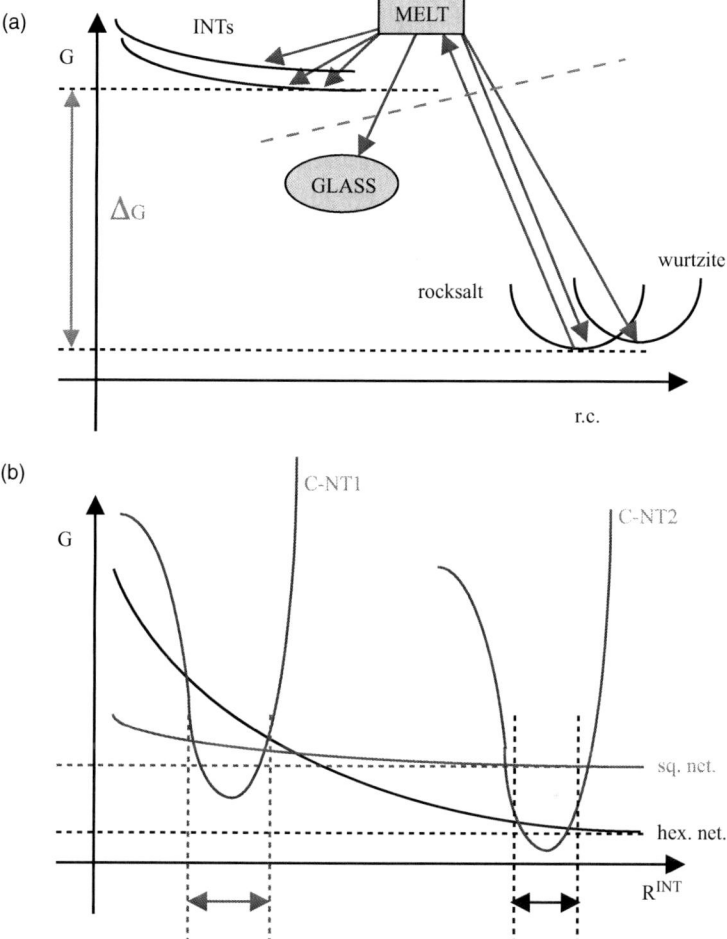

Figure 10.9 (a) Schematic representation of the energy landscape for the INT formation for molten salts filling C-NTs. The thermodynamically stable rocksalt and wurtzite crystals are melted at high temperature to generate the molten salt (remaining arrows). The arrow pointing up indicates a selection of pathways accessible to the system on cooling, including the formation of both hexagonal net- and square net-based INTs. The insertion of C-NTs into the system effectively filters a large section of the landscape (effectively precluding crystallisation back to the bulk crystals or glassy states) as indicated by the long dashed line. (b) Magnified section of the energy landscape in (a) highlighting the role of the C-NTs in "selecting" the INT structures. The INT energies for a potential that forms square nets at low INT radius and hexagonal nets at high radius are shown. A relatively low-radius C-NT (labelled C-NT1) selects low-radius INTs and hence favours the formation of square net-based structures, despite the hexagonal sheets being thermodynamically favoured over the square nets. A high-radius C-NT (labelled C-NT2) selects INTs of a higher radius and hence favours the formation of hexagonal net-based structures. The double-headed arrows represent the approximate range of the INT radii selected by each C-NT.

Table 10.1 lists the INT structures that can be formed around these radii. For hexagonal INTs the INT radius is given by $R = (a_0/2\pi)(n^2 + m^2 + nm)^{(1/2)}$. The INTs are formed within a well-defined range of radii. Furthermore, the distribution of INTs formed can be largely predicted from the radius difference with respect to the confining C-NT using a Boltzmann distribution. The relatively simple nature of this dependence arises from the lack of morphology dependence of the hexagonal net-based INTs on morphology. In contrast, Table 10.2 lists the morphologies of the hexagonal net structures that can be formed at low radius, with the INT radius given by $R = (a_0/2\pi)(n^2 + m^2)^{(1/2)}$. Of these, the $(2,2)_{sq}$ and $(3,2)_{sq}$ are found to form within a $(10,10)$ C-NT, whilst the $(3,1)_{sq}$ INT, which has a radius intermediate between those observed, does not form. The key difference with respect to the hexagonal tubes is that the energies

Table 10.1 Possible folding morphologies for the hexagonal net-based INTs expressed in terms of their respective $(n,n)_{hex}$ chiral vectors. The INT radius is given by $R = (a_0/2\pi)(n^2 + m^2 + nm)^{(1/2)}$ with the (n,m) dependence shown here. The horizontal lines indicate the lower and upper limits of the INTs observed to form in a $(12,10)$ C-NT.

(n,m)	$n^2 + m^2 + nm$
(4,2)	28
(5,1)	31
(6,0)	36
(4,3)	37
(5,2)	39
(6,1)	43
(4,4)	48
(7,0),(5,3)	49

Table 10.2 Possible folding morphologies for the hexagonal net-based INTs expressed in terms of their respective $(n,n)_{sq}$ chiral vectors. The INT radius is given by $R = (a_0/2\pi)(n^2 + m^2)^{(1/2)}$ with the (n,m) dependence shown here. The bold figures indicate the two INT morphologies observed to form within a $(10,10)$ C-NT.

(n,m)	$n^2 + m^2$
(1,1)	2
(2,1)	5
(2,2)	**8**
(3,1)	10
(3,2)	**13**
(4,0)	16
(4,1)	17
(3,3)	18

of the square net-based INTs are themselves dependent upon the folding morphology. As a result, high-energy INTs, such as the $(3,1)_{sq}$, are effectively precluded.

The energy landscape may also be used to rationalise the mechanism of formation of the square net-based INTs. The filling profiles shown in Figure 10.1 correspond to a systematic chain-by-chain build up of INT structure. Figure 10.10 shows a section of the underlying energy landscape for this potential. The effect of adding the C-NT/INT interaction energy (highlighted in the figure) is to thermodynamically discriminate against larger radii INTs (where the INT structure is pushed against the C-NT inner repulsive wall). The transition between these structures can be understood in terms of thermo-dynamically driven "hopping" between successive energy basins on this land-scape. Here, the mechanism follows a "chain-by-chain" motif in which successive (charge neutral) M-X-M-X . . . chains are added, giving the order $(1,1)_{sq} \rightarrow (1 \times 3)_{sq} \rightarrow (2,2)_{sq} \rightarrow (3,2)_{sq}$. Three points are worthy of specific comment. First, the $(2,2)_{sq}$ and $(3,2)_{sq}$ minima are close in energy, reflecting the fine balance between their respective radii and the morphology dependence of the folding energy. As a result, the transformation to the $(3,2)_{sq}$ structure, which occurs from the $(2,2)_{sq}$ basin, is not strongly thermodynamically driven.

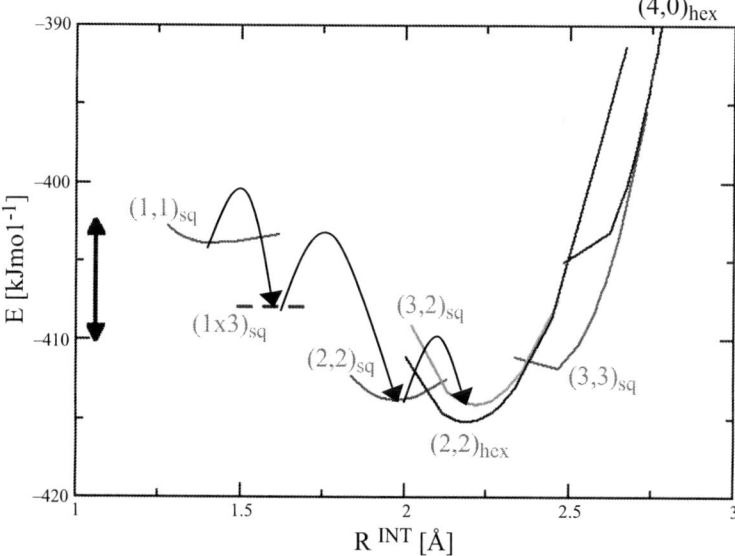

Figure 10.10 The dynamic energy landscape for the filling of the (10,10) C-NT. The INT morphologies are indicated on the figure. The arrows highlight the "chain-by-chain" basin-hopping mechanism by which the C-NT fills. The energy scale for the INTs is given by the left ordinate, whilst that for the C-NT/INT interaction is given by the right ordinate. The thermal energy at the filling temperature ($T = 800$ K) is indicated by the double arrow.

To further support the idea of basin "hopping" at these thermal energies (at 800 K, RT is around $7\,kJ\,mol^{-1}$ highlighted in Figure 10.10 we note that a number of "reverse hops" (transformations in which the number of chains in the INT structure falls) are observed on the simulation timescale. In these cases both the transformations $(1,1)_{sq} \leftrightarrow (1 \times 3)_{sq}$ and $(1 \times 3)_{sq} \leftrightarrow (2,2)_{sq}$ are observed. Secondly, the minimum for the $(2,2)_{hex}$ INT structure lies at a lower energy than either the corresponding $(2,2)_{sq}$ or $(3,2)_{sq}$ INT minima. However, a mechanism to transform from a square planar to a hexagonal structure involves the breaking of bonds, which equates to a relatively high activation barrier for this process compared with the "chain-by-chain" building mechanism to give the $(3,2)_{sq}$ structure. As a result, the square-plane-based structures appear kinetically stable over hexagonal INTs of similar radii. Thirdly, the effect of the encasing C-NT interaction can be seen for the larger radii INTs on this landscape. As a result, the $(3,3)_{sq}$ INT becomes inaccessible on the simulation timescales as, although only marginally thermodynamically unfavourable with respect to the $(3,2)_{sq}$ INT, the insertion of a sixth ion chain involves too high an activation barrier to be accessed.

10.6 Conclusions

In this chapter, the formation of inorganic nanotubes of stoichiometry MX by the filling of carbon nanotubes has been discussed. The use of relatively simple interatomic potential models makes the calculations computationally tractable and, as a result, allows for the simulation of the relatively long length- and timescales required to observe the filling of the carbon tubes directly. These observations allow for information regarding the formation mechanisms to be obtained (information that is difficult to obtain directly from experiment). Energy-minimisation calculations, initiated from ideal INT structures obtained by folding the respective planar nets, has allowed the underlying phase diagrams to be established and further allows the INT energetics to be rationalised in terms of the interatomic interactions. Finally, the formation of INTs can be rationalised by reference to the energy landscape. The carbon tubes act as effective landscape filters both in terms of effectively removing the bulk crystalline and liquid states and in terms of selecting the INTs themselves.

References

1. H. W. Kroto, J. R. Heath, S. C. O'Brien, R. F. Curl and R. E. Smalley, *Nature*, 1985, **318**, 162–163.
2. P. M. Ajayan and T. W. Ebbesen, *Rep. Prog. Phys.*, 1997, **60**, 1025–1062.
3. S. Iijima, *Nature*, 1991, **354**, 56–58.
4. D. S. Bethune, C. H. Klang, M. S. de Vries, G. Gorman, R. Savoy, J. Vasquez and R. Beyers, *Nature*, 1993, **363**, 605–607.
5. M. L. Cohen, *Mater. Sci. Eng.*, 2001, **C15**, 1.

6. L. Scheffer, *et al.*, *Phys. Chem. Chem. Phys.*, 2002, 2095.
7. O. Tal, *et al.*, *Chem. Phys. Lett.*, 2001, **344**, 434.
8. M. Hershfinkel, *et al.*, *J. Am. Chem. Soc.*, 1994, **116**, 1914.
9. M. Remškar, *et al.*, *Appl. Phys. Lett.*, 1996, **69**, 351.
10. G. L. Frey, *et al.*, *Phys. Rev. B*, 1999, **60**, 2882.
11. P. J. Ziemann and A.W. Castleman Jr., *J. Chem. Phys.*, 1991, **94**, 718.
12. M. Wilson, *J. Phys. Chem.*, 1997, **B101**, 4917.
13. T. W. Ebbesen, *J. Phys. Chem. Solids*, 1996, **57**, 951.
14. A. R. Ubbelohde, *The Molten State of Matter*, John Wiley & Sons, Chichester (1978).
15. D. A. Britz and A. N. Khlobystov, *Chem. Soc. Rev.*, 2006, **35**, 637.
16. J. Sloan, A. I. Kirkland and J. L. Hutchison, *et al.*, *Comptes Rendus Physique*, 2003, **4**, 1063–1074.
17. A. N. Khlobystov, D. A. Britz and G. A. D. Briggs, *Acc. Chem. Res.*, 2005, **38**, 901.
18. S. C. Tsang, Y. K. Chen, P. J. F. Harris and M. L. H. Green, *Nature*, 1994, **372**, 159–162.
19. P. M. Ajayan, T. W. Ebbesen, T. Ichihashi, S. Iijima, K. Tanigaki and H. Hiura, *Nature*, 1993, **362**, 522–525.
20. P. M. Ajayan, T. Ichihashi and S. Iijima, *Chem. Phys. Lett.*, 1993, **202**, 384–388.
21. P. M. Ajayan, O. Stephan, Ph. Redlich and C. Colliex, *Nature*, 1995, **375**, 564–567.
22. Y. K. Chen, M. L. H. Green and S. C. Tsang, *Chem. Commun.*, 1996, 2489–2490.
23. J. Sloan, J. Cook, A. Chu, M. Zwiefka-Sibley, M. L. H. Green and J. L. Hutchison, *J. Solid State Chem.*, 1998, **140**, 83–90.
24. J. Sloan, D. M. Wright, H.-G. Woo, S. R. Bailey, G. Brown, A. P. E. York, K. S. Coleman, J. L. Hutchison and M. L. H. Green, *Chem. Commun.*, 1999, 699–700.
25. J. Sloan, M. Terrones, S. Nufer, S. Friedrichs, S. R. Bailey, H.-G. Woo, M. Rühle, J. L. Hutchison and M. L. H. Green, *J. Am. Chem. Soc.*, 2002, **124**, 2116–2117.
26. J. Sloan, M. C. Novotny, S. R. Bailey, G. Brown, C. Xu, V. C. Williams, S. Friedrichs, E. Flahaut, R. L. Callendar, A. P. E. York, K. S. Coleman, M. L. H. Green, R. E. Dunin-Borkowski and J. L. Hutchison, *Chem. Phys. Lett.*, 2000, **329**, 61–65.
27. R. R. Meyer, J. Sloan, R. E. Dunin-Borkowski, A. I. Kirkland, M. C. Novotny, S. R. Bailey, J. L. Hutchison and M. L. H. Green, *Science*, 2000, **289**, 1324–1326.
28. J. Sloan, S. J. Grosvenor, S. Friedrichs, A. I. Kirkland, J. L. Hutchison and M. L. H. Green, *Angew. Chem. Int. Ed.*, 2002, **41**, 1156.
29. E. Philp, J. Sloan, A. I. Kirkland, R. R. Meyer, S. Friedrichs, J. L. Hutchison and M. L. H. Green, *Nature Mater.*, 2003, **2**, 788–791.
30. S. Friedrichs, R. R. Meyer, J. Sloan, A. I. Kirkland, J. L. Hutchison and M. L. H. Green, *Chem. Commun.*, 2001, 929–930.

31. S. Friedrichs, A. I. Kirkland, R. R. Meyer, J. Sloan and M. L. H. Green, *Electron Microscopy and Analysis*, 2004, 455.

32. S. Friedrichs and M. L. H. Green, *Z. Metallkd.*, 2005, **96**, 419.

33. S. Friedrichs, U. Falke and M. L. H. Green, *ChemPhysChem*, 2005, **6**, 300.

34. E. Flahaut, J. Sloan, S. Friedrichs, A. I. Kirkland, K. S. Coleman, V. C. Williams, N. Hanson, J. L. Hutchison and M. L. H. Green, *Chem. Mater.*, 2006, **18**, 2059.

35. R. Carter, J. Sloan, A. I. Kirkland, R. R. Meyer, P. J. D. Lindan, G. Lin, M. L. H. Green, A. Vlandas, J. L. Hutchison and J. H. Harding, *Phys. Rev. Lett.*, 2006, **96**, 215501.

36. G. Brown, S. R. Bailey, J. Sloan, C. Xu, S. Friedrichs, E. Flahaut, K. S. Coleman, M. L. H. Green, J. L. Hutchison and R. E. Dunin-Borkowski, *Chem. Commun.*, 2001, 845.

37. M. Hulman, H. Kuzmany, P. M. F. J. Costa, S. Friedrichs and M. L. H. Green, *Appl. Phys. Lett.*, 2004, **85**, 2068.

38. S. Y. Hong, R. Popovitz-Biro, G. Tobias, B. Ballesteros, B. G. Davis, M. L. H. Green and R. Tenne, *Nano. Res.*, 2010, **3**, 170.

39. S. Y. Hong, *et al.*, *Nature Mater.*, 2010, **9**, 485.

40. J. S. Bendall, A. Ilie, M. E. Welland, J. Sloan and M. L. H. Green, *J. Phys. Chem. B*, 2006, **110**, 6569–6573.

41. A. Ilie, J. S. Bendall, D. Roy, E. Philp and M. L. H. Green, *J. Phys. Chem. B*, 2006, **110**, 13848–13857.

42. E. L. Sceats, J. C. Green and S. Reich, *Phys. Rev. B*, 2006, **73**, 125441.

43. E. L. Sceats, M. L. H. Green, A. I. Kirkland and J. C. Green, *Chem. Phys. Lett.*, 2008, **466**, 76–78.

44. N. Kuganathan and J. C. Green, *Chem. Commun.*, 2008, 2432.

45. E. Bichoutskaia and N. C. Pyper, *Chem. Phys. Lett.*, 2006, **423**, 234–239.

46. X. Y. Zhang, X. H. Yan and Y. R. Yang, *Physica E*, 2010, **42**, 1896–1900.

47. D. Q. Fang, A. L. Rosa and R. Q. Zhang, *et al.*, *J. Phys. Chem. C*, 2010, **114**, 5760–5766.

48. D. B. Zhang, T. Dumitrica and G. Seifert, *Phys. Rev. Lett.*, 2010, **104**, 065502.

49. M. Wilson and P.A. Madden, *J. Am. Chem. Soc.*, 2001, **123**, 2101.

50. M. Wilson, *J. Chem. Phys.*, 2002, **116**, 3027.

51. M. Wilson, *Chem. Phys. Lett.*, 2002, **366**, 504–509.

52. M. Wilson, *Nano Lett.*, 2004, **4**, 299–302.

53. M. Wilson, *Chem. Phys. Lett.*, 2004, **397**, 340.

54. M. Wilson, *J. Chem. Phys.*, 2006, **124**, 124706.

55. M. Wilson and S. Friedrichs, *Acta. Crystallogr. A*, 2006, **62**, 287.

56. M. Wilson, *Faraday Discussions*, 2007, **134**, 283.

57. C. L. Bishop and M. Wilson, *Mol. Phys.*, 2008, **106**, 1665–1674.

58. C. L. Bishop and M. Wilson, *J. Mater. Chem.*, 2009, **19**, 2929.

59. C. L. Bishop and M. Wilson, *J. Phys.: Condens. Matter*, 2009, **21**, 115301.

60. G. G. Tibbetts, *J. Cryst. Growth*, 1983, **66**, 632.

61. M. Baldoni, S. Leoni, A. Sgamellotti, G. Seifert and F. Mercuri, *Small*, 2008, **3**, 1730.

62. E. Bichoutskaia and N.C. Pyper, *J. Phys. Chem. B*, 2006, **110**, 5936.
63. E. Bichoutskaia and N.C. Pyper, *J. Chem. Phys.*, 2008, **129**, 154701.
64. D. Frenkel and B. Smit, *Understanding Molecular Simulation*, Academic Press, San Diego (2001).
65. M. Allen and D. Tildesley, *Computer Simulation of Liquids*, Oxford University Press, Oxford (1987).
66. W. H. Press B. P. Flannery S. A. Teukolski W. T. Vetterling, *Numerical Recipes*, Cambridge University Press, Cambridge 1992.
67. M. J. L. Sangster and M. Dixon, *Adv. Phys.*, 1976, **23**, 247.
68. A. J. Stone, *Theory of Intermolecular Forces*, Oxford University Press, Oxford, 1996.
69. M. Born and J.E. Mayer, *Z. Phys.*, 1932, **75**, 1.
70. P. A. Madden and M. Wilson, *Chem. Soc. Rev.*, 1996, **25**, 339.
71. A. Aguado, L. Bernasconi, S. Jahn and P. A. Madden, *Faraday Discuss.*, 2003, **124**, 171.
72. W. A. Steele, *J. Phys. Chem.*, 1978, **82**, 817.
73. J. Tersoff, *Phys. Rev. B*, 1988, **37**, 6991.
74. R. Saito, G. Dresselhaus and M. S. Dresselhaus, *Physical Properties of Carbon Nanotubes*, Imperial College Press, London 1998.
75. M. W. Finnis, *Acta. Metall. Mater.*, 1992, **40**, S25–S37.
76. M. W. Finnis, R. Kaschner, C. Kruse, J. Furthmüller and M. Scheffler, *J. Phys. Condens. Matter*, 1995, **7**, 2001.
77. J. I. Siepmann and M. Sprik, *J. Chem. Phys.*, 1995, **102**, 511–524.
78. M. García-Hernández, P.S. Bagus and F. Illas, *Surf. Sci.*, 1998, **409**, 69–80.
79. S. T. Cui, *Mol. Phys.*, 2006, **104**, 2993.
80. S. K. Reed, O. J. Lanning and P. A. Madden, *J. Chem. Phys.*, 2007, **126**, 084704.
81. T. R. Gingrich and M. Wilson, *Mol. Phys.*, 2010, **500**, 178.
82. D. H. Robertson, D. W. Brenner and J. W. Mintmire, *Phys. Rev. B*, 1992, **45**, 12592.
83. D. Sánchez-Portal, E. Artacho, J. M. Soler, A. Rubio and P. Ordejon, *Phys. Rev. B*, 1999, **59**, 12678.
84. D. J. Wales, *Energy Landscapes*, Cambridge University Press, Cambridge 2004.
85. G. T. Pickett, M. Gross and H. Okuyama, 2000, **85**, 3652.

CHAPTER 11

Native and Irradiation-Induced Defects in Graphene: What Can We Learn from Atomistic Simulations?

JANI KOTAKOSKI AND ARKADY V. KRASHENINNIKOV

Division of Materials Physics, University of Helsinki, P.O. Box 42 (Pietari Kalmin katu 2), 00014 Helsinki, Finland

11.1 Introduction

Graphene – the ultimately thin honeycomb-like membrane consisting only of sp^2-bonded carbon atoms – has been one of the most attractive research subjects since monolayer graphene flakes became available for experiments.[1] Indeed, the yearly number of publications that mention graphene (see Figure 11.1) demonstrates an exponential growth. A considerable part of these publications are theoretical studies, and the general consensus is that at the moment experimental understanding lags behind theory. This is because graphene has been simulated by the computational materials-science community well before it was experimentally discovered. In fact, the electronic properties of this material were addressed for the first time more than 60 years ago.[2] The reason for the early interest arose from the fact that the most stable carbon allotrope, graphite, is formed by stacking a large number of graphene layers on top of each other to obtain the bulk structure kept together by weak van der Waals forces between the individual layers. Thus, to understand the

RSC Theoretical and Computational Chemistry Series No. 4
Computational Nanoscience
Edited by Elena Bichoutskaia
© Royal Society of Chemistry 2011
Published by the Royal Society of Chemistry, www.rsc.org

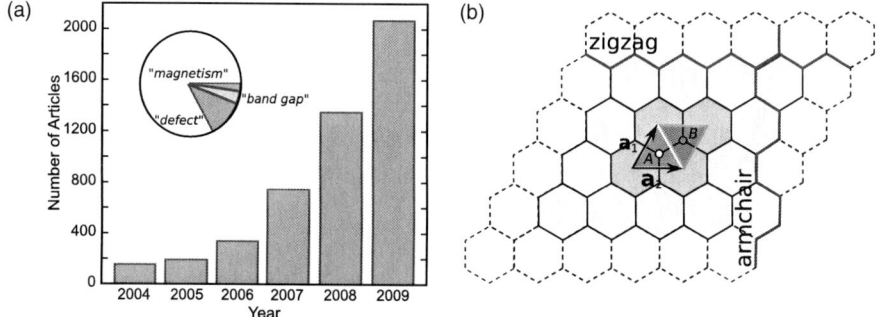

Figure 11.1 Number of publications per year (a) which are found with the ISI Web of Knowledge (published by Thomson Reuters) search at http://isiknowledge. com/ with search term *"graphene"*, since the first report on experiments on monolayer graphene in 2004. The piechart shows how many of those publications were associated with search terms *"magnetism"*, *"bandgap"* or *"defect"*, the last of which was combined with other related keywords. Note that some publications are likely to be found in more than one category. Graphene honeycomb lattice (b) with the two triangular sublattices *A* and *B* marked along with the lattice vectors $\mathbf{a}_1$ and $\mathbf{a}_2$, as well as zigzag and armchair lattice directions in the structure.

physics and chemistry of graphite, one must first study graphene as the prototype material.

Humankind has used graphite in paints for centuries because the layers can easily be detached from one another and used to mark surfaces with a dark colour. More recently, graphite has been utilised in pencils and as a lubricant for the same reason. In contrast to the weak interlayer interaction, the intralayer sp^2 carbon bonds are among the strongest bonds known in nature – even stronger than the sp^3-bonds of diamond. Carbon is famous for the most complicated chemistry possible for any element, which has – as an extreme example – allowed the evolution of life on Earth through the DNA molecule. It has four valence electrons that occupy the 2s (two electrons), and 2p orbitals. In compounds, carbon exists in different *sp*-hybridisations, such as sp^1, sp^2 and sp^3, where the superscript denotes the number of 2p orbitals involved in bonding. Of these, sp^1 leads to a 1D chain, sp^2 to 2D graphene and sp^3 to 3D diamond. The left-over orbitals do not take part in the bonding. Surprisingly, graphite was also found to exhibit peculiar electronic properties (*e.g.* conductivity similar to that of a poor metal, negative temperature coefficient). When the rise of quantum mechanics made it possible to understand the electronic structure of materials, scientists set out to find the underlying reason for these properties. The early work of Wallace[2] applied a tight-binding (TB) model to the π electrons of the sp^2-bonded graphene monolayer, and showed that this material is a *zero-bandgap semiconductor* (or in other words, a *semimetal*) with unusual linearly dispersing electronic excitations reminiscent of relativistic Dirac electrons. However, the role of electron–electron interactions in graphene is still a subject of intensive research. The latest developments in the theory of the electronic structure of graphene can be found in a recent review by Castro Neto *et al.*[3]

However, not only has the academic interest been the driving force for intensive research of the electronic properties of graphene. Our time is sometimes referred to as the Information Age, and we indeed rely heavily on the ever-increasing computational power. This has been made possible due to the success of materials science in miniaturising silicon-based electronic components (mainly transistors) and increasing the speed of their operation. As the atomic scale is being approached, it has become evident that new materials are needed to retain the computational speed increase and circumvent the fundamental size-related problems. Graphene and other nanoscale carbon materials, with their extraordinary properties,[3,4] have been proposed for this purpose.

An important aspect regarding electronic properties is the effect of disorder. This is particularly relevant to graphene grown with the chemical vapor deposition (CVD) method. The CVD graphene is known to contain a large number of dislocations and disordered areas, some of which may originate from the growth processes, and some from imaging the structure[5] in a transmission electron microscope. In addition to native, pre-existing defects, defects can also be introduced *via* irradiation (*e.g.* in outer space, nuclear reactors) or created in a chemically harsh environment (*e.g.* by oxidation). Impacts of energetic particles such as electrons or ions typically give rise to formation of atomic defects in solids, sometimes making the material unusable for a particular application. Historically, the necessity of understanding irradiation effects in carbon materials emerged during the 1950s with the appearance of nuclear power plants in which the irradiation-induced degradation of graphitic components was an important issue (see ref. 6 for an example). ISI Web of Knowledge lists 4 research articles with search terms "*graphite*" and "*irradiation*" during the 1950s, and 56 in the 1960s. After this, the number steadily increased until the early 1990s, after which it exploded to reach 1054 during the first decade of this century. The exponential growth in the number of publications occurs at the same time as the appearance of carbon-nanotube publications (carbon nanotubes were brought to common knowledge in 1991[7]). During all these decades, electron irradiation has also been used to study the radiation-damage effects in graphitic materials.

Despite the detrimental effects irradiation has on the target, beams of energetic particles can also be used for tailoring material properties. The most obvious example is ion implantation technology customarily employed in the semiconductor industry.[8] Computer simulations have shown[9,10] that it is possible to use ion irradiation to replace carbon atoms by boron and nitrogen *via* ion irradiation to either dope or functionalize a carbon nanotube. Experiments have confirmed these results.[11,12] In graphene, due to the lack of curvature, this process can be expected to be even easier. Experiments have also demonstrated that the morphology of graphene can be changed in a controllable manner using electron irradiation.[13–15] While structural defects are known to alter the electronic properties of carbon nanomaterials,[16,17] this opens up the possibility for creating nanoscale devices for electronics from graphene.

In general, nanomaterials have a drastically different response to radiation with respect to the bulk counterparts because of their dimensions; an energetic particle typically passes through a nanostructure without depositing all of its energy into the system. Moreover, the displaced target atoms customarily become ejected out from the nanostructure instead of contributing to a collision cascade. Thus, both the primary and secondary collisions in nanostructures differ from those in bulk. Especially in the case of a monatomic layer, such as graphene, the conventionally used models for estimating irradiation damage in solids fail, as demonstrated in ref. 18. The difference in deposited energies to nanostructures and bulk under irradiation means that using irradiation in a useful manner to modify nanostructures is possible with less damage as compared to bulk structures. Despite a few recent studies,[19–22] no experiments have yet revealed ion-irradiation effects in graphene at the microscopic level. Therefore, computational work is needed as guidance for the later experiments. Also, the microscopic processes occurring under electron irradiation of graphene have remained unknown, as will be explained in this chapter.

In this chapter, we give an overview of recent theoretical progress in our understanding of native and irradiation-induced defects in graphene with a particular stress on the theoretical results. We first discuss the observed electron- and ion-bombardment mediated effects in graphene, and outline the experimental techniques used to characterize these effects. Then, computational methods employed to simulate irradiation of graphene are briefly introduced. We then describe the variety of defects that exist in this material either naturally or due to irradiation. Finally, we explain how irradiation can be used to modify the properties of graphene, and why atomistic simulations play a crucial role in understanding the active microscopic processes.

11.2 Experimental Evidence for Defects in Graphene

11.2.1 Transmission Electron Microscopy (TEM) and Defects Created by the Electron Beam

Discussing recent scientific progress in understanding the physics of sp^2-bonded carbon, and especially graphene, without considering high-resolution transmission electron microscopy (HR-TEM) is – if not meaningless – at least disingenuous. HR-TEM is currently the ultimate device for atomic-scale imaging. It offers a two-dimensional projection of the target material – with an atomic resolution if the device is aberration corrected. Clearly, for such a device, the optimal target material would be a two-dimensional atomically thin membrane, *e.g.* graphene. For such a target it is possible to see each individual atom if the local atomic configuration under the electron beam remains stationary during the exposure, as shown in Figure 11.2. This property can also be used in utilising graphene as the thinnest possible TEM grid material for HR-TEM

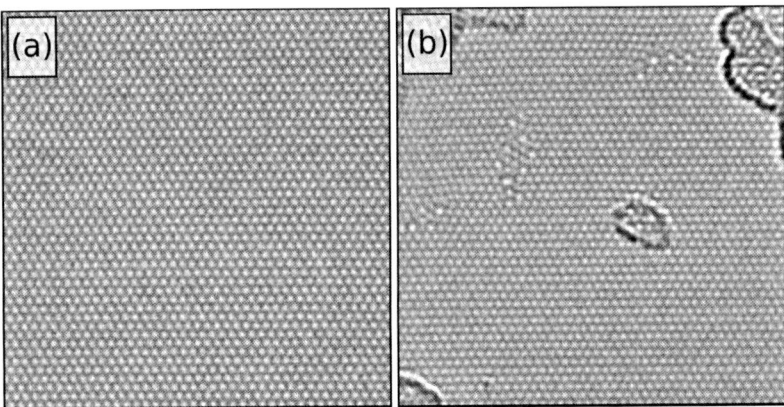

Figure 11.2   HR-TEM images of graphene. When the acceleration voltage is below
the threshold for defect production, no damage is created during imaging
(a). Brief exposure of graphene to high-energy electron irradiation
creates a set of complex defect structures built from a random set of
polygons (b) with a locally slightly lowered density due to ejected atoms.
Images courtesy of Jannik Meyer, University of Ulm, Germany.

experiments; when carbon atoms in the lattice are stationary, their positions
can be extracted from the TEM images completely.[14] The stability of the atomic
structure under the beam (its radiation hardness) depends on possible
pre-existing defects, environment (for example foreign molecules present in the
TEM vacuum chamber), and the acceleration voltage of the TEM. When
displacements of target atoms occur due to electron impacts, it is impossible to
follow the motion of the atoms *in situ* because of the difference between
the timescale of displacement events (10^{-21} s) and exposure times (1 s).[23] The
first HR-TEM images of graphene have revealed both the perfect honeycomb
structure as well as various defects (Figure 11.3).[13–15,24]

In the case of graphene, electron irradiation clearly cannot be used to
introduce such drastic structural changes as has been demonstrated for other
carbon nanomaterials (*e.g.* using carbon nanotubes and onions as pressure
vessels by decreasing their diameter under a TEM beam[25,26]). However, there
are ways to also alter the morphology of graphene with a TEM. For example,
reorganisation of edges of large holes has been shown (Figure 11.4).[15] With
electron irradiation, under conditions where graphene cannot compensate the
loss of atoms *via* bond rotations, creation of holes and deriving monatomic
carbon chains between two such holes has been demonstrated.[27,28] Overall,
energetic electrons will cause significant changes to the atomic network after a
sufficient dose by introducing a randomised polygon structure in the place of
the honeycomb lattice. Typical result of continuous electron irradiation with
a high enough acceleration voltage (~ 100 kV) is shown in Figure 11.2(b).
Curiously, most atoms remain three-coordinated with sp^2-hybridisation.
Hence, the structure also remains truly two-dimensional. In the defected

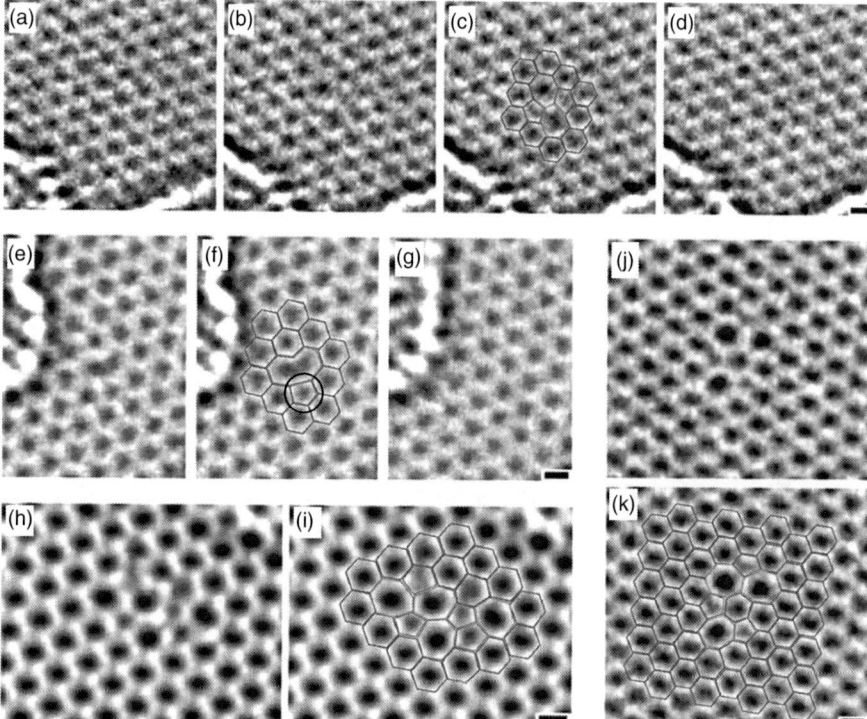

Figure 11.3 Metastable defects found in HRTEM image sequences. (a–d) Stone–Wales (SW) defect: (a) unperturbed lattice before appearance of the defect, (b) SW defect (c) same image with atomic configuration superimposed, (d) relaxation to unperturbed lattice (after ca. 4 s). (e–g) Reconstructed vacancy: (e) original image and (f) with atomic configuration; a pentagon is indicated. (g) Unperturbed lattice, 4 s later. (h and i) Defect image and configuration consisting of four pentagons and heptagons. Note the two adjacent pentagons. (j and k) Defect image and configuration consisting of three pentagons and three heptagons. This defect returned to the unperturbed lattice after 8 s. In spite of the odd number of 5–7 pairs, this is not a dislocation core (it is compensated by the rotated hexagon near the centre of the structure). All scale bars are 2 Å. From ref. 14.

regions some atoms are missing and the lattice has accommodated the locally lowered density by reorganising the atoms. The kinetic energy that allows for the reorganisation is provided by the electron beam *via* knock-on collisions with target atoms.

11.2.2 Defects Produced by Ion Irradiation

Impacts of ions can do much larger modifications to a graphene lattice than what is possible with electrons. This is mainly because of two major differences

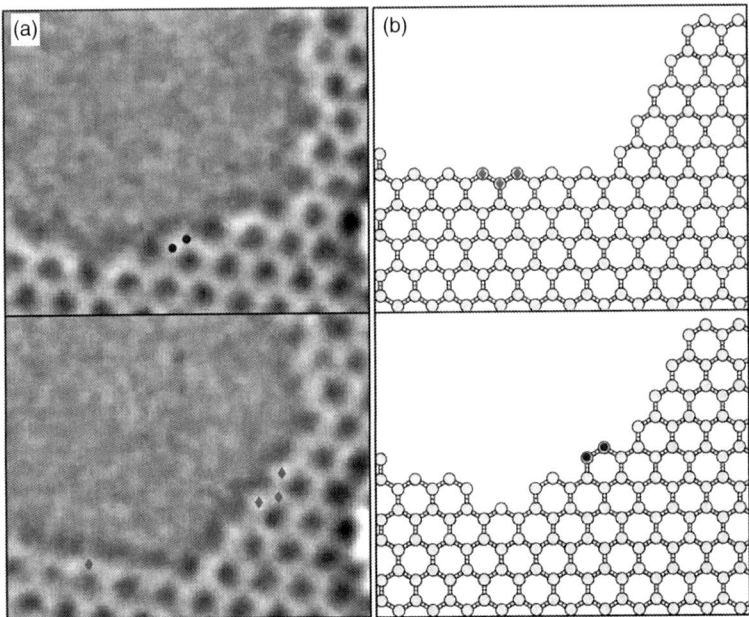

Figure 11.4 Edge reconfiguration. (a) Conversion of an armchair edge (top) to a zigzag edge (bottom). The two atoms marked as dots in the upper frame are gone in the lower frame, where four new carbon atoms are indicated as diamonds. The 7-hexagon armchair edge is transformed into a 9-hexagon zigzag edge with a 60° turn. The transformation occurs due to migration of atoms along the edge. (a) Similar behaviour is observed in the kinetic Monte Carlo simulation of hole growth, where three zigzag atoms (diamonds, top) disappear and two armchair atoms (dots, bottom) appear. From ref. 15.

between electron and ion irradiation. First, ions are much more massive than electrons, which allows for larger momentum transfer. Also, the space angle available for scattering is not limited in the case of ions for the same reason. Secondly, by choosing very low irradiation energies, and carefully selecting the ion species used in the process, it is possible to alter the local chemistry *via* introducing impurities. Obviously, electron irradiation can never alter the chemistry of a target system in this way.

Ion irradiation is an attractive method for a systematic study of the role of disorder on the properties of a material, because irradiation doses and hence the amount of disorder can be easily controlled. A two-dimensional target material brings in an additional advantage; very few (if any) ions will get trapped in the target material if this is intentionally avoided by selecting a high enough irradiation energy (above 100 eV). Already, quite a few experiments on ion irradiation of graphene have been carried out.[19–22,29–32] To estimate the ion-irradiation-induced disorder in graphene, Raman spectroscopy,[20,22] atomic force microscopy,[21,22] local mobility measurements[29] and atomic-resolution scanning tunneling microscopy[19] have been used, as described below.

11.2.3 Experimental Identification of Irradiation-Induced Defects

There are several experimental techniques that can be used to detect defects in carbon nanostructures in general and specifically in graphene. These techniques include scanning tunneling microscopy (STM)[19,33] and TEM (as explained above), which can be combined with transport measurements, micro-Raman[20] and other spectroscopic techniques to get the signal from specific regions of the sample. Since the defects introduced with an electron beam in a TEM are typically directly analysed from the produced TEM images, the main focus here is on defects produced *via* ion irradiation.

In Figure 11.5, an example of an STM measurement of a defected area on a graphite surface (*i.e.* top-most graphene layer of graphite) after Ar$^+$ irradiation is shown. Panel (a) displays four monovacancies (one missing atom) that each introduce a triangular-shaped signal in spatial variation of the tunneling current. Local density of states [Figure 11.5(d)] displays a peak density for a vacancy at the Fermi level indicating a flat defect state in the electronic band structure. Similar measurements, also with Ar$^+$ ions, have been carried out for graphene.[19] In these measurements, the defects were found to induce disorder in the hopping amplitudes in addition to acting as scattering centres for

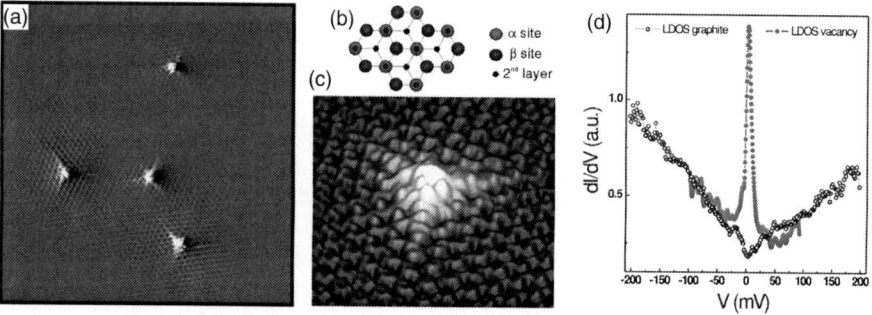

Figure 11.5 (a) $17 \times 17\,\text{nm}^2$ STM topography, measured at 6 K, showing the graphite surface after the Ar$^+$ ion irradiation. Single vacancies occupy both α and β sites of the graphite honeycomb lattice. Sample bias: $+270\,\text{mV}$, tunneling current: 1 nA. (b) Schematic diagram of the graphite structure. (c) 3D view of a single isolated vacancy. Sample bias: $+150\,\text{mV}$, tunneling current: 0.5 nA. (d) STS measurements of the LDOS induced by single vacancy and of graphite. Open circles correspond to dI/dV spectra measured on pristine graphite and solid circles correspond to dI/dV spectra measured on top of the single vacancy, showing the appearance of a sharp resonance at E_F. dI/dV measurements were done consecutively at 6 K with the same microscopic tip. From ref. 33.

electrons. The most important consequence of the disorder is the substantial reduction in the Fermi velocity, revealed by bias-dependent imaging of electron-density oscillations near defect sites, unlike the sharp peak at the Fermi level for a monovacancy [Figure 11.5(d)]. In the experiment reported in ref. 19, the atomic structure of the defects was not identified. The observed hillocks in the STM images may have originated not only from defects in graphene, but also from defects in the substrate under the graphene. Moreover, most defects were likely rather produced by atoms sputtered from the substrate rather than from the initial impacts by Ar^+ ions because the defect density appears higher than what would be expected from the irradiation dose (5×10^{11} ions/cm^2).

Another commonly used technique for estimating defect concentration of carbon nanostructures is Raman spectroscopy.[34,35] It should be pointed out that this technique can hardly be used to detect individual defects, as it provides information averaged over the finite laser spot area (typically hundreds of nm). In Figure 11.6, the measured Raman spectra are shown for a graphene sheet irradiated with 500-eV Ne ions on top of a SiO_2 substrate.[20] The increase in the D band is an indication of the appearing defects, when compared to the pristine structure. Defect scattering was found to give a conductivity proportional to charge-carrier density, with mobility decreasing as the inverse of the ion dose. Defected graphene appeared to be insulating at low temperatures. The results were explained by ion-irradiation-induced localised defect states in the electronic band structure. Although a comprehensive data set was produced on the effects of disorder on electronic transport in graphene, the microscopic nature of the produced defects in the reported experiments remains a mystery.

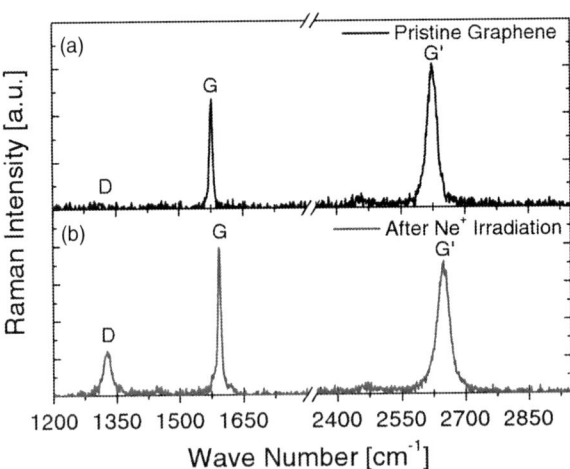

Figure 11.6 Raman spectra (wavelength 633 nm) for (a) pristine graphene and (b) graphene irradiated by 500-eV Ne ions at a dose of 5×10^{12} ions/cm^2. Note increase of the intensity of the D band associated with defects. From ref. 20.

In another recent study,[22] a combination of atomic force microscopy (AFM) and Raman spectroscopy was used to assess the effects of a substrate layer and graphene on defect production. 500-keV carbon irradiation of monolayer, bilayer and multilayer graphene on a SiO_2 substrate was carried out. At such high energies, electronic stopping dominates as the energy-transfer method over nuclear stopping, which governs the process at lower energies. The ratio between the D peak and the G peak in monolayer graphene was found to be higher than that for bilayer or multilayer structures (Figure 11.7). This indicates a higher amount of disorder in the monolayer structure, and the importance of the environment (the substrate) on defect production rate. AFM results also demonstrated that at fluences higher than $5 \times 10^{13}\,\text{cm}^{-2}$ the morphology of monolayers becomes fully conformed to that of the substrate, while graphene ripples are still present on bilayers and multilayers.

Graphene flakes have also been irradiated with 0.4–0.7-MeV protons on Si/SiO_2 substrates.[21] Again, the irradiated samples were investigated both by Raman spectroscopy and AFM. No increase in D-band intensity was reported. This is due to the fact that graphene becomes essentially transparent for protons with such high energies. AFM probing revealed bubble-like features which appeared at sample surface after the irradiation. This result was explained by agglomeration of gas molecules (most likely O_2 released from the SiO_2 substrate). This result also demonstrates the gas-holding ability of graphene – even after high-energy proton irradiation.

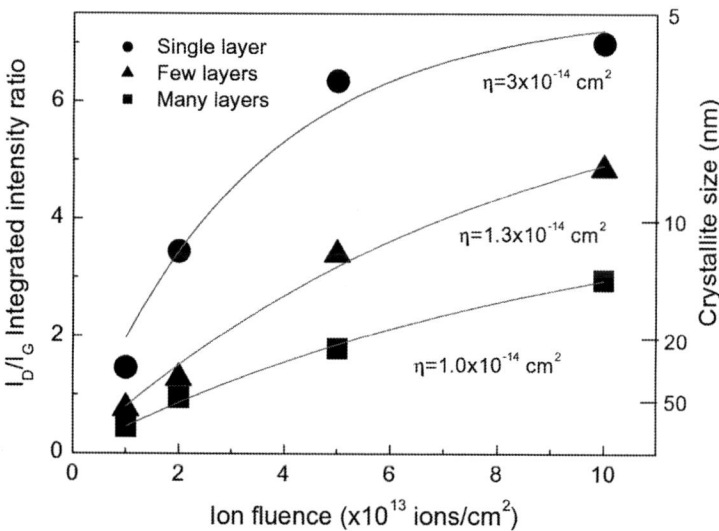

Figure 11.7 Ratio of the D peak and G peak intensities (I_D/I_G) as a function of the ion fluence v for a monolayer, a bilayer and a multilayer of graphene. From Ref. 22.

Overall, despite the interesting insights into the influence of defects on graphene properties obtained from the first irradiation experiments on graphene, no microscopic picture of the damage-production mechanism and defect types has been obtained experimentally. Even the interpretation of STM images is not straightforward, as various defects in graphitic systems can give rise to peaks in the local density of states near the Fermi energy.[36–38] Therefore, in order to understand the microscopic mechanisms that are active during the formation of these defects, atomistic simulations are needed. This is partly due to the much shorter times for defect production as compared to the exposure times in a TEM, and partly due to the fact that other techniques do not reveal the exact atomic configuration of the defected graphene.

11.3   Computational Methods

Although the main focus of this chapter is not on the computational methods used for graphene simulations but rather on giving an overview of the results obtained with them regarding defects in graphene, we feel that a short description of the methods is helpful for the reader at this point, at least for the sake of defining the terms used later in this chapter. For a more thorough discussion of the methods and their specific use in simulations of carbon systems we refer the reader to the recent review articles on the subject.[39,40] Below, we will briefly describe the different approximations used in simulating electron or ion irradiation of nanostructures. The methods are presented in the order of increasing accuracy (and thus transferability, but at the expense of increasing computational cost). We omit here the statistical methods that can be employed to model events with macroscopic timescales (such as defect diffusion) because they can only be used to model the system after the actual irradiation event, and even then only when more accurate simulations have been first performed to map the possible events and relevant energetics. We stress, however, that these methods give useful insight when relating short-timescale simulation results to the experiments, where annealing effects inevitably take place. An example of such a method for carbon nanostructures is the kinetic Monte Carlo model applied for defect migration on carbon nanotubes in refs. 41 and 42.

Molecular dynamics (MD) simulations solve numerically the equations of motion for a set of interacting atoms with known initial coordinates $\mathbf{r}_i$ and velocities $\mathbf{v}_i$.[43] These methods have been of great use in modeling ion-irradiation events in solids, and particularly in interpreting or predicting experimental results. Atomic interactions – both between two target atoms and between the ion and target atoms – can be described at different levels of sophistication: empirical potentials (EP), tight binding (TB) or density functional theory (DFT) models are typically employed. All these models can also be used in static calculations to understand energetics (and, in the case of quantum-mechanical models, electronic properties) of native and irradiation-induced defects.

MD simulations of radiation effects are normally used to provide insight into qualitative mechanisms that cannot be directly assessed with experiments. However, several amendments are typically necessary when MD is used in out-of-equilibrium simulations. For radiation processes the most typically applied modifications are electronic stopping as a frictional force[44] (used with EP, can typically be omitted for carbon nanostructures where knock-on damage governs the ion irradiation process), realistic high energy repulsive interaction[45] (if all electron DFT is not used) and adaptive time step (Δt), which both ensures that the Δt is not too large when energetic ions or atoms are present and that no computational time is wasted with a too short Δt when the system has cooled down.[44] It is also customary to apply one of several temperature-control methods (see *e.g.* refs. 46–48) to atoms at periodic boundaries to model heat dissipation in a much larger target structure. For a description of a typical simulation setup for irradiation simulations, we again refer to the recent review article on ion- and electron-irradiation effects in nanostructured materials.[40] If the reader is particularly interested in ion irradiation of graphene, we refer to the (so far) only MD study on the subject.[18]

11.3.1 Empirical Potentials (EP)

Empirical (or analytical) potentials are a set of equations that give the energy of the system as a function of the positions of the atoms. For each, a set of parameters is fitted to reproduce experimental results, or results obtained with methods of higher accuracy (typically DFT data). During the years, several interaction models have been developed for carbon. The most widely used one is the hydrocarbon potential by Brenner,[49] and its extensions.[50,51] For metals, models such as the embedded-atom method are widely employed.[52,53] For covalently bonded materials, bond-order formalism, as implemented in Tersoff-like potentials (including the Brenner potential),[49,54,55] has proven to be a very good approximation. The major drawback of the empirical potentials is the fact that they have been fitted to a set of parameters that have been obtained for certain reference structures (typically in equilibrium). Thereby, one can never be sure how well these methods describe situations that have not been included in the fitting database. Nevertheless, empirical potentials have played an important role in increasing the understanding of ion irradiation in bulk materials, as well as in carbon nanostructures.

11.3.2 Tight Binding (TB)

A step towards higher accuracy from EP is the tight binding (TB) model. TB is the simplest quantum mechanical method in atomistic simulations. In this approach the Schrödinger equation is solved for electrons moving in a field of atom cores, but the exact Hamiltonian is replaced by a parametrised matrix. Parameters are most often obtained by fitting to the experimental or DFT data. Typically, atomic-like basis sets are used so that the Hamiltonian has the same symmetry properties as the atomic orbitals. A nonorthogonal self-consistent charge TB method[56,57] with parameters derived from DFT calculations (a second-order expansion of the Kohn–Sham total energy in DFT with respect to charge density fluctuations) has been successfully applied to model electron-irradiation effects in covalently bonded materials such as carbon nano-structures.[58–60] DFT-based TB (DFTB) has also been used to model electron impacts in *h*-BN and BN nanotubes (along with carbon structures),[61] but, most likely due to ionic bonding, the BN results have caused much confusion when compared to the experimental results. Recent first principles calculations seem to resolve this issue.[62] For carbon materials this problem does not exist. Because fitting to the experimental data is avoided in DFTB, it has a much greater transferability than the conventional TB approaches. This makes it well suited for simulating irradiation events.

11.3.3   Density Functional Theory (DFT)

The DFT approach (a good overview of the formalism and its implementation is offered in ref. 63) relies on two theorems by Hohenberg, Kohn and Sham. The theorems state that: (1) the ground state energy of a nondegenerate electronic state is a unique functional of its density; (2) this energy can be found by variation of the universal density functional with respect to the charge density. Hence, explicit treatment of the wavefunction of the system is not required. Instead, charge density is enough to find the ground-state energy. This reduces the parameter space in the optimisation from $3N$ (where N is the number of ions and electrons in the system) to 3 (spatial coordinates). However, the exact density functional is not known and must be approximated.

DFT and other quantum-mechanical methods have a high accuracy, but they are computationally demanding. This limits the studied systems to a few hundred atoms and simulation times to picoseconds. Hence, most irradiation-related problems are out of reach for these methods. DFT describes well the atomic structure of defects in many materials, and it has been successfully used for simulating behaviour of various defected systems. This makes DFT indispensable in understanding the structures after the irradiation event.

11.3.4   Time-Dependent DFT (TD-DFT)

Conventional MD simulations are based on the Born–Oppenheimer approximation, which assumes that electrons are always in the ground state when the dynamics of the atoms is considered. For most cases, this is a valid assumption. However, when the velocities of the ions (atoms) approach the Fermi velocity v_F of the material (in graphene $v_F = 8 \times 10^5$ m/s, or $c/300$), this approximation becomes invalid. For ion irradiation, this limit would be reached at irradiation energy of 3 keV for a proton as the projectile. In cases such as this, a combination of time-dependent DFT and classical MD for ions[64,65] can be used to obtain microscopic insight into the interaction of energetic particles with target atoms, because the dynamics of the electronic system and that of the ions are treated on the same footing in real time. This method has also been applied in simulating proton collisions with carbon nanostructures.[66]

11.3.5 Summary of the Simulation Methods

Due to the differences in the computational cost of the methods, they can be used to model events at different timescales and systems of different size. MD simulations within EP can be used to understand the microscopic processes during the ballistic phase of an irradiation event. In graphene, the probability for creating collision cascades is drastically lowered from that of bulk structures because of the two dimensional structure, but they nonetheless occur when the ion displaces a target atom with a trajectory in the direction of the graphene plane. In fact, the conventionally used binary collision model (in which the exact atomic structure of the target material is not explicitly described) fails in describing ion irradiation of graphene due to the absence of collision cascades, as described in ref. 18. After the ballistic phase, MD can also describe the formation of heat spikes, their thermalisation, formation of a sound/shock wave and how it spreads beyond the regime of the ballistic collisions. MD can also describe the nature of the defects produced. However, the major drawback of the MD method lies in the description of the defect structures and energetics. Quantum-mechanical methods (DFTB, DFT) can provide a much more reliable picture of defect properties. As was mentioned above, these methods are limited by the system sizes and timescales they can handle. However, combining results of EP MD simulations and DFTB/DFT calculations makes it possible to reach an accurate understanding on the actual defect production occurring during ion or electron irradiation of graphene and other carbon nanostructures.

11.4 Theoretical Analysis of Defects in Graphene

More than 11% of the studies listed by ISI Web of Knowledge for a search with keyword "*graphene*" also featured words related to defects (*e.g.* "*defect*", "*dislocation*", "*vacancy*") (see Figure 11.1). This gives an idea of the interest of the scientific community on defects in this material. As mentioned above, defects are interesting both by themselves because, *e.g.* intrinsic defects appear in graphene used in experiments (especially in CVD grown graphene), and because they offer a possibility for tailoring the electric and magnetic properties of graphene and graphene-based electronic devices *via* controlled introduction of defects. Moreover, with respect to irradiation effects, it is clearly impossible to completely understand the irradiation response of any material without precise microscopic knowledge of the defect formation mechanisms.

The multitude of carbon allotropes with a variety of shapes and sizes is due to the ability of carbon to exist in various hybridisations. This also means that carbon networks can reorganize the atomic structure like no other material can do.

New bonds restructure the lattice around defects by creating modified but coherent networks that retain many of the original properties of the structure. This leads to a basically unlimited number of ways to accommodate extra and missing atoms in graphene. Carbon nanomaterials are also well known for their ability of self-healing during annealing.

In the following, we will turn our focus on both natural (intrinsic) defects and defects that can form due to electron or ion irradiation. The main stress will be on computational work in accordance with the topic of this book. We will first discuss point defects, such as vacancies and carbon adatoms, including topological defects, *e.g.*, *Stone–Wales* defects. We will also address the evolution of point defects by briefly touching upon impurity atoms in graphene. The actual processes leading to the irradiation-induced defects are discussed later. We will also discuss line defects such as dislocation lines.

11.4.1 Point Defects

Two of the most often considered defect structures in carbon nanomaterials are a monovacancy and a carbon adatom. Adatoms play the role of interstitials in graphene because there is not enough space for an interstitial to fit into the two-dimensional structure. In principle, an additional atom could fit into the open hollow of one of the hexagons, but this is energetically so much unfavoured that it will never happen in practice. Another important example is the so-called Stone–Wales defect formed by rotating a C–C bond.

11.4.1.1 Mono- and Divacancies

When one carbon atom is simply removed from the graphene lattice, an immediate Jahn–Teller distortion[67–69] occurs to saturate the dangling bonds of two of the three undercoordinated carbon atoms [see Figure 11.3(f) and Figure 11.8]. This leads to a (5–9) defect structure in which one dangling bond remains in the middle of the 9-membered carbon ring. Contrary to metals, these reconstructions are a very prominent feature of vacancies in carbon nanomaterials, as demonstrated for example by the high pressure created by removing atoms from carbon nanotubes[25] at high temperatures, so that the atomic bonds can efficiently reconstruct. The formation energy for a monovacancy in graphene is about 7.5–7.7 eV, as estimated with DFT calculations both within local density approximation (LDA)[68] and the generalised gradient approximation (GGA).[70,71] Formation energy is here defined as

$$E_f = E_d(N - n) + n\mu_C - E(N), \qquad (11.1)$$

where $E_d(N - n)$ is the total energy of the defected structure composed from $N - n$ atoms, $\mu_C = E(N)/N$ is the chemical potential for carbon, $E(N)$ the total energy of the pristine system with N atoms and n the number of removed atoms when creating the vacancy. Note that E_f defines the energy difference between the defected system and the pristine one. It does not describe the energy needed to actually produce the defect – which is higher than E_f – because of a kinetic barrier in between the two states. The migration barrier for monovacancies has been estimated to be around 1.3–1.7 eV in graphene.[68]

After removing the undercoordinated atom of a monovacancy, a new bond is formed between the two new undercoordinated atoms, similar to what happens for a monovacancy (see Figure 11.8). Thus, we now get a (5-8-5) defect that is conventionally considered divacancy structure of carbon nanomaterials. Because of the saturation of all dangling bonds in the structure, we have again a sp^2-bonded carbon network (although with slightly distorted bond angles) where all atoms have three neighbours. In that way, graphene (and other sp^2-bonded carbon structures) are self-healing under irradiation. The saturation of

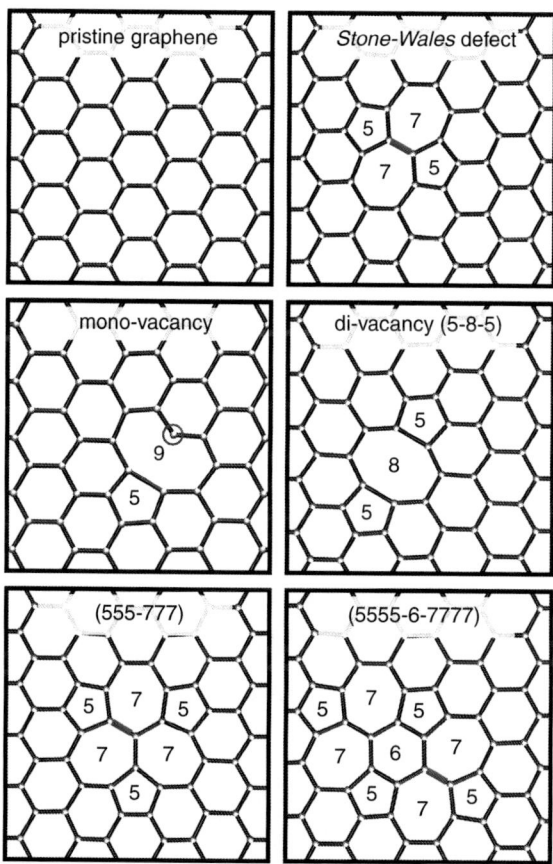

Figure 11.8 Different simple defect structures in graphene. A *Stone–Wales* defect is obtained from pristine graphene by rotating one bond. In a similar manner, (555-777) is obtained from (5-8-5) and (5555-6-7777) from (555-777) by rotating another bond. The undercoordinated atom of the monovacancy structure is marked with a circle.

the dangling bonds also leads to curious energetics of graphene vacancies since the divacancy formation energy is ca. 7.28 eV (or 3.64 eV/n where n is the number of missing atoms, as given by DFT GGA calculations with PBE parametrisation,[72,73] earlier DFT LDA value is 8.7 eV[68]) *i.e.* similar to that of a monovacancy. Divacancies have been thought to remain immobile in graphene due to a migration barrier of 7 eV.[68] However, taking into account the bond-rotation process, they can actually travel under an electron beam. This matter is discussed in more detail below in the context of electron irradiation.

11.4.1.2   *Multivacancies*

As there is no way of removing an odd number of carbon atoms in such a way that the undercoordinated atoms could saturate the dangling bonds, it is clear that all energetically favoured vacancy structures have an even number of missing atoms. This is similar to what was noticed for carbon nanotubes earlier.[74] The main difference between graphene and nanotubes is that in nanotubes a local diameter change assists in accommodating the locally lowered atomic density. A recent study[75] considered some simple multivacancy structures in graphene, and concluded that these defects could appear in nanoporous graphene with up to 1% concentration. However, the atoms in graphene can easily rearrange, which opens up the possibility for more complex polygon networks than what can happen in nanotubes or what was considered in ref. 75. Nanoporous graphene has also been considered as a membrane for gas separation.[76]

An interesting example of complex polygon networks is the previously mentioned hæckelite structure[77] that consists of pentagons and heptagons arranged as a group of three pairs (555-777). When an individual (555-777) defect[78] is incorporated into graphene sheet (see Figure 11.8), it is easy to notice that it accommodates two missing atoms. More interestingly, this structure can be obtained from the (5-8-5) defect by simply rotating one bond. As the local strain is reduced by transforming the octagon and three hexagons to one pentagon and three heptagons, this defect has lower formation energy than the (5-8-5) divacancy [3.31 eV/n (DFT GGA); a similar value has been reported in the literature[78,79]]. Moreover, Lee *et al.*,[80,81] showed using a TB model combined with MD, that a (5-8-5) divacancy can be formed from two separated vacancies *via* migration, and that the divacancy can transform to a (555-777) *via* bond rotation during the dynamical simulations. However, as the barrier for bond rotation is estimated to be at least 4–10 eV (depending on the local atomic configuration),[82] temperatures above 3000 K were needed to initiate these transformations during times reachable in the simulations.

Jeong *et al.* did an extensive comparison of the energetics of dislocations and combinatorial (555-777) defect structures and noted that the (555-777) defects are energetically favoured when eight or less atoms are missing, whereas dislocations would be preferred after this.[83] However, (555-777) and (5-8-5) are not the only possible divacancy structures that can occur in graphene.

By rotating a nearby bond, a structure with four pentagons, four heptagons and one rotated hexagon (by 30°) in the middle of the grain boundary of alternating pentagons and heptagons is formed (5555-6-7777)[79] (see Figure 11.8). Although this defect has a higher formation energy (ca. 3.35 eV/n) than the (555-777) defect, the energy difference is minor when compared to the (5-8-5) divacancy. In fact, in graphene nanoribbons (5555-6-7777) has been reported to have a lower formation energy than (555-777).[79]

It turns out that the (5555-6-7777) can be considered as the elementary multivacancy structure in graphene. The vacancy structure can be extended by adding these structures right next to each other so that pentagons of one overlap with the heptagons of the other. A rotated hexagon is obtained at each such polygon in the lattice. When more and more defects are added (and atoms thus removed), the structure starts to form a small misoriented grain in the graphene superlattice. For four and six missing atoms, it is possible to arrange these defects so that the formation energies (2.86 eV/n and 2.70 eV/n, as given by DFT GGA, respectively) are clearly below those for dislocations or combinatorial (555-777) defects. There is no reason why this could not be extended to larger multivacancy structures.

As a conclusion, there is a multitude of ways to arrange carbon atoms to form a polygon network to compensate for a certain number of missing atoms. However, the lowest-energy multivacancy structures in graphene are formed by kernels of rotated hexagons that are accommodated to the zigzag edges of the surrounding graphene superlattice by a chain of alternating pentagons and heptagons. This notion is supported by the experimental results of extended electron irradiation, which typically yields defects of this type (see Figure 11.2).

11.4.1.3 Carbon Adatoms

Carbon interstitials have played an important role in explaining radiation damage in graphite. These defects can easily appear after ion or electron impact, because of the empty space between the separate graphene layers. In principle, for each created monovacancy there exists an interstitial carbon atom somewhere else in graphite. Therefore, irradiating graphite will lead to formation of Frenkel pairs.[84] Adatoms also have an interesting role in self-healing of defected carbon nanotubes.[42] Again because of the two-dimensional structure, Frenkel pairs are far less frequently created in graphene, as displaced atoms are typically sputtered away (instead, vacancies will appear). As was mentioned above, no interstitial atoms in the traditional sense exist in graphene. Instead, any additional atoms in graphene structures remain on top of the structure as adatoms.

Although creation of vacancies is less likely to produce adatoms in graphene than interstitial atoms in graphite, adatoms can still appear under certain experimental conditions. The reason for this is that graphene samples always contain carbon absorbates from which a TEM beam can sputter atoms that get trapped on the graphene monolayers. When following the evolution of defect structures under TEM, one sometimes observes a surprising increase in the number of carbon atoms at the defect, which is due to travelling carbon adatoms. The minimum energy configuration for carbon adatoms is the bridge structure in which the atom occupies a position on top of a carbon–carbon bond, as was shown recently.[85,86] The migration barrier in graphene has been estimated to be ca. 0.47 eV[85] (earlier estimates were even lower[87]), which means that they are highly mobile.

Carbon adatoms have also been suggested to play a role in catalysing bond rotations[82] and self-healing processes,[88] and when joined as dimers to lead to *inverse Stone–Wales* defects and extended graphene bubble structures with locally introduced slight curvature.[89,90] However, bubbles of this kind remain to be observed experimentally. Some interest in carbon adatoms on sp^2-bonded carbon structures has also resulted from the indications for introducing magnetism in these systems *via* additional atoms.[85,91]

11.4.1.4   Impurities on Graphene

Impurity atoms can hardly be considered as *native* defects in graphene. However, we want to stress that a considerable body of research has been carried out on this topic. On the one hand, metal atoms and clusters play a major role in the catalytic activity during growth of carbon nanomaterials, and on the other hand, they have a major influence on the properties of these structures.[26] The chemisorbed adatoms on graphene either position themselves on top of a hexagon, on top of a carbon bond (*e*-type) or on top of a carbon atom (*s*-type).[92] Adatoms have been widely discussed in the context of magnetism (see *e.g.* refs. 93–95), but this has been far from the only subject of adatom studies.[93,94,96–104]

In addition to metal atoms, since B and N are the natural dopants in carbon systems, much interest has emerged on doping carbon nanomaterials with boron and nitrogen. For example, MD simulations[9,10] have shown that it is possible to get up to 40% of the impinging ions into the substitutional position in carbon nanotubes by selecting the irradiation energy appropriately. Later experiments confirmed that N irradiation of nanotubes indeed leads to substitutional doping.[11,12] Although the mentioned studies were made with carbon nanotubes, there is no fundamental physical reason why this would not work for graphene. Furthermore, due to the lack of curvature, one can expect even higher substitution probabilities for graphene than what was found for nanotubes.

Another topic that has gathered much attention regarding carbon nano-materials is their ability to bind hydrogen in order to use them in hydrogen-storage applications, to solve yet another problem that humankind is facing at the moment. So, perhaps ironically, carbon nanomaterials are considered to be a part of the solution for putting forward so-called *carbon free technology*, in which the energy production is no longer based on fossil fuels. Since also this topic is beyond native defects and only remotely related to radiation damage in graphene, we simply list here in an exemplary way a few recent studies touching upon the subject (see *e.g.* refs. 96, 99, 105–107). Some work has also been done on hydrogen-related defects and their properties.[108–110]

11.4.1.5 Topological Defects

The so-called *Stone–Wales* defect[111] is probably the most famous defect in carbon nanostructures (see Figure 11.8). It can be formed by selecting one bond in the hexagonal lattice and rotating it by 90° to transform four hexagons to two pentagons and two heptagons (5-7-5-7). Thus, the system is still composed of the "correct" number of atoms, but the atoms are arranged in a "wrong" way, so that we deal with topological disorder.

The formation of the SW defect is assisted by local curvature so that the formation energies are lower in fullerenes and carbon nanotubes than in graphene.[112–114] It is believed that the graphene sheet with a SW defects remains flat, although a recent study[115] utilising highly accurate quantum-mechanical calculations suggested that the formation of the defect could introduce a slight curvature in the lattice. The *Stone–Wales* defects have also been observed during TEM experiments [see Figure 11.3(c)]. Although it has the lowest formation energy of any defect in carbon nanostructures ($E_f \approx$ 5.8 eV in graphene according to quantum Monte Carlo calculations;[115] the DFT GGA value is close to 4.6 eV), it is so high that the room-temperature equilibrium concentration should be too low to explain the frequent observations. As explained below, these defects are actually introduced during the observation under a TEM at acceleration voltages starting from ca. 80 kV *via* knock-on collisions between an electron and a single target atom.

In general, bond rotation is the other important way for altering graphene morphology in addition to removing atoms. Since bond rotations can change the hexagonal structure into a network of alternating pentagons, hexagons and heptagons (or even tetragons and octagons, although they have been much less frequently considered), this allows for a way to drastically change the local atomic structure but retaining the sp^2 bonding and thus flat – or almost flat – carbon membrane.

11.4.2 One Dimensional Defects: Dislocations and Grain Boundaries

A glide dislocation monopole in graphene is formed by a pentagon–heptagon pair so that the shortest possible dislocation involves either a *Stone–Wales* defect or an *inverse Stone–Wales* defect, as shown in ref. 89. Dislocations of different lengths – involving different numbers of missing atoms – are illustrated in Figure 11.9. In their study,[83] Jeong and coworkers showed that dislocations are energetically favoured over so-called hæckelite structures[77,80,81] when each structure has the same number of missing atoms ($N > 8$), as mentioned above in the context of multivacancy structures. In addition to accommodating missing atoms, extended dislocation structures serve as grain boundaries when they occur between two graphene grains with differing orientations.

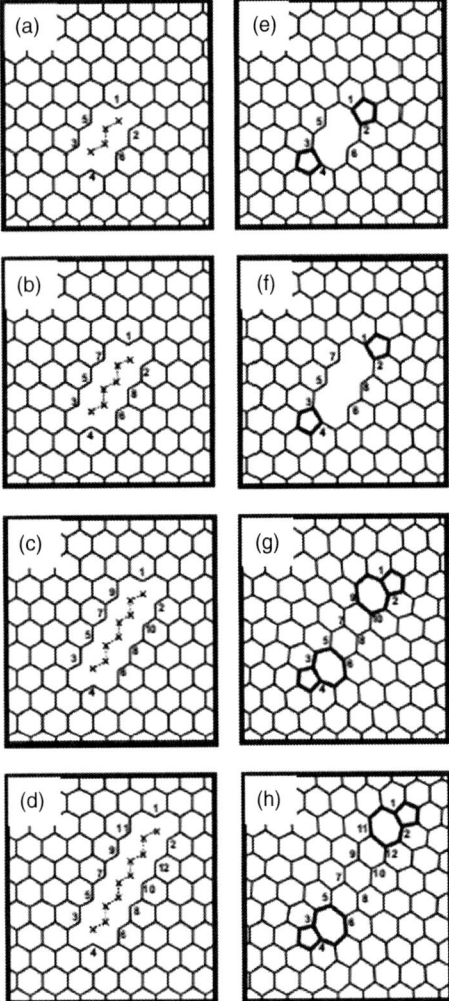

Figure 11.9 The initial geometries [(a)–(d)] and the relaxed geometries [(e)–(h)] of graphenes containing four [(a) and (e)], six [(b) and (f)], eight [(c) and (g)], and ten [(d) and (h)] vacancy units. Crosses and dashed lines in (a)–(d) indicate the positions of carbon vacancies and zigzag chains of carbon vacancies, respectively. The bold lines in (e)–(h) indicate pentagon and heptagon rings formed by the structural relaxation. From ref. 83.

A set of advanced ways of organising the dislocation monopoles in a graphene superlattice was recently considered by Yazyev and Louie.[116] They developed a systematic approach for constructing atomic structures of topological defects in graphene, and were able to find grain-boundary structures that exhibit extremely low formation energies (see Figure 11.10 for example structures). The authors note that none of the considered atomic arrangements exhibited either electronic states at the Fermi level or magnetic moments.

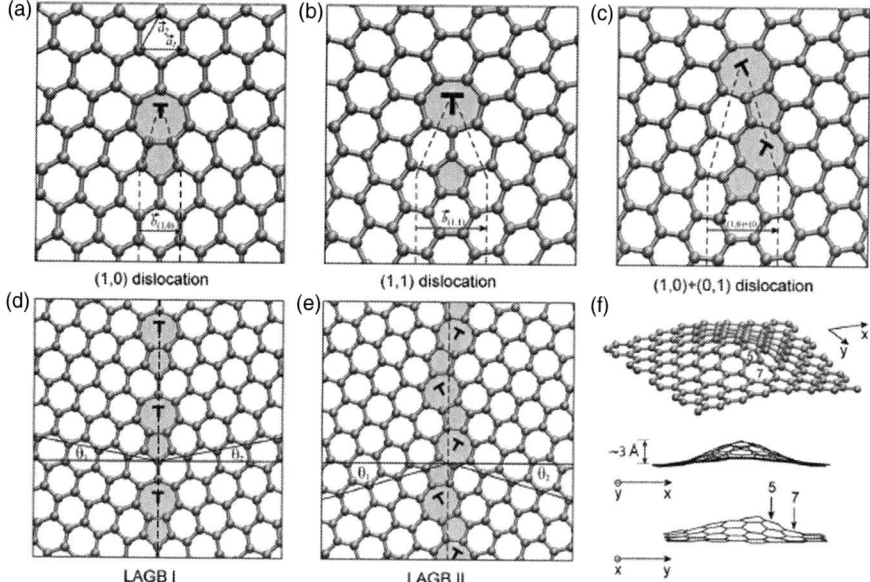

Figure 11.10 (a–c) Atomic structures of (1,0) and (1,1) dislocations, and a (1,0) + (0,1) dislocation pair, respectively. The dashed lines delimit the introduced semi-infinite strips of graphene originating at the dislocation core. Non-6-membered rings are shaded. (d,e) Atomic structures of the $\theta = 21.8°$ (LAGB I) and the $\theta = 32.2°$ (LAGB II) symmetric large-angle grain boundaries, respectively. The dashed lines show the boundary lines and the solid lines definite angles θ_1 and θ_2. (f) Buckling of the graphene layer due the presence of a (1,0) dislocation. From ref. 116.

With respect to the atomic configurations, one must keep in mind that in reality there are several other issues that play a role in determining the actual grain boundaries in addition to equilibrium thermodynamics of graphene. For example, when graphene is grown on a nickel substrate, the mismatch between graphene lattice and the underlying Ni(111) surface leads to the formation of a grain boundary consisting of alternating sets of octagons and pentagon pairs (see Figure 11.11).[17] Another way of obtaining this structure would be to align divacancies in graphene to the zigzag orientation. In contrast to the above-mentioned grain boundaries, this atomic arrangement was reported to display occupied electronic states at the Fermi level and thus to have metallic nature.

11.5 Irradiation-Induced Defects

All the discussed defects can be produced, at least in principle, by electron or ion irradiation of graphene. In order to fully understand why particular defects are more commonly encountered in the experiments, it is not sufficient to simply study the structure and energetics of the defects, but also the microscopic effects responsible for their production must be known.

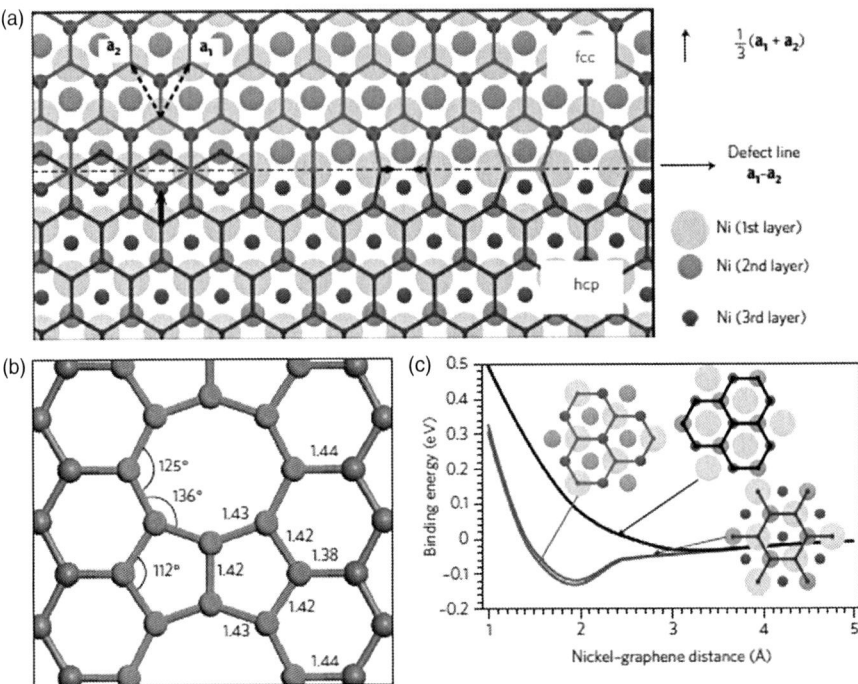

Figure 11.11   Structural model and schematic formation of an extended one-dimensional defect in graphene. Two graphene half-lattices, with unit-cell vectors $\mathbf{a}_1$ and $\mathbf{a}_2$ (shown as dashed arrows), are translated by a fractional unit-cell vector $1/3(\mathbf{a}_1 + \mathbf{a}_2)$, indicated by the vertical vector (solid arrow) (a). The two half-lattices can be joined along the $\mathbf{a}_1 - \mathbf{a}_2$ direction, indicated by the horizontal vector, without any unsaturated dangling bonds, by restructuring the graphene lattice. The domain boundary can be constructed as shown, by joining two carbon atoms, indicated by the two arrows, along the domain boundary line. This reconstructed domain boundary forms a periodic structure consisting of octagonal and pentagonal carbon rings. The underlying Ni(111) structure illustrates how the extended defect is formed by anchoring two graphene sheets to a Ni(111) substrate at slightly different adsorption sites. If one graphene domain has every second carbon atom located over a fcc hollow site and the other domain over a hcp-hollow site, then the two domains are translated by $1/3(\mathbf{a}_1 + \mathbf{a}_2)$ relative to one another. The calculated adsorption energies for these two domains are very similar, but both are lower in energy than a third possible adsorption configuration with all carbon atoms on hollow sites, as shown in c. The DFT relaxed geometry of the defect structure, including bond lengths (in Å) and bond angles, is shown in (b). From ref. 17.

11.5.1   Electron Beam-Driven Morphological Changes

The most accurate methods suitable for modeling electron beam damage in solids (DFT, TDDFT) cannot explicitly describe collisions of energetic electrons and ions in a solid. However, the damage created by energetic electrons can still be computed. Instead of launching an electron towards the structure in question, we can limit ourselves to studying what happens *after the fact*, *i.e.* after the electron–ion collision has happened. To model the processes occurring in a TEM, it is sufficient to know the acceleration voltage used in the experiment. Then, it can be relativistically calculated how much momentum (and thus, kinetic energy) an electron can transfer (as a maximum) to a target atom:[23]

$$T_{max} = \frac{2ME(E + 2mc^2)}{(m + M)^2 c^2 + 2ME},$$ (11.2)

where M is the mass of the target ion and m that of an electron, E the energy of the relativistic electron, and c the speed of light. If we now assign this kinetic energy to a certain target atom and follow the time evolution of the system, we can understand how certain TEM conditions will alter the morphology of the target material. However, in order to understand all events that can occur, one obviously must also consider energy transfer below the maximum ($T < T_{max}$), and at all possible scattering angles.

The most important parameter is the displacement threshold T_d, which denotes the limiting kinetic energy above which the target atom will be displaced from its initial position. In pristine graphene this leads to a production of a single vacancy. In the case of a pre-existing vacancy, an additional atom will be sputtered. A thorough study was recently carried out within the DFTB approach by Zobelli and coworkers[61] in which T_d was calculated as a function of the scattering angle in carbon and BN nanotubes as well as in graphene (see Figure 11.12; note the different notation for kinetic energy) and hexagonal BN monolayers. Although their results for BN have caused some confusion in the interpretation of later

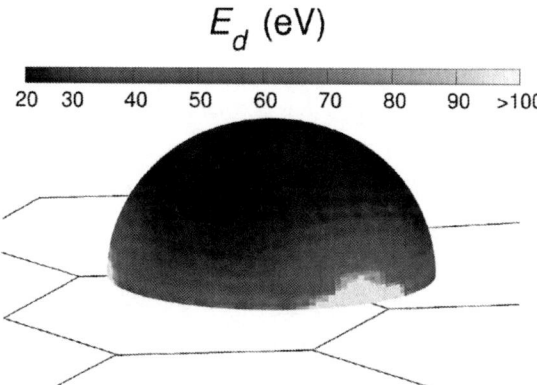

Figure 11.12 Three-dimensional representation of the map of the emission threshold function $E_d(\delta, \gamma)$ for a carbon atom in a graphitic layer as a function of the spherical coordinates δ and γ. The scale indicates the emission energy values from 20 eV to more than 100 eV. The sphere indicates the emission direction for the ejected carbon atom. The sphere is centred on the initial position of the targeted C atom. From ref. 61.

experiments[117,118] it was still possible to explain the experimental results for the lowest energies used. Later, DFT calculations[62] showed that indeed the displacement thresholds were underestimated, probably because of problems in describing the charge transfer in BN systems within the DFTB model. In general, as discussed in ref. 61, the theoretical T_d values can be expected to be 10–20% higher than the experimental ones, because only the ground-state dynamics is considered in the calculations. The beam-induced excitations may reduce the bonding energy of the atoms in the lattice, thus reducing the threshold energy. However, as the effect is expected to be systematic, one can assume that the obtained thresholds are overestimations, and simply correct for this effect by reducing the threshold value to match the reference experimental data.

In order to calculate the actual displacements occurring under the electron beam, one can use the equation for the displacement cross-section for a target atom within the McKinley–Feshbach formalism for Coulomb scattering[119] as[23]

$$
\sigma = \frac{4Z^2 E_R^2}{m^2 c^4} \pi a_0^2 \left(\frac{1-\beta^2}{\beta^4}\right) \times \left\{ 1 + 2\pi\alpha\beta \left(\frac{T_d}{T_{max}}\right)^{1/2} \right.
$$
$$
\left. - \frac{T_d}{T_{max}} \left[1 + 2\pi\alpha\beta + (\beta^2 + \pi\alpha\beta)\ln\left(\frac{T_{max}}{T_d}\right)\right]\right\},
$$

(11.3)

where Z is the atomic number of the target atom, E_R the Rydberg energy (13.6 eV), a_0 the Bohr radius (5.3×10^{-11} m), $\beta = v/c$, and $\alpha = Z/137$. Displacement rate becomes thus

$$
p = \sigma j, \tag{11.4}
$$

where j is the beam current.

The McKinley–Feshbach formalism assumes a uniform displacement threshold distribution over the relevant space angles. Although this at first might seem an inappropriate approximation noting that the T_d values, as reported in ref. 61, vary from close to 20 eV up to more than 100 eV (the DFT GGA value is ca. 22.2 eV, or 17.8–20.0 eV with a 10–20% correction corresponding to an acceleration voltage of 89.5–99.6 kV), one must also remember that due to the mass difference between an electron and a carbon atom, the possible scattering angles are all close to the direction parallel to the electron beam. Within these angles, as seen in Figure 11.12, the variations in T_d are rather small, and thus applying the McKinley–Feshbach formalism is justified.

After introducing the McKinley–Feshbach formalism, we now apply it for estimating electron beam damage in solids. As mentioned, the typical view has been that T_d is sufficient to describe the process. If energies higher than T_d are considered, it is obvious that the displaced carbon atom can introduce more structural changes in graphene. This obviously requires that the displacement occurs in a direction along the graphene plane, as noticed by Yazyev et al.[84] For example, they observed formation of *Stone–Wales* defects when a carbon atom is displaced into some inplane directions with $T > 30$ eV. This and

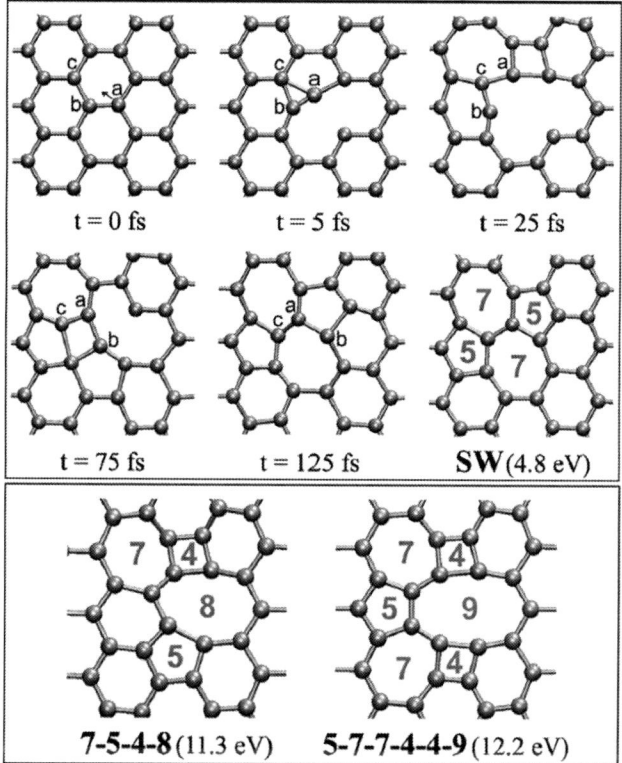

Figure 11.13 Top: The mechanism of formation of a Stone–Wales defect upon inplane knock-on displacement. The carbon atoms involved in the rearrangement are marked with letters. Bottom: Atomic structures of 7-5-4-8 and 5-7-7-4-4-9 topological arabic numbers indicate non-hexagonal ring. From ref. 84.

another example of their findings are shown in Figure 11.13. The remarkable reorganisation ability of carbon when the local atomic structure is disturbed is evident from this figure.

However, the inplane displacement with $T > T_d$ cannot account for the reorganisation that has been observed in graphene under continuous bombardment with electrons, as shown in Figure 11.2(b), because in a typical TEM experiment graphene is set perpendicular to the electron beam for the best possible imaging conditions. Therefore, the local heating effect of the electron beam has sometimes been thought to cause the reorganisation events, as displayed in the high-temperature simulations by Lee *et al.*[80,81] This explanation fails for two reasons: (1) graphene is an excellent heat conductor, which means that any produced heat would be quickly dispersed to the whole structure and (2) single collisions between an electron and an ion would not bring about such heating as what would be required for providing enough energy for overcoming the 4–10 eV barriers for bond rotations.

The answer to this puzzle lies again in the extraordinary ability of carbon to reorganize when its atomic structure is perturbed. When simulations are carried out for momentum values (the momentum transferred from the electron to the ion) right below T_d at various space angles close to the electron-beam direction, three outcomes are found. The most typical event is a failed ejection of an atom that leads back to pristine graphene. The second most typical event is the formation of a Frenkel pair where the atom can almost escape the graphene sheet, but still gets attached to it after it has moved away from its original position (Figure 11.14). Due to low migration energy barrier, the atom can easily find its way back to recombine with the corresponding vacancy. The third possibility is a bond rotation, which occurs when the local atomic structure is perturbed, when the displaced atom either (1) is moved out from its original position, but doesn't have enough time to relax back before the atom returns (Figure 11.15), or (2) circles around one of its neighbours due to the kick by the

Figure 11.14 Formation of a Frenkel pair due to electron irradiation. As the atom remains close to the formed vacancy, it can easily migrate back and the defects can recombine. The DFTB MD result.

electron and drops back next to its original position thus forming the *Stone–Wales* defect (see Figure 11.16).

These different processes each have a different kinetic energy threshold, similar to T_d, which defines the corresponding rate p at which they occur under a certain TEM beam. Therefore, by changing the acceleration voltage of the TEM, which defines the processes that can occur and corresponding p, one can select the ratios at which different processes occur. Below the low energy limit, nothing happens to graphene, whereas at the other extreme (high energy and/or high dose) large holes are created. In between, the knock-on collision-driven bond rotations are active and play an important role in accommodating the locally lowered atomic density by producing complex polygon networks. The low T_d for displacing any dangling bond atoms in graphene (13.5 eV, or 10.8–12.2 eV with a 10–20% correction) is so low that they are effectively

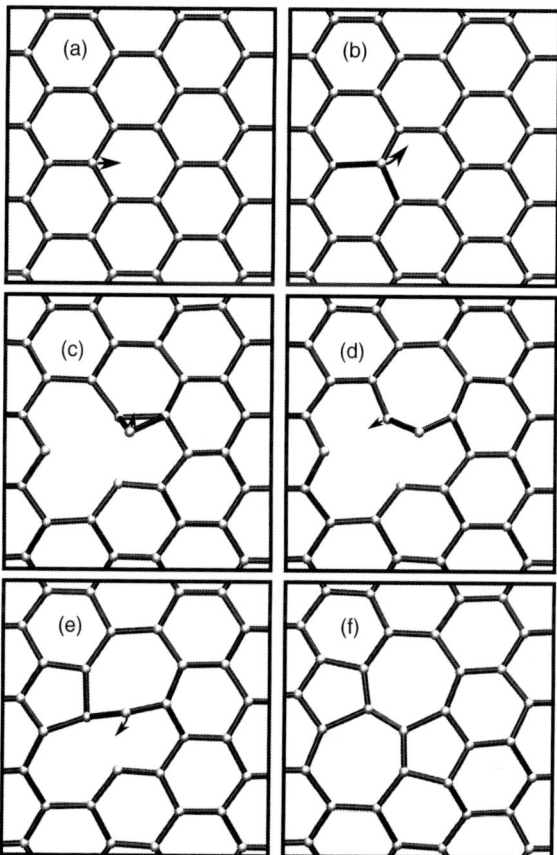

Figure 11.15 Formation of a *Stone–Wales* defect *via* the "*direct*" mechanism in which the displaced atom simply falls back to its original position, but pushes another atom during the process to cause a bond rotation. The DFTB MD result.

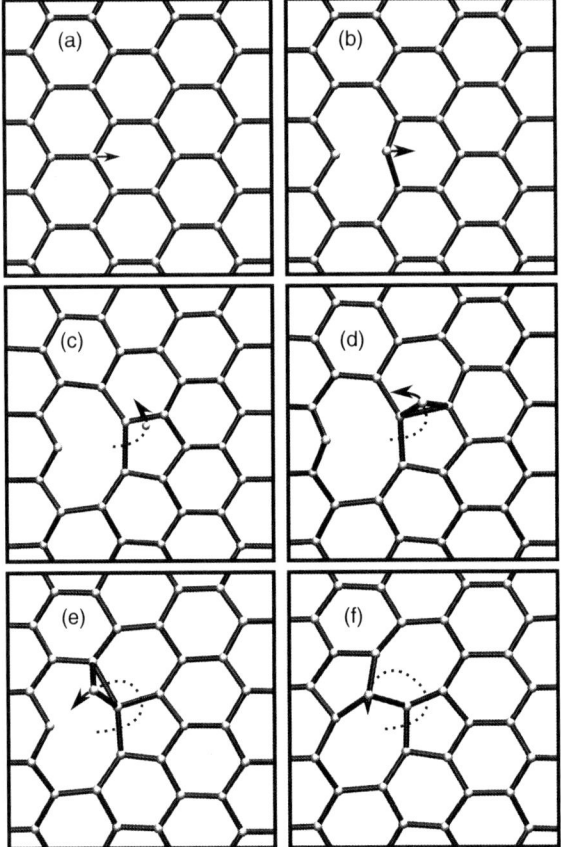

Figure 11.16   Formation of a *Stone–Wales* defect *via* the "*circle*" mechanism in which the displaced atom circles its neighbour and initiates the bond rotation. The DFTB MD result.

ejected already at low acceleration voltages (50.0–62.8 kV), which means that undercoordinated carbon atoms are rarely observed in the complex polygon networks.

11.5.2   Ion Irradiation-Induced Defect Production

Since ions are much heavier than electrons and can thus transfer larger momentum to the target, ion irradiation has more profound effects on solids than electron irradiation. As mentioned before, the energy transfer to the electronic system of the target structure is not treated properly within the classical MD, but this is only relevant to simulations of graphene irradiation for high energies for the impinging ions. However, within the TDDFT formalism this effect is treated properly as demonstrated in ref. 66.

In a recent study,[18] noble-gas ion irradiation of graphene was simulated using the empirical potential by Brenner[49] to describe the carbon–carbon interaction amended with a correct description at short interatomic distances[45] and the universal repulsive potential for carbon–noble-gas ion interactions.[45] To check for the validity of this method, graphene T_d was calculated by the Brenner interaction model (22.04 eV), and a surprisingly good agreement with DFT value (22.2 eV) was noted. Also the minimum kinetic energy needed for an Ar ion to cause an ejection of a carbon atom in graphene was calculated for both EP model and DFT GGA dynamical calculations. The obtained values were 32.33 eV and 32.74 eV, respectively. A good agreement was also noted for the penetration barrier for the noble-gas ions through graphene. Again, a good agreement between EP MD and DFT MD (see Figure 11.17) was seen. From these comparisons it was concluded that the used EP method adequately describes the major features of the ion-irradiation process of graphene.

From the results of the EP MD simulations, it was found that the produced defects can be grouped into the following categories: monovacancies, divacancies, amorphous regions, Frenkel pairs and *Stone–Wales* defects. The last two contributed only a minor fraction to the bulk of the observed defects. Amorphisations included larger defected areas consisting of randomised polygon networks covering areas of up to 60 Å^2, but typically involving only one or two ejected carbon atoms. Any defects more complex than divacancies created in graphene are due to secondary recoils of a displaced carbon atom in the target plane.

Overall, the general trends for defect production in graphene under ion irradiation are as follow. If the scattering angle for the carbon atom is close to the initial velocity of the ion, a monovacancy is produced [Figures 11.18(a)–(d)] unless the ion hits right in the middle of a carbon–carbon bond and has such a

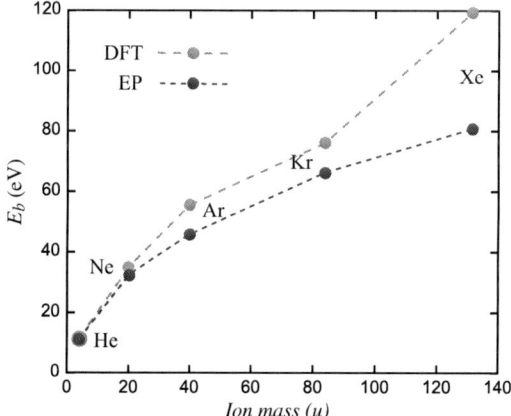

Figure 11.17 Comparison between DFT (GGA) and EP on the penetration barriers of various noble-gas ions through a middle of a hexagon in graphene sheet as a function of the ion mass.

velocity that the collision cross-section between the ion and the two closest atoms is large enough to sputter both atoms and create a divacancy [Figures 11.18(e)–(h)]. Otherwise, if a carbon atom is ejected, depending on the exact scattering angle and transferred momentum, the carbon atom can initiate ejection of another carbon atom [Figures 11.18(i)–(l)] or a larger inplane collision cascade that will lead to an amorphisation event [Figures 11.18(m)–(p)]. An inplane collision will only occur at high irradiation energies, because it requires that the target atom is essentially immobile during the time it interacts with the ion; symmetric interaction with respect to the graphene plane

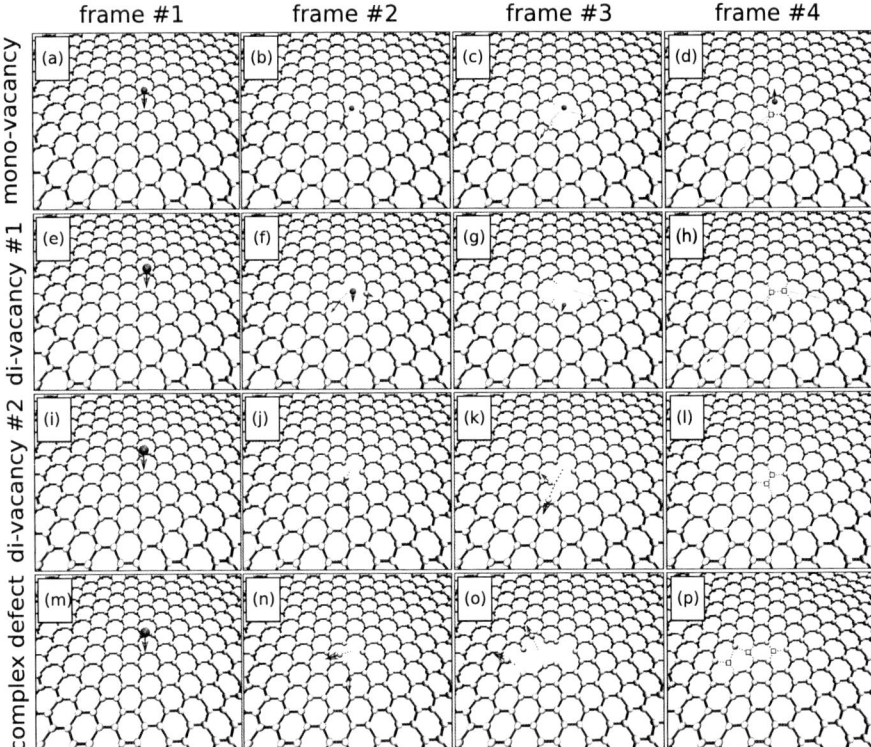

Figure 11.18   Defect creation processes during an ion-irradiation event. A mono-vacancy is always produced by a direct knock-on displacement by the incoming ion (a–d). For divacancies, two formation mechanisms exist: direct displacement of both two atoms by low-energy ions (e–h) and higher-energy inplane displacement of one atom that knocks out the second atom (i–l). All higher-order defects are created by the inplane displacements as shown in the example of a complex defect (m–p). Missing atoms are marked with squares in the last frame for each defect. Note that the lattice reconstructions have not had enough time to occur in these snapshots (and do not customarily appear in EP simulations, which needs to be taken into account when analysing the EP MD results).

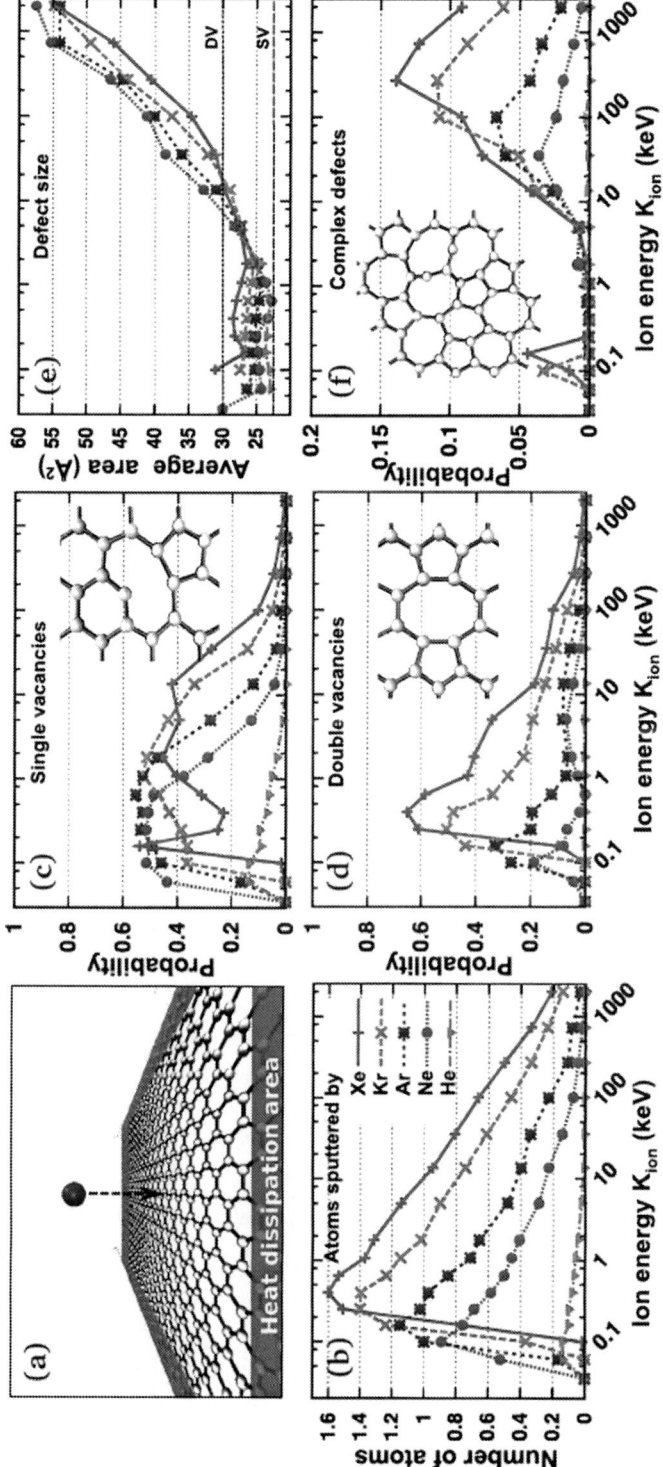

Figure 11.19 Production of defects in graphene under ion irradiation as revealed by EP MD. (a) Simulation setup. (b) Number of sputtered atoms per ion impact as a function of ion energy. [(c) and (d)] Probability for single- and double-vacancy formation as a function of ion energy. The insets show the atomic structures of the reconstructed vacancies. (e) Average area covered by a single defect (when formed) – typically still an sp^2-bonded network of carbon atoms. The areas corresponding to a SV and DV are marked. (f) Probability for creating defects other than SV/DV (except FP/SW), see the inset for an example. From ref. 18.

will result in a displacement in a direction perpendicular to the velocity of the ion.

Probabilities for producing various defects are presented for all considered ions as functions of irradiation energy in Figure 11.19 along with a schematic illustration of the simulation setup and examples of the defect structures. By carefully selecting ion species and energy, it is possible to select such parameters from the data that for example 50% monovacancies and 50% divacancies will be produced with no additional defects. Varying this ratio, it is possible to study the effects these defects have on the electronic properties of graphene. One interesting aspect is the fact that when the energy is increased, for example for a Xe ion, the total defected area will be smaller than what would be produced at lower energies with the same irradiation dose (for 2-MeV Xe the defected area becomes 30% of that of energies close to 0.5 keV). This opens up the possibility of using graphene in ion-beam analysis methods as a membrane for irradiating volatile targets or targets that cannot be put into a vacuum chamber (like pieces of art or living cells), especially when high-energy ions can be used. Also, since the monovacancies dominate at the lower energies, where also the total damaged area is larger, it is evident that energies close to the monovacancy production maximum would be ideal for instance for cutting graphene with a focused ion beam (FIB).

The above results were obtained for ions oriented perpendicular to the graphene sheet. An additional parameter to control the defect production is the irradiation angle, bringing the number of parameters to three (ion mass m, energy K_{ion} and angle ψ). By varying the angle, one can shift the energies both at which the penetration of the ion through the graphene sheet occurs and when graphene becomes transparent for the ions (Figure 11.20). The angle variation also gives some control on the defect types that will be created since by tilting graphene the projected atomic density under irradiation increases.

Clearly, the defect production in graphene under ion irradiation is a rather complex problem even for the nonbonding noble gases. As was shown above, atomistic simulations will provide such insight into the relevant processes that cannot be obtained using any current experimental methods. However, much work still remains to be done on this matter. In particular, determining the conditions at which ion irradiation can be used to dope graphene with various elements such as boron, nitrogen or

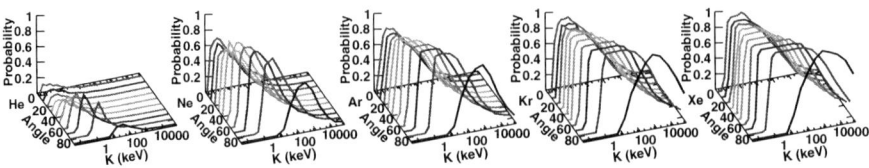

Figure 11.20   Defect-creation probability of graphene under ion irradiation as a function of both irradiation angle ψ and energy K. Image courtesy of Ossi Lehtinen, University of Helsinki, Finland.

magnetic metals will be crucial if this method will be used in nano-engineering graphene with impurity atoms.

11.6 Conclusions and Outlook

Understanding the structure and energetics of native defects in graphene is crucial since they may govern the electronic properties of the material. In particular, CVD-grown graphene is known to contain a large number of dislocations and grain boundaries. In addition to pre-existing defects, some defects will inevitably be produced during the preparation of the material for applications. Moreover, defects can also be introduced deliberately into the structure in order to modify or study the properties of the system. Although the history of radiation damage studies in graphitic materials dates back to the post-World-War-II era, many of the microscopic processes that occur during the radiation process have remained unknown, especially in the context of monolayer graphene, which became available for experiments only in the middle of the last decade (2004).

Effects of continued electron irradiation can be considered as collective sampling of separate elastic collisions between an electron and a target atom, each of which – when having any effect – will either (1) eject the atom, (2) create a Frenkel pair where the displaced atom remains in the system as an adatom on graphene and can easily migrate back, or (3) cause a bond rotation. The actual rates for each of the events depends on the acceleration voltage *i.e.* kinetic energy of the electrons, and can be estimated within the McKinley–Feshbach formalism for Coulomb scattering. At the low-energy limit, electron irradiation has no effect on graphene, whereas at the high-energy/high-dose limit atoms are simply sputtered. Curiously, in between these two extremes, graphene can accommodate the loss of atoms by reorganising to form a complicated network of nonhexagonal rings *via* bond rotations.

Ion irradiation has a much stronger effect on graphene because of the larger momentum transfer between the ion and a target atom. When the transferred energy exceeds the displacement threshold, and the scattering angle of the target atom is close to the initial velocity of the ion, a monovacancy is always produced. When an ion hits in between two carbon atoms, which are nearest neighbours, and with a sufficient collision cross-section, a divacancy is directly produced. All other defects, and in fact also some divacancies, are created *via* secondary processes that are initiated by an inplane displacement of the carbon atom. These only occur at high irradiation energies.

Atomistic simulations are the only way to obtain detailed knowledge on microscopic events occurring during a radiation process of graphene. As graphene is the ultimate thin membrane with its thickness of only one atomic layer, the conventional knowledge of radiation damage in solids must be revised for this special case. Moreover, since carbon displays more flexibility in its ability to form molecules and rearrange bonds than any other element, many of the occurring processes are highly nontrivial. This opens possibilities to use atomistic simulations on radiation damage of carbon nanomaterials for designing experiments and methods that can be used for nanoengineering these exciting materials for possible uses in nanoelectronics, or perhaps as hydrogen storage.

To sum up, in this chapter we presented an overview of the latest theoretical and experimental data on defects in graphene. We showed that atomistic computer simulations carried out at different levels of sophistication ranging from empirical potential to first-principle techniques have proven to be an indispensable tool in the interpretation of the experimental results and in getting the microscopic understanding of the types of defects present in this material. We analysed the structure of not only native but also irradiation-induced defects, and gave a detailed account of how the defects influence graphene properties. We also discussed at length the mechanisms of defect formation under ion and electron irradiation and pointed out new avenues for defect-mediated engineering of graphene-based structures and materials.

Acknowledgements

We wish to acknowledge Jannik Meyer and Ossi Lehtinen for contributing images and unpublished results to this chapter. Our highest gratitude is due to the Academy of Finland, which has provided us funding through several projects, and to Center for Scientific Computing, Espoo, Finland, for generous grants on computer time through the grand challenge program IECBN09.

References

1. K. S. Novoselov, A. K. Geim, S. V. Morozov, D. Jiang, Y. Zhang, S. V. Dubonos, I. V. Grigorieva and A. A. Firsov, Electric field effect in atomically thin carbon films, *Science*, 2004, **306**(5696), 666–669.
2. P. R. Wallace, The band theory of graphite, *Phys. Rev.*, 1947, **71**(9), 622–634.
3. A. H. Castro Neto, F. Guinea, N. M. R. Peres, K. S. Novoselov and A. K. Geim, The electronic properties of graphene, *Rev. Mod. Phys.*, 2009, **81**(1), 109–162.
4. P. Avouris, Z. Chen and V. Perebeinos, Carbon-based electronics, *Nature Nano*, 2007, **2**(10), 605–615.
5. H. Cao, Q. Yu, R. Colby, D. Pandey, C. S. Park, J. Lian, D. Zemlyanov, I. Childres, V. Drachev, E. A. Stach, M. Hussain, H. Li, S. S. Pei and Y. P. Chen, Large-scale graphitic thin films synthesized on ni and transferred to insulators: Structural and electronic properties, *J. Appl. Phys.*, 2010, **107**(4), 044310.
6. S. B. Austerman and J. E. Hove, Irradiation of graphite at liquid helium temperatures, *Phys. Rev.*, 1955, **100**(4), 1214–1215.
7. S. Iijima, Helical microtubules of graphitic carbon, *Nature*, 1991, **354**, 56–58.
8. J. W. Mayer and S. S. Lau, *Electronic Materials Science For Integrated Circuits in Si and GaAs*, MacMillan, New York, 1990.

9. J. Kotakoski, A. V. Krasheninnikov, Y. Ma, A. S. Foster, K. Nordlund and R. M. Nieminen, B and n ion implantation into carbon nanotubes: Insight from atomistic simulations, *Phys. Rev. B*, 2005, **71**(20), 205408.

10. J. Kotakoski, J. A. V. Pomoell, A. V. Krasheninnikov and K. Nordlund, Irradiation-assisted substitution of carbon atoms with nitrogen and boron in single-walled carbon nanotubes, *Nucl. Instrum. Methods Phys. Res. B*, 2005, **228**(1–4), 31–36. Proceedings of the Seventh International Conference on Computer Simulation of Radiation Effects in Solids.

11. C. Morant, J. Andrey, P. Prieto, D. Mendiola, J. M. Sanz and E. Elizalde, Xps characterization of nitrogen-doped carbon nanotubes, *Phys. Status Solidi*, 2006, **203**, 1069–1075.

12. F. Xu, M. Minniti, P. Barone, A. Sindona, A. Bonanno and A. Oliva, Nitrogen doping of single-walled carbon nanotubes by low energy ion implantation, *Carbon*, 2008, **46**(11), 1489–1496.

13. A. Hashimoto, K. Suenaga, A. Gloter, K. Urita and S. Iijima, Direct evidence for atomic defects in graphene layers, *Nature (London)*, 2004, **430**, 870.

14. J. C. Meyer, C. Kisielowski, R. Erni, M. D. Rossell, M. F. Crommie and A. Zettl, Direct imaging of lattice atoms and topological defects in graphene membranes, *Nano Lett.*, 2008, **8**(11), 3582–3586.

15. C. O. Girit, J. C. Meyer, R. Erni, M. D. Rossell, C. Kisielowski, L. Yang, C.-H. Park, M. F. Crommie, M. L. Cohen, S. G. Louie and A. Zettl, Graphene at the Edge: Stability and Dynamics, *Science*, 2009, **323**(5922), 1705–1708.

16. G. Gómez-Navarro, P. J. De Pablo, J. Gómez-Herrero, B. Biel, F. J. Garcia-Vidal, A. Rubio and F. Flores, Tuning the conductance of single-walled carbon nanotubes by ion irradiation in the Anderson localization regime, *Nature Mater.*, 2005, **4**, 534.

17. J. Lahiri, Y. Lin, P. Bozkurt, I. I. Oleynik and M. Batzill, An extended defect in graphene as a metallic wire, *Nature Nano*, 2010, **5**, 326–329.

18. O. Lehtinen, J. Kotakoski, A. V. Krasheninnikov, A. Tolvanen, K. Nordlund and J. Keinonen, Effect of ion bombardment on a two-dimensional target: atomistic simulations of graphene irradiation, *Phys. Rev. B*, 2010, **81**, 153401.

19. L. Tapasztó, G. Dobrik, P. Nemes-Incze, G. Vertesy, Ph. Lambin and L. P. Biró, Tuning the electronic structure of graphene by ion irradiation, *Phys. Rev. B*, 2008, **78**(23), 233407.

20. J.-H. Chen, W. G. Cullen, C. Jang, M. S. Fuhrer and E. D. Williams, Defect scattering in graphene, *Phys. Rev. Lett.*, 2009, **102**(23), 236805.

21. E. Stolyarova, D. Stolyarov, K. Bolotin, S. Ryu, L. Liu, K. T. Rim, M. Klima, M. Hybertsen, I. Pogorelsky, I. Pavlishin, K. Kusche, J. Hone, P. Kim, H. L. Stormer, V. Yakimenko and G. Flynn, Observation of graphene bubbles and effective mass transport under graphene films, *Nano Lett.*, 2009, **9**(1), 332–337.

22. G. Compagnini, F. Giannazzo, S. Sonde, V. Raineri and E. Rimini, Ion irradiation and defect formation in single layer graphene, *Carbon*, 2009, **47**(14), 3201–3207.

23. F. Banhart, Irradiation effects in carbon nanostructures, *Rep. Prog. Phys.*, 1999, **62**, 1181–1221.

24. M. H. Gass, U. Bangert, A. L. Bleloch, P. Wang, R. R. Nair and A. K. Geim, Free-standing graphene at atomic resolution, *Nature Nano*, 2008, **3**(11), 676–681.

25. L. Sun, F. Banhart, A. V. Krasheninnikov, J. A. Rodríguez-Manzo, M. Terrones and P. M. Ajayan, Carbon nanotubes as high-pressure cylinders and nanoextruders, *Science*, 2006, **312**, 1199.

26. L. Sun, A. V. Krasheninnikov, T. Ahlgren, K. Nordlund and F. Banhart, Plastic deformation of single nanometer-sized crystals, *Phys. Rev. Lett.*, 2008, **101**(15), 156101.

27. C. Jin, H. Lan, L. Peng, K. Suenaga and S. Iijima, Deriving carbon atomic chains from graphene, *Phys. Rev. Lett.*, 2009, **102**(20), 205501.

28. A. Chuvilin, J. C Meyer, G. Algara-Siller and U. Kaiser, From graphene constrictions to single carbon chains., *New J. Phys.*, 2009, **11**(8), 083019 (10pp).

29. F. Giannazzo, S. Sonde, V. Raineri and E. Rimini, Irradiation damage in graphene on SiO_2. probed by local mobility measurements, *Appl. Phys. Lett.*, 2009, **95**(26), 263109.

30. S. H. M. Jafri, K. Carva, E. Widenkvist, T. Blom, B. Sanyal, J. Fransson, O. Eriksson, U. Jansson, H. Grennberg, O. Karis, R. A. Quinlan, B. C. Holloway and K. Leifer, Conductivity engineering of graphene by defect formation, *J. Phys. D: Appl. Phys.*, 2010, **43**(4), 045404.

31. K.-J. Kim, H. Lee, J. Choi, H. Lee, M. C. Jung, H. J. Shin, T.-H. Kang, B. Kim and S. Kim, Surface property change of graphene using nitrogen ion, *J. Phys. Condens. Matter*, 2010, **22**(4), 045005(4pp).

32. K. W. Lee, H. Kweon, J. J. Kweon and C. E. Lee, Electron spin resonance study of proton-irradiation–induced defects in graphite, *J. Appl. Phys.*, 2010, **107**(4), 044302.

33. M. M. Ugeda, I. Brihuega, F. Guinea and J. M. Gómez-Rodríguez, Missing atom as a source of carbon magnetism, *Phys. Rev. Lett.*, 2010, **104**(9), 096804.

34. A. C. Ferrari, J. C. Meyer, V. Scardaci, C. Casiraghi, M. Lazzeri, F. Mauri, S. Piscanec, D. Jiang, K. S. Novoselov, S. Roth and A. K. Geim, Raman spectrum of graphene and graphene layers, *Phys. Rev. Lett.*, 2006, **97**(18), 187401.

35. A. C. Ferrari, Raman spectroscopy of graphene and graphite: Disorder, electron–phonon coupling, doping and nonadiabatic effects, *Solid State Commun.*, **143**(1–2), 47–57. Exploring graphene - Recent research adcances.

36. A. Tolvanen, G. Buchs, P. Ruffieux, P. Groning, O. Groning and A. V. Krasheninnikov, Modifying the electronic structure of

semiconducting single-walled carbon nanotubes by Ar^+ ion irradiation, *Phys. Rev. B*, 2009, **79**(12), 125430.

37. A. V. Krasheninnikov, K. Nordlund, M. Sirviö, E. Salonen and J. Keinonen, Formation of ion irradiation-induced atomic-scale defects on walls of carbon nanotubes, *Phys. Rev. B*, 2001, **63**, 245405.

38. A. V. Krasheninnikov, Predicted scanning microscopy images of carbon nanotubes with atomic vacancies, *Solid State Commun.*, 2001, **118**, 361–365.

39. D. Tománek, Carbon-based nanotechnology on a super computer, *J. Phys. Condens. Matter*, 2005, **17**, R413.

40. A. V. Krasheninnikov and K. Nordlund, Ion and electron irradiation-induced effects in nanostructured materials, *J. Appl. Phys.*, 2010, **107**(7), 071301.

41. J. Kotakoski, A. V. Krasheninnikov and K. Nordlund, Kinetic Monte Carlo simulations of the response of carbon nanotubes to electron irradiation, *J. Comp. Theor. Nanosci.*, 2007, **4**(6), 1153–1159.

42. Y. Gan, J. Kotakoski, A. V. Krasheninnikov, K. Nordlund and F. Banhart, The diffusion of carbon atoms inside carbon nanotubes, *New J. Phys.*, 2008, **10**.

43. M. P. Allen and D. J. Tildesley, *Computer Simulation of Liquids*, Oxford University Press, Oxford, England, 1989.

44. K. Nordlund, J. Keinonen and A. Kuronen, Effect on the interatomic Si–Si-potential on vacancy production during ion implantation of Si, *Phys. Scr.*, 1995, **T54**, 34–37.

45. J. F. Ziegler, J. P. Biersack and U. Littmark, *The Stopping and Range of Ions in Matter*, Pergamon, USA, 1985.

46. H. J. C. Berendsen, J. P. M. Postma, W. F. van Gunsteren, A. DiNola and J. R. Haak, Molecular dynamics with coupling to external bath, *J. Chem. Phys.*, 1984, **81**(8), 3684.

47. S. Nose, A molecular-dynamics method for simulations in the canonical ensemble, *Mol. Phys.*, 1984, **52**(2), 255–268.

48. S. Nose, A unified formulation of the constant temperature molecular-dynamics methods, *J. Chem. Phys.*, 1984, **81**(1), 511–519.

49. D. W. Brenner, Empirical potential for hydrocarbons for use in simulating the chemical vapor deposition of diamond films, *Phys. Rev. B*, 1990, **42**, 9458–9471.

50. S. J. Stuart, A. B. Tutein and J. A. Harrison, A reactive potential for hydrocarbons with intermolecular interactions, *J. Chem. Phys.*, 2000, **112**(14), 6472–6486.

51. D. W. Brenner, O. A. Shenderova, J. A. Harrison, S. J. Stuart, B. Ni and S. B. Sinnott, A second-generation reactive empirical bond order (REBO) potential energy expression for hydrocarbons, *J. Phys.: Condens. Matter*, 2002, **14**, 783–802.

52. M. S. Daw, S. M. Foiles and M. I. Baskes, The embedded-atom-method: a review of theory and applications, *Mater. Sci. Rep.*, 1993, **9**, 251–287.

53. C. I. Kelchner, D. M. Halstead, L. S. Perkins, N. M. Wallace and A. E. DePristo, Construction and evaluation of embedding functions, *Surf. Sci.*, 1994, **310**, 425–435.

54. J. Tersoff, Empirical interatomic potential for carbon, with applications to amorphous carbon, *Phys. Rev. Lett.*, 1988, **61**(25), 2879.

55. K. Albe, K. Nordlund and R. S. Averback, Modeling the metal-semi-conductor interaction: Analytical bond-order potential for platinum-carbon, *Phys. Rev. B*, 2002, **65**(19), 195124.

56. D. Porezag, Th. Frauenheim, Th. Köhler, G. Seifert and R. Kaschner, Construction of tight-binding-like potentials on the basis of density-functional theory: Application to carbon, *Phys. Rev. B*, 1995, **51**(19), 12947–12957.

57. M. Elstner, D. Porezag, G. Jungnickel, J. Elsner, M. Haugk, Th. Frauenheim, S. Suhai and G. Seifert, Self-consistent-charge density-functional tight-binding method for simulations of complex materials properties, *Phys. Rev. B*, 1998, **58**(11), 7260–7268.

58. F. Banhart, J. X. Li and A. V. Krasheninnikov, Carbon nanotubes under electron irradiation: Stability of the tubes and their action as pipes for atom transport, *Phys. Rev. B*, 2005, **71**, 241408(R).

59. A. V. Krasheninnikov, F. Banhart, J. X. Li, A. S. Foster and R. M. Nieminen, Stability of carbon nanotubes under electron irradiation: Role of tube diameter and chirality, *Phys. Rev. B*, 2005, **72**, 125428.

60. T. Loponen, A. V. Krasheninnikov, M. Kaukonen and R. M. Nieminen, Nitrogen-doped carbon nanotubes under electron irradiation simulated with a tight-binding model, *Phys. Rev. B*, 2006, **74**(7), 073409.

61. A. Zobelli, A. Gloter, C. P. Ewels, G. Seifert and C. Colliex, Electron knock-on cross-section of carbon and boron nitride nanotubes, *Phys. Rev. B*, 2007, **75**, 245402.

62. J. Kotakoski, C. H. Jin, O. Lehtinen, K. Suenaga and A. V. Krasheninnikov, Knock-on damage in hexagonal boron nitride monolayers, *Phys. Rev. Lett.*, 2010.

63. R. Martin, *Electronic Structure*, Cambridge University Press, Cambridge, UK, 2004.

64. O. Sugino and Y. Miyamoto, Density-functional approach to electron dynamics: Stable simulation under a self-consistent field, *Phys. Rev. B*, 1999, **59**(4), 2579–2586.

65. O. Sugino and Y. Miyamoto, Erratum: Density-functional approach to electron dynamics: Stable simulation under a self-consistent field, *Phys. Rev. B*, 1999, **59**, 2579; *Phys. Rev. B*, 2002, **66**(8), 089901.

66. A. V. Krasheninnikov, Y. Miyamoto and D. Tomanek, Role of electronic excitations in ion collisions with carbon nanostructures, *Phys. Rev. Lett*, 2007, **99**, 016104.

67. R. H. Telling, C. P. Ewels, A. A. El-Barbary and M. I. Heggie, Wigner defects bridge the graphite gap, *Nature Mater.*, 2003, **2**, 333–337.

68. A. A. El-Barbary, R. H. Telling, C. P. Ewels, M. I. Heggie and P. R. Briddon, Structure and energetics of the vacancy in graphite, *Phys. Rev. B*, 2003, **68**(14), 144107.
69. P. O. Lehtinen, A. S. Foster, Y. Ma, A. V. Krasheninnikov and R. M. Nieminen, Irradiation-induced magnetism in graphite: A density functional study, *Phys. Rev. Lett.*, 2004, **93**, 187202.
70. L. Li, S. Reich and J. Robertson, Defect energies of graphite: Density-functional calculations, *Phys. Rev. B*, 2005, **72**(18), 184109.
71. A. V. Krasheninnikov, P. O. Lehtinen, A. S. Foster and R. M. Nieminen, Bending the rules: Contrasting vacancy energetics and migration in graphite and carbon nanotubes, *Chem. Phys. Lett.*, 2006, **418**, 132–136.
72. J. P. Perdew, K. Burke and M. Ernzerhof, Generalized gradient approximation made simple, *Phys. Rev. Lett.*, 1996, **77**, 3865.
73. J. P. Perdew, K. Burke and M. Ernzerhof, Erratum: Generalized gradient approximation made simple, *Phys. Rev. Lett.*, 1997, **78**, 1396.
74. J. Kotakoski, A. V. Krasheninnikov and K. Nordlund, Energetics, structure, and long-range interaction of vacancy-type defects in carbon nanotubes: Atomistic simulations, *Phys. Rev. B*, 2006, **74**(24).
75. J. M. Carlsson and M. Scheffler, Structural, electronic, and chemical properties of nanoporous carbon, *Phys. Rev. Lett.*, 2006, **96**(4), 046806.
76. De en Jiang, V. R. Cooper and S. Dai, Porous graphene as the ultimate membrane for gas separation, *Nano Lett.*, 2009, 4019–4024.
77. H. Terrones, M. Terrones, E. Hernández, N. Grobert, J.-C. Charlier and P. M. Ajayan, New metallic allotropes of planar and tubular carbon, *Phys. Rev. Lett.*, 2000, **84**(8), 1716–1719.
78. R. G. Amorim, A. Fazzio, A. Antonelli, F. D. Novaes and A. J. R. da Silva, Divacancies in graphene and carbon nanotubes, *Nano Lett.*, 2007, **7**(8), 2459–2462.
79. R. Y. Oeiras, F. M. Araujo-Moreira and E. Z. da Silva, Defect-mediated half-metal behaviour in zigzag graphene nanoribbons, *Phys. Rev. B*, 2009, **80**(7), 073405.
80. G. D. Lee, C. Z. Wang, E. Yoon, N. M. Hwang, D. Y. Kim and K. M. Ho, Diffusion, coalescence, and reconstruction of vacancy defects in graphene layers, *Phys. Rev. Lett.*, 2005, **95**, 205501.
81. G.-D. Lee, C. Z. Wang, E. Yoon, N.-M. Hwang and K. M. Ho, Vacancy defects and the formation of local haeckelite structures in graphene from tight-binding molecular dynamics, *Phys. Rev. B*, 2006, **74**(24), 245411.
82. C. P. Ewels, M. I. Heggie and P. R. Briddon, Adatoms and nanoengineering of carbon, *Chem. Phys. Lett.*, 2002, **351**, 178–182.
83. B.W. Jeong, J. Ihm and G.-D. Lee, Stability of dislocation defect with two pentagon–heptagon pairs in graphene, *Phys. Rev. B*, 2008, **78**(16), 165403.
84. O. V. Yazyev, I. Tavernelli, U. Rothlisberger and L. Helm, Early stages of radiation damage in graphite and carbon nanostructures: A first-

principles molecular dynamics study, *Phys. Rev. B*, 2007, **75**(11), 115418.

85. P. O. Lehtinen, A. S. Foster, A. Ayuela, A. V. Krasheninnikov, K. Nordlund and R. M. Nieminen, Magnetic properties and diffusion of adatoms on a graphene sheet, *Phys. Rev. Lett.*, 2003, **91**, 017202.

86. A. V. Krasheninnikov, K. Nordlund, P. O. Lehtinen, A. S. Foster, A. Ayuela and R. M. Nieminen, Adsorption and migration of carbon adatoms on carbon nanotubes: Density-functional *ab initio* and tight-binding studies, *Phys. Rev. B*, 2004, **69**, 073402.

87. M. Heggie, B. R. Eggen, C. P. Ewels, P. Leary, S. Ali, G. Jungnickel, R. Jones and P. R. Briddon, Ldf calculations of point defects in graphites and fullerenes, *Electrochem. Soc. Proc.*, 1998, **98**, 60.

88. L. Tsetseris and S. T. Pantelides, Adatom complexes and self-healing mechanisms on graphene and single-wall carbon nanotubes, *Carbon*, 2009, **47**(3), 901–908.

89. M. T. Lusk, D. T. Wu and L. D. Carr, Graphene nanoengineering and the inverse Stonethrower–Wales defect, *Phys. Rev. B*, 2010, **81**(15), 155444.

90. M. T. Lusk and L. D. Carr, Nanoengineering defect structures on graphene, *Phys. Rev. Lett.*, 2008, **100**(17), 175503.

91. R. Singh and P. Kroll, Magnetism in graphene due to single-atom defects: dependence on the concentration and packing geometry of defects, *J. Phys. Condens. Matter*, 2009, **21**(19), 196002.

92. V. V. Cheianov, O. Syljuasen, B. L. Altshuler and V. Fal'ko, Ordered states of adatons on graphene, *Phys. Rev. B*, 2009, **80**(23), 233409.

93. B. Uchoa, L. Yang, S.-W. Tsai, N. M. R. Peres and A. H. Castro Neto, Theory of scanning tunneling spectroscopy of magnetic adatoms in graphene, *Phys. Rev. Lett.*, 2009, **103**(20), 206804.

94. H. Johll, H. C. Kang and E. S. Tok, Density-functional theory study of FE, CO, and NI adatoms and dimers adsorbed on graphene, *Phys. Rev. B*, 2009, **79**(24), 245416.

95. E. J. G. Santos, D. Sánchez-Portal and A. Ayuela, Magnetism of substitutional CO impurities in graphene: Realization of single π vacancies, *Phys. Rev. B*, 2010, **81**(12), 125433.

96. C. Ataca, E. Akturk, S. Ciraci and H. Ustunel, High-capacity hydrogen storage by metallized graphene, *Appl. Phys. Lett.*, 2008, **93**(4), 043123.

97. K.T. Chan, J. B. Neaton and M.L. Cohen, First-principles study of metal adatom adsorption on graphene, *Phys. Rev. B*, 2008, **77**(23), 235430.

98. A. H. Castro Neto and F. Guinea, Impurity-induced spin-orbit coupling in graphene, *Phys. Rev. Lett.*, 2009, **103**(2), 026804.

99. W. S. Kim, S. Y. Moon, S. Y. Bang, B. G. Choi, H. Ham, T. Sekino and K. B. Shim, Fabrication of graphene layers from multiwalled carbon nanotubes using high dc pulse, *Appl. Phys. Lett.*, 2009, **95**(8), 083103.

100. D. W. Boukhvalov and M. I. Katsnelson, Destruction of graphene by metal adatoms, *Appl. Phys. Lett.*, 2009, **95**(2), 023109.

101. T. O. Wehling, M. I. Katsnelson and A. I. Lichtenstein, Impurities on graphene: Midgap states and migration barriers, *Phys. Rev. B*, 2009, **80**(8), 085428.

102. O. U. Akturk and M. Tomak, Bismuth doping of graphene, *Appl. Phys. Lett.*, 2010, **96**(8), 081914.

103. W. Zhang, L. Sun, Z. Xu, A. V. Krasheninnikov, P. Huai, Z. Zhu and F. Banhart, Migration of gold atoms in graphene ribbons: Role of the edges, *Phys. Rev. B*, 2010, **81**(12), 125425.

104. M. Klintenberg, S. Lebègue, M. I. Katsnelson and O. Eriksson, Theoretical analysis of the chemical bonding and electronic structure of graphene interacting with group IA and group VIIA elements, *Phys. Rev. B*, 2010, **81**(8), 085433.

105. S. Casolo, O. M. Lovvik, R. Martinazzo and G. F. Tantardini, Understanding adsorption of hydrogen atoms on graphene, *J. Chem. Phys.*, 2009, **130**(5), 054704.

106. K. Xue and Z. Xu, Strain effects on basal-plane hydrogenation of graphene: A first-principles study, *Appl. Phys. Lett.*, 2010, **96**(6), 063103.

107. H. McKay, D. J. Wales, S. J. Jenkins, J. A. Verges and P. L. de Andres, Hydrogen on graphene under stress: Molecular dissociation and gap opening, *Phys. Rev. B*, 2010, **81**(7), 075425.

108. L. Pisani, B. Montanari and N. M. Harrison, A defective graphene phase predicted to be a room temperature ferromagnetic semiconductor, *New J. Phys.*, 2008, **10**(3), 033002 (9pp).

109. Z. Liu, S.-K. Joung, T. Okazaki, K. Suenaga, Y. Hagiwara, T. Ohsuna, K. Kuroda and S. Iijima, Self-assembled double ladder structure formed inside carbon nanotubes by encapsulation of $H_8Si_8O_{12}$, *ACS Nano*, 2009, **3**(5), 1160–1166.

110. J. A. Vergés and P. L. de Andres, Trapping of electrons near chemisorbed hydrogen on graphene, *Phys. Rev. B*, 2010, **81**(7), 075423.

111. A. J. Stone and D. J. Wales, Theoretical studies of icosahedral C_{60} and some related species, *Chem. Phys. Lett.*, 1986, **128**, 501–503.

112. Y. Miyamoto, A. Rubio, S. Berber, M. Yoon and D. Tománek, Spectroscopic characterization of Stone–Wales defects in nanotubes, *Phys. Rev. B*, 2004, **69**, 121413(R).

113. E. Ertekin, D. C. Chrzan and M. S. Daw, Topological description of the Stone–Wales defect formation energy in carbon nanotubes and graphene, *Phys. Rev. B*, 2009, **79**(15), 155421.

114. T. Shimada, D. Shirasaki and T. Kitamura, Stone–Wales transformations triggered by intrinsic localized modes in carbon nanotubes, *Phys. Rev. B*, 2010, **81**(3), 035401.

115. J. Ma, D. Alfe, A. Michaelides and E. Wang, Stone–Wales defects in graphene and other planar sp^2-bonded materials, *Phys. Rev. B*, 2009, **80**(3), 033407.

116. O. Yazyev and S. Louie, Topological defects in graphene: dislocations and grain boundaries, *Phys. Rev. B*, 2010, **81**(19), 195420.

117. C. Jin, F. Lin, K. Suenaga and S. Iijima, Fabrication of a freestanding boron nitride single layer and its defect assignments, *Phys. Rev. Lett.*, 2009, **102**(19), 195505.

118. J.C. Meyer, A. Chuvilin, G. Algara-Siller, J. Biskupek and U. Kaiser, Selective sputtering and atomic resolution imaging of atomically thin boron nitride membranes, *Nano Lett.*, 2009, **9**(7), 2683–2689.

119. W.A. McKinley and H. Feshbach, The coulomb scattering of relativistic electrons by nuclei, *Phys. Rev.*, 1948, **74**(12), 1759–1763.

CHAPTER 12

The Atomic-, Nano-, and Mesoscale Origins of Graphite's Response to Energetic Particles

MALCOLM I. HEGGIE AND CHRISTOPHER D. LATHAM

Department of Chemistry and Biochemistry, University of Sussex, Falmer, Brighton, BN1 9QJ, UK

12.1 Introduction and Background

Ably assisted by Miss Yardley, and Messrs Astbury and Wood, J. D. Bernal determined the structure of graphite in 1924 by the then new technique of X-ray diffraction.[1] He describes the arrangement of carbon atoms as being inplane, hexagonal nets, alternately superposed so that half of the atoms in each net lie opposite the vertices of the hexagons of their neighbours (labelled α) while the remainder (labelled β) are opposite the centres of the hexagons, as shown in Figure 12.1. It immediately provided an explanation for the pronounced basal cleavage plane and trigonal hexagonal symmetry of graphite crystals, and its softness and flexibility compared with diamond. Bernal also sought to reconcile the structure of graphite with Bohr's atomic theory for carbon atoms. From this he deduced that the arrangement and nature of the electron orbitals would be different from those in diamond, resulting in the different thermal, electrical, and optical properties known at that time. He also proposed that that the electrical conductivity of graphite should be anisotropic, being greater in the basal plane than perpendicular to it, and hoped that measurements might be performed to resolve the matter. Bernal's correct model has subsequently both guided and

RSC Theoretical and Computational Chemistry Series No. 4
Computational Nanoscience
Edited by Elena Bichoutskaia

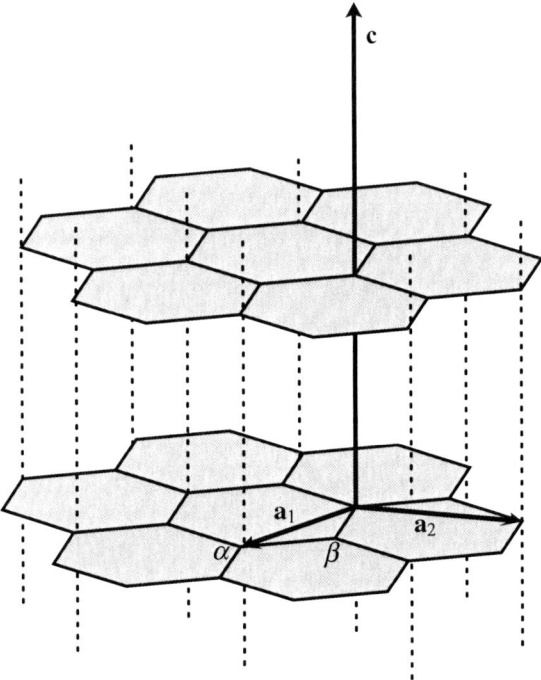

Figure 12.1   The structure of graphite depicting the lattice vectors $\mathbf{a}_1$, $\mathbf{a}_2$, and $\mathbf{c}$, and α and β atoms. A note for readers whose background is in carbon nanotubes (CNT): the usual convention for the angle between basal lattice vectors $\mathbf{a}_1$ and $\mathbf{a}_2$ is $120°$ in graphite, rather than $60°$ for CNTs.

misguided models of how graphite responds to external perturbations, particularly those caused by exposure to radiation from energetic particles.

The main interest here concerns the use of graphite as a neutron moderator and reflector material in nuclear fission reactors. This dates back to the original, experimental Chicago Pile-1 constructed in 1942. Many more reactors built since then have used graphite for the same purpose, including ones in operation now. Graphite may also be used in designs proposed for the future. It works by slowing down, from energies of order 2 MeV, the neutrons released in a nuclear fission chain reaction, trapping them as a gas-like "atmosphere" within the reactor core with thermal energies, at which the nuclear reactions are at their optimum. From the outset, it was realised that "the collision of neutrons with the atoms of any substance placed in the pile will cause displacement of these atoms" that would have consequential effects on the properties of materials exposed to radiation within a reactor.[2] Experiments, already in progress then, would be necessary to determine these effects.

One such effect is radiation-induced dimensional change. When a crystal of graphite receives a high dose of energetic particle radiation, it can expand along the c-axis by an amount well in excess of one third of its original length.[3] An explanation for this phenomenon was proposed in the 1960s by Iwata and

Suzuki.[4] As described by Bernal, graphite has a layered structure. Each energetic particle entering a crystal knocks many atoms into the spaces between the layers, leaving behind vacant lattice sites within the sheets. The interstitial atoms migrate and aggregate, initially forming C_2 molecules, and eventually forming new sheets in between the existing ones. Hence, the crystal expands in the c-direction, perpendicular to the sheets. The original layers gradually shrink owing to the loss of atoms from them. Unfortunately, this appealingly simple and attractive model, it will be shown, is flawed. The picture that emerges instead, is one where, in addition to the point-defect processes, the sheets behave more like pieces of fabric that crinkle, ruck, and eventually fold over upon themselves into structures resembling a nanosized version of some geological formations (recumbent folds). The role of point defects within the crystal is also extremely important, since they modify its properties, and large amounts of energy—called Wigner energy—are associated with their formation.

The purpose of many early nuclear reactors was to produce plutonium for atomic weapons rather than for the generation of energy. These reactors were simply an air-cooled matrix of graphite blocks and uranium fuel, similar to the original Chicago Pile-1, except on a much larger scale. In them, fissile plutonium-239 is created by neutron capture from the common, but nonfissile, isotope uranium-238, where the neutrons are provided by fission of uranium-235 atoms. Crucially, these reactors operated at a relatively low temperature, not much above ambient. Over time, point defects and their associated Wigner energy accumulated in the graphite. Reactor operations included occasional annealing of the graphite by allowing the reactor temperature to rise by reducing cooling. This process released the Wigner energy, and at least partially restored the physical properties of the graphite blocks. Unfortunately, the dangers associated with this procedure were not fully appreciated, resulting in the accident at the Windscale plant in 1957.[5] Prior to the accident, annealing cycles had not released all the Wigner energy, which was also distributed unevenly within the graphite. Positive feedback led to an uncontrollable release of heat. Although the graphite did not catch fire, the temperature rose sufficiently to ignite the uranium-metal fuel. It is estimated that, consequently, about 4–8 GBq of radioactive material were emitted into the environment.

Subsequent experiments on graphite irradiated and annealed over a wide range of temperatures from cryogenic to several thousand Kelvin exhibit complex behaviour, as shown by Kelly's review.[6] When graphite irradiated at cryogenic temperatures is annealed, the onset of energy release, changes in electrical properties, thermal properties, crystal dimensions, compared with the unannealed material, begins from about 50 K, and is well established by 100 K. The release of energy does not occur smoothly: it follows an irregular pattern with several large peaks, and many smaller ones up to about 600 K (ref. 7). The two largest peaks are at 240 K and 475 K. By 600 K, the properties of the material are close to virgin graphite.[8] Further release occurs at higher temperatures and more smoothly for higher-dose and higher-temperature irradiations.[9] For the low-temperature, low-dose cases, although the interlayer spacing d_{002} and density of irradiated graphite is restored by annealing, crystals

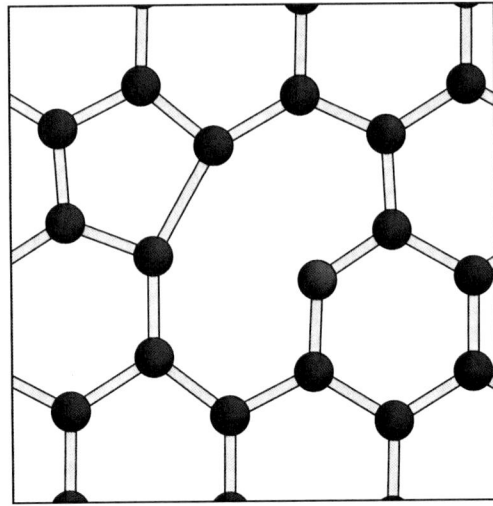

Figure 12.2   The structure of an isolated monovacancy in graphite.

still exhibit dimensional change that is stable to 600 K. Thus, the expansion in the *c*-direction is compensated by shrinkage in the basal plane, as is also the case for irradiations above 550 K.

In order to maximise their thermodynamic efficiency, nuclear reactors used for energy production are operated at the highest feasible temperatures, well in excess of 500 K, above which no appreciable accumulation of Wigner energy occurs in graphite and they are not cooled by air, but by carbon dioxide. However, the problem of radiation-induced dimensional change, and a related phenomenon, which tends to mitigate these changes (radiation-induced creep), still occurs. There are also concerns in some quarters about how to dispose of the graphite from reactors when they are decommissioned at the end of their lives, given that the exact atomic and microstructural changes after spending several decades in a reactor core are not perfectly known. Even when the problem of high radiation levels is ignored, access to these places in order to make measurements is extremely difficult. Moreover, graphite is much less amenable to experimental investigation than materials such as silicon or diamond. It is polycrystalline, highly defective, and contains many impurities. The "bricks" of graphite used in nuclear reactors are manufactured by pyrolysis from hydrocarbon sources such as pitch or tar of some form; hence, residues of the precursors are inevitably present. They are also composite in nature, with a hierarchy of structure from the initial coke particles down to the pyrolysed binder phase, with perhaps different degrees of graphitisation. Crystal grain sizes are normally no larger than the micrometre scale, and their orientations are deliberately engineered to be as random as possible to give isotropic behaviour. For scientific investigation, where purity and simplicity is needed, the highest-quality synthetic material is used. It is known as highly oriented (or ordered) pyrolytic graphite (HOPG), where the crystallographic *c*-axes of the crystals exhibit relatively little ($<1°$) misalignment, while still being poly-crystalline. In general, the still large crystallite misorientation in basal directions is unimportant, since the behaviour of graphite is nearly isotropic in these directions. Small single crystals can be found in Nature with dimensions of order a few millimetres; however, they are still difficult to work with. Even in the most ideal case, there is the fundamental problem that optical, infrared, and microwave radiation cannot penetrate the interior of graphite crystals; thus, it is not possible to conduct the kind of spectroscopic investigations that have revealed the nature of defects in semiconductor crystals in exquisite detail. These circumstances raise the importance of using theory to calculate the properties and behaviour of graphite.

In this short review, we shall confine ourselves to discussing point defects and line defects, and not considering planar defects, such as grain boundaries. It is certain that an initially ordered crystallite will progressively disorder with neutron irradiation and that disorder will include turbostratic disorder. This implies at the very least the introduction of twist boundaries, whose structure and energetics have been studied.[10]

12.2 Using Density Functional Theory to Calculate the Properties of Graphite

The theoretical methods upon which the calculations described in the present work are based are well documented elsewhere; however, it is worth reviewing a few points that are applicable to graphite. Central to density functional theory (DFT) is the Hohenberg–Kohn theorem. This allows us to reduce the $3N$ coordinates needed to describe N electrons with the Schrödinger equation to the three spatial dimensions required to describe the charge-density distribution with the Kohn–Sham equation. Within this formalism, the full potential of the system is replaced with an equivalent potential, generated by the charge-density distribution $n(\mathbf{r})$, in which each electron moves.

Some approximations are then needed to make further progress to a solution. The good news is that for carbon these mainly work very well. Possibly the most troublesome, though, is the energy term arising from the effects of quantum-mechanical exchange and correlation, $E_{xc}[n(\mathbf{r})]$. This term, and the potential associated with it v_{xc}, are unknown. In the first instance, they can be replaced with their calculated values for a homogeneous electron gas of the same density at each point in space. This is the local density approximation (LDA), or the local spin density approximation (LSDA) when spin polarisation is taken into account. Its main effect, where we are concerned, is that binding and cohesive energies tend to be slightly overestimated, and interatomic distances are usually underestimated. However, in practice these errors and any consequences arising from them are relatively small.

In the case of graphite, an additional problem concerns the weak interlayer bonding. This is often attributed to "van der Waals" forces. This may be correct insofar as the term includes all weak forces between chemical moieties, but is often misinterpreted as London dispersion forces alone, and these are not included in the LDA. Moreover, the long-range behaviour of v_{xc} for the LDA is not well described. Hence, it might be thought that the LDA cannot reproduce properly the properties and behaviour of graphite. Charlier *et al.* have demonstrated that the weak interlayer bonding in graphite is, in fact, accounted for correctly by the overlap of the partially filled carbon *p*-orbitals projecting perpendicular to each graphene sheet.[11,12] Furthermore, the contribution from v_{xc} is essential: without it the equilibrium interlayer distance expands by 26%. Thus, the interlayer bonding in graphite is mainly accounted for by its band structure, rather than the dynamical phenomenon of London dispersion forces. This will be expanded upon later.

Clearly, the electron-density distribution in graphite is far from homogeneous; therefore it is reasonable to suppose that including contributions due to the density gradient in $E_{xc}[n(\mathbf{r})]$ might improve matters. This is called a generalised gradient approximation (GGA). Several varieties of GGAs exist, constructed with different levels of complexity and application in mind. For most systems GGAs reduce the overbinding of the LDA. Unfortunately, in the case of graphite, GGAs greatly overcompensate the overbinding to the extent that there is no net interlayer binding. Thus, for graphite, GGAs do not yield a stable structure, and appear not to provide any useful advantage over the LDA. Perhaps this might be expected, since GGAs are normally constructed for situations where bonding is strong.

Model systems are typically constructed in either clusters or supercells. The cluster formalism, uses a selected section of crystal containing up to a few hundred atoms including the defect of interest. It resembles a large molecule. Since a structure such as this necessarily has a large surface relative to its volume, interactions between the defect and the surface is a concern. Furthermore, it is also necessary to passivate the surface, usually using hydrogen atoms, otherwise host atoms at the surface will introduce an unwanted spectrum of energy levels into the electronic system, owing to the presence of unpaired electrons where bonds have been cut to create the surface. The supercell formalism exploits the periodicity of crystals by taking a defect model and placing identical images of itself ad *infinitum* in all directions, thereby creating a lattice of defects. In this case, the main concern is whether the defect images interact significantly with one another. The periodic boundary conditions of such a system also place geometric constraints on the models that can be constructed. For example, extended defects such as dislocations must always occur in opposite pairs, which inevitably enhances interactions for a given cell size.

An important quantity for both theory and experiment to estimate is the defect formation energy. In general terms, this is defined as,

$$E_{\mathrm{f}}(q) = E_d(q) + q\mu_{\mathrm{e}} - \sum_i n_i \mu_i, \qquad (12.1)$$

where $E_{\mathrm{f}}(q)$ is the formation energy of a defect in charge state q, and $E_d(q)$ is the total energy of a system containing a defect d that comprises n_i atoms, molecules or formula units of type i each having chemical potential μ_i. The chemical potential of i is defined to be the Gibbs free energy per particle. For condensed matter at zero temperature and pressure, the entropy and pressure contributions to the Gibbs free energy can be neglected. Hence, μ_i is the total energy per species. These chemical potentials represent reservoirs from which atoms are taken to create the state d. They are normally assumed to be a standard state that has the minimum μ_i for each i. The correct choice is not always obvious, particularly where more than one chemical species is involved. For example, it may be a compound such as an oxide, nitride, carbide, and so on. The quantity μ_e is the electron chemical potential with respect to the top of the valence band of the pure material. Since graphite is semimetallic, defects in it never carry any net electric charge; hence, in this case the term $q\mu_e = 0$. Nevertheless, in some special circumstances the system could become semiconducting, in which case charged defects can exist.

The $1s$ electrons in carbon atoms occupy states that have energies much lower than those for the valence electrons. As such, these electrons to not contribute in any significant way to chemical bonding in all forms of carbon or its compounds. Thus, it is usual to replace the core electrons and the nuclear potential with a pseudopotential, rather than including them explicitly in the calculation. Since methods based on DFT normally scale in terms of computational resources needed as N^3, this represents a significant saving. While this approximation works extremely well for carbon, the same is not always true for other elements. This may be a problem when dealing with impurities in graphite.

Valence orbitals are normally described by some kind of basis functions such as Gaussians, plane waves, or wavelet functions for both the charge density and the electronic wavefunction. Many choices are available. It is always necessary to perform checks to ensure that the results are not affected by the various parameters available. In supercell calculations, this includes the density of the grid employed to sample the Brillouin zone in reciprocal space. As a rule, metallic systems require a denser mesh than semiconducting or insulating ones to achieve results that are converged with respect to mesh density. In this respect, graphite is typical. However, the anisotropy in its electrical conductivity means that a less-dense mesh is needed in the c-direction than in the basal plane. An additional problem for graphite concerns the band crossing at $\mathbf{k}$. Here, the band structure has finite density of states (as given by the reciprocal of the derivative of the total energy E_{total} with respect to $\mathbf{k}$), and is rather intricate; hence, the total energy is highly sensitive to the Brillouin-zone sampling near $\mathbf{k}$. It is crucial to be aware of this point, and consider carefully whether it affects the results of a calculation. Further details are explained in the next section on bonding and elastic parameters.

12.3 Bonding and Elastic Parameters

The overall energetics of line defects are controlled by the elastic constants of the material. The elastic constants of graphite have been calculated many times. There has not as yet been exact agreement between calculation and measurement, but DFT, especially with the LDA, gets close. The commonly perceived difficulty is that DFT does not contain the dynamic correlation between electron densities that typifies the London dispersion forces. These forces dominate intermolecular interactions for closed-shell molecules without electric multipoles. This perception has impeded acceptance of DFT calculations for graphite, in the face of rather good agreement with experiment. Recalling that integration over the Brillouin zone is usually replaced with a weighted summation over special **k**-points, the density of these points needs to be exceptionally high and include **k** in order to reproduce the interlayer binding. If a clue were needed for the nature of the interlayer interaction this is one. The two degenerate wavefunctions at the **k**-point are rather special, as pointed out in the seminal work by Novoselov and collaborators.[13] They each exist on different sublattices, and have zero density on the other. This property is known as "pseudospin" The phenomenon is strikingly similar to the nonbonding orbital typical of odd alternant polycyclic aromatic hydrocarbons.[14] This orbital in Hückel theory has the same energy as the atomic p_z orbitals of the basis, and it only has weight on every other carbon atom. Such odd alternants are π-radicals. Similarly, the degenerate wavefunctions at the **k**-point of graphene are "nonbonding" in Hückel theory—they have the same energy as the atomic basis—and might be regarded by a chemist as π-biradicals. Thus, the binding of graphene sheets can be looked upon as covalent bonding between these biradicals.

An early attempt to describe and calculate this band-structure effect, including its effect on elastic constants was made by Green and Spain.[15] It was shown by Charlier *et al.*[11,12] that the LDA gave a good description of graphite, reinforced by Schabel and Martins[16] and by Girifalco and Hodak,[17] who showed this description was acceptable below 15% strain. Kolmogorov and Crespi[18] demonstrated that any correction to LDA was relatively insensitive to shear (and hence LDA gave a sensible value for C_{44}).

Savini and Heggie[19] showed that the agreement between LDA calculation and experiment was substantial and also put forward the possibility that the graphene layers could buckle when compressed. Beyond a very small compressive strain, this buckling replaces the quadratic dependence of energy on strain with a linear one (*i.e.* a constant force, rather than a force proportional to strain, typical of Hooke's Law). This modified the elastic constants, for which they coined the term mesoscale, since a natural wavelength of buckling of about 7 nm emerged. This work has not yet been verified and published, so here we present their classical LDA elastic constants, which are in rather good agreement with experiment, as shown in Table 12.1.

There is clear scope for improvement in C_{33} and in C_{13}, which could potentially come from either the buckling effect, or inclusion of dispersion forces in some manner. Similar agreement with experiment has been reported by Mounet and Marzari,[21] who chose to emphasize their GGA results obtained by imposing the experimental value of c/a, and also demonstrated good reproduction of thermal properties.

Although there have been attempts to incorporate dynamical correlation into a density functional these have not been as successful for graphite as one might

Table 12.1 The elastic parameters for graphite calculated using the LDA by Savini and Heggie,[19] and a summary of experimental results from Cousins and Heggie.[20]

	C_{11}	C_{12}	C_{33}	C_{44}	C_{13}
Savini and Heggi	1105	182	30.5	4.8	− 2.3
Cousins and Heggie	1060 ± 20	180 ± 20	36.5 ± 1	5.05 ± 0.35	7.9 ± 3.5

hope. The interlayer stiffness, C_{33}, for example is significantly underestimated at one third of the experimental value in a prominent study.[22]

12.4 Point Defects

When a crystal of graphite is struck by a sufficiently energetic particle, such as a neutron in a nuclear reactor, then a cascade of events follows, which creates Frenkel defects in its wake. The threshold energy needed to displace an atom irreversibly in this system is about 30 eV (refs. 6 and 23). Fission of uranium-235 atoms yield neutrons with energies up to 20 MeV wih a mean energy of 2.0 MeV, and these typically impart an energy of order 10^5 eV to the primary knock-on atom (PKA) when they strike a graphite crystal.[24] Further details of the collision cascade arc given in ref. 25, pages 4800–4801. Subsequently, when the conditions allow migration of lattice vacancies or self-interstitial atoms, Frenkel defects may vanish, or aggregates of interstitials or vacancies form, and thereby release energy. DFT can be used to calculate the properties and behaviour of the various point defects involved in these processes, where the principal entities created initially are lattice vacancies, self-interstitials, Frenkel defects, vacancy pairs, and di-interstitials.

Post-cryogenic-irradiation annealing experiments seemed to suggest that the interstitial activation energy was very low ($\lesssim 0.1$ eV), allowing the atoms to move relatively unhindered between the layers.[26] Later work revised this figure upwards to 0.33 eV (refs. 27 and 28), but this still implies that self-interstitial atoms are mobile at well below room temperature. In the early model, the net volume per interstitial atom is given as 3.3 times the volume per atom in ideal graphite, free from defects.[8] Lattice vacancies were supposed to be essentially immobile species, with an activation energy for migration of over 3 eV (ref. 26). Calculations based on DFT show that, in fact, the opposite is true. This misunderstanding in part arose from (understandable) ignorance of the structures for these defects. Once this ignorance was abolished, it became much easier to see why interstitials are relatively immobile species, and that vacancies have low barriers to migration. Experiments still lag behind theory here; however, the most recent measurements for vacancies support the models based on DFT calculations.

12.4.1 Lattice Vacancies in Graphite

When one carbon atom is removed from an otherwise perfect crystal of graphite to form an "ideal" lattice vacancy, the defect structure possesses

D_{3h} symmetry, with a threefold axis through its centre (Figure 12.2). According to a simple "defect molecule" description, this structure has a doubly degenerate $^2E'\sigma$ state occupied by one electron. Consequently, the structure can lower its energy by the Jahn–Teller effect.[29] The true picture, in fact, is somewhat more complex than this. It has been established by calculations based on the LDA that, in a graphene sheet, a partial transfer of charge occurs from the σ- to the π-system, which is reversed when the symmetry is lowered by the Jahn–Teller effect.[30] Two of the three initially equivalent atoms pair up to make one side of an approximately pentagonal ring neighbouring the vacancy.[30–33] This structure has C_{2v} symmetry. The odd, unpaired atom lies on the mirror plane. This plane passes through the centre of the reconstruction bond, and is oriented perpendicular to it. Although there is some disagreement among these calculations over the extent to which the unpaired atom is displaced out of the basal plane, it is generally agreed that the formation energy is $E_f \approx 7$–8 eV. This is close to the value deduced from experiments on graphite, where $E_f = 7.0 \pm 0.5$ eV (ref. 26). It is also agreed in the more recent calculations that a monovacancy in a graphene sheet carries a magnetic moment $M \approx 1.0$–1.1 μ_B. However, whether this remains true in a graphite crystal—where the interlayer interactions may affect the outcome—is uncertain.[34]

Theory has also provided information about the dynamical properties of vacancies in graphite. The reconstruction bond can form in any one of three equivalent positions. Calculations show that it can switch easily between the three orientations, experiencing a barrier of only 0.13 eV in the process.[30] This means that reorientation occurs with a frequency of about 65 GHz at room temperature, and 30 kHz at 77 K, according to the calculations. It explains why microscopic observations of defects attributed to isolated monovacancies exhibit threefold symmetry,[35] in apparent contradiction to the C_{2v} symmetry of the ground state structure: reorientation is so fast that images are simply an average view.[30]

The rate at which isolated monovacancies in graphite migrate is very much slower than their reorientation; nevertheless, it is now known to be significantly faster than was originally believed. Indications that the long-established value for the energy barrier to migration at $E_a \approx 3.1 \pm 0.5$ eV was incorrect emerged from theory based on the LDA, which has consistently yielded lower values over recent times: 1.6 eV (ref. 36), 1.7 eV (ref. 30), 1.01 eV (ref. 37), and 1.3 eV (ref. 33). Direct observation of vacancies by scanning tunneling microscopy (STM) finally resolved the matter, finding that $E_a \approx 0.9$–1.0 eV, and may even be made lower in some circumstances.[38] This result means that vacancies in graphite become mobile at room temperature. The process responsible for the 3.1-eV barrier found in the earlier experiments is a mystery, for which a resolution will be discussed later.

Given that vacancies can move relatively easily, it is natural to ask what happens when they meet. There are two components to this problem: what are the equilibrium structures for pairs and aggregates of vacancies, and what are the dynamics of their formation? Concerning the first part of the problem, this can be further divided into two categories, namely intralayer defects (two or more vacancies on a single graphene sheet), and interlayer defects (vacancy pairs on neighbouring graphene sheets). Theory finds that there are many structures for pairs of vacancies that are bound, *i.e.* the total energy of the composite defect is less than two times the energy of an isolated monovacancy. Remarkably, the lowest-energy divacancy aggregates have a smaller formation energy than a single monovacancy; thus, they are more stable than a mono-vacancy, and will, therefore, have a higher equilibrium concentration (although we note that thermal equilibrium rarely pertains).

The most stable divacancy is the so-called Haeckelite structure divacancy, or 555-777 defect, as shown in Figure 12.3. As the latter designation suggests, this defect has three fivefold and three sevenfold rings of carbon atoms, arranged around the atom at its centre, thereby giving it D_{3h} symmetry (the same as a monovacancy). The name Haeckelite was suggested by Terrones *et al.*,[39] owing to the resemblance of the structure to some observed in the mineral skeletons radiolaria microorganisms, observed by the German biologist Ernest Haeckel. The binding energy E_b of Haeckelite structure divacancy, with respect to two isolated monovacancies, is calculated to be 8.50 eV for the LSDA, and 8.61 eV

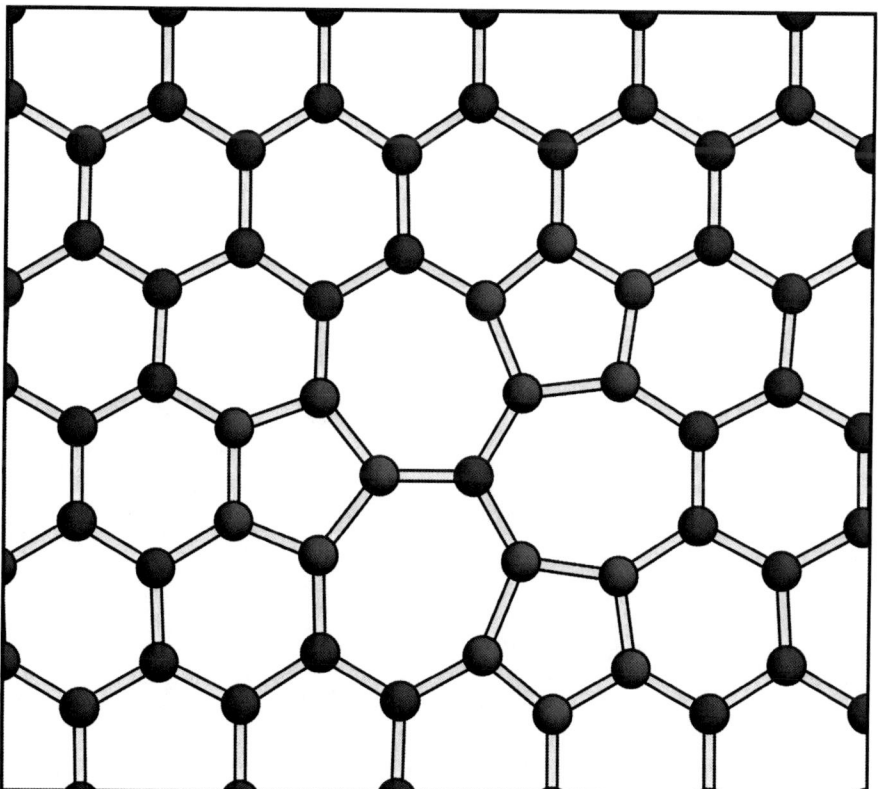

Figure 12.3 The Haeckelite structure divacancy in graphite.

for the GGA.[37] Its energy is also about 0.9 eV lower than a simple nearest-neighbour, $\alpha\beta$ divacancy (illustrated in Figure 12.4). This is consistent with an earlier result for the nearest-neighbour divacancy where $E_b \approx 7.7$ eV calculated using the LDA.[40] Calculations show as well that there is a significant energy barrier of about 5.2 eV required to convert a nearest neighbour divacancy into a Haeckelite structure defect.[37]

The process of vacancy coalescence, inevitably proceeds through third-neighbour divacancies, of which there are two possible structures, as illustrated in Figures 12.5 and 12.6. Both of these structures are metastable: their respective total energies for the *trans* (Figure 12.5) and *cis* (Figure 12.6) forms are 5.1 eV and 5.8 eV higher than the $\alpha\beta$ state, and their binding energies are 3.5 eV and 2.8 eV (ref. 41). The *trans* form of the defect has an acetylene-like central bond, about 1.2 Å long, which, together with the two reconstruction bonds the opposite ends of the defect, enhances the stability of the structure. Ewels *et al.* suggests that these defects, owing to their stability, present significant barriers to defect aggregation.[42] This view is confirmed by simulations of the vacancy coalescence process, where the activation energy to transform

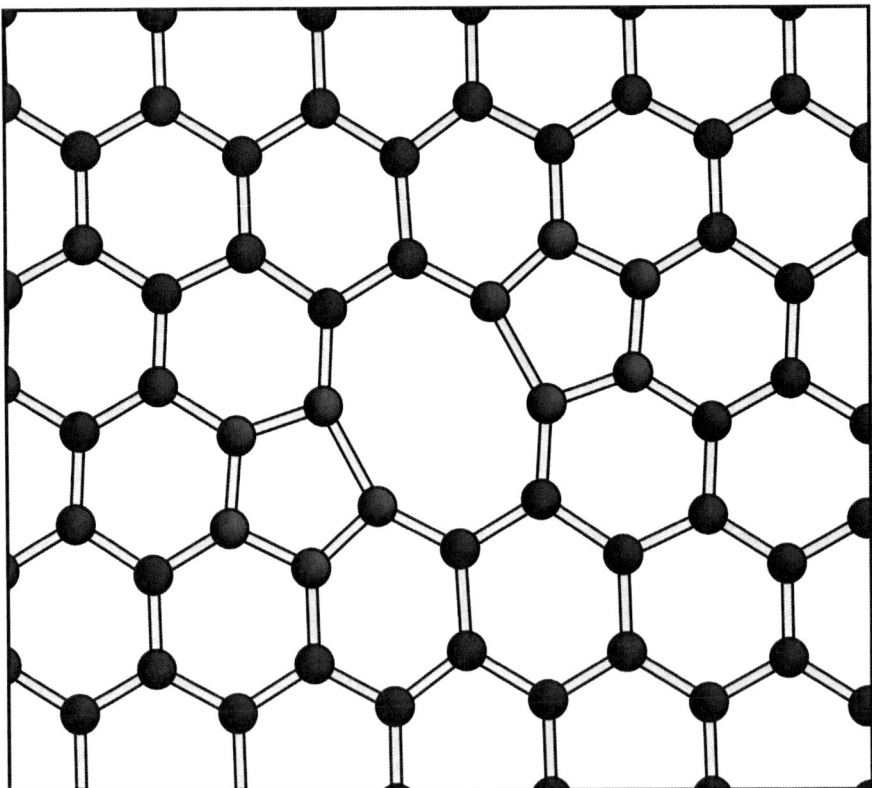

Figure 12.4   The structure of a simple, nearest-neighbour, $\alpha\beta$ divacancy in graphite.

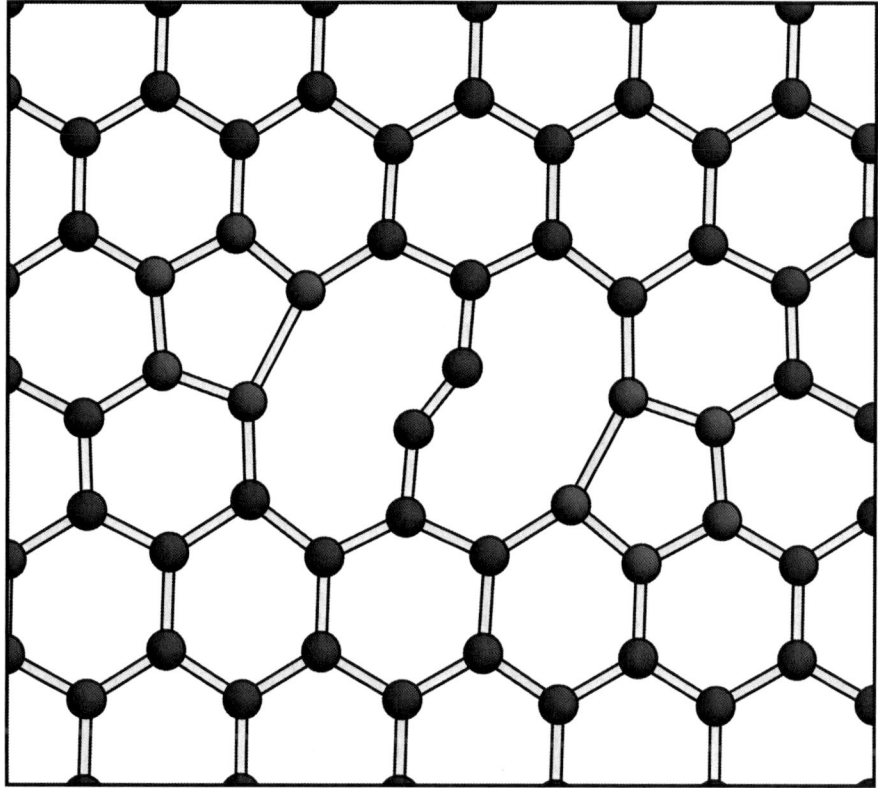

Figure 12.5 Third-neighbour *trans* divacancy in graphite.

the *trans* divacancy into a nearest-neighbour divacancy is about 1.5 eV, according to the LDA.[37] The *cis* divacancy, being slightly less stable, is likely to experience a slightly lower barrier to coalescence into the $\alpha\beta$ state.

Vacancies encountering one another on neighbouring graphene sheets in graphite can form bound pairs, according to calculations using the LDA by Telling *et al.*[40] It is found that these defects are stabilised by crosslinking bonds between the graphene sheets. A range of binding energies from 1.4 eV to 1.9 eV is found, depending on the details of the calculations and the structural configuration. In the right circumstances, such as where graphite locally adopts AA-stacking of the layers, the stability of these defects is enhanced, yielding $E_b \approx 3$ eV. Telling *et al.* also conclude that the activation energy for dissociation of interplanar vacancy pairs is 3.2–3.6 eV. This is close enough to the historically accepted figure of 3.1 $\pm$ 0.5 eV for vacancy migration to suggest that this may in fact be the process that was observed in the experiments.

Thus, the process of vacancy aggregation and coalescence releases a significant amount of energy during the process of annealing, and involves several energy barriers that may be associated with some of those observed by Iwata and others, described in Kelly's review.[6]

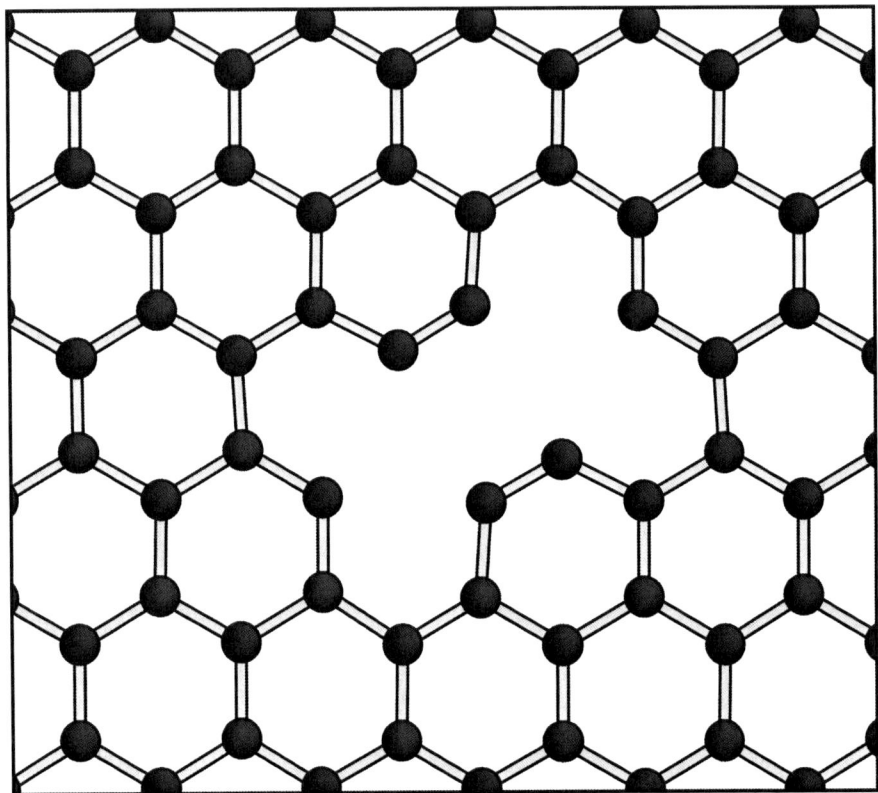

Figure 12.6   Third-neighbour *cis* divacancy in graphite.

12.4.2   Self-Interstitial Atoms in Graphite

Self-interstitial atoms in graphite have never been observed directly. Consequently, our knowledge from experiment about the properties and behaviour of interstitial atoms is largely based on circumstantial evidence, inference, and assumptions. Much of this work was done long before *ab initio* simulations based on DFT were widely available. These have revealed that the nature of interstitial atoms in graphite is very different from what was previously the majority belief. Unfortunately, the old beliefs carry so much momentum that the results of *ab initio* theory have been slow to gain acceptance in the graphite-radiation-damage research community.

In order to explain the phenomenon of dimensional change in irradiated graphite, it was *assumed* that interstitial atoms and pairs of interstitial atoms in graphite exist as nearly free entities, able to migrate easily within the spaces between the layers, forming loose aggregates.[4] As such, the interstitial atoms do not form covalent bonds with the host. Each one occupies a volume equivalent to 3.3 atomic volumes, thereby accounting for the observed expansion in the *c*-direction of irradiated graphite crystals.[8] Attractive though this model and its

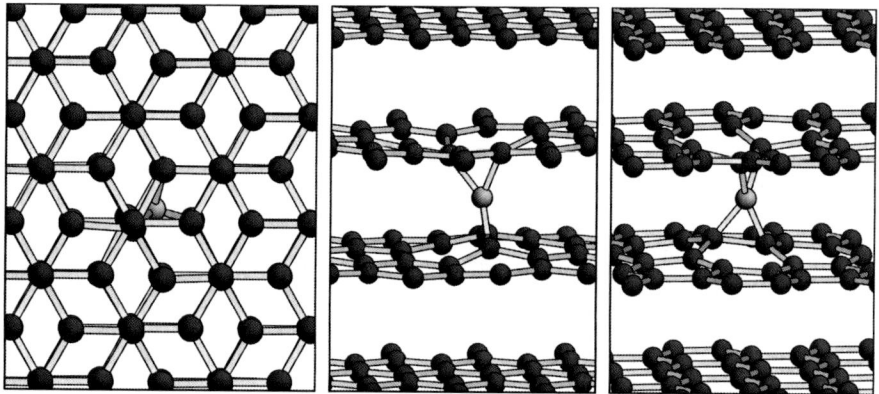

Figure 12.7   The spiro-interstitial in graphite (from Latham *et al.*[43]).

subsequent refinements might have been, it has several serious deficiencies.[43,44] Specifically, the assumptions on which the model is based are not supported by theory, which predicts that self-interstitial atoms in graphite are covalently bonded to the host,[40,45–47] and generate no appreciable expansion.[43]

The structure of an isolated self-interstitial defect in hexagonal graphite is illustrated in Figure 12.7. This is known as a spiro-interstitial owing to the resemblance of its core to a spiropentane molecule.[45] The defect has two nearly equilateral triangles of single bonds, arranged with their planes at 60° to each other, such that it possesses C_2 symmetry.[43] This is a relatively stable local minimum on the energy surface, when compared with a free atom having a formation energy of the cohesive energy of graphite. The formation energy of a spiro-interstitial is estimated to be about 5.85 eV (ref. 43). The defect imposes a significant shear stress along its [1010] axis on the adjacent graphene layers to which it is bonded. When the layers are moved by half a bond length in this direction, the total energy of the defect is significantly lower, implying that the defect density will be higher at the cores of partial basal dislocations where this shear strain prevails.[40,43,46]

The almost-single-bond length of the four covalent bonds that the interstitial atom makes to the host implies that the activation energy for migration is relatively high. In one case, it is estimated to be at least 1.4 eV (ref. 46). For net migration to occur in the basal plane, not only must the interstitial atom move along $\langle 1120 \rangle$ directions between second-neighbour α atoms, the (1010) plane on which the atom lies must also undergo reorientation around the *c*-axis between the neighbouring pairs of α atoms, as illustrated in Figure 12.8. A more detailed investigation of the reaction path gives a range of values for the energy barrier to migration from 1.2 eV to 1.7 eV, depending on the magnitude of a correction for van der Waals interactions applied.[47] According to these calculations, the reaction passes through several intermediate steps. Starting from a spiro-interstitial state, a migrating carbon atom moves first to a metastable grafted-interstitial state bridging a C_α–C_β bond of the host (Figure 12.9), then to a

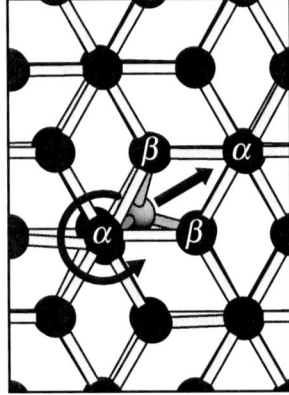

Figure 12.8   Reorientation and migration of an interstitial atom in graphite.

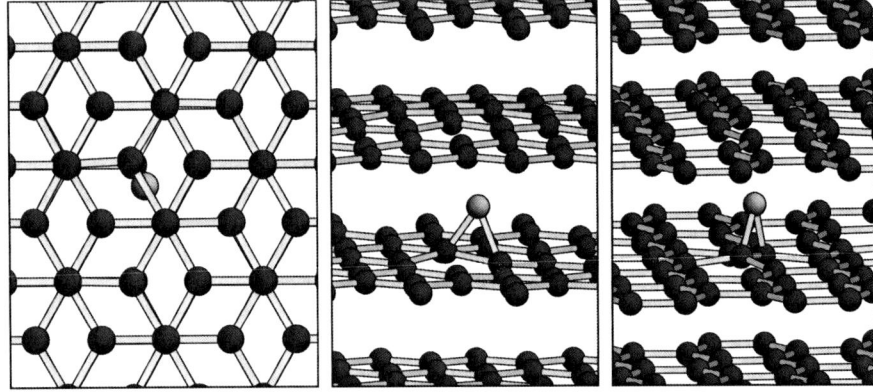

Figure 12.9   The grafted-interstitial in graphite.

metastable β split-interstitial state at the halfway point (Figure 12.10), before progressing to a neighbouring spiro-interstitial site *via* the corresponding grafted state.[47] Notice that exchange of a migrating interstitial atom with a host atom in the intermediate β split-interstitial state allows self-diffusion to occur along the *c*-direction equally easily as in the basal plane. This could account for the peculiar observation that, although graphite is highly anisotropic, the activation energy for self-diffusion is isotropic.[26] For this part of the reaction the activation energy is about 0.5 eV. Apparently, Ma interpreted this to equate with the 0.33 eV barrier observed in experiments[27,28] and claimed reasonable agreement. However, unless we are mistaken, this appears to be no more than an attempt to force agreement with the historical view of interstitial migration, since to invoke migration between metastable states in this way, rather than between ground states, is difficult to understand.[44]

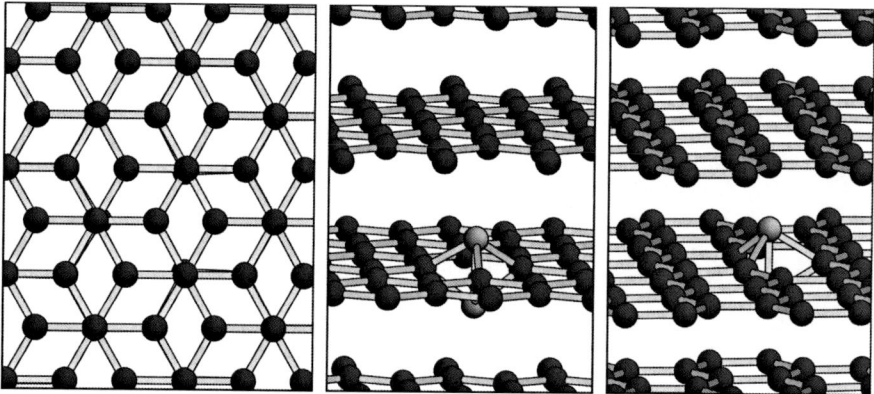

Figure 12.10 The β split-interstitial in graphite.

One clue from experiment in support of a high activation energy for interstitial migration is provided by transmission electron microscope observations by Reynolds and Thrower[48] of interstitial prismatic dislocation loops, which estimate the activation energy for loop growth is about 1.25 eV. One interpretation of this result by Brown *et al.*[49] is that the interstitial carbon atoms are trapped by boron impurities; however, as they note themselves, the concentration of boron in fission-reactor graphite is intentionally very small, which presents some difficulties for their model. They also conclude that the kinetics of loop formation are inconsistent with Iwata and Suzuki's model for irradiation-induced dimensional change, which involves small clusters of interstitials containing 2–4 carbon atoms. Instead, Heggie *et al.*[50] suggest that the 1.25-eV activation energy measured by Reynolds and Thrower is in fact that for self-interstitial migration. Nevertherless, this may not be the whole story, since the structure and dynamics of basal dislocations—where the concentration of self-interstitials is locally enhanced—is likely to affect this energy.

Pairs of interstitial atoms form several covalently bound structures that have binding energies of up to about 3 eV with respect to isolated self-interstitial atoms, according to calculations employing the LDA.[43] This implies that the activation energy for migration of interstitial pairs in graphite could be this large. The two lowest-energy forms of interstitial pairs are shown in Figures 12.11 and 12.12. These two structures have nearly equal formation energies. All of the complexes reported produced very little expansion of the graphite host. Furthermore, it is important to note that, unlike the historic model, these calculations do not contain any assumptions about whether isolated interstitial atoms or interstitial pairs form freely floating entities, or are covalently bonded to the host; they are *only* conjugate-gradient geometry optimisations starting from various initial positions.[44] The algorithm employed only progresses down in energy, and never traverses any energy barrier with respect to the atomic coordinates.

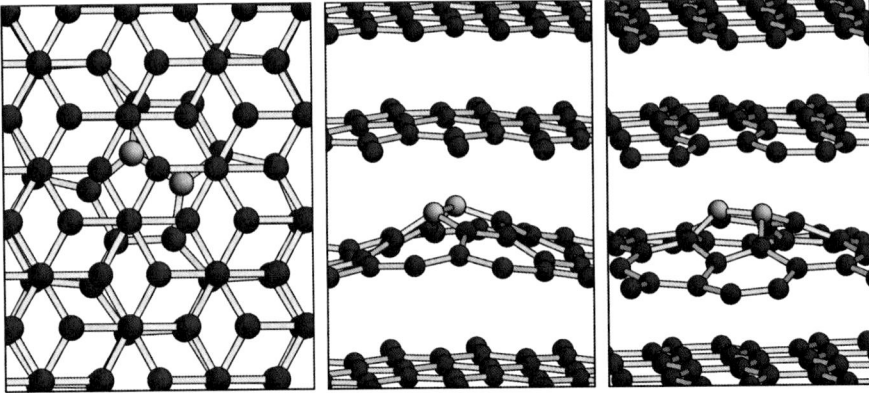

Figure 12.11   The grafted, intralayer bridge complex in graphite (from Latham et al.[43]).

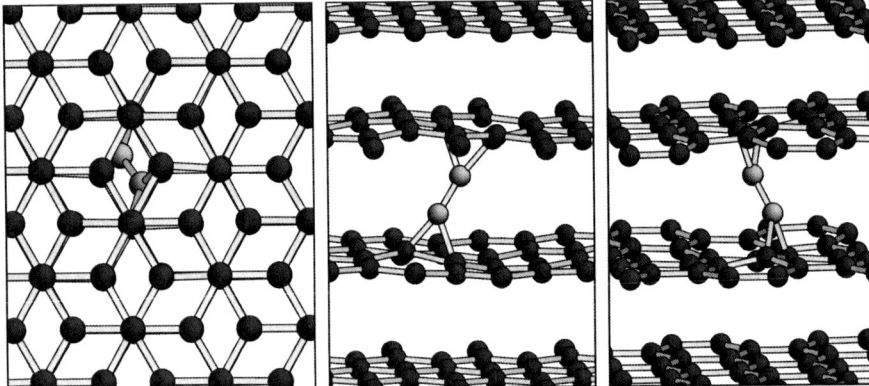

Figure 12.12   The twin-triangle interlayer bridge complex in graphite (from Latham et al.[43]).

12.4.3   Frenkel Defects in Graphite

Careful calorimetric measurements of the stored energy in electron-irradiated graphite by Mitchell and Taylor led them to conclude that the large Wigner energy release peak at about 475 K is neither caused by recrystallisation, nor by the aggregation of interstitial C_2 molecules—as was advocated by others at the time—but by recombination of Frenkel defects.[24] This yields about 14 eV per displaced atom, rather than about 4–5 eV had the atoms been in the form of C_2 molecules, they argue. They also deduce from their experiments that the recombination process occurs over a range of energy barriers from 0.2 eV to 1.38 eV, and estimate that the formation energy of a Frenkel defect is 14.4 eV.

Calculations by Ewels *et al.* have validated this idea.[51] In a graphite crystal, a lattice vacancy can occur on either an α or a β site, and metastable configurations exist for interstitial atoms standing at the very edge of both types of

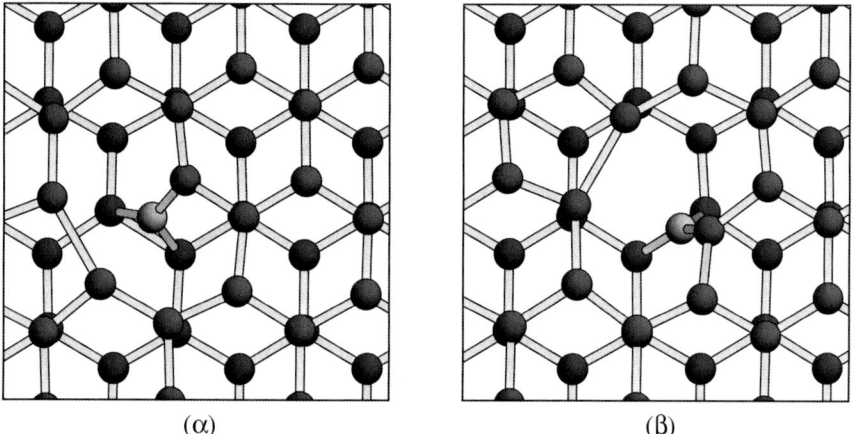

Figure 12.13 The α and β intimate Frenkel defects in graphite.

vacancy. In both cases, the interstitial atom has a Y-shaped arrangement of bonds to it, with the arms attached to the graphene layer opposite the vacancy, and the leg anchored in the layer containing the vacancy. See Figure 12.13. These defects have formation energies that are estimated to be about 2.9 eV less than that for a widely separated Frenkel defect ($E_f = 13.7$ eV). The calculated barrier to recombination is 1.3 eV, and thus is close to the upper limit of the range found by Mitchell and Taylor. The fact that this barrier exists, provides further evidence to support the concept that interstitial atoms in graphite are relatively immobile species.

In experiments performed by Asari *et al.*,[52] they observed two distinct processes occuring during the annealing of graphite that had been irradiated with 3-keV He$^+$ ions. These were found to have activation energies of 0.89 ± 0.10 eV and 1.8 ± 0.3 eV, which they attribute, respectively, to interstitial atoms migrating to lattice vacancies and annihilating them, and to aggregation of vacancies diffusing within graphene sheets. In later work,[53] they reported a slightly higher activation energy of 0.98 ± 0.10 eV for the fast process, but were puzzled by the activation energy for "vacancy clustering", which was much lower than expected. However, they were working on the prevailing assumption at the time that interstitials are relatively mobile species, and vacancies migrate much more slowly. If, as theory indicates, the opposite is true, then a more likely explanation for what they observed in these experiments was vacancies migrating to interstitial atoms for the fast process, and interstitials forming bound pairs for the slow process.

12.5 Line Defects

Line defects are characterised by the direction of the line **1** and a quantity that describes their topological properties. Imagine a cut is made from one of the surfaces and terminates on the line. If extra material is inserted into (or removed from) this cut a topological defect is created: a wedge disclination for an added (removed) wedge or an edge dislocation for an added (removed) slab. The important quantity then, is the angle of the wedge or the thickness of the slab. The latter being known as the edge component of the Burgers vector **b**.

Other topological defects can be created without addition or removal of material: for example the sides of the cut may suffer a relative translation or rotation, creating, respectively, a twist disclination and a screw dislocation. For the latter, the translation is the screw component of the Burgers vector and is parallel to the line **1**.

For each of these defects the lattice can repair itself, healing the cut surfaces from insertion or removal, by rebonding that forces elastic deformation of the surrounding material. The rebonding is effective only if the atoms of the cut surfaces can match up, which, given the periodicity of the lattice can happen for Burgers vectors equal to lattice vectors and for angles $2\pi/n$ about rotational symmetry axes, where n is the order of the axis. The rebonding takes on a different character or is frustrated close to the dislocation line (within a "core radius", r_c, which is normally a bond length or two). This is the core of the defect. It is the interesting bit for structure, motion and energy and has been the subject of first-principles investigation for twenty years. Insofar as it is possible to assign it an energy, that energy is labelled E_c.

The role of such defects in plastic deformation was postulated independently by Orowan, Polanyi and Taylor in 1934, and developed by J. M. Burgers.[54] Plastic deformation is achieved by crystal planes slipping over one another. The plane over which they slip is called the glide plane. The slip is not simultaneous all over the glide plane, and a dislocation line is the boundary between the slipped and unslipped regions. The notion of inserting or removing material, which was used to introduce the concept of a dislocation, is not necessary: the extra material can, and in practice always does, result from slip. For example, Figure 12.14 shows the principle of a half-plane gliding in from the surface to produce an edge dislocation. Dislocations in graphite were amongst the first to be identified in the electron microscope,[55,56] and their theory was intensively studied,[57,58] providing a basis for the understanding described here.

Single dislocation lines cannot end in the material—they must either begin and end at a surface or form a loop. The Burgers vector for an isolated

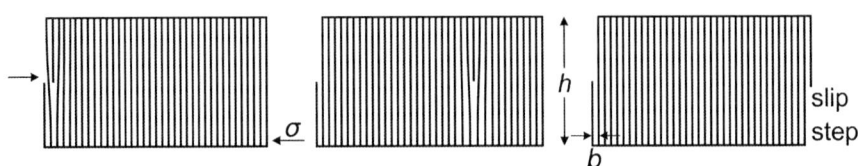

Figure 12.14   Schematic of a theoretical action of an applied stress, σ, to introduce a dislocation (left), move it through crystal (middle) until it exits (right), leaving slip step, b, on the surface. The elastic deformation has been ignored. The resulting strain on the crystal of height, h, is b/h.

dislocation is constant along the line (or loop). Where dislocations meet, the vector sum of Burgers vectors must be zero.

The quantitative elastic theory of these topological defects was formulated in the late 1940s and 1950s, giving us: (1) the atomic positions outside the dislocation core and hence, the boundary conditions for the core; (2) ability to calculate the force on a dislocation due to an applied stress (a configuration force); (3) an appreciation of the energies of dislocations, including their divergences. It yields the energy per unit length (E) of a dislocation threading a cylinder of material radius R (eqn (12.2)). This theory is only valid for an elastic continuum. It breaks down at dimensions close to the interatomic distance; thus, as described previously, a cylinder of radius r_c is excluded from the calculation, and its energy is represented by a core energy E_c.

$$E = E_c + \left(\frac{Kb^2}{4\pi}\right)\ln\left(\frac{R}{r_c}\right),\qquad(12.2)$$

where K is known as the energy factor, and b is the magnitude of the Burgers vector. For a dipole of dislocations separated by a distance d it is usual to consider the strain energy within a radius $R = d$, allowing that the more distant strain fields of the dislocations cancel. Likewise, in a network of dislocations, the energy of a dislocation line is estimated using R comparable with an interdislocation distance. If this is a matter of micrometres, then the quantity $\ln(R/r_c)$ can be of order 10, and E can be of order hundreds of eV per Å. This is an enormous energy compared with a typical point defect energy (1–10 eV), and it is obvious that an isolated dislocation in an infinite medium ($R = \infty$) has infinite energy—dislocations are thus not equilibrium defects, unlike point defects.

However, they can be created out of equilibrium by particle radiation. Imagine a fission neutron. It scarcely interacts with matter—it has no charge, so no long-range interactions with the electrons and nuclei, and its interactions are limited to direct collisions with them. Conservation of momentum means that its interactions with the more massive nucleus is meaningful, especially for fission neutrons whose energy spectrum has a mean of about 2 MeV). Nuclei are small compared with the size of the atom, and the neutrons have a mean free path of centimetres. Thus, the neutron collisions punch a carbon nucleus seemingly from "out of nowhere" The displaced carbon atom (the primary knock-on atom or PKA) cannons away with a mean energy of 50 keV, strongly interacting with the material around it. In addition to the historically acknowledged displaced atoms (Frenkel defects), the energy and momentum of the PKA may also be dissipated by forming and extending dislocations. This process is clearly seen in an electron microscope in a visualisation procedure, "diffraction contrast", which mimics with electrons the technique devised by Lang for X-rays.[59] A dislocation necessarily leads to bending of planes to accommodate the extra half-plane. The bending is largest near the core. Many angles of tilt between the regular crystal planes of the lattice and electron

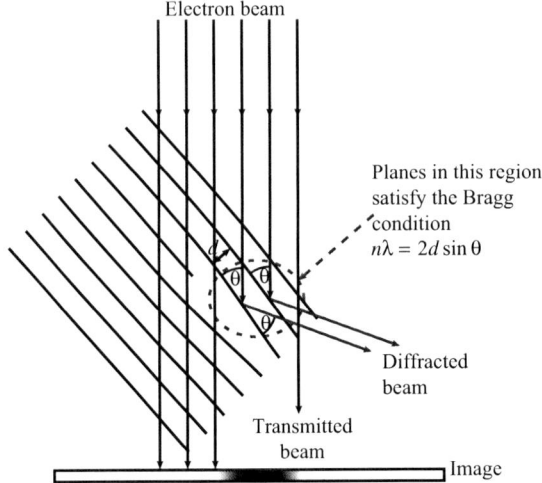

Figure 12.15 Schematic diagram showing how dislocations produce "diffraction contrast" in an electron microscope. When the crystal is tilted appropriately, Bragg reflection occurs only at atomic planes that are bent by a dislocation core.

beam allow efficient transmission and a bright field in the image. However, a judicious choice of angle can satisfy the Bragg diffraction condition *only* for the bent planes around the core. Such planes satisfying the Bragg condition diffract electrons away from the transmitted beam, resulting in a dark line on the image. Conversely, when the crystal as a whole satisfies the Bragg condition – creating a dark field – dislocations do not, and they show up as bright lines. See Figure 12.15, noting that the lines representing atomic planes in the picture are not necessarily graphene sheets, but could be any applicable planes in the crystal.

   In the early stages of irradiation (10^{17} to 10^{18} n cm^{-2}) or nominally 10^{-3} to 10^{-4} displacements per atom (dpa), the increases in dislocation density can be clearly seen.[60] The present authors have not seen such images at much higher doses, but it seems reasonable that the diffraction contrast condition, normally giving an image spread over 20–50 nm, requires low dislocation densities. Although we note that there is a weak beam technique that enabled approximately 5 nm separation of glide partial dislocations in silicon.[61] However, for the time being it does not seem unreasonable to suppose that the same processes occurring at these low dislocation densities continues to higher densities, albeit clouded by interactions and cooperative phenomena.

12.5.1 Dislocations and Disclinations

Before we go on to consider dislocations in detail, some remarks on disclinations follow. They are a fascinating field of study, but in a conventional solid their independent existence is limited by a prohibitive relationship between energy per unit length, wedge angle, and cylinder radius, R (varying as R^2). It is interesting to note that removing a wedge of material and then reinserting the same wedge displaced in the vacant region can be equivalent to removing a slab of material. Thus, in creating a positive and negative wedge disclination close to each other (a wedge disclination dipole) we have created an edge dislocation. The negative 60° wedge disclination in graphite with its axis through a graphene hexagon removes of one of the vertices of the hexagon making its core a pentagon. The positive 60° wedge disclination core is similarly a heptagon.

Heptagon–pentagon pairs are thus dislocations. They are "perfect" dislocations in the sense that **b** is a lattice vector, **a**. They glide in a prismatic plane (whose normal is given by $\mathbf{b} \times \mathbf{1}$, where **1** is **c**. In the Fujita–Izui classification[62,63] they are "nonbasal" dislocations and their two forms were discussed by Jenkins[64] and Thomas and Roscoe.[65] First-principles calculations of such dislocations in graphene have been performed[66] identifying their glide planes (Figure 12.16) as either cutting through the zigzags (labelled the glide version) or through the centres of hexagons (labelled shuffle). The glide set (pentagon–heptagon pair) is the ground state and the shuffle set, where the shared bond has trapped an interstitial, is metastable, featuring a two-fold coordinated carbon atom.

The energy barrier to dislocation glide is called the Peierls barrier of the first kind. DFT calculations show that it is enormous for the glide dislocation (7.6 eV) making it immobile at all but the highest of temperatures, while for the shuffle set it is a more accessible 2.2 eV (ref. 66). The glide set dislocation will trap an interstitial atom with an energy of 2.3 eV, and then it is as if the interstitial catalyses motion, in a similar fashion to the autocatalysis effect invoked earlier for fullerenes.[67,68] Radiation damage, which steadily creates interstitials, may well mobilize these dislocations at reactor temperatures, and there has been some early discussion that shuffle dislocations in neighbouring layers could bond to each other[69] through the undercoordinated carbon atoms. However, the major role these dislocations have been thought to play is as the

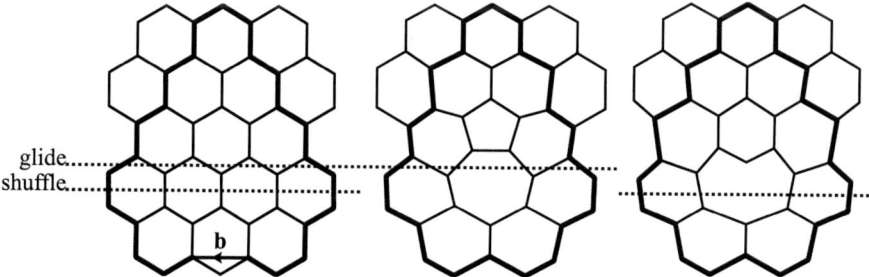

Figure 12.16 Illustration of glide planes (dotted), Burgers circuit (bold), and Burgers vector, **b**, for glide (centre) and shuffle (right) set nonbasal **b** = **a** dislocations.

dislocations that result from the terminations of lines of healed vacancies that contract the basal plane.[25,70] Later in this section it will be seen that buckling of the layers, through defects that bridge between layers,[40] could provide a better explanation, even though dismissed earlier.[70]

More recent studies of these nonbasal dislocations in graphene, using similar approaches but in a grain-boundary-style superlattice,[71] or using a discretised theory of the dislocation strain field[72] have examined the electronic and magnetic properties. A very recent study invoked the dislocation in a "realistic" model of amorphous graphene.[73] In connection with the subject of this review, we note that there are good reasons to believe neutron irradiation does not amorphize graphite under reactor conditions.[74]

Clearly there is an added complication for these nonbasal dislocations in graphite compared with graphene, in that they can penetrate more than one layer, but this has not so far been treated, to the best of our knowledge. Recall that, in the absence of junctions and surfaces, dislocations must form continuous loops with a constant Burgers vector. Thus, a pair of nonbasal dislocations of opposite sign must be part of a loop completed by basal dislocations.

Glide is the easy movement for an edge dislocation, but it can only occur on the glide plane defined in orientation by $\mathbf{b} \times \mathbf{1}$, and in position by the axis (roughly the centre of the shared pentagon–heptagon bond). Addition of an interstitial or removal of a vacancy from a "glide" dislocation forms the "shuffle", moving the axis (and glide plane) from the zigzag to hexagon centres, a distance of $a\sqrt{3}/4$. Further similar addition recreates the "glide" dislocation and a displacement of $a\sqrt{3}/2$. This is the process of climb, the name indicating the energetic difficulty of point defect emission or the statistical difficulties of point-defect absorption.

The pentagon–heptagon pair is a topological defect in the lattice. The graphene lattice is hexagonal with a basis of two atoms, so the basal lattice vectors connect an atom with its second neighbour. Imagine a wide and closed circuit jumping between second-neighbour atoms around the dislocation axis (left of Figure 12.16). If the same circuit were performed about a bond centre in perfect graphene, then it would not close (centre of Figure 12.16). Taking a right-handed circuit about the axis, the Burgers vector is the vector that closes the circuit from the finish to the start (**FS-*RH*** convention[54]). This "Burgers" circuit demonstrates that the pentagon–heptagon pair is not a point defect and cannot spontaneously revert to two hexagons and thus perfect graphene.

A dipole of dislocations, a pentagon–heptagon pair next to an inverted pair, on the other hand, can revert to four hexagons, since a Burgers circuit around both does close. This has become known as a Stone–Wales defect (or 5775 defect) but was first discussed by Dienes,[75] and then more recently by Kaxiras and Pandey[36] as a metastable state in the course of directed or concerted exchange of two carbon neighbours in graphite and graphene. The barriers for creation and annihilation of this defect are, however, high (9.2 eV and 4.4 eV, respectively[46]). It appears quite likely that this defect could arise during neutron irradiation, or as damage anneals,[51] and indeed in addition to other point defects with varied ring statistics.[76]

Considering the wedge disclination and how it might move, imagine a pentagon–heptagon pair gliding towards a pentagon such that the heptagon collides with it, annihilates itself and the pentagon. What remains is a pentagon displaced from the original. Thus, a disclination may move by either absorption or emission of edge dislocations, but these are very energetic species, and they must glide a long way for a single displacement of the pentagon. They tend to be ruled out of bulk radiation damage, but they can be formed in the electron microscope near the edge of nanometric graphene pieces.[77] Before leaving disclinations, we should remark that the prohibitive energy of the strain field arising from a disclination in a large piece of material (large R) does not apply to a disclination in a sheet, like graphene. In a layer, large strains can be accommodated by curvature, which is a symmetry breaking distortion going beyond second-order elasticity theory normally used to derive stresses and strains around line defects. Thus, carbon nanocones, carbon-nanotube caps and fullerenes are rather stable entities, even though they contain 1–5, 6, and 12 pentagons, respectively.

12.5.2 Prismatic Edge Dislocations

Returning to the theme of dislocations, the prismatic edge dislocation is where the notional inserted plane is part of a graphene layer. The perfect prismatic edge dislocation has a basal line direction (*e.g.* **a**) and **b** = **c**. Thus, the insertion comprises two graphene sheets stacked AB fashion. Where they terminate, the resulting bonds can reconstruct pairwise, perfectly connecting the sheets, and yielding a fold in the graphite. This can happen without any side effects, preserving AB stacking, only for a line direction along **a**, *i.e.* zigzag edges. For $1 = 2\mathbf{a}_1 + \mathbf{a}_2$, *i.e.* armchair edges, a translation along the edge of half a C–C bond length is needed. For other directions there is a penalty of rotational disorder between the graphene sheets to be paid. In order for a nanotube-like fold to form, one or other sheet must be rotated. This could perhaps be the origin of the rotational disorder arising from radiation damage (and the inverse process of rotational ordering that removes turbostratic disorder during graphitisation).

In modelling dislocations in an infinite solid, a supercell with periodic boundary conditions is standard, implying an array of dislocations. A single dislocation could not satisfy these boundary conditions, so a dipole is used, often in the glide configuration. This is shown in Figure 12.17(a), which represents the extra part plane as the upright of a **T** symbol, and the glide plane as the crosspiece. The climb dipoles (Figures 12.17(b) and (c)) can be arranged to include or exclude the added part planes, making them, respectively, interstitial or vacancy dipoles. Depending on the supercell geometries, the glide and climb forms of dipole can be interchangeable, as becomes obvious by inspection of the resulting superlattice. Note that, as well as optimising geometries for any

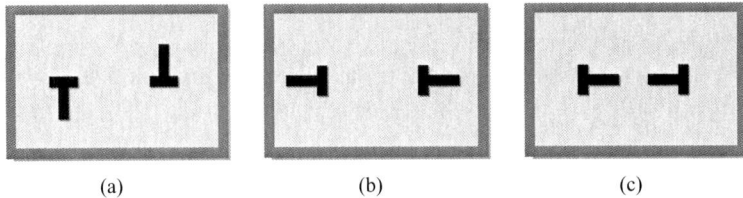

(a)   (b)   (c)

Figure 12.17   Dislocation dipoles. The leg of the **T** symbol indicates where extra
material has been inserted into a crystal, ending at the crosspiece. (a) a
glide dipole; (b, c) climb dipoles, which are (b) vacancy type, or
(c) interstitial type.

of these cells, superlattice vectors must be optimised. According to elastic
theory[78] one of the superlattice vectors is modified from those for crystalline
graphite by a vector $\mathbf{b}(W/L)$, where the supercell length is L, and the dislocation
separation is W.

Making an interstitial dipole of prismatic edge dislocations (Figure 12.17(b))
creates a flattened nanotube-like structure which, if small enough, becomes a
nanotube.[79] A vacancy dipole comprises two folds neighbouring a void region.
The properties of the folds have been given earlier.

These dipoles are infinite, parallel dislocations representing ribbons of extra
or missing material within the lattice. In place of a ribbon we could consider a
disk, for which the diagrams of Figure 12.17 are cross-sections through the
diameter of the disk. In this case, there are dislocation loops of interstitial or
vacancy nature: the first tending to form embedded fullerenoid structures upon
relaxation when small and the second, a wormhole.[80] The latter has been
mooted as a possible radiation defect.[80]

The Burgers vector can be reduced to $\mathbf{c}/2$ giving a single-layer insertion or
removal, but this is not a lattice vector and the dislocation is thus partial. The
stacking of graphene layers is labelled A, B or C, each differing from its
neighbour by an inplane displacement along a C–C bond. This kind of shear
between layers can convert A to B, or B to C, or C to A. Insertion of either an
A- or a B-layer implies one AA-stacked pair, which is high in energy (18 meV
$\mathrm{\mathring{A}}^{-2}$).[25] Thus, the lowest-energy state is achieved by a C-plane. ABC or
rhombohedral stacking corresponds to a stacking fault in AB graphite,[25] but it
is a very low energy defect (0.053 meV $\mathrm{\mathring{A}}^{-2}$). We know of no reports of isolation
of a rhombohedral graphite crystal. An interstitial dipole of these dislocations
is a ribbon of graphene embedded in graphite. Insertion of a C-layer, giving an
ABABCABAB sequence has three bad stackings (ABC, BCA and CAB), and is
expected to be unstable against passage of a basal dislocation (see later)
between the B- and C-layers that generates a sequence ABABABCBC with only
one (ABC). See Figure 12.18.

It is doubtful whether such prismatic partial edge dislocations could glide,
but they can climb by absorption or emission of interstitials or vacancies.
They can also slip in the lattice by the passage of a dipole of basal dislocations
(see later).

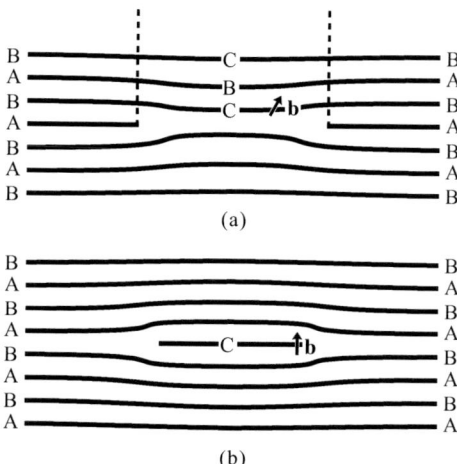

Figure 12.18   Schematic illustration of prismatic dislocations in graphite: (a) vacancy type, where **b** is not parallel to **c**; (b) interstitial type, where **b** is parallel to **c** (by CDL after Amelinckx and Delavignette[81]). In the first case the prismatic dipole has combined with a basal edge dislocation dipole, in the region between the broken lines. If this basal dipole were to expand, then the four top layers would all be BCBC stacking. Likewise, for the interstitial type, passage of a basal dislocation over the layer C would convert the top three layers to CBC.

Reducing the infinite ribbon of inserted or removed graphene to a disk creates a dislocation loop (which can be either interstitial, *i.e.* from insertion, or vacancy, from removal). Baker and Kelly were the first to prove radiation damage gave rise mostly to these interstitial loops, formed by the accretion of interstitials.[82] The nucleation process, which, if homogeneous, starts from the encounter of bonding of two interstitials, is temperature and flux dependent. The best evidence and measurements on these loops come from high-temperature damage, or damage annealed to high temperature (about 1200 K).

A simple treatment of these $b = c/2$ dislocations from first principles proceeded from AA-stacked graphite.[83] Where the line direction is along **a**, the graphene edge in the core is zigzag and the undercoordinated carbon atoms bond with single bonds into a line of carbon atoms in a neighbouring plane, converting them to sp^3 hybridisation and fourfold coordination. For lines along $2\mathbf{a}_1 + \mathbf{a}_2$ (*i.e.* armchair edges), this does not happen, because undercoordinated carbon atoms neighbour each other, and can improve their stability by formation of an extra π-bond, lending the edge C–C bond some triple bond character (Figure 12.19(b)). Employing AA-graphite is an artifice that makes graphite $b = c/2$ dislocations perfect, but AA is unstable, if computationally convenient.

In the same study, the $b = c/2$ screw dislocation (Figure 12.19(c)) turns out to be very stable for three reasons. First, the elastic anisotropy of graphite conspires to make its energy factor, which is equal to the basal shear modulus, C_{44}, and hence about 5 GPa (Table 12.1) the least of all dislocations by an order of magnitude. Secondly, the calculated Peierls potential is very high ($2.8\,eV/\mathring{A}$). Thirdly, the core pierces hexagon centres, converting hexagons into single helices, with only modest bond stretches of 2.1%. Although the elasticity theory predictions for the strain field indicate the dislocation does not visibly perturb the crystal-structure projection, calculations suggest the contraction of the core hexagon might be visible. It also appears that the screw tends to pull the layers into AA-alignment, otherwise it must weave between hexagons in AB-layers, and any lengthening of a dislocations leads to an increase in energy. The supercells used were of dipoles, and since there is no "added part plane" for the screw, and **b** is parallel to **1**, there is no defined glide plane, so the complications of Figure 12.17 do not arise. However, we note that a dipole of screw dislocations is the same topology as the ramp in a multistorey car park. Curiously, even though elasticity theory dictates that the energy of a dislocation line varies as the square of the magnitude of the Burgers vector, dislocations of several hundred **c** have been observed in graphite,[84] and an explanation has been offered for their existence,[79] involving the cooperative formation of AA-stacking.

12.5.3 Basal Dislocations

Turning now to basal dislocations, which glide between layers, they typify the reasoning that led to the idea of a dislocation. Shearing two layers past each other is resisted by an elastic force dictated by Hooke's law and the basal shear modulus, C_{44}. Given the periodicity of the lattice along a high-symmetry direction, and assuming the potential energy varies sinusoidally with this period, the amplitude of the potential can be deduced if its second differential with respect to shear, at zero shear strain, is set equal to C_{44}. Thus, this method for the theoretical shear strength of a crystal, when applied to graphite basal shear, gives a strength of 380 MPa (ref. 85), close to 340 MPa given by Frenkel's rigid-sphere model,[86] and two or three orders of magnitude larger than measured shear strengths. Basal dislocations are responsible for this discrepancy, because they allow shear to proceed progressively across the basal plane, as the dislocation glides across it, rather than all at once.

In representing the basal dislocation, Telling suggested the fact of the extra material planes on one side of the glide plane be represented as a perturbation in the planarity of the layers (Figure 12.20). For a dislocation, its elastic energy varies as the square of the Burgers vector. The total Burgers vector, **b**, must be conserved, but where a low-energy stacking fault exists, a dislocation can split into two partials, $\mathbf{b}_1$ and $\mathbf{b}_2$, separated by the fault. While $\mathbf{b} = \mathbf{b}_1 + \mathbf{b}_2$, for the dissociation, the sum of the squares of **b** for the partials is smaller, giving an energy reduction that can pay for the formation of stacking fault, and a repulsive force between the partials (the Peach–Koehler force). The energy per unit area on the ribbon of stacking fault causes a constant force from a surface-tension effect, which tends to oppose the repulsion of the partials. Where the forces balance should give the equilibrium separation of the partials, which is observed to be 50–150 nm, as summarised by Telling and Heggie,[85] where the range comes from expected variations with dislocation direction (since the

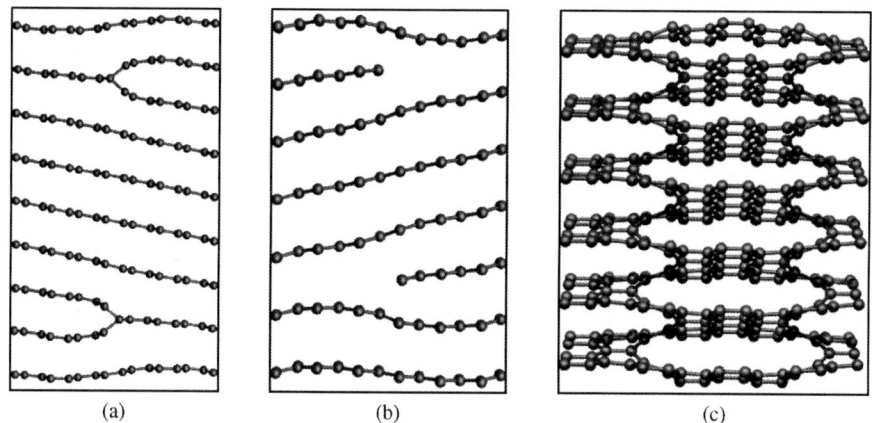

Figure 12.19 Dipoles of prismatic **c**/2 dislocations in AA-graphite: (a) zigzag edge; (b) armchair edge; (c) screw (from Suarez-Martinez.[79])

Peach–Koehler force varies with the different mix of edge and screw character), and other unidentified sources.

Basal dislocations have been invoked in the history of radiation damage but only in the limited sense of being pinned by interstitials, giving, for example, the pinning–unpinning model for radiation creep,[87] and also explaining how the measured values for C_{44} for rather perfect natural and synthetic crystals is an order of magnitude smaller than what is now accepted as the crystal value (Table 12.1). The principle behind this is thought to be the reversible motion of basal dislocations. It is possible to deduce important properties of dislocations from a simple analytical model based on the first-principles calculation of the potential surface arising from basal shears (the gamma surface).[85] From this one concludes that the stress required to move basal dislocations (their Peierls stress) is effectively zero, which appears compatible with reversibility of basal

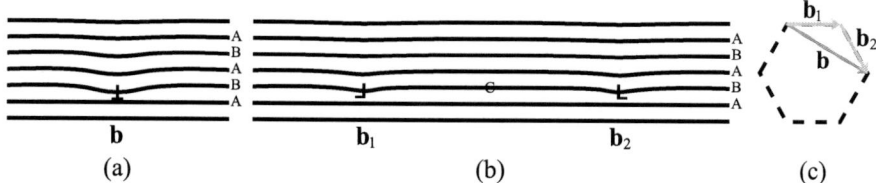

Figure 12.20 Basal dislocations: (a) perfect, (b) dissociated and (c) conservation of Burgers vector upon dissociation such that $\mathbf{b} = \mathbf{b}_1 + \mathbf{b}_2$. Capital letters label stacking arrangements. Dissociation causes a B-layer to convert to C, giving an intrinsic stacking fault. Note that the partial Burgers vectors correspond to C–C bond directions and length, $\mathbf{b}_1$ being pure edge (90°), $\mathbf{b}_2$ being mixed (30°) character.

glide. It also confirms that basal dislocation cores are very wide ($\approx 4.3\,$nm), and puts forward the view that if basal dislocation segments were limited to 50 μm lengths, then the stress required to operate dislocation multiplication mechanisms could be the cause of the measured strength of graphite.

12.5.4 Buckle, Ruck, and Tuck

There is clear evidence from diffraction contrast on low neutron dose specimens that dislocation density increases and that closed loops can appear[60] during irradiation. Work by one of us (MTH) has consistently pointed to a wider role for basal dislocations in radiation damage[50,88] demonstrating two sources for dimensional change based on basal dislocations and their interactions with each other and with point defects. Two concepts are involved. First, graphene layers in the graphite buckle when the dislocation core is constrained by the presence of bridging defects such as spiro-interstitials.[50] Secondly, that basal dislocations may interact to cause folds in the layers creating so-called "ruck and tuck" defects.[88]

Starting from the discovery that point defects often give rise to bonds between layers,[40] it became apparent that these bonds could pin layers together that were of different length. Stretching or compressing C–C bonds is very difficult, but if a layer is compressed, then it may buckle easily, and in this way match up with the shorter layer to which it is pinned. The layers either side of the glide plane in a basal edge dislocation differ in length by the Burgers vector. If the pinning points are separated by less than the natural width of the basal dislocation, this buckling causes a marked expansion in the c-axis dimension.

As usual, a dipole structure must be used. Figure 12.21 shows an array of basal dislocations built from a glide dipole cell, which is indicated by a dashed line. A glide dipole of basal dislocations in a cell much smaller than the core width is bound to be unstable and will annihilate. However, in this example the structure is held in place by bridging defects (spiro-interstitials in this case), which prevent the shear that would be necessary for annihilation. There is pronounced expansion in the c-dimension, as is found with radiation damage below 250 °C. This formation must collapse when the interstitials or other bridging defects become mobile, which is presumably around 250 °C.

Note that pushing basal dislocations together also leads to confinement, and this can easily be arranged in a calculation as in Figure 12.22, which arises from a multiple climb dipole-type supercell (c.f. Figure 12.19). Note the wavelengths of buckling in these examples are somewhat artificial, arising from limitations of the cell size used. These simulations, which employ the Universal Force Field implemented within the CERIUS^2 package, were performed by I. Suarez-Martinez.

When edge dislocations of opposite sign glide together (i.e. a dipole collapses) they annihilate: their extra part-planes combine to form a complete plane, and thus a region of perfect graphite is the product. There is a tremendous release of energy, recalling that the line energies are high.

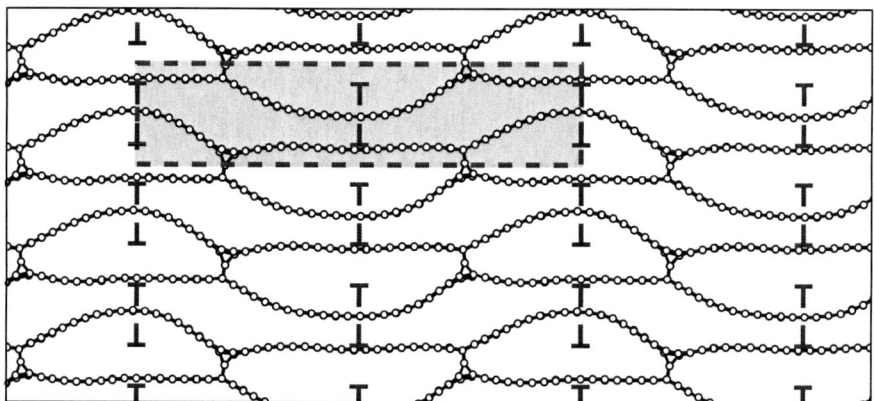

Figure 12.21 An array of basal dislocations in graphite confined by spiro-interstitials.

If they are not gliding on the same interlayer gallery, then there is an excess or a deficit of material compared with perfect graphite. In a layered material, this excess or deficit is confined to the layer or layers between the annihilating dislocations. Thus, the excess causes a lengthened layer and this lengthening can be most easily accommodated by rucking up and tucking the ruck in,

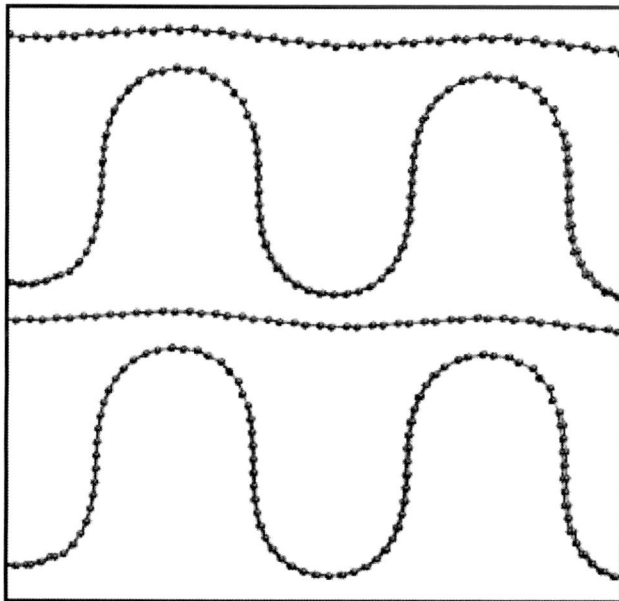

Figure 12.22 An array of multiple climb dipoles of basal edge dislocations in graphite, each gliding on neighbouring interlayer galleries (by I. Suarez-Martinez).

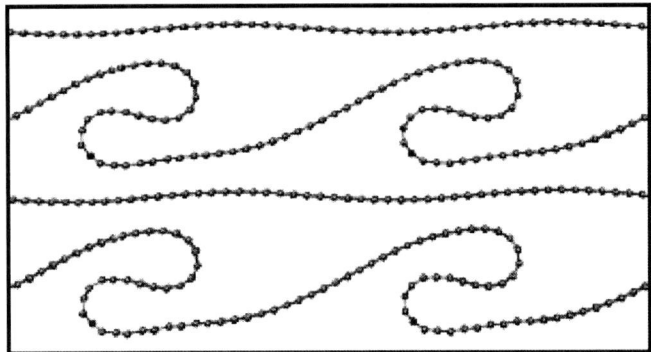

Figure 12.23   An array of ruck and tuck defects in graphite (by I. Suarez-Martinez).

making a "ruck and tuck" defect (Figure 12.23). A nomenclature has been suggested[88] in which the extra material is specified as if it were a nanotube, using the chiral index notation $(n, m) \times k$ where k denotes the number of layers involved. This example is of a single plane defect $(k = 1)$ in a rather small orthorhombic cell. In principle, a corresponding deficit of material cases can lead to breaking of one or a number of layers.

Figures 12.21–12.23 are clearly caricatures of radiation damaged material, which will neither be so regular, nor would defects such as the ruck and tuck be so close to one another under most conditions, but they would likely have separations and widths of tens of nanometres to micrometres. The folds, once created, can be extended by the passage of a basal dislocation (Figure 12.24) that climbs as it passes the defect, depositing a small part of its "added part-plane" With each passage, the crystal dimensions are altered according to the Lehto–Öberg correction[78] for the supercell, treating each fold as a perfect prismatic edge dislocation. It is fascinating to observe that basal glide can lead to growth in the c-axis dimension, and that a pile-up of basal edge-dislocation dipoles can cause (or rather be equivalent to) a dipole of perfect prismatic edge dislocations. The way this occurs seems entirely consistent with the phenomenology of damage at high temperatures and doses,[25] adding to the well-established mechanism of prismatic loop formation, and the possible aggregation of vacancies into lines.

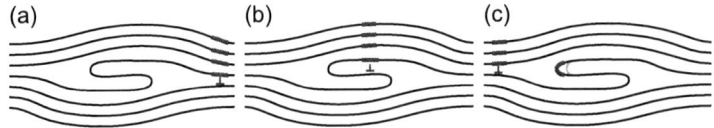

Figure 12.24   Schematic illustration showing how glide of a basal dislocation from right to left can deposit material into ruck and tuck defect, demonstrating that there is a continuous process of extension of the defect once it has been created (by CDL after I. Suarez-Martinez[79]).

Dislocation simulations in a supercell require dipoles, and those dipoles have always produced lattice expansion for the sizes that have been employed so far. The size of cell studied, or even the study of single dislocations, can be greatly increased by moving from an atomistic calculation to one in which the layers are treated as flexible membranes that stick to each other with a weak potential. This is the approach taken by Bo Yang, Michel Barsoum, and collaborators. Initial indications are that the core of the basal dislocation does cause an opening of the interlayer space.[89] Such a result could not be anticipated by the continuum theory of dislocations based on second-order elasticity, but might possibly arise from higher-order theories. Clearly a large, atomistic and, preferably first-principles, calculation of the structure and energy of basal dislocations is required.

12.6 Summary

Compared with diamond, graphite is far less amenable to experimental investigation of defects present within the bulk of the material. Consequently, advances in our understanding of the response of graphite to energetic particle irradiation—which affects its properties at every scale—rely heavily on *ab initio* models of defects within the atomic realm. Without these calculations it would not be possible to test the validity of various ideas about the nature of radiation damage in graphite. Indeed, theory has shown that the early assumptions about the properties and behaviour of vacancies and interstitial atoms in graphite, turns out to be completely the opposite of what was originally believed, until relatively recently in the more than half a century of research into the subject.

The science of line defects is by and large a rather specialised field, and substantial developments in theory were made more than fifty years ago. This is now a very interesting time, where the peculiarities of layered materials are offering new challenges to the theory of dislocations and their solution promises a better understanding of immensely important practical problems such as radiation damage in graphite, not to mention new avenues in the nano-engineering of materials.

If the reader is stimulated to read beyond this update on the science of graphite and radiation damage, there is no doubt that Brian Kelly's book "The Physics of Graphite"[6] remains one of the most thorough sources for its day.

Acknowledgements

Thanks for provision of funding and resources to:

- British Energy (part of EdF).
- HPC-Europa2.
- CSC, Finland.
- NSC, Sweden.

The views expressed here are the authors' own, and not necessarily those of our sponsors.

References

1. J. D. Bernal, *Proc. R. Soc. Lond., Ser. A*, Dec 1924, **106**, 749.
2. E. P. Wigner, *J. Appl. Phys.*, Nov 1946, **17**, 857.
3. B. T. Kelly, *Dimensional Changes in Graphite and the Thermal Expansion Coefficient,* IAEA, Vienna, 2000, pp. 49–84.
4. T. Iwata and H. Suzuki, in *Radiation Damage in Reactor Materials – 1962*, Proceedings series, Vol. 99 (International Atomic Energy Agency, Vienna, 1963) pp. 565–579.
5. L. Arnold, *Windscale 1957: Anatomy of a Nuclear Accident*, 3rd edn Palgrave Macmillan, Basingstoke 2007.
6. B. T. Kelly, *Physics of Graphite,* Applied Science Publishers, Barking and New Jersey, 1981.
7. L. Bochirol and E. Bonjour, *Carbon*, in French with English summary 1968 (Oct), **6**, 661.
8. T. Maeta, H. Iwata and S. Okuda, *J. Phys. Soc. Jpn.*, Dec 1975, **39**, 1558.
9. B. T. Kelly, B. J. Marsden, K. Hall, D. G. Martin, A. Harper, A. Blanchard, *Irradiation Damage in Graphite due to Fast Neutrons in Fission and Fusion Systems*, TECDOC, Vol. 1154 (IAEA, Vienna, 2000).
10. J.-M. Campanera, G. Savini, I. Suarez-Martinez and M. I. Heggie, *Phys. Rev. B*, Jun 2007, **75**, 235449.
11. J. Charlier, X. Gonze and J. Michenaud, *Carbon*, 1994, **32**, 289.
12. J. Charlier, X. Gonze and J. Michenaud, *Europhys. Lett.*, Nov 1994, **28**, 403.
13. K. S. Novoselov, A. K. Geim, S. V. Morozov, D. Jiang, M. I. Katsnelson, I. V. Grigorieva, S. V. Dubonos and A. A. Firsov, *Nature*, Nov 2005, **438**, 197.
14. J. N. Murrell, S. F. A. Kettle and J. M. Tedder, *The Chemical Bond,* John Wiley and Sons, 1978.
15. J. F. Green and I. L. Spain, *Phys. Rev. B*, May 1975, **11**, 3935.
16. M. C. Schabel and J. L. Martins, *Phys. Rev. B*, Sep 1992, **46**, 7185.
17. L. A. Girifalco and M. Hodak, *Phys. Rev. B*, Feb 2002, **65**, 125404.
18. A. N. Kolmogorov and V. H. Crespi, *Phys. Rev. B*, Jun 2005, **71**, 235415.
19. G. Savini and M. I. Heggie, in *Carbon 2007 Proceedings* (Seattle, Washington, 2007).
20. C. S. G. Cousins and M. I. Heggie, *Phys. Rev. B*, Jan 2003, **67**, 024109.
21. N. Mounet and N. Marzari, *Phys. Rev. B*, May 2005, **71**, 205214.
22. H. Rydberg, M. Dion, N. Jacobson, E. Schröder, P. Hyldgaard, S. I. Simak, D. C. Langreth and B. I. Lundqvist, *Phys. Rev. Lett.*, Sep 2003, **91**, 126402.
23. F. Banhart, *Rep. Prog. Phys.*, Apr 1999, **62**, 1181.
24. E. W. J. Mitchell and M. R. Taylor, *Nature*, Nov 1965, **208**, 638.
25. R. H. Telling and M. I. Heggie, *Philos. Mag.*, Nov 2007, **87**, 4797.

26. P. A. Thrower and R. M. Mayer, *Phys. Status Solidi A*, May 1978, **47**, 11.
27. K. Niwase, *Phys. Rev. B*, Dec 1995, **52**, 15785.
28. K. Niwase, *Phys. Rev. B*, Sep 1997, **56**, 5685.
29. C. A. Coulson, M. A. Herraez, M. Leal, E. Santos and S. Senent, *Proc. R. Soc. Lond., Ser. A*, Aug 1963, **274**, 461.
30. A. A. El-Barbary, R. H. Telling, C. P. Ewels, M. I. Heggie and P. R. Briddon, *Phys. Rev. B*, Oct 2003, **68**, 144107.
31. P. O. Lehtinen, A. S. Foster, Y. Ma, A. V. Krasheninnikov and R. M. Nieminen, *Phys. Rev. Lett.*, Oct 2004, **93**, 187202.
32. P. O. Ma and, Y. Lehtinen, A. S. Foster and R. M. Nieminen, *New J. Phys.*, Jun 2004, **6**, 68.
33. A. V. Krasheninnikov, P. O. Lehtinen, A. S. Foster and R. M. Nieminen, *Chem. Phys. Lett.*, Jan 2006, **418**, 132.
34. R. Faccio, H. Pardo, P. A. Denis, R. Yoshikawa Oeiras, F. M. Araújo-Moreira, M. Veríssimo-Alves and A. W. Mombrú, *Phys. Rev. B*, Jan 2008, **77**, 035416.
35. J. G. Kushmerick, K. F. Kelly, H.-P. Rust, N. J. Halas and P. S. Weiss, *J. Phys. Chem. B*, Mar 1999, **103**, 1619.
36. E. Kaxiras and K. C. Pandey, *Phys. Rev. Lett.*, Dec 1988, **61**, 2693.
37. G.-D. Lee, C. Z. Wang, E. Yoon, N.-M. Hwang, D.-Y. Kim and K. M. Ho, *Phys. Rev. Lett.*, Nov 2005, **95**, 205501.
38. J. I. Paredes, P. Solis-Fernández, A. Martínez-Alonso and J. M. D. Tascón, *J. Phys. Chem. C*, May 2009, **113**, 10249.
39. H. Terrones, M. Terrones, E. Hernández, N. Grobert, J.-C. Charlier and P. M. Ajayan, *Phys. Rev. Lett.*, Feb 2000, **84**, 1716.
40. R. H. Telling, C. P. Ewels, A. A. El-Barbary and M. I. Heggie, *Nature Mater.*, May 2003, **2**, 333.
41. A. A. El-Barbary, *First Principle Characterisation of Defects in Irradiated Graphitic Materials*, PhD thesis. School of Life Sciences, University of Sussex (2005).
42. C. P. Ewels, M. I. Heggie, A. El-Barbary, A. Gloter, R. H. Telling, J. P. Goss, and P. R. Briddon, in *Proceedings of Carbon 2004*, edited by W. Hoffman, American Carbon Society (American Carbon Society, Edwards, California, 2004) p. G013, available online, http://acs.omni-booksonline.com/data/papers/2004_G013.pdf.
43. C. D. Latham, M. I. Heggie, J. A. Gámez, I. Suárez-Martínez, C. P. Ewels and P. R. Briddon, *J. Phys.: Condens. Matter*, Oct 2008, **20**, 395220.
44. C. D. Latham, G. L. Haffenden, M. I. Heggie, I. Suarez-Martinez and C. P. Ewels, *Phys. Rev. B*, 2010, **82**, 056101.
45. M. I. Heggie, B. R. Eggen, C. P. Ewels, P. Leary, S. Ali, G. Jungnickel, R. Jones, and P. R. Briddon, in *Fullerenes: Chemistry, Physics, and New Directions*, Recent Advances in the Chemistry and Physics of Fullerenes and Related Materials, Vol. 6, (ed.) K. M. Kadish and R. S. Ruoff, The Electrochemical Society, Pennington NJ, 1998, pp. 60–67.
46. L. Li, S. Reich and J. Robertson, *Phys. Rev. B*, Nov 2005, **72**, 184109.

47. Y. Ma, *Phys. Rev. B*, Aug 2007, **76**, 075419.
48. W. N. Reynolds and P. A. Thrower, *Philos. Mag.*, Sep 1965, **12**, 573.
49. L. M. Brown, A. Kelly and R. M. Mayer, *Philos. Mag.*, Apr 1969, **19**, 721.
50. M. I. Heggie, I. Suárez-Martínez, G. Haffenden, G. Savini, A. El-Barbary, C. Ewels, R. Telling, and C. S. G. Cousins, in *Management of Ageing Processes in Graphite Reactor Cores*, Vol. 309, (ed.) G. B. Neighbour, The Royal Society of Chemistry, Cambridge, UK, 2007, pp. 83–90.
51. C. P. Ewels, R. H. Telling, A. A. El-Barbary, M. I. Heggie and P. R. Briddon, *Phys. Rev. Lett.*, Jul 2003, **91**, 025505.
52. E. Asari, M. Kitajima, K. G. Nakamura and T. Kawabe, *Phys. Rev. B*, May 1993, **47**, 11143.
53. E. Asari, M. Kitajima and K. G. Nakamura, *Carbon*, Nov 1998, **36**, 1693.
54. J. P. Hirth and J. Lothe, *Theory of Dislocations,* 2nd edn, John Wiley and Sons, New York, 1982.
55. A. Grenall, *Nature*, Aug 1958, **182**, 448.
56. I. M. Dawson and E. A. C. Follett, *Proc. R. Soc. Lond., Ser. A*, Dec 1959, **253**, 390.
57. S. Amelinckx, *The Direct Observation of Dislocations*, Solid State Physics, Vol. Supplement 6, Academic Press, New York, 1964.
58. S. Amelinckx, P. Delavignette, and M. Heerschap, in *Chemistry and Physics of Carbon*, Vol. 1, (ed.) P. L. Walker, Marcel Dekker, New York, 1965, Chap. 1, pp. 1–71.
59. A. R. Lang, *Acta Crystallogr.*, Mar 1959, **12**, 249.
60. P. A. Thrower, in *Chemistry and Physics of Carbon*, Vol. 5, (ed.) P. L. Walker, Marcel Dekker, New York, 1969, Chap. 3.
61. I. L. F. Ray and D. J. H. Cockayne, *Proc. R. Soc. Lond., Ser. A*, Dec 1971, **325**, 543.
62. F. E. Fujita and K. Izui, *J. Phys. Soc. Jpn.*, Feb 1961, **16**, 214.
63. K. Izui and F. E. Fujita, *J. Phys. Soc. Jpn.*, May 1961, **16**, 1032.
64. G. M. Jenkins, in *Chemistry and Physics of Carbon*, Vol. 11, (ed.) P. L. Walker, Marcel Dekker, New York, 1973, pp. 189–242.
65. J. M. Thomas and C. Roscoe, in *Chemistry and Physics of Carbon*, Vol. 3, (ed.) P. L. Walker, Marcel Dekker, New York, 1968, Chap. 1, pp. 1–44.
66. C. P. Ewels, M. I. Heggie and P. R. Briddon, *Chem. Phys. Lett.*, Jan 2002, **351**, 178.
67. B. R Eggen, C. D. Latham, M. I. Heggie, R. Jones and P. R. Briddon, *Synth. Met.*, Feb 1996, **77**, 165.
68. B. R Eggen, M. I. Heggie, G. Jungnickel, C. D. Latham, R. Jones and P. R. Briddon, *Science*, Apr 1996, **272**, 87.
69. G. M. Jenkins, *Carbon*, Feb 1969, **7**, 9.
70. B. T. Kelly, *Nature*, Jul 1965, **207**, 257.
71. O. V. Yazyev and S. G. Louie, *Phys. Rev. B*, May 2010, **81**, 195420.
72. A. Carpio, L. L. Bonilla, F. de Juan and M. A. H. Vozmediano, *New J. Phys.*, May 2008, **10**, 053021.

73. V. Kapko, D. A. Drabold and M. F. Thorpe, *Phys. Status Solidi B*, May 2010, **247**, 1197.
74. B. T. Kelly, *J. Nucl. Mater.*, Jul 1990, **172**, 237.
75. G. J. Dienes, *J. Appl. Phys.*, Nov 1952, **23**, 1194.
76. O. V Yazyev, I. Tavernelli, U. Rothlisberger and L. Helm, *Phys. Rev. B*, Mar 2007, **75**, 115418.
77. A. Chuvilin, U. Kaiser, E. Bichoutskaia, N. A. Besley and A. N. Khlobystov, *Nature Chem.*, Jun 2010, **2**, 450.
78. N. Lehto and S. Öberg, *Phys. Rev. Lett.*, Jun 1998, **80**, 5568.
79. I. Suárez-Martínez, *Theory of Diffusion and Plasticity in Layered Carbon Materials*, DPhil thesis, School of Life Sciences, University of Sussex (May 2007).
80. E. R. Margine, A. N. Kolmogorov, D. Stojkovic, J. O. Sofo and V. H. Crespi, *Phys. Rev. B*, Sep 2007, **76**, 115436.
81. S. Amelinckx and P. Delavignette, *Phys. Rev. Lett.*, Jul 1960, **5**, 50.
82. C. Baker and A. Kelly, *Philos. Mag.*, Apr 1965, **11**, 729.
83. I. Suarez-Martinez, G. Savini, G. Haffenden, J.-M. Campanera and M. I. Heggie, *Phys. Status Solidi C*, Jul 2007, **4**, 2958.
84. F. H. Horn, *Nature*, Oct 1952, **170**, 581.
85. R. H. Telling and M. I. Heggie, *Philos. Mag. Lett.*, Jul 2003, **83**, 411.
86. J. Frenkel, *Z. Phys. A*, in German. 1926 (Jul), **37**, 572.
87. B. T. Kelly and A. J. E. Foreman, *Carbon*, Apr 1974, **12**, 151.
88. M. I. Heggie, I. Suarez-Martinez, G. Savini, G. L. Haffenden and J. M. Campanera, *Radiation damage in graphite-a new model*, IAEA, Vienna, 2010.
89. M. Barsoum, private communication.

Subject Index